The Rise of the Wave Theory of Light

Jed Z. Buchwald

The Rise of the Wave Theory of Light

Optical Theory and Experiment in the Early Nineteenth Century

The University of Chicago Press Chicago and London

Jed Z. Buchwald is associate professor in the Institute for the History and Philosophy of Science and Technology at the University of Toronto. He is the author of *From Maxwell to Microphysics,* also published by the University of Chicago Press.

The University of Chicago Press, Chicago 60637
The University of Chicago Press, Ltd., London

Printed in the United States of America

98 97 96 95 94 93 92 91 90 89 54321

Library of Congress Cataloging in Publication Data

Buchwald, Jed Z.
The rise of the wave theory of light : optical theory and experiment in the early nineteenth century / Jed Z. Buchwald.
p. cm.
Bibliography: p.
Includes index.
ISBN 0-226-07884-1
ISBN 0-226-07886-8 (pbk.)
1. Light, Wave theory of—History—19th century. I. Title.
QC403.B83 1989
535′.13′09—dc19 88-18647
CIP

To the serene lady, the demon, and the imp

Contents

Part 2 Fresnel, Diffraction, and Polarization

Part 3 Controversy and Unification

Appendixes

Acknowledgments

I have discussed aspects of this material with many people over the years. Among them I especially thank Geoffrey Cantor, Bernard Cohen, Stillman Drake, Caspar Hakfoort, Naum Kipnis, Ole Knudsen, Martin Klein, Tom Kuhn, Alan Shapiro, Dan Siegel, and Tom Whiteside. They will certainly not agree with every one of my claims (or with one another), but they helped me reach them in their several ways.

Material from two of my articles in *Archive for History of Exact Sciences* has been incorporated in this book, by permission of Springer-Verlag: "Optics and the Theory of the Punctiform Ether," 21(1980:245–78, and "Experimental Investigations of Double Refraction from Huygens to Malus," 21(1980):311–73. Elements of part 2 are reproduced from my article "Fresnel and Diffraction Theory," which appeared in *Archives Internationales d'Histoire des Sciences* 33(1983): 36–111, by permission of the editor.

Introduction

> By the genius of Young and Fresnel the wave-theory of light was established in a position so strong that henceforth the corpuscular hypothesis was unable to recruit any adherents among the younger men.
>
> Sir Edmund Whittaker,
> *A History of the Theories of Aether and Electricity*

Since the middle of the 1830s it has been customary to distinguish the scientists involved in the debates leading to the rise of the wave theory of light as either emissionists or wave theorists. The emissionists, as they called themselves (or "corpuscularians" in Whittaker's terminology), conceived of light as a sequence of rapidly moving particles that are subject to forces exerted by material bodies. Wave theorists thought of light as a regular sequence of spreading disturbances in a universally present medium that they termed the ether. Whittaker's remarks reflect William Whewell's early and influential opinion that the wave theory came to dominance by exposing the comparative inadequacy of the emission theory.

Historians have long recognized that the emission theory was able to explain contemporary experiments, even though the explanations were usually highly unsatisfactory in the eyes of wave theorists. Where the emission theory was unavoidably weak compared with its opponent was in generating quantitative results. By about 1830 quantitative deduction had become extremely desirable, and so the replacement of the emission theory by the wave theory was, one might argue, more a function of a change in the canons of what a theory must do than of its failing abysmally to explain some new experiment.

On the whole this view has a great deal to recommend it, because there can be no doubt that the concept of waves in the ether did replace the idea of particles of light and that the wave theory *is* more powerful than the emission theory. The young Cambridge mathematicians who rapidly extended the theory in the 1830s were attracted to it for a number of reasons, but a principal one was that waves could be used to obtain results, whereas it seemed clear by then that this could not be done with light particles. By the end of the 1830s only a few diehards remained committed to the emission theory, which had almost completely vanished from the journals of science.

My principal goal here is to alter the historical terms in which these events have been laid out for nearly a century and a half. Although we will indeed see that waves replaced light particles, a deeper process also took place at about the same time, a process that only partially coincides with the first one. While waves in the ether were replacing particles of light as tools of explanation, wave fronts were also beginning to replace rays as tools of analysis. This second process was at least as difficult for many people as the first one because it required them to alter many fundamental optical concepts concerning the nature of a *ray* and its relation to a *beam*.

To put the point crudely and inadequately, many people had immense difficulty understanding that the *wave front* is irreconcilable with the concept of an *isolatable ray*—that rays must be abandoned entirely as physical objects in their own right in order to deploy the mathematical apparatus of the wave theory. Even those who were willing to forgo further commitment to the emission theory (and there were several by the 1820s) often refused to admit that one cannot use rays in the old ways. In the emergence of the wave theory to dominance, I will argue, one finds a profound dichotomy in understanding as well as in conception: a dichotomy between the most elementary images of the nature of a ray, on the one hand, and a dichotomy between the physical models of light as particles and light as waves in the ether, on the other.

The first dichotomy is a particularly subtle one. Even today, in historical retrospect, it is difficult to uncover in all its aspects because it appears most strikingly in problems that remain hard to quantify. Diffraction, which Fresnel did tame (and which I will discuss in detail), hardly raises the issue at all. But polarization, especially those aspects of it that escape exact representation, immediately implicates the dichotomy concerning the nature of a ray that forms a major part of my discussion. Consequently, much of my history will concentrate on the issues raised by polarization and the optics of crystals. These issues may seem to be peripheral ones—mere asides to the main story—because they do not directly concern the grand historical drama that culminated in Fresnel's great triumph in diffraction: his winning a Paris Academy prize in 1819 for an empirically successful quantification of the phenomenon. To give them, as I do, nearly equal place with diffraction would be a gross distortion were this true.

But it is not. What the modern eye focuses on in retrospect may not be, indeed very often *will not be,* what the contemporary eye saw. Very few scientists at the time thought that optical theories stood or fell with their ability to quantify diffraction. Indeed, diffraction caused little stir before 1819 because it did not seem related to the subject of most intense contemporary interest, which was polarization. In granting Fresnel a prize for diffraction, the judges of the Paris Academy were, they felt, conceding comparatively little.

The signal importance of polarization to the historical events arises primarily from its close association with deep-seated views on the nature of rays and beams of light. As understood (I shall argue) by most optical scientists before about 1830, a *beam* of light consists of a collection of objects called *rays*. These objects, it was assumed, possess individual identities and so can be counted, the intensity of a beam being measured numerically by the number of rays it contains. The ray, it was further thought, has an inherent *asymmetry* about its length. Think of it as a stick with a crosspiece nailed to it at right angles. Given the direction of the ray, the orientation of the crosspiece in a plane at right angles to the ray determines the ray's asymmetry.

According to this way of thinking *polarization,* the central topic of experiment and theory in the 1810s, does not characterize the individual rays in a beam but only the beam as a collection of rays. One cannot, that is, speak of the polarization of a single ray. This has no meaning because a single ray, as it were, has its crosspiece permanently nailed on and so is always just as asymmetric as it can ever be. Instead, *polarization* as a category applies only to collections of rays, to *beams,* in the following manner. If the asymmetries of the rays in a given beam point randomly in many directions, then the beam is, in contemporary parlance, *unpolarized.* If, on the other hand, one can group the rays in a beam into a number of sets each of whose elements shares a common asymmetry, or even if this can be done only for a certain portion of the rays in the beam, then the beam is said to be *partially polarized* (sometimes *partially unpolarized*). If all the rays have the same asymmetry, then the beam is just *polarized.* Accordingly, the polarization of a beam is altogether a matter of degree, of the numbers of rays that are aligned in various directions. And so this property of light is very closely tied to instrumental techniques: if a polarization detector is comparatively insensitive, a particular beam may prove to be unpolarized; but if the device is very sensitive, the same beam might just as reasonably prove to be partially polarized. Since the category is not absolute—it is a matter of grouping rays into sets whose properties can be detected—both statements would be correct.

This instrumentally based conception is difficult indeed to reconcile with the understanding that was developed by Fresnel and that we hold today. According to his (and our) idea, at a given instant and at a given point on the wave front, every wave of light possesses an asymmetry that lies in the front itself. Rays of light are mathematical abstractions; they are the directions joining the center of the wave to the front itself. In a purely analytical sense, one can link a ray to the asymmetry in the front at the point of the front that the ray contains. Accordingly, in the wave theory one can say that a *ray* is polarized, if one wishes, because polarization refers to the asymmetry at a point in the front, and to each such point there corresponds only one ray. But—and this is a signal characteristic of the wave theory—the rays, being mere lines, cannot be counted. A beam of light is therefore not a collection of rays; it cannot be dissected into independent parts.

The effects of this primitive difference between the ray and the wave theories reveal themselves in almost every situation that alters polarization. Consider, for example what happens when a beam of light reflects and refracts. Suppose that this beam is polarized in the sense of ray theory. Then it consists of a set of rays all of whose asymmetries point more or less in the same direction. When we ask what happens to the polarization when this group is reflected and refracted, we are asking what happens to the asymmetries of the rays. We want, that is, to know what new directions the asymmetries can point in after reflection and refraction, and how many rays will have asymmetries that point nearly in each of these new directions. We do not need to know what happens to the asymmetry of any given ray, because the measurable phenomena depend only on how many rays have a given asymmetry—they do not depend on whether the asymmetry of some particular ray does or does not retain its original direction. Once we know what asymmetries occur, and how many rays have approximately the allowed asymmetries, we can calculate what can be seen in a polarization detector, because these devices merely select rays according to their asymmetries.

The wave theory provides a picture of reflection completely different from this one. In it we begin with a wave striking the surface at a given instant. The wave intersects the surface in a line, and at each point along that line there is a direction, in the plane of the wave itself, that specifies the wave's polarization at that point. When we ask what happens to the polarization on reflection, we are asking what happens to each of these directions. *That is, here we are dealing with what happens to the individual ray, not with collections of rays.* We must accordingly introduce an utterly different conception of a *beam,* one that will specify what our devices can detect.

In the wave theory a beam must be thought of as a very small segment of a front instead of as a set of rays. So small, in fact, that every point on the segment must have the same polarization. The vector that represents in direction this polarization may also represent in magnitude the amplitude of the oscillation at every point of that small frontal segment. When the beam is reflected, the amplitude's components in and perpendicular to the plane of incidence are altered in different amounts, and these alterations determine the polarizations as well as the intensities of the reflected and refracted light.

One might therefore put one aspect of the difference between the wave and ray conceptions in the following way. The wave theory calculates the change in polarization of an individual ray, which it confounds with a beam. No ray theorist ever claimed to be able to calculate what happens to an individual ray, which cannot be confounded with a beam; rather, he concerned himself with the redistribution of the rays in a beam in respect to their asymmetries. Accordingly the two theories differ as well in their attitudes toward what detectors reveal about the beam. A ray theorist at once admits that a beam's polarization is a matter of degree—of how many rays have asymmetries with *approximately* some given direction, and of how many such directions there are. A wave theorist disagrees. Since he confounds the beam with the ray, the category of polarization is for him very nearly absolute: it involves a property of the unique ray in question. Whether or not a detector provides evidence of polarization depends almost entirely on the detector rather than the state of the beam—whereas for the ray theorist it is a question of detector accuracy in relation to beam structure.

There is a second aspect to the difference as well, one that also emerges from the fact that ray theorists think of polarization as a property of collections of rays whereas wave theorists think of it as a property of the individual ray. The major and also nearly absolute distinction in the wave theory is between the polarized and the unpolarized. For each of these two categories there are subcategories. The wave theory postulates two kinds of unpolarized light: *partially unpolarized* and *completely unpolarized,* or natural, light. There are three kinds of polarized light: *elliptical, circular,* and *plane,* as diagramed below.

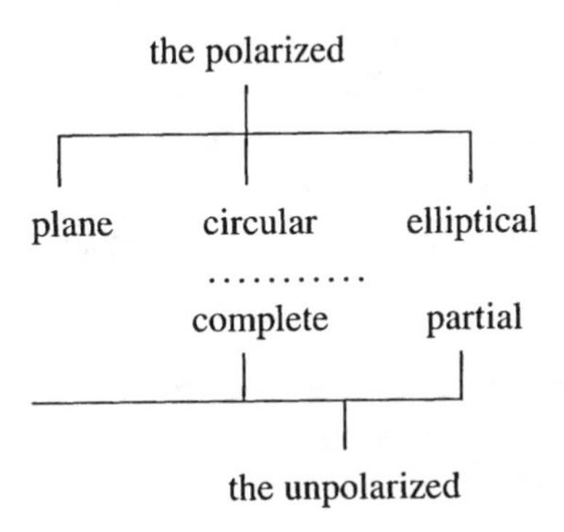

Neither a reflecting mirror nor an analyzing crystal can—used singly—distinguish between partially polarized and elliptically polarized light, or between completely unpolarized light and circularly polarized light. Only carefully constructed combinations of these devices can bring out the differences, all of which depend on a fundamental physical distinction: that every type of polarized light can be decomposed into orthogonal components that have fixed phase and amplitude differences, whereas this cannot be done for either of the two kinds of unpolarized light.

This goes to the heart of the essential opposition between ray theory and wave theory, which exists in part because the concept of coherence is foreign to ray theory. In the latter one cannot easily introduce a distinction between circularly polarized and unpolarized, or between elliptically polarized and partially unpolarized, because polarization concerns only how much light is thrown into either of the two beams in an analyzing crystal or into the reflected and refracted beams by a mirror. Further connections between the rays in the beam are of no consequence here and so there is, according to ray theory, no physical difference *at all* between these kinds of light. It would be difficult indeed for ray theory to produce a distinction without deeply altering its long-standing conviction that polarization concerns collections rather than individuals.[1]

We shall see many times that the vast gulf between these concepts made it extremely difficult for scientists who did not think about polarization in the same way to communicate their experimental results to one another without leaving an inevitable residuum of ambiguity. This ambiguity could, and often did, lead to controversies that ended in frustration and anger—in particular to a striking public confrontation between Arago and Fresnel, on the one hand, and Biot, on the other. Biot had worked extensively on chromatic polarization, the major topic of optics in France during the 1810s. This extremely influential phenomenon, discovered by Arago, involves the production of colors when polarized light that has passed through thin slices of certain crystals cut in certain ways is then either passed through an analyzing crystal or else reflected at the polarizing angle from a mirror. Biot, Arago, and almost everyone else except Young and Fresnel assumed implicitly that in this phenomenon the asymmetries of the rays in the original polarized beam are selectively rotated to new azimuths depending upon the color that corresponds to the particular ray. (Because in this theory rays are selected by the apparatus, I shall—adopting terminology introduced, as we shall see, by Thomas Young—call those who thought about light in this way *selectionists,* and the theory they espoused *selectionism.*)

Biot built a quantitative theory founded on the assumption that rays can be grouped into sets with common asymmetries. When Fresnel developed a theory based on interference, he and Biot fought strongly and frustratingly over whether the formulas Fresnel deduced were much the same as Biot's. I say frustratingly because Fresnel did not see that Biot's quantitative work depended exclusively on the chromatic selection of rays, whereas Biot did not admit that Fresnel's work was incompatible with ray counting. Accordingly, Fresnel attacked Biot for not having properly used emission principles, which Biot had not needed here, and Biot reacted in honest and rather angry puzzlement. Neither he nor Fresnel pinpointed the major issue; if it had once been brought into the open, they would have realized that their most basic concepts were so different as to prohibit almost any kind of discussion at all.

David Brewster provides a second compelling example of this failure to construe experiments and formulas in compatible ways. About 1830 he wrote several articles on what he termed elliptical polarization and on the structure of partially polarized and natural light. Brewster claims not to need any hypothesis about the nature of light. He tries to make use of formulas and parameters that are derived from the wave theory in a way that avoids the theory proper. The result in this case is almost complete nonsense, because he creates a hodgepodge that, despite his claim to have avoided theory, mixes the tacit assumption that rays can be treated as individuals with concepts that have little meaning outside the wave theory proper. And so Brewster never thoroughly appreciated the most fundamental requirement of the wave theory, that rays cannot be treated as things. Nor was he alone. Neither Biot nor Poisson appreciated this point, and even Arago, who early promoted Fresnel's work, had difficulties with it. But there is also another side to this coin. It was apparently very hard for wave theorists to see that selectionists often needed only to count rays rather than to rely directly on the intricate details of particles and forces, as Fresnel's commentary on Biot's work nicely shows.

Although this difference was most significant for polarization, it was important in a very different way for diffraction. It has been argued (Kipnis 1984) that until the early 1830s almost everyone, including those who were competent to understand the mathematics involved, thought of diffraction entirely as a process in which rays influence one another. There were two major ways to think like this. The less sophisticated way simply ignored Huygens's principle and reverted to some variant of Fresnel's and Young's original method, in which only two rays are compared with one another for their path differences. The more sophisticated way (empirically as well as mathematically) considered Huygens's principle to be an extension of this procedure—to be, that is, solely a method for dealing with more than two rays. Both alternatives are, in my sense of the term, selectionist because both implicitly assume that the only proper subject of optics is the ray of light.[2]

This blurring of the boundaries between the two conceptions greatly aided the dissemination of Fresnel's work, because it made the principle of interference comprehensible as a newly discovered property of a well-known thing, the optical ray. Had this not been possible I strongly doubt that, despite the empirical power of his formulas, Fresnel's work would have so readily won the prize offered by the Paris Academy in 1818. Both Laplace and Poisson were on the prize commission, and both of them thought this way about Fresnel's formulas. And yet Poisson admitted early on that he felt Fresnel's diffraction experiments posed grave difficulties for the emission theory. Poisson, like Brewster and many others, implicitly distinguished the emission theory from pure ray counting by, on occasion, voicing doubts about, or taking an agnostic attitude toward, light particles while never abandoning the physical ray.[3]

The persistence of this way of thinking through the early 1830s (and even beyond in several cases) means that one cannot treat the emerging dominance of the wave theory exclusively as a change from one well-understood though empirically barren scheme, based on particles and forces, to another more fruitful scheme based on waves in the ether. One must also try to capture the effects of the more delicate and difficult distinctions between rays and waves, distinctions that usually remained just below the surface of scientific discourse, subtly affecting its texture and tone. For example, David Brewster's increasing anger at being shunted to the scientific periphery cannot be fully understood without recognizing that he remained honestly convinced that much of his work did not require the emission theory and was ipso facto independent of hypothesis. He was wrong about its unhypothetical character, but few if any of the wave theorists after 1830 recognized precisely why he was wrong, though they certainly knew that he was. Brewster was not simply an old partisan of an outmoded theory who was rejected by a younger, highly trained professional elite devoted to new ideas—though he was at least that. He, like many others, was by the 1830s an outsider in an even more profound sense because he simply could not understand why his work no longer compelled attention.

I have chosen to concentrate in substantial part on this issue of the nature of the ray, rather than upon the conflicting physical models that embody the different conceptions of light, because it seems to me that here we can focus directly upon a very deep and, at the time, ill-understood change, one that occasioned resistance to the wave theory even by those who no longer directly advocated emission principles. To some the arithmetic of ray counting was axiomatic; to others, later, the method of continuous analysis required by the wave theory was equally axiomatic. These were not differences that could easily be recognized, unlike the difference between a particle and a wave.

Nevertheless, one must not carry this new emphasis too far. Although I shall attempt to show that it does cast new light upon the early history of the wave theory, still the traditional emphasis upon physical principles also captures something of equal importance: the contemporary concern of many scientists to integrate their quantitative analyses with acceptable physical models, whether based on particles or on the ether. If one could not do so—if a theory seemed utterly incapable of integration with such principles, then it stood no chance at all. This was why, for example, John Herschel went to a good deal of effort to show that the ether could, at least conceivably, accommodate selective absorption: he had no quantitative theory for the phenomenon, but he had to provide (in effect) an existence proof. Similarly, it was necessary for ray theorists not merely to use the principle of interference but to explain how it could conceivably work even were light to consist of particles.

The kind of theoretical difference I have tried to explore here through the example of ray and wave theory seems to me pervasive in the history of physics. In electromagnetism I have argued that what one thinks of as perhaps the most elementary concept—the idea of electric charge—was so profoundly different in Britain and on the Continent in the last quarter of the nineteenth century that it was very difficult for physicists to communicate fruitfully with one another. This was even harder in that the intricate equations required by electromagnetism often masked the conceptual difficulties until, well into a deduction or a description of experiment, an impasse was reached. Although the existence of a special British conception of charge was not so entirely forgotten as was selectionism, nevertheless the scope and nature of the idea were rapidly lost after about 1905.

These two cases illustrate something that has characterized controversies in physics since at least the time of Newton. Often arguments between physicists seem to revolve endlessly or to go off in apparently arbitrary directions, in the end petering out with no clear resolution, until at some time it becomes clear that the argument, unresolved though it may have been, is no longer of interest to anyone except a few people who are by then on the periphery of the physical community. When one finds irresolute controversies, one may also find that the ostensible, the explicitly recognized differences are not the only ones at issue, and that other differences run so deep that they have precluded mutual understanding. Over time one way of thinking becomes pervasive and the other becomes perhaps not incorrect, but certainly irrelevant. Ray theory, for example, became irrelevant well before it became incorrect, and the electron replaced the British conception of charge before the latter became difficult to reconcile with experiment. And yet in neither case did anyone explicitly recognize what it was that had been replaced. In electromagnetism not only did Lorentz's or Larmor's fields and particles replace ether models, but electrons also replaced field discontinuities. In optics not only did ether waves replace light particles, but wave fronts also replaced physical rays.

I remarked in my previous book, on Maxwellian electrodynamics, that reconstructing dead theories presents a profound difficulty. Since they may be utterly different in their fundamental conceptions from current views, it is hard to discuss them without adopting the original language or else producing inevitable inconsistencies by using familiar, modern terminology. In the case of electromagnetism the problem was compounded by the intricate mathematics (forming a language of its own, and one that changed over time) that all these theories require. Here we are not quite so troubled by technical difficulties of this sort, because the mathematics is not so complicated as the intricate vector manipulations of electrodynamics. Consequently scientists of the time were not usually plagued by difficulties in grasping one another's mathematics, whereas the electrodynamics of the 1880s and 1890s presented many problems of this sort. Nevertheless, there remained a gaping conceptual impasse between ray theory and wave theory.

My goal, as it was in my previous book, will therefore be to lead the reader past modern preconceptions, but without entirely abandoning a modern perspective, to an understanding of deep foundations—here those of optics just before and just after the development of the wave theory. And again I must ask for patience and indulgence. In order to penetrate to the core of the issues, we must examine a number of exemplary problems and discuss at some length the arguments of the time that directly illuminate these issues. Accordingly, I have divided the book into three major parts. The first part, chapters 1 through 4, concentrates on ray optics, and especially on the development of polarization theory on that basis by Malus, Arago, and Biot. Readers who have not encountered such things as chromatic polarization before may find the going a bit difficult, particularly in chapters 3 and 4, but this is in substantial part because I shall be attempting to bring out the quantitative essense of a now utterly forgotten theory.

The second major part, chapters 5 through 8, concentrates almost exclusively on Fresnel, first on his development of diffraction theory and then on his problems in abandoning earlier conceptions of polarization. Here, in chapters 7 and 8, the wrenching difficulty of grasping not merely an outmoded idea, but someone else's efforts to move beyond it, will be manifest. Finally, in the third part (chapters 9 through 12) I turn to the conflict between ray theorists and wave theorists, to Fresnel's "final unification," and (briefly) to the spread of the theory, including the interesting misunderstandings about it that were common until well into the 1830s. I intend the Introduction to serve as a sort of Baedeker for this structure when, as may happen, the landmarks become unfamiliar. I will on occasion repeat in different words, or elaborate at greater length, many of the points introduced here, perhaps rambling backward over what may by then seem familiar ground. However, only in this way can I hope to give a proper feeling for the physics of the time.

Until recent years most histories of physics have concentrated almost entirely on theory, with experiment appearing only at the edges, as a sort of boundary condition. Yet in most cases the development of theory and experiment are so tightly interwoven, often in the work of a single person, that they cannot easily be separated. Experiment constrains and is constrained and motivated by theory; theory constructs the world by specifying what is accessible to experiment. Consequently, to concentrate one's history on theory not only misses half the story but inevitably obscures what the theory was about.

I do not intend by this to embrace, or even to suggest, the notion that phenomena are in any meaningful sense *created* by experiment. On the contrary, I firmly believe that whatever happened in one instance will always happen in similar instances. But one of the things the integrated whole of theory and experiment does is to specify the kinds of situations that count as *similar instances*, and this can be profoundly difficult to do. For example, in measuring polarized light, is it the same thing—are the "instances" sufficiently similar—to use the eye and an analyzing crystal as it is to use some sort of sensitive intensity-measuring device (a photometer) and a crystal? Modern theory—the wave theory—says there is no essential difference between the two instances. The photometer merely increases one's accuracy and reveals nothing truly new. But ray theory not only insists on a difference, it implies that the results obtained with the photometer will have a much deeper theoretical significance than those that can be obtained with the unaided eye: for they provide the numbers of rays in bundles that must exist but cannot be seen by the unaided eye.

This example is only one of several to be explored below that indicate the symbiosis between theory and experiment in the early history of the wave theory. Indeed, many of the chapters contain an experimental kernel or kernels that are permeated by, and in turn permeate, developing theory. As a guidepost for the reader, I list them here:

Chapter 1 Constraints are imposed by naked-eye observations with doubly refracting crystals.
Chapter 2 Malus's polarizing mirrors detect ray symmetry.
Chapter 3 Arago fails to use Malus's symmetry detectors in the proper way to probe beam structure.
Chapter 4 Biot succeeds where Arago failed.
Chapter 5 Fresnel fabricates a point source of light to examine the spacing of regions where two light rays meet.
Chapter 6 A demand for higher accuracy and theoretical consistency pushes theory to integral computations, thereby forging a thoroughly new structure.
Chapter 7 Traditional conceptions of the role of the analyzing device compel Fresnel to envision wave combination as though the analyzer were always present.
Chapter 8 Reinterpretation of the functioning of the analyzer permits the formulation of new theory.
Chapter 9 A conflict arises over what is being measured.
Chapter 11 What experiment fails to distinguish, theory at first confounds.
Chapter 12 Brewster presents a creative amalgam of old theory, new experiment, and wave categories.

In major part because of this close connection between theory and experiment, it may be difficult for readers who are not well versed in optics to follow certain parts of the argument. To mitigate the difficulty—but not remove it entirely, since that is simply not possible—I have tried to include enough schematic diagrams to suggest the nexus between theory and experiment. Few of the diagrams are therefore actual images of a device. They are meant to suggest the structure of the device in the context of contemporary theory rather than literally to depict a piece of apparatus.

PART 1

Selectionism

1 The Optical Ray

Pick up a piece on experimental optics written after about 1810 and compare it with another written before the middle of the eighteenth century. You will at once find at least three striking differences. The older piece, if it contains any mathematics to speak of, employs geometry. The newer piece uses algebra. The older piece at most provides a few numbers, and it almost certainly does not contain much discussion of the experiment's accuracy. The newer piece provides tables and a reasonably detailed discussion of accuracy.

These are major changes, and they had a powerful impact on optics. Until algebra replaced geometry, for example, there was little chance that Huygens's pulse theory could be applied to complicated situations, because the geometric constructions required were often too intricate to use for computation. And until the format of experimental reports became reasonably standardized, until data were presented and analyzed in some detail, there was a great deal of room for competing formulas. The rapid developments in optics that we shall examine required analytic methods and elaborate experimental reports, and these did not appear for the most part until the third quarter of the eighteenth century.

Here I will prepare the ground for these developments by first examining the development of a new concept of *ray* during the seventeenth century. Then, after briefly discussing the older, competing view, I will turn to what was, or even could be, done with this new view at the time. I will concentrate especially on alternative quantifications for crystal optics, since that was the area in which the new concepts appeared in sharpest relief and in which the change in the quantitative and experimental structure of optics is most clearly apparent. This will bring us to the verge of a rapid series of developments in France shortly after the turn of the century that virtually created a new experimental optics and that were closely bound to a particular understanding of the physical nature of a ray.

1.1 The Ray and Medium Theories

Proposition 3 of Euclid's *Optics* asserts that "from every object there is a distance at which it is no longer seen." To prove this fact of, we would today say, visual perception, Euclid remarks that at some distance from the eye "none of the visual rays" that proceed from the eye to the object "will fall upon it," because the object[1] will lie entirely between[2] neighboring rays. For Euclid, it seems, the "ray" of light was a distinct physical object; one could in principle even count rays. The concept of a ray of light underwent many changes between antiquity and the seventeenth century as phenomena that were later attributed to the eye as a perceiving organ were distinguished from phenomena that depend solely upon the physical nature of light.[3] Nevertheless until the seventeenth century the ray, in its several mutations, was generally thought of as the physical foundation of light.

During the 1600s several mechanical theories of light emerged in which the ray was displaced from its foundational role but retained much of its former physical reality.[4] All these theories presume that a mechanical substratum fills the universe and that light is some sort of disturbance in this substratum or medium. In 1644 Thomas Hobbes developed a theory on this basis in which the ray still retained a distinctly physical character by insisting that it must occupy a finite volume—though for Hobbes the ray proper is not a thing but a "path through which the motion from the luminous body is propagated [in Hobbes's theory instantaneously] through the medium" (Shapiro 1974, 148, quoting from Hobbes). Indeed, only Christiaan Huygens and Ignace Pardies before him did not continue to grant the ray some sort of physical, rather than purely mathematical, reality during this period.[5]

The complex development during the second half of the seventeenth century of medium theories of light culminated in Huygens's *Traité de la lumière*. Huygens reduced the ray to a geometrical construct—in fact to a line—as Pardies had before him, but he went even further. Pardies defined the ray as the normal to the spherically expanding pulse that he took to constitute the physical basis of light. Huygens did not define the ray in this way. Instead, he introduced a more fundamental principle from which the direction of the ray can be deduced, and which is equivalent to Pardies's definition only in certain circumstances.

Huygens introduced his new principle in this way:

> That each particle of matter in which a wave spreads, ought not to communicate its motion only to the next particle which is in the straight line drawn from the luminous point [this was Pardies's assumption], but that it also imparts some of it necessarily to all the others which touch it and which oppose themselves to its movement. So it arises that around each particle there is made a wave of which that particle is the centre. (Huygens 1950 [1690], 19)[6]

Huygens defines the ray in terms of these secondary pulses that together reconstitute at any moment the primary pulse of light: the rays are lines joining the original center of disturbance to the points of the secondaries that touch the tangent common to them all. If the pulses are spheres, this is no different from Pardies's definition of the rays as the normals to the fronts. But, and this is the signal characteristic of Huygens's theory, if the pulses are not spheres then the rays will be oblique to the front. And if they are oblique, as we shall see they are in a certain crystal, then Snel's law will not hold.

Huygens's theory accordingly insists on a deep distinction between the elementary physics of light, which resides in his secondary pulses, and the optical ray, which is a mathematical construct defined with respect to a surface, the front, which is itself a construct (of secondary pulses). To accept Huygens's principle requires abandoning the idea that a ray has much intrinsic physical significance, and very few people were willing to do so.[7]

The failure of the principle to gain adherents is hardly surprising in light of the fact that, as a method solely for explaining the rectilinear propagation of light, it has nothing to offer that surpasses, for example, Pierre Ango's assertion (in his account of Pardies's theory) that front "parts" move only along the normals to the front.[8] Nor does it explain Snel's law any better. The true power of the principle arises only when the usual law of refraction does not hold—when, that is, the secondary pulses are not spheres, as Huygens argued they will not be in the crystal Iceland spar.

The peculiar optical behavior of Iceland spar was first described by Erasmus Bartholin in 1669. In crude description, the crystal doubles light, producing two images, one of which remains stationary while the crystal rotates about a normal to the facet, whereas the other image seems to revolve about the stationary one. One gets a better understanding of what occurs from Huygens's description (see fig. 1.1). Cover the surface ABCD of the crystal, leaving only a small hole at K. Then, Huygens remarked, two rays will be formed from a single ray that strikes normally along IK. One of the two, KL, goes straight through the crystal in the ordinary way, without deviation, whereas the other or "extraordinary" ray, KM, will lie in the crystal's principal section[9] and is inclined to KL; at emergence it is refracted back along the facet normal. From this it is simple to understand the appearance of a point marked at L on the base of the crystal. L will be seen, with the eye at I, by the extraordinarily refracted ray RI and by the usual or "ordinary" ray KI. Observations of this sort, Huygens remarked, led him to conclude that there are always two refractions, one of which follows the usual rule though the other does not. That is, one of the rays obeys Snel's law, but the other cannot be fitted to the law even if it is assigned a different index of refraction.

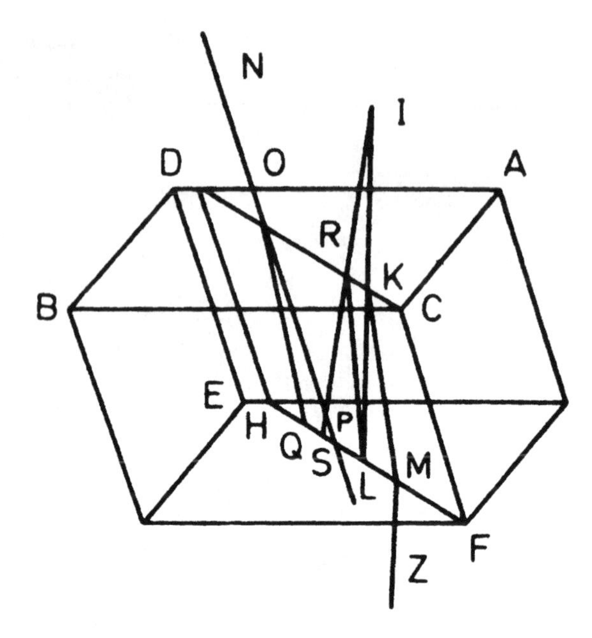

FIG. 1.1 Ray paths in double refraction.

Since Snel's law follows only from spherical secondaries, Huygens knew that the crystal had to propagate pulses of this shape as well as pulses of some other shape. He conjectured that the new pulses were spheroidal and proceeded to test the conjecture as best he could (on which see appendix 1). And here Huygens's principle is indispensable, because only it can determine the ray paths that are produced by nonspherical secondaries.

The pulse front, which even in Pardies's theory was more of an accessory to the rays than a primary interest (Shapiro 1974, 214), must now be at the center of attention. Yet this is required, empirically, only by refraction in a particular crystal (as far as was then known). Few scientists at the time were prepared even to adopt Huygens's geometry for determining the refraction (to the extent that either the construction or Huygens's principle proper was well understood) for such a limited purpose. This changed markedly a hundred years later when new experimental evidence that seemed strongly to confirm Huygens's geometry combined with certain developments in France to precipitate rapid changes in optics.

1.2 The Ray in Newton's *Opticks* and the Emission Theory

Isaac Newton always thought of light rays as individual physical objects, though he was careful to construct his early theory of color in a way that did not presume this definition (Shapiro 1975). We shall be concerned, however, not with developments during the seventeenth century, but with the early nineteenth century in France, where Newton's *Opticks,* not his articles or his *Lectiones opticae,* was a standard source. The *Opticks* defines the ray in a way that unquestionably requires that it be a discrete thing:

> Defin. I. *By the Rays of Light I understand its least Parts, and those as well Successive in the same Lines, as Contemporary in several Lines.* For it is manifest that Light consists of parts, both Successive and Contemporary; because in the same place you may stop that which comes one moment, and let pass that which comes presently after; and in the same time you may stop it in any one place, and let it pass in any other. For that part of Light which is stopp'd cannot be the same with that which is let pass. The least Light or part of Light, which may be stopp'd alone without the rest of the Light, or propagated alone, or do suffer any thing alone, which the rest of the Light doth not or suffers not, I call a Ray of Light. (Newton 1704)

This amounts to defining a ray as an atom of light.

By the 1790s in France Newton's definition of a ray was treated as essentially synonymous with an emission theory of light. Indeed, we shall frequently see that French physicists rarely distinguish between a "ray" and a "molecule of light" (though in the eighteenth century the word "molecule" connoted only a smallest part). There was, among them, a considerable looseness of expression in that they tended to use the same word—"ray"—to refer to a collection or bundle of rays (though on occasion they do use the word "bundle"), and they never explicitly distinguished between successive and collateral rays—between rays that follow one another and rays that lie next to one another. However, this looseness of expression does not reflect any conceptual confusion, because the essential requirement for French optics was that the basic physical elements of light be isolatable, individual objects.[10]

During the eighteenth century there was considerable development and criticism of the emission theory, and alternatives to it were also proposed that retained the physical identity of the ray without assuming light to consist of particles on which matter exerts Newtonian forces (Cantor 1983). Indeed, Young's and Fresnel's early accounts of interference were still based upon interactions between rays, and both of them (as well as Euler, who may not have known it) ignored Huygens's principle as an explanation for rectilinear propagation, instead relying on the requirement that a wave passing through a narrow slit will not diverge very much, making it behave quite like a ray.[11] Before Fresnel's adoption of Huygens's principle in 1818 and, even more important, his work on polarization and double refraction, the ray continued to have a central function even in the wave theory.[12]

Until well into the 1830s the old understanding of a ray persisted among many physicists who, though they might no longer insist on the emission theory, nevertheless did not adopt the wave theory. David Brewster, for example, defined a ray in the following way in 1831:

> Light moves in straight lines, and consists of separate and independent parts called rays of light. If we admit the light of the sun into a dark room through a small hole, it will illuminate a spot on the wall exactly opposite to the sun.—the middle of the spot, the middle of the hole, and the middle of the sun, being all in the same straight line. If there is dust or smoke in the room, the progress of the light in straight lines will be distinctly seen. If we stop a very small portion of the admitted light, and allow the rest to pass, or if we stop nearly the whole light, and allow only the smallest portion to pass, the part which passes is not in the slightest degree affected by its separation from the rest. The smallest portion of light which we can either stop or allow to pass is called a *ray of light.* (Brewster 1831, 12)

Until the last sentence Brewster's account could apply to the wave theory, because he writes only of parts; but the last sentence presumes that there are "smallest" parts, and this can be true only if rays are isolatable, discrete objects—and yet in this book Brewster (reluctantly) treated the wave theory as superior to the emission theory, remarking: "Each of these two theories of light is beset with difficulties peculiar to itself; but the theory of undulations has made great progress in modern times, and derives such powerful support from an extensive class of phenomena, that it has been received by many of our most distinguished philosophers" (Brewster 1831, 118).

1.3 The Transformation of Experiment

The years between Newton's death in 1727 and the foundation in the 1790s of the Parisian Ecole Polytechnique present many difficulties for the historian of optics. Unlike the last quarter of the seventeenth century or the first quarter of the nineteenth, these six-odd decades did not give birth to major unifying optical theories that are capable of exact expression. They are eclectic years, during which no single system of optics was widely accepted, despite the growing influence by midcentury, even outside Britain, of Newton's *Opticks*. Of even greater significance for understanding the fundamental changes that took place in the early 1800s, very little quantitative work, either theoretical or experimental, was accomplished. By far the majority of texts concerned with optics were purely qualitative. Furthermore, until well after midcentury the approaches to, and standards of, those experiments that were done scarcely differed from what they had been in the 1690s. By about 1810 all this had changed. Optical experiments had by then achieved unprecedented accuracy and were reported in meticulous, exacting detail. Of equal importance, they were linked to formulas that derived from a theory capable of exact expression.

There were two immediate stimuli to this transformation in optics. First, the foundation of the Ecole Polytechnique produced, by the early 1800s, a group of men with a common mathematical background and an intense concern with accurate measurement. Second, precisely because of this existing concern, several among them were prepared to recognize a challenge to their theoretical beliefs when they became aware of certain experimental results concerning indexes of refraction obtained early in the decade by the English chemist William Hyde Wollaston—even though Wollaston's own work did not differ markedly from eighteenth-century standards.

To appreciate the magnitude of the changes that had taken place by 1810, consider figure 1.2, which concerns double refraction, the area in optics around which the changes in experimental method coalesced. Here I have graphed the distance between the ordinary and the extraordinary rays in the principal section of the crystal as a function of the distance of the ordinary ray from the normal, for four laws: Huygens's (using both his parameters and modern ones), Haüy's (1788), La Hire's (1710), and Newton's. (Hereafter I shall call this distance the "amplitude of aberration," following Haüy's terminology.) The differences between them are striking. La Hire's law nowhere approaches accuracy, while Newton's does so only at very small incidences. Haüy's formula works reasonably well up to about 45° of incidence. Nevertheless Newton, La Hire, and Haüy were each convinced that their respective formulas worked well. This was possible, in the cases of Newton and La Hire, not because the experimental techniques of the time were inadequate to reveal the inaccuracies, but because there were no commonly accepted standards of experimental reporting demanding that formulas be pushed as far as technique permitted. In Haüy's case the problem was the technique itself, for the approach he used could not in fact differentiate between his formulas and Huygens's. It is worth spending a few moments on La Hire and Haüy in order to appreciate by contrast the very large change in standards and technique that had taken place by the end of the century.

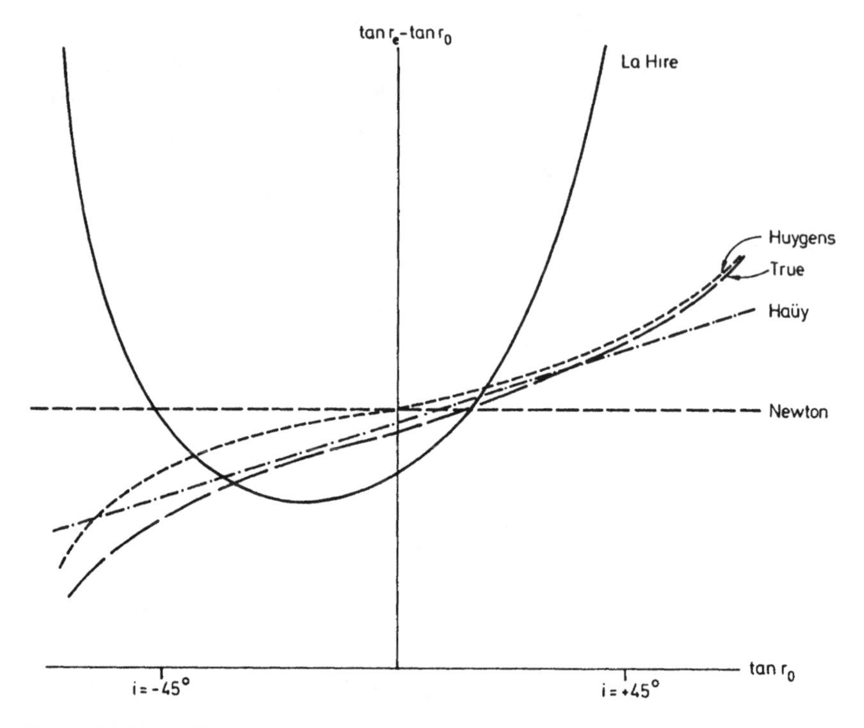

FIG. 1.2 The amplitude of aberration.

1.4 Confronting Formulas with Experiment

Philippe de La Hire was familiar with both Huygens's *Traité* and Newton's *Opticks,* though he took little from either except a rather loose concern to establish quantitative relations. He did not set out to test Huygens's construction or Newton's rule. He simply ignored both and offered instead an alternative that did less violence than either to the usual laws of optics but that the common observational technique of the time could easily have proved inaccurate.

La Hire argued, in effect, that nothing new was really taking place optically. Indeed, he continued to believe that Snel's law, considered solely as a relation between an angle of incidence and one of refraction, continued to hold. What did change was the line from which the angles must be measured. La Hire hypothesized that this line must be the one along which an incident ray is not doubly refracted. With respect to it, La Hire calculated the index to be as 4.5 to 3. Computing the refractions in the principal section for this law,

I find that the predicted distances of the refractions on the base of a 39 mm crystal differ from the true distances by at least 0.5 mm below 20° of incidence, and by about 1 mm over the next 10°. The accuracy of the technique used by Huygens (and probably almost everyone else until Wollaston) is better than 0.5 mm, so La Hire could easily have found his law highly inaccurate—provided that he could actually have calculated the position of the refraction from it.

This is not an insignificant task, since the formulas that locate the refraction on the crystal base can be obtained only by projecting the rays and the fixed axis onto it. To do so requires some rather intricate geometry, followed by extensive computation, even for the principal section.[13] Not even Huygens had actually examined a series of incidences to test his construction; La Hire did not do any more than Huygens had, and he probably did considerably less.

At least until well past the middle of the century, there was little if any notion that formulas should be confronted systematically with experiment. Huygens used experiment primarily to determine parameters. La Hire and Newton did the same. One might attribute this lack of efforts to test formulas to contemporary views on the relation between theory and experiment, which no doubt differed in many respects from those that became common after about 1800. But however different these views may have been, neither Huygens, Newton, nor La Hire ever questioned that a formula must be adequate to the demands of experiment.

The connection between scientists' methodological views on theory and reality may not be close to their laboratory practice, though I would not, of course, deny that some link between the two may exist. Scientists in the laboratory do not usually decide consciously to apply this or that method derived from considerations of the metaphysical links between science and the world. For the most part they do what they think their colleagues (or their intended audience) will generally accept, at least from the standpoint of experimental method. Huygens certainly did not expect anyone to quarrel with the way he determined his parameters, and in fact no one did—for they proceeded in almost exactly the same way.

The fundamental requirement that formulas must connect directly to experiment did not alter between the beginning and the end of the century. What did change was the structure of experimental reports. At the beginning of the century it was enough to compute a few parameters and report that the formula seemed to work pretty well. By the early years of the nineteenth century it was essential to use the experimentally determined parameters to examine the formula's range of accuracy. Almost no one in 1700 gave explicit estimates of the accuracy to be expected with a given apparatus and formula; by about 1810, in France at least, estimates of accuracy (though often weak) were considered essential components of experimental papers. One of the factors in this change was the replacement of geometry by algebra; this made it possible to carry out calculations with a minimum of effort and to avoid errors that could, and did, arise because of the complex geometric structure that was usually necessary to capture what can be encompassed in a few algebraic formulas.

1.5 Computing with Geometry

Our third eighteenth-century formula for double refraction, invented by the crystallographer René-Just Haüy, indicates that this sea change in experimental practice did not occur all at once, that parts of it emerged before others. Haüy, like his predecessors, reported very few numbers and gave no estimates of accuracy. Moreover, his experimental arrangement was not substantially more sophisticated than theirs—like them, he used a pen, a ruler, a piece of paper, and a crystal of reasonable height. Furthermore, he did not think it necessary—if he thought of the point at all—to report numerical results or carefully specify his observational accuracy and its effect on the formulas he tested. But quite unlike his predecessors, Haüy attempted systematically to confront formulas with experiment.

Haüy directly measured the quantity I have graphed in figure 1.2—the distance on the base of the crystal between the ordinary and the extraordinary rays, or what Haüy called the "amplitude of aberration." He used an old, and extremely simple, observational technique in an attempt to satisfy the emerging contemporary standards for confronting formulas with experiment.

Haüy marked a series of points on a piece of paper at diverse distances from one another, though on the same line, and he placed the crystal over the points in such a manner that the principal section contained the line. He then observed the points two at a time through the crystal, looking always in the principal section. Four images were in general visible. But if the distances of two of the points were varied, a distance between them could always be found such that only three images were visible (see fig. 1.3). What happens is this. Let n be so marked that it is seen in ordinary refraction by the ray ok (set normal to the crystal for simplicity) and by the extraordinary ray om. Let r′ be so marked that it is seen by the ordinary ray ol and by the extraordinary

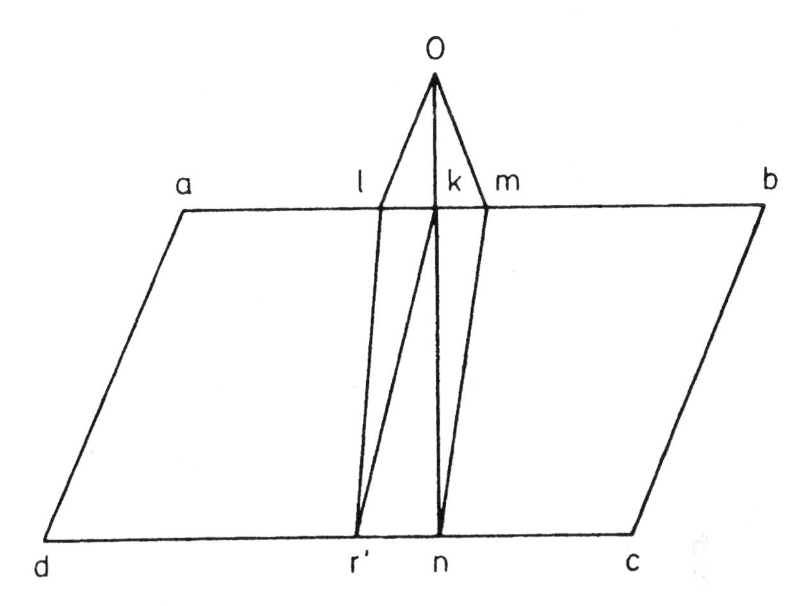

FIG. 1.3 Haüy's test of Newton's law.

ray ok. The ordinary image of n is then coincident with the extraordinary image of r′. The ordinary image of r′ and the extraordinary image of n appear on opposite sides of point k, directly above n. Consequently one sees only three images.

With this Haüy could at once test Newton's law, which asserts that the amplitude of aberration should be independent of the angle of incidence. That, he reported, is easily seen not to be true. Given the points r′, n marked just as described, whenever the eye is placed in the principal section it should see only three images at all inclinations, according to Newton's law. But as the eye moves toward b the images separate in such a fashion that the extraordinary image of r′ is displaced away from the ordinary image of n in the direction of d. That is, the amplitude of aberration is decreased for rays inclined toward b and increased for rays inclined toward a.

The separation Haüy observed is quite simple to see, and it can also be calculated. For example, using Huygens's construction with modern parameters and Snel's law, I find that, for a 39 mm crystal, the image separation along the crystal base reaches 2 mm at 30° of incidence (see Buchwald 1980b, 371–72). By combining a completely traditional experiment with a clever new observational technique, Haüy disposed of Newton's formula.

Haüy then turned to Huygens and La Hire; he claimed that their formulas also fail, but that one he himself invented works very well. That claim must astound in view of figure 1.2, which shows Haüy's own formula to be considerably worse than Huygens's even using Huygens's own parameters. What probably happened here strikingly reveals Haüy's incomplete emergence from earlier standards as well as showing how the provenance of a formula can influence the way an experimenter deals with it.

To understand what occurred, begin with Haüy's own formula. In figure 1.4 ∠gan equals the deviation of a normally incident ray (abcd is a principal section). From g draw, at ∠qgd equal to 30°, the line segment qg to an. Then from the point e where the ordinary ray ke from an incident ray ik strikes the base, draw the segment er parallel and equal to qg. Finally, from the point of incidence k draw a line kr through r. That line is the extraordinary ray according to Haüy. From this he easily deduced that the distance el on the base between the ordinary and the extraordinary refractions equals gn + (em/

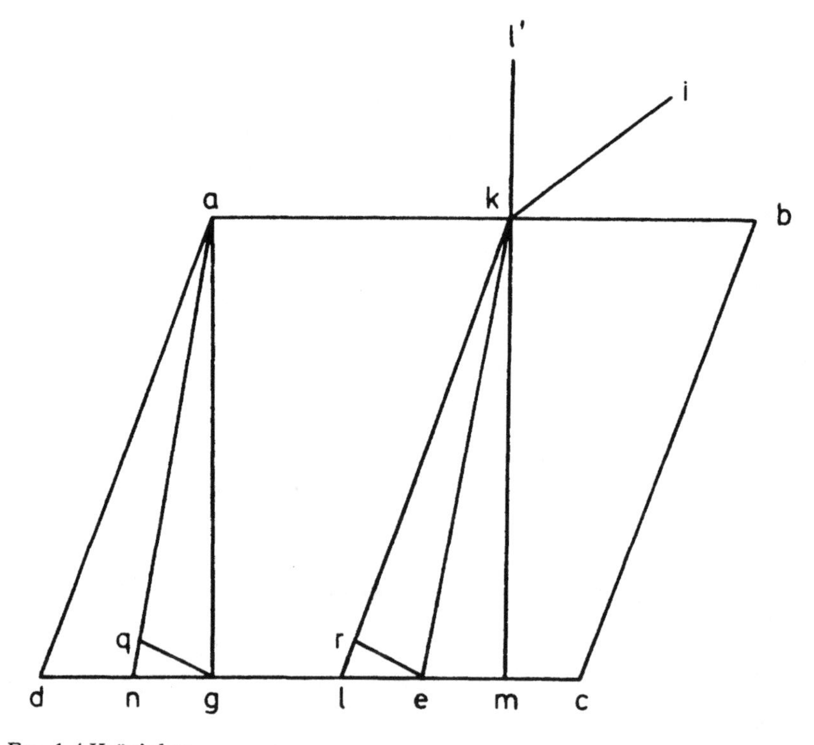

FIG. 1.4 Haüy's law.

$\sqrt{243}$). Consequently if δ is the normal deviation ($\angle$nag), then Haüy's law is equivalent to the algebraic formula:

$$\tan(r_e) = \tan(\delta) + \left[1 + \frac{\text{em}}{\sqrt{243}}\right]\left[\frac{\sin(i)}{\sqrt{n^2 - \sin^2(i)}}\right]$$

Here i is the angle of incidence; r_e is the angle of extraordinary refraction, and n is the index of ordinary refraction.

But Haüy did not formulate his law algebraically, and this led to several errors, the first of which concerns the relation between it, Newton's rule, and Huygens's construction.[14] Haüy, like Huygens, proceeded exclusively in terms of ratios, and this makes computations tedious. Yet computation is precisely what one needs here: the only way to test the formulas is to compute the positions of rays on the crystal base, and that is a task much better adapted to algebra than to geometry. Until algebraic, rather than synthetic, methods came into common use, there was a practical barrier to many systematic comparisons with experiment: the large amount of computation that was often necessary to obtain even a single result. In the case of Huygens's construction proper, it was actually impossible to perform any computations at all outside the principal section and its normal plane because the constructive, geometric approach produced a formula for the measurable coordinates only in these two planes.

The difficulties posed by the use of geometric methods where algebraic ones are more appropriate appear most strikingly in Haüy's attempt to show that La Hire's formula fails as badly as Newton's. To minimize computations, he tried to use the results of his test of the Newton formula. In figure 1.5, point m lies directly below the point of incidence on the crystal facet; the x axis lies in the principal section. Consider a ray that strikes above m at a given incidence but in any plane; its ordinary refraction will always strike the base of the crystal on a circle e^1e^2e of a radius equal to the product of the crystal height by the tangent of the angle of ordinary refraction, because for a fixed incidence the angle of refraction is the same for all planes of incidence. The incident ray, when prolonged to the base, always strikes along a larger circle h^1h^2h of a radius equal to the crystal height by the tangent of the angle of incidence. La Hire's fixed axis strikes the base at point s, which lies at a distance from m equal to the product of the height by the tangent of the axial inclination.

The incident ray, the extraordinary ray, and the fixed axis lie, according to

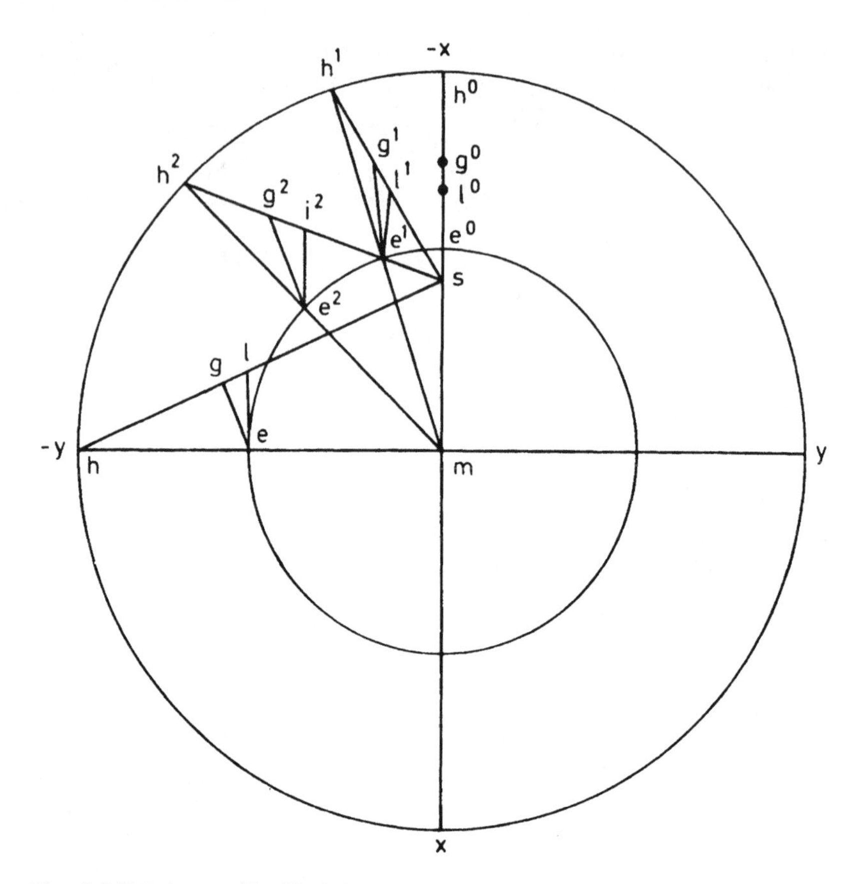

FIG. 1.5 Haüy's test of La Hire's law.

La Hire's formula, in the same plane. Consequently the points g^1g^2g of the base struck by the extraordinary rays from incident ones that, prolonged to the base, strike at h^2h^2h must lie somewhere along the lines h^1s, h^2s, hs. Precisely where they lie is fixed by La Hire's formula, but this would be immensely hard to compute without expressing it algebraically. Haüy accordingly proposed to show by geometric reasoning that the assumption of a common plane is by itself untenable, and for this he thought he could use experiments done entirely in the principal section, in which case the positions of g^1g^2g need not be known precisely.

The amplitudes of aberration for the four incidences in the figure are e^0g^0, e^1g^1, e^2g^2, eg where e^0g^0 corresponds to a ray incident in the principal section and striking the crystal base, when prolonged, at h^0. In general, Haüy continued, the aberration will not be parallel to mx (the principal section), but for incidences in that section it will be. Suppose then that the aberrations are as in the figure, and consider the line segments el, e^2l^2, e^1l^1, which are parallel to mx by construction. From the figure, e^2l^2 is equal to $(ms/m^2h^2)e^2h^2$. Since similar relations hold for e^1l^1 and el, and since mh^0, mh^1, and mh are equal to one another, it follows that e^0l^0, e^1l^1, e^2l^2, el are also equal to one another. Haüy concluded that, since the aberration must ultimately become parallel to mx, it will finally become equal to the constants e^0l^0, which are parallel to mx.

Precisely the same situation obtains for rays that, coming from the $-x$ region, strike the base in the $+x$ region, that is, the aberration of a ray incident in the principal section at the same angle as the ray which, prolonged, strikes at h^0, but on the opposite side of the normal at m, must be equal to the constants e^0l^0. Consequently, Haüy concluded, according to La Hire's formula rays of equal but opposite incidences in the principal section produce the same aberrations. But his experiments with Newton's formula had clearly shown that the aberration decreases as the eye moves from above $+x$ to above $-x$. La Hire's law must therefore be erroneous.

Haüy was wrong, though not about the inadequacy of La Hire's law. Consider, for example, a 39 mm crystal and 35° of incidence. According to Haüy's argument the constants e^0l^0 should be 5.56 mm. A direct calculation using the algebraic translation of La Hire's law yields 7.12 mm. For an incidence of $-35°$ in the principal section, the aberration rises markedly to 10.35 mm, far indeed from Haüy's 5.56 mm. How could he have been so badly mistaken?[15]

Geometry was the problem. Consider the points g^1g^2g as generated by the motion of a point along the line sh^0 as the line rotates successively with the plane of incidence to the points h^1h^2h. According to Haüy's construction, when the rotating line lies in the principal section, the moving point lies at l^0. In fact it lies at g^0. In other words, Haüy erroneously assumed that, because e^1g^1, and so forth, ultimately become parallel to e^0l^0, they must also become equal to it. He was betrayed by his limit procedure, in that he failed to realize that e^1g^1 could become parallel to mx by collapsing the triangles $e^1l^1g^1$ as well as by causing g^1 to approach l^1. The complex approach imposed by geometric reasoning vitiated Haüy's comparison of the law with experiment.

Haüy, I remarked above, felt his own law was superior to Huygens's construction, which he found "inaccurate" at higher incidences. Here we have an instance in which not the kind of mathematics used, but the method of observation that Haüy, and most everyone before him, employed was inadequately matched to the formulas being tested. Haüy's claim also reflects the incomplete development, even as late as 1788, of standards for reporting experiments designed to compare a set of formulas with one another.

Haüy implied that at lower incidences—he did not write how low—Huygens's construction worked reasonably well. But when he tried higher incidences he found it failed. Now the surprising fact, which I have ascertained by experiment, is that Haüy was correct. In this kind of experiment, where the eye must judge image coincidences, every law, including Huygens's with modern parameters, fails to represent the amplitude of aberration for a large crystal beyond about 55° of incidence. At and beyond this angle, the images reach coincidence before they should, and they remain together as the incidence increases. The effect may be due to a complicated phenomenon that involves the finite size of the eye's pupil.

Haüy did, then, have reason to reject Huygens's law, but he had as much reason to reject his own, since it fails in precisely the same way Huygens's does.[16] But it seems highly probable that Haüy discovered the empirical failure of Newton's and Huygens's laws before he invented his alternative. His replacement overcomes the deficiencies of Newton's and works as well as Huygens's at low incidences. I think it quite likely that Haüy never bothered to test it where he already knew Huygens's law failed.

No tables of experimental data, no list of the results of computation appear in Haüy's article. He disposes of Newton easily, La Hire incorrectly, and Huygens inadequately by citing only general properties. There are few numbers. Of his own alternative he remarked only that "it always appeared to me that the result of observation was consistent with that of calculation" (Haüy 1788, 47). Yet Haüy was very sensitive to the necessity of minimizing observational errors, and he certainly sought to compare the empirical adequacies of the four laws (including his own) he had available to him.

Haüy's work delineates a boundary in the history of experimental optics. It recognizes the need to compare the empirical worth of various formulas, whereas both Newton and La Hire had essentially ignored alternatives to their own proposals and Huygens had made no attempt to test his construction over several incidences. Haüy strongly felt that a series of tests were essential, and he performed a fair number. But in other, deeper respects his work remains closer to his predecessors' than to his successors'.

Like his predecessors, Haüy preferred to work with geometric, rather than algebraic expressions. This means that instead of inserting terms into an algebraic formula, one must construct a solution through a series of proportions. There are two effects. First, the amount of labor involved is usually much greater for the constructive method than for the algebra. As a result very few examples are ever worked through, and lists of results are hardly likely to be included. Second, regularities that may be easily found in the algebra are often hidden by the geometry, because the route from proposition to proof must go through a construction. Haüy's errors would have been much more difficult to make had he used algebraic formulas.[17]

A major part of the sea change in optical experiment that occurred during the last part of the 1700s involved the replacement of geometric with algebraic formulations. We shall presently see that the major stimulus to the exceedingly rapid growth of quantitative optics in the early 1800s was Malus's translation of Huygens's construction into algebra. It would be only a slight exaggeration to claim that algebra had to replace geometry before the systematic confrontation of formulas with experiment could become common—or at least they had to occur in tandem. Haüy felt the need to effect the comparison, but his mathematical tools were too cumbersome to do so easily. Haüy no doubt applied constructive methods incorrectly, but the methods themselves ill fit the task he set himself.

Haüy struggled with traditional methods in another respect as well. He clearly recognized that he had to find quantities to observe that could easily (given the mathematical limitations he labored under) distinguish between the formulas. And he succeeded in doing so by recognizing the amplitude of aberration as a quantity he could observe quite directly and about which the several laws differ. But like his predecessors, Haüy was content with naked-eye observations, and these were inadequate to his task. Moreover Haüy would perhaps have realized the inadequacy of his technique had he undertaken a series of measurements at many incidences and compared the empirical results with those of calculation—since then he would have seen how badly his, as well as Huygens's, law begins to fail at higher incidences. Instead, he tried one or two high incidences that he apparently computed using only Huygens's construction; finding that it failed, he turned to his own, which he did not subject to the same test. This is precisely what no one was likely to do because the constructive method made it so difficult.

1.6 Embodying a Formula in an Instrument

There was one way to avoid the computational problems raised by constructive methods without turning to algebra, and that was to embody the geometric solution in an instrument. Before 1802 the several methods used to determine indexes of refraction depended upon measuring distances in one way or another, with the result that a calculation—and so a construction—remained necessary to determine the index (Brewster 1813). In 1802 the

English chemist William Hyde Wollaston bypassed the computational difficulties posed by geometric methods by creating a device that incarnated the construction.

It was a well-known consequence of Snel's law that, for any pair of contingent media, the maximum refraction occurs when the ray incident from the more refractive medium coincides with the surface of separation at its incidence (or, in realistic terms, scrapes it). The sine of the angle of refraction is then equal to the reciprocal of the index. That maximum angle is vice versa the angle of internal incidence beyond which a ray is not refracted past the interface but is reflected back into the more refractive medium; that is, it is the angle of total internal reflection. Where, specifically, a prism has (relative to air) an index of n^1, and its external medium has a lesser index of n, then the sine of the angle of total internal reflection for the prism is equal to n/n^1. If, therefore, one has a prism of known index, it is possible to construct the index of the external medium by measuring the angle of total reflection. The brilliance of Wollaston's technique was that it eliminated the construction by using the eye only to judge the presence or absence of light rather than to mark distances.

In Wollaston's diagram (fig. 1.6), A is a square prism of known index n^1; B is a substance of unknown, smaller index n. If a ray cb enters B at nearly grazing incidence to the plane of separation between A and B, then it will be refracted at ∠hbd, the maximum of refraction, which is the angle of total internal reflection for a ray incident in A upon B. Consequently sin(∠hbd) is

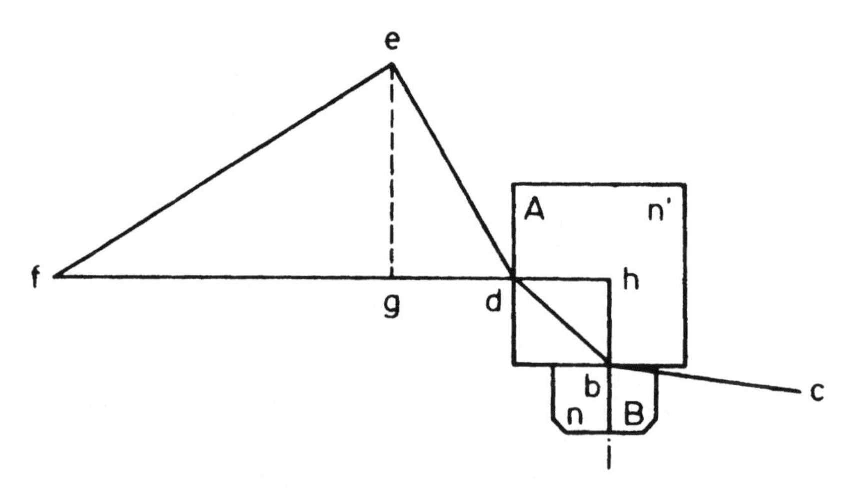

FIG. 1.6 Wollaston's ray paths.

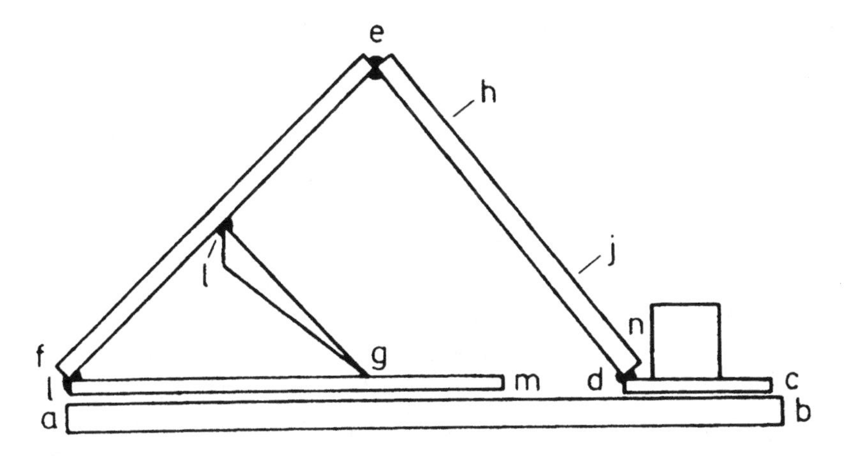

FIG. 1.7 Wollaston's device.

equal to n/n^1, and ∠cbi is very nearly 90°, from which it follows that ef/gf is in the same ratio as n^1 to n.[18]

This can be embodied in an instrument. Build a device in which ef, ed remain always the same, while ∠fen can be varied until a ray seen along ed marks the limit of total reflection. Wollaston's device (fig. 1.7) does just that. In it ab, dc, en, ef, and lm are separate wooden boards that are hinged to one another at f, e, and d; l′g is a wedge of length $\frac{1}{2}$ef that is hinged to ef at the latter's midpoint. By construction the movable point g remains vertically below e whatever the value of ∠fen. Wollaston cut ef to a length of 15.83 inches, and he cut ed 10 inches long, so that the index of the prism was 1.583 (hence made of flint glass). He must then have marked a scale, calibrated in hundredths of an inch, along board lm such that point g of the wedge pointed to the corresponding value of n, equal to gf/ed, for a given ∠efg. No calculation of any kind needed to be done to measure an index of refraction.

At some point Wollaston put a piece of Iceland spar in the device. He obtained a confusing series of "indexes" that depended upon the orientation of the spar relative to the plane of incidence, and he apparently asked Thomas Young for an explanation (Young 1809, 339). Young referred him to Huygens's *Traité*. Wollaston reacted favorably to its account of double refraction:

> The optical properties of Iceland spar have been so amply described by Huygens, in his *Traité de la Lumière,* that it could answer little purpose to make any addition to those which he enumerated. But, as the law to which he has reduced the oblique refraction occasioned by it, could not be verified by former meth-

> ods of measurement, without considerable difficulty, it may be worth while to offer a new and easy proof of the justness of his conclusion. For, since the theory by which he was guided in his inquiries, affords (as has lately been shown by Dr. Young) a simple explanation of several phenomena not yet accounted for by any other hypothesis, it must be admitted that it is entitled to a higher degree of consideration than it has in general received. (Wollaston 1802b, 381)

At once Wollaston faced the daunting task of applying Huygens's construction. He read the *Traité* with enough care to obtain from it the geometric solution I have called the "law of proportions" (appendix 1) and to see that this solution could be interpreted in terms of his device. In Wollaston's diagram (fig. 1.8), the plane of incidence FRO intersects the Huygens spheroid in a diameter FO that lies on the surface of the crystal. FTO is the section of the spheroid by the plane of refraction. This plane contains CT, the conjugate to FO, as well as FO itself. The refraction CI of an incident ray RC is determined by the law of proportions: VR/EI = N/FC. When set to measure the angle at which the top of the crystal just becomes visible, Wollaston's device detects a ray that has, in effect, emerged from the crystal at grazing internal incidence. Consequently FC = EI, and VR/EI = VR/RC = sin(i)/1, where i is the angle at which the ray emerges into the glass prism (∠hbd in fig. 1.6). Hence sin(i) is N/FC. Since one reads off the value of gf/ed directly from the scale, Wollaston at once had (N)(ef)/(ed)(FC). He took the reciprocal as the relative value of FC, since N and ef are permanently fixed.

The device, then, could measure a radius of Huygens's spheroid as easily

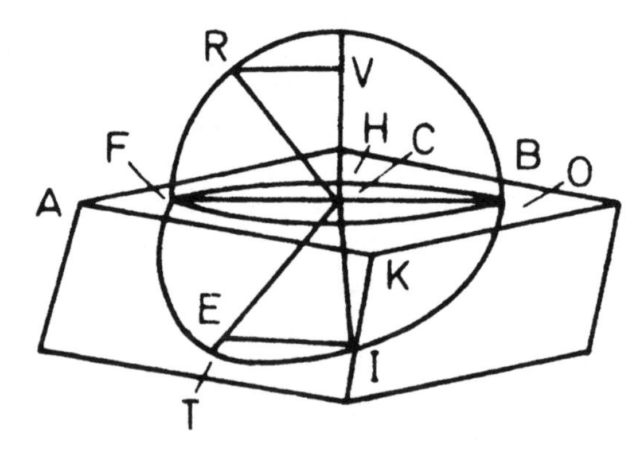

FIG. 1.8 What Wollaston measured.

as an index of refraction, and Wollaston carried out a series of experiments designed to test the law of proportions. We need not examine them (see Buchwald 1980b for details) to see a problem immediately. Wollaston's device necessarily fixes the angle of incidence for a given arrangement, so that one can only test the variation of FC as the crystal's principal section alters with respect to the plane of incidence. Wollaston did just that, and in two experiments he found excellent agreement. In one of these the inclination was 50°57′30″; in the other it was 39°2′30″. What Wollaston failed to realize was that the law of proportions is invalid except in the principal section or its normal. He had demonstrated that a law that should not work nevertheless does.

Geometry was again the problem. Huygens did not, and probably could not, provide a constructive method that permitted computation outside the principal section or its normal plane. Wollaston needed a method adaptable to his device, and he read Huygens far enough to find one—the law of proportions. Quite probably Wollaston perused the *Traité* looking for a geometric method and stopped when he found one, not realizing that it is generally inapplicable to his device.

Yet he found good agreement between the law and experiment. The reason is that the difference between the law of proportions, used in any plane, and the proper law is actually less than the experimental accuracy of Wollaston's device, which one can show to have been between 20′ and 30′ in the incidence i.[19] But one can also show that the law of proportions deviates from the true law by almost exactly this same amount.

Wollaston had certainly improved the instrumental aspects of double-refraction measurements, but in fact it did him little good. Since he did not, and probably could not, translate Huygens's construction into algebra, his device was actually useless. But the great complexity of the construction, combined with Wollaston's desire to find a way of using his device to test it, left him entirely unaware of the problem.

1.7 The New Optics

1.7.1 War, Plague, and Scientific Creativity

On 19 December 1808 Pierre-Simon de Laplace announced to the First Class of the Paris Institute that Etienne-Louis Malus, a graduate of the Ecole Polytechnique, had confirmed Huygens's construction for double refraction (*Procès-verbaux* 1804–7, 631–32). Malus's experimental work contrasts starkly with that of his predecessors. Where they constructed solutions, he applied analytic geometry (which he had learned from Gaspard Monge);

where they haphazardly mentioned a few numbers, he provided tabular lists and strove to take account of inaccuracies in measurement. And where (with the exception of Wollaston) they relied on the simplest of experimental arrangements, he built and used highly accurate devices. Malus established new standards for experimental work in optics, standards it rapidly became essential to satisfy if one's work was to be taken seriously.

Born in 1775, Malus took the examination for the Mézières school for engineers in 1793 (Arago 1857, 2:118). The school was "suppressed," however, and Malus became for a time a manual laborer working on field fortifications. Then, Arago records:

> M. Lepère, engineer of roads and bridges, who was directing a part of these constructions, having remarked certain peculiar and unexpected arrangements in the manner in which the soldiers executed the excavations and raised the mounds, was desirous to learn the origin of these practices; they pointed out to him the man who had indicated these as the means best suited to attain the desired end with the least possible fatigue. A few moments' conversation showed the engineer that he had found in the humble labourer of the 15th Battalion of Paris a superior man; and he accordingly sent him to the "Ecole Polytechnique," which had just been founded. (Arago 1857, 2:118–19)

During the decade after his graduation from the Ecole, Malus fought and engineered in Egypt and elsewhere in the Middle East. In Syria he contracted the plague and witnessed bloody massacres. His journal powerfully describes the pillage of Jaffa by French troops to which he was witness:

> The enemy was overthrown, discouraged, and retired, after a sharp firing of musquetry from the houses and forts of the city; they kept their ground, however, at some points, and continued their fire for an hour. During this time the soldiers, scattered through all parts, killed men, women, children, old persons, Christians, and Turks;—everything that bore the human form was the victim of their fury.
>
> The tumult of carnage, the broken doors, the houses shaken by the noise of the firing and of arms, the cries of the women, the father and child overthrown one on the other, the violated daughter on the corpse of her mother; the smoke of dead bodies burned in their garments which had been set on fire, the smell of blood, the groans of the wounded, the cries of the conquerors disputing together over the spoils of their expiring prey, infuriated soldiers

> responding to the cries of despair by exclamations of rage and redoubled blows; lastly, men satiated with blood and gold, falling down in mere weariness on the heaps of corpses;—such was the spectacle which this unfortunate city presented until night. (Arago 1857, 2:124)

The dead were revenged by plague. "A burning fever, and violent pains in the head," Malus wrote, "forced me to seek repose. . . . Only one man in twelve escaped. St. Simon arrived in Egypt and came to see me; he was then in perfect health, in two days afterwards he was dead. . . . I was now alone, without strength, without help, without friends. . . . Two men of the corps of sappers undertook the care of me, and they perished one after the other." Put on board a ship, whose captain soon died of plague, Malus felt suddenly "relieved from suffocation." He recovered, though not without effects on his health that were probably responsible for his death a decade later.

At Lesbieh he recuperated, surrounded by plague victims and witness to scenes of almost incredible horror. "One wretched woman, of whom I had taken care because she was absolutely deserted, begged of me the evening of her death to give a piastre to the gravediggers, that she might be preserved from becoming a prey to the jackals. I fulfilled her wish, and caused them to bury her at the extremity of the plain where the dead were deposited." One can imagine Malus's relief when he left what he must certainly have thought was his deathbed for a posting where "we encamped in huts whose walls and roofs were composed of palm-leaves interwoven." In that hut he composed his first memoir on optics.

Malus's frame of mind when he turned to optics was very different from Arago's or Fresnel's a decade later. Arago had to justify his appointment to the Institute; Fresnel early on wished to make a major "discovery." Half dead from the effects of the plague, with vivid memories he found hard to forget, Malus embraced optics as a refuge, not as a new battleground. He sought not to dispute but to elaborate and to improve upon concepts that the chemistry lectures at the Ecole had imparted, to prove, in particular, that light is a compound of caloric and oxygen (Arago 1857, 2:135). This contrasts remarkably with Fresnel's first work, which sought to prove precisely the opposite—that light cannot possibly be a compound substance. Arago remarks Malus's "perfect initiation" in the "general principles" and "details" of contemporary chemistry, in which he was not alone since most Ecole graduates for at least the next decade had similar knowledge.

Malus's chemical discussion of light, which was never published, appeared before book 10 of Laplace's *Mécanique céleste,* which effectively established an orthodox understanding of the emission theory. Malus likely had had very little if any direct contact with Laplace before about 1807; his adherence to the emission theory therefore resulted primarily from chemical considerations rather than physical or mechanical ones. We shall see that Fresnel's rejection of the emission theory derived from chemical considerations very like Malus's. Much that Malus found persuasive, Fresnel later found unacceptable. Both men were polytechnicians; both were mathematically adept, though hardly creative; both were extremely good experimentalists; both developed new, quantitative theories; both had very poor health (Malus because of the plague, Fresnel from childhood). And both began their optical researches during periods of enforced retirement from military action. But Malus was retired by plague, and for him optics was a refuge from the horrors of war, death, and disease. Fresnel was retired by politics because he refused to support the return of Napoleon during the hundred days. For him unorthodox positions in optics were aspects of a rebellious nature that appeared early in his life and that sought renown.

1.7.2 The General Problem of Optics

We do not know precisely what occupied Malus in optics between his recovery from the plague and 1807, when he first read a memoir to a Parisian audience. A letter dated 5 October 1800 indicates, according to Arago, that Malus was "occupied theoretically with that most important meteorological question, the distribution of heat in different climates" (Arago 1857, 2: 138)—an appropriate topic in view of his sojourn in the Middle East as well as the role of heat (caloric) in the structure of light particles.[20] Malus did not have direct, continuing contact with Paris until 1809—indeed, he founded the science society at Lilles, where he was stationed—so that his work emerged essentially in isolation (very unlike the later work of Arago and Biot, or even Fresnel's after 1815).

Malus's isolation may have been something of an advantage in that he was able to distinguish two aspects of optics from one another in a very clear way, which would have been difficult had he worked from the beginning in Paris. With the publication of book 10 of the *Mécanique céleste,* Laplace had firmly directed his contemporaries' attention to the physics of short-range forces (Fox 1974). This had a number of significant effects, but for our purposes the most immediate one was to conflate two things that Malus implicitly kept separate: the substantial existence of rays was not distinguished from the hypothesis that rays are composed of particles.

Despite his deep-seated belief in the emission theory, the first work that Malus read to a Parisian audience abstracted from light's physical structure. Instead, Malus distilled the essence of optics as he understood it into a single goal:

> The rays that emanate from a luminous point in a medium of uniform density may be regarded as a system of right lines passing through this point. When these rays encounter the surface of a body that reflects or refracts them, their mutual disposition experiences different modifications from which arise all the phenomena of optics. (Malus 1810, 216)

In this first study Malus expertly deployed methods that he had learned from the mathematician Gaspard Monge while at the Ecole (Chappert 1977, chaps. 4–5—though Malus did not mention Monge) to establish the proposition (subsequently known as Malus's theorem) that all the rays in a bundle that emanate from a given point are, after a single reflection or refraction, normal to some surface.[21] Laplace, Joseph-Louis Lagrange, Monge, and Sylvestre Lacroix examined the paper, concluding:

> To apply thus, without any limitation on its generality, calculation to phenomena;—to deduce, from a single consideration of a very general kind, all the solutions which before were only obtained from particular considerations,—is truly to write a treatise on analytical optics, which, considering the whole science in a single point of view, cannot but contribute to the extension of its domain. (Arago 1857, 2:139)

Though Malus's work here did not depend in any way upon the existence of physical rays, still the goal of the essay was to determine the behavior of rays, not the surface to which the rays are normal after a single refraction or reflection. That surface was an analytical device without physical significance that depended entirely upon the laws of reflection and refraction. If these laws were different, then Malus's deduction would have to be redone entirely and might not work. Malus never, for example, attempted to see whether such a surface would exist given Huygens's construction. Indeed, even after Pierre-Charles Dupin demonstrated in 1816 that Malus's analysis could be amended to show that such a surface exists after any number of reflections or refractions, the conclusion remained purely mathematical since the surfaces had no physical significance (see Chappert 1977, chap. 5 on Dupin).

Nevertheless, in concentrating on the properties of ray collections, or bundles, Malus in a striking way set the physics of the ray apart from the ray itself. Furthermore, the program he defined could be, and was, applied not only to the directions of rays but to other properties they might share as elements of a given bundle. Like their "mutual dispositions," these properties depend in no direct way upon ray physics, though we shall see that they do require that rays have substantial existence (which was not required for Malus's theorem).

Of the four commissioners who reported on Malus's first work, only Laplace was not exclusively a mathematician, and there can be little doubt they considered Malus's results entirely analytical in significance. Yet not seven months after presenting this work to the Institute's First Class (20 April), Malus presented another paper (16 November) that recurred directly to the physics of rays and so to the kinds of questions that had initially engaged him. It seems very likely that he undertook this investigation at the urging of Laplace, with whom he probably had contact about the time he read his first paper (or perhaps shortly after). This second paper concerned Wollaston's device for measuring indexes of refraction.

1.7.3 Problems with Wollaston's Device

Despite the war, both of Wollaston's articles were translated into French and published in 1803 in the *Annales de Chimie*. Laplace's first major excursions into the physics of short-range forces had appeared in 1805 (atmospheric refraction) and 1806 (capillarity). Further, in the recently published book 10 of the *Mécanique céleste* Laplace had extended his considerations to establish formulas for the refractive indexes of opaque as well as transparent bodies. These formulas were, as we shall see in a moment, not reconcilable with the structure of Wollaston's device. Laplace almost certainly asked Malus to look into the question, perhaps even suggesting to him that Wollaston's device could not be used for opaque bodies.

On emission principles Wollaston's device is useless for opaque substances. The device depends upon the occurrence at a certain angle of total internal reflection within a prism of glass that abuts the substance whose index one wishes to measure. By emission principles total internal reflection must begin at a certain fixed angle of internal incidence at which the square of the vertical component of the particle's velocity is precisely equal to the vis viva the attraction of the medium abstracts from it. The particle will, at this incidence and for a range of angles beyond it, necessarily emerge from the medium but will curve back into it. The maximum distance it achieves from the interface is small but finite. At some higher incidence the particle does not emerge at all from the medium.

Both these incidences can be calculated as functions of the loss in vis viva, which Laplace did, and Malus reiterated the formulas in his paper (Malus 1807, 511; see Chappert 1977, 58–59). Suppose the incident velocity from within is U; then the vertical component at angle ϑ of incidence is $U \cos \vartheta$. If $U^2\cos^2\vartheta$ is equal to v^2, which is the amount constantly taken away from the vis viva, then the particle will reach its maximum possible penetration into the boundary layer between the prism and the substance that abuts it. Laplace calculated that if ϑ is so large that $U \cos \vartheta$ is less than $v/\sqrt{2}$, then the particle will not emerge from the prism at all.

A salient characteristic of internal reflection, even with opaque bodies as the bounding medium, is that the reflected rays have, excepting dispersion, the same color as the incident rays; that is, the opaque bounding medium does not impose its color upon the rays internally reflected in Wollaston's prism. If a ray penetrates the opaque medium it must therefore be completely absorbed. We cannot, then, use v/U as the cosine of the angle at which internal reflection begins when we are measuring opaque substances; we must instead use $v/U\sqrt{2}$. Only the first angle ($\cos^{-1}[v/U]$) follows from Snel's law.[22] But the indexes Wollaston gave for opaque substances were obtained directly from his scale in precisely the same way he obtained indexes for transparent bodies—from Snel's law. Laplace almost certainly noticed what he saw as Wollaston's incorrect assumption and set Malus the task of testing the second formula ($\cos^{-1}[v/U\sqrt{2}]$).

The problem this at once posed for Malus is that Wollaston's device is constructed on the assumption that the angle of total internal reflection is the one implied directly by Snel's law. It embodies the law. One cannot use the device to test the very thing its structure presupposes unless—as Wollaston had thought to be the case with double refraction—one can reinterpret the device's structure in a way consistent with an alternative law. But that is impossible when opaque substances bound the prism in the apparatus, because the internal reflection must then begin past the point at which (were the body transparent) the emergent ray would parallel the interface—and that condition (which is still satisfied by double refraction) must be fulfilled because the very structure of Wollaston's device depends upon it.

Consequently Malus had to utilize a different method of observation. Instead of using Wollaston's device, Malus dispensed with its wooden framework (which embodies the angle implied by Snel's law) and instead directly measured the angle at which total internal reflection begins. He described his method:

> To measure the angles under which the reflection begins, I employed an instrument composed of a platform of ground glass and a vertical shaft armed with a sight that could be raised at will and that carried a vernier marked in tenths of millimeters.
>
> After melting several drops of wax on a prism, at the extremity of one of its faces, I applied the other part of that face to the glass platform. I measured exactly with the sight the tangent of the angle the visual ray formed with the vertical at the moment when the wax disappeared, and knowing the refractive force of the prism and the angle between its planes, I concluded from them the angle of incidence on the face to which the body is applied. (Malus 1807, 513)

Malus used wax because it could be made liquid and transparent by heating or solid and opaque by cooling, and because he could test the substance at different densities, also by heating. Laplace's formula can then be applied quite directly. Suppose the (greater) speed within the prism is U, while the (lesser) speed within the opaque body is V; the "index" n of the opaque body in relation to that of the prism will be U/V. Suppose we observe the beginning of total internal reflection at angle ϑ. The light particle's speed will be decreased by an amount v. According to Laplace's formulas, the reflection begins when $U \cos \vartheta$ equals αv, where α is 1 for transparent, and $\frac{1}{2}$ for opaque, substances. Consequently we have the following formula for the index:

Laplace's Formula

$$\frac{1}{n^2} = 1 - \frac{1}{\alpha} \cos^2\vartheta$$

Of course, to test the formula we must do more than merely compute indexes, since we have no way to know whether our values are correct when the body is opaque. This is why Malus experimented at different densities. According to Laplace's reasoning, the refractive power of a substance ($n^2 - 1$) should be proportional to its density. Malus accordingly calculated the ratios of power to density at different temperatures and listed the results in four tables. Of particular importance, he also provided an estimate of the error inherent in his observational method:

> Since results of this kind are not susceptible of absolute rigor, I calculated for the different cases the limit of observational errors. I recognized in this way that an error of a minute in the determination of the angles of the prism brought a difference of only 0.00026 (twenty-six hundred-thousandths) in the ratio of the sine

> of incidence to the sine of refraction, and that a similar error in the observation of the angles of the broken ray in the same element a difference of only 0.0004 (four ten-thousandths). I had reason to think that the errors would not be very considerable, because my observations were made on objects situated at more than two thousand meters and were repeated with different instruments. (Malus 1807, 514)

The provision of error estimates, combined with tabular data, characterizes all of Malus's work, as well as those papers in French optics after him that achieved widespread recognition. The results were indeed striking, since the refractive powers determined in this way were accurate to the third decimal place—and, to this accuracy, had the same proportion to density at all temperatures measured in both the solid and liquid states.

This was not the first time refractive powers had been measured in order to test their proportionality to density. In 1806 Biot, assisted by Arago, had at Laplace's instigation measured the indexes of several gases at different densities (See Frankel 1972, 187–98). They had used a highly accurate device to measure ray angles (Borda's repeating circle; see below) and had taken great care to eliminate errors, which was especially difficult to do because they used a hollow glass prism. Malus's memoir was no doubt closely patterned on theirs—in both cases Laplace probably had a direct influence on the experiments and analysis. Malus, then, was working in an existing tradition, but one he pushed very far by concentrating directly upon the estimation of error and the extensive testing of a formula in several circumstances. Though embryonic in this first work, these characteristics rapidly matured in his study of double refraction.

1.7.4 A New Experimental Optics

Malus won the prize competition for double refraction announced on 21 December 1807 by confirming Huygens's construction. We do not have extensive details concerning the events that led up to the competition or precisely what occurred during the year between its announcement and Malus's award. Shortly after Malus's death in 1812 Jean-Baptiste-Joseph Delambre remarked that Malus's *Traité,* which contained his ray analysis, "brought attention to a phenomenon that had occupied Huygens and Newton, double refraction" (Cited by Frankel 1972, 231; Delambre 1812). But double refraction was not mentioned in the published version of the *Traité,* and more to the point, the *Traité* contained very little of direct use for it. More than likely Malus's interest in the phenomenon was not an outgrowth of a desire to apply the ray methods of the *Traité* elsewhere, but rather reflects Laplace's acute interest in the subject by the middle of 1807.

Laplace was no doubt pleased that Malus had been able to bypass Wollaston's device, and to confirm Laplace's formula, in measuring the indexes of opaque bodies. Malus read his paper to the First Class on 16 November, and the prize competition was announced a little more than a month later. In the interim Laplace may have realized that Malus's argument cannot apply to Wollaston's use of the device for double refraction because Iceland spar is transparent. Consequently, it is more than likely that the issue of double refraction came to the forefront of Laplace's attention only in the late fall of 1807 at the earliest—before that time he may have thought in a vague way that Wollaston's device would be as useless for double refraction as for opaque substances. Malus's argument clearly showed this was not the case.[23]

Malus began experimenting early in January. 'When the Institute* (*January 1808) called physicists' attention to double refraction," he wrote, "[a subject] that differences of opinion among the greatest geometers had rendered inconclusive, I began by observing and measuring a long series of phenomena on natural and artificial faces of Iceland spar" (Malus 1811b, 503). Malus must certainly have recognzied that Wollaston's apparatus remains entirely worthless for testing Huygens's construction until the internal reflection implied by the latter can be calculated. It therefore seems very probable that Malus translated the construction into algebra—a step that transformed contemporary optics—precisely to check Wollaston's two confirming measurements. Using Wollaston's parameters, he would then have found, no doubt to his great surprise, that theory and experiment differ by at most 15′.

He did not go on to use Wollaston's device in other planes of incidence because there would have been little reason to doubt that it works generally, since the planes Wollaston used were arbitrary. But Malus could still have questioned whether the complete formula works, because Wollaston's device confirms only that the denominator in Malus's formula has a certain form. The only way to confirm the full formula is to measure refractions for various angles of incidence in given planes of incidence, and Wollaston's device cannot do so.

The events that led to Malus's experiments began, then, with a translation of Huygens's construction into algebra sometime during the mid- to late fall of 1807 (see appendix 2). Having effected the translation, Malus calculated the predicted angles for Wollaston's two confirming experiments; here he certainly found excellent agreement between theory and experiment. He then apparently communicated the result to Laplace (this is perhaps what Lacroix was confusedly referring to in the report on Malus's *Traité*), who would have been both stunned and skeptical. Laplace then probably urged Malus to undertake a comprehensive series of experiments, whose importance would be underwritten by offering a prize. Realizing that Wollaston's device was incapable of testing refractions at arbitrary incidences, Malus recurred to apparatus in common scientific use in France that could be adapted for double refraction.

Since he could not use internal reflection, Malus had to measure angles of incidence directly. For this he could use Borda's repeating circle, previously used for much the same purpose a few years earlier by Biot and Arago (see fig. 1.9). Because of its accuracy this device distinguishes Malus's experiments in double refraction from all previous work.[24]

To observe the extraordinary refraction at various angles of incidence,

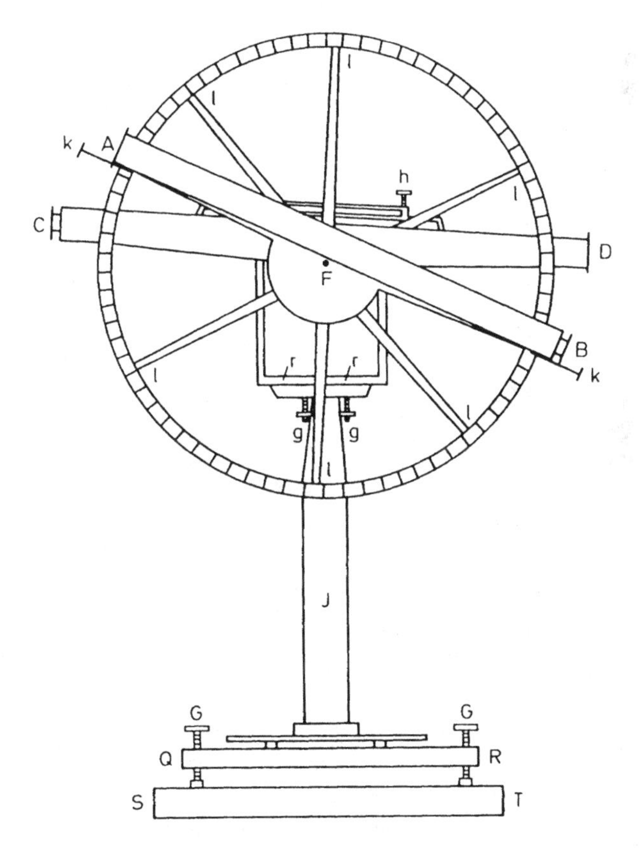

FIG. 1.9 Borda's repeating circle.

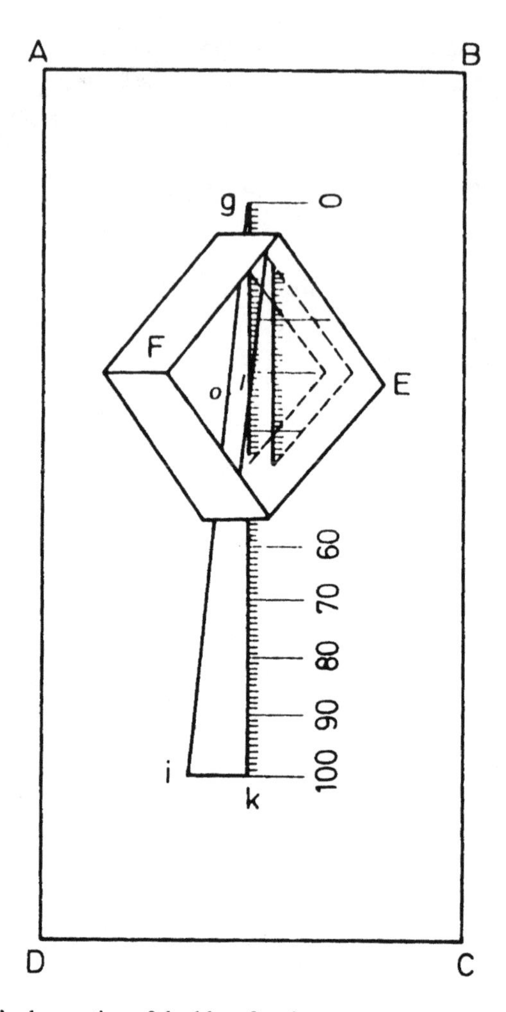

FIG. 1.10 Malus's observation of double refraction.

Malus adopted Haüy's procedure without mentioning him and decided to observe the amplitude of aberration (see fig. 1.10). In the figure, ABCD is a plate of "whitened copper" on which a scale 10 cm long and divided into millimeters is inscribed. At k, line ik is drawn 1 cm long and perpendicular to gk; i and g are joined. Then if a point l is marked along gk, and the length of the perpendicular to gk at l and drawn to ig is ol, we have ol equal to gl tan (∠igk), or gl/10. This relationship gave Malus a very high accuracy indeed in measuring the aberration.

To observe aberrations in the principal section, the crystal was placed over the scale with its principal section normal to gk and with its upper solid acute angle at E. Sighting from a point, for example, directly above gk, the extraordinary image of ig is displaced toward gk. By moving the crystal along gk and keeping the eye perpendicularly above it, a point l can be found where the ordinary image of gk (which is undisplaced at normal incidence) intersects the displaced, extraordinary image of ig. At this point, therefore, the distance between ig and gk equals the product of the tangent of the angle of normal deviation by the height. Similarly, if one sights at any angle (looking from E toward F) and the crystal is moved until the extraordinary image of ig intersects the ordinary image of gk at, say, point l on gk, then the distance cc′ between ig and gk will be the aberration and equals gl/10. The distance gl can be read directly off the scale gk. Since Malus could estimate gl to the nearest 0.2 mm, he had the aberration cc′ to the nearest hundredth of a millimeter, affording a highly accurate test.

To use this device requires what no one before Malus had possessed in double refraction experiments: the ability to set or measure angles of incidence accurately with the Borda repeating circle. In most of his experiments Malus began with a set angle of incidence. He placed the repeating circle with its plane perpendicular to the scale gk. The copper plate ABCD was set on a level ground-glass surface (perhaps the same one he had used in his experiments on opaque indexes), and he then moved it over the glass, with the crystal in place on it, until the intersection of the images appeared along the edge of a copper rectangle with which Malus replaced the movable glass. (Occasionally Malus first set the aberration and then turned the rectangle until he observed image coincidence; i.e., he occasionally measured the incidence for a set refraction.) Once the intersection was found, all he had to do was observe the point on the scale gk where it occurred. Measuring the aberration in a plane perpendicular to the principal section was harder to do, since it has two components, but this presented no unusual difficulties (for details, see Buchwald 1980b, 359–60).

For the principal section Malus measured the aberration at ten different incidences (±10°, ±20°, ±30°, ±45°, ±60°), and each measurement was itself the average of repeated observations. For the normal plane he measured at five incidences (10°–50°). Finally, to test the dependence of the refractions on the angle between the facet and the optic axis, he used cut crystals and measured at five or seven incidences.

Malus's penetrating experimental acumen was matched only by the nearly unprecedented method he used to test the formulas. Instead of doing several experiments to compute the necessary parameters and then using these to test the remaining measures, Malus took a much more difficult, but vastly more reliable, route (for details see Buchwald 1980b, 361–63). In an elaborate analysis, he used groups of experiments to calculate sets of values for the parameters, and he then averaged the sets. Using these results, he computed the aberration for each of the experiments and compared those numbers with observation. The results of this elaborate exercise in the minimization of experimental error were remarkable, showing an agreement between formula and measurement of better than 1 percent.

Malus's work here amounted to the foundation of a new experimental optics. No one before had achieved such high accuracy in these kinds of measurements; no one before had carried the use of intricate algebraic formulas in conjunction with experiment to such a high art. And perhaps of greatest importance, no one before Malus had so painstakingly sought to eliminate errors by averaging and then to use the accurate parameters that resulted to compare a large tabular series of data with a formula.

1.7.5. Deriving Huygens's Construction from Least Action

Malus's work must have proceeded rapidly, because at the end of January 1808 Laplace already possessed his translation of Huygens's construction—even though, according to Malus, he had begun to experiment only several weeks earlier.[25] The immediate problem Malus's experiments posed was that the formulas derived from Huygens's pulse theory of light, which would seem at first to make them incompatible with the emission theory.

Both Laplace and Malus probably realized independently of one another that the difficulty was only apparent, for the following reasons.[26] Huygens's account requires that the path taken by a ray in traveling from point A to point B must be a minimum—that Pierre de Fermat's principle of least time must hold. He had demonstrated in the *Traité* that Fermat's principle of least time for rays follows, assuming that the angles of incidence and refraction are in the ratio of the respective velocities [Huygens 1950, 42–45]. Laplace realized at once, in his words, "the identity of [Fermat's principle of least time] and the manner in which Huygens envisages the refraction of light" (Laplace 1878–1912, 12:282). Laplace actually generalized Huygens's demonstration. He did not, as Huygens had, begin with rays for which the sines of incidence and refraction are in the ratio of the velocities. Instead, he showed that the principle holds for any shape of wave surface by considering directly the paths followed by different points of an incident plane front in reaching simultaneously the tangent plane to the surface in refraction. (In Huygens's demonstration the front played no role.) Laplace, in other words, demonstrated that the basic requirement of Huygens's account—that all points of a given front must simultaneously reach the refracted front—implies Fermat's principle for rays.

Indeed, Laplace felt Huygens's requirement and Fermat's principle were "identical" (Laplace 1878–1912, 12:273), by which he meant that anything that can be deduced from a given wave surface can also be obtained from the principle of least time when one knows the velocity of the ray as a function of direction. This marks a subtle and important change from the way a pulse theorist would have considered the principle. To the latter, ray behavior follows from the behavior of the front. Laplace, and Malus, thought physically in terms of rays, not fronts. Laplace had demonstrated that Fermat's principle always holds for rays in Huygens's theory. Consequently it had to be possible to find a velocity—the velocity of the ray—that, when inserted in the principle of least time, would then yield the behavior of the ray as a function of direction without explicitly considering the front itself. (That velocity would necessarily differ from the speed at which the front advances, though the one can be deduced from the other.)

Now in the case of Snel's law the difference between the pulse and emission theories when we limit considerations to the behavior of rays, including their velocity, amounts essentially to this: that light slows down according to the former and speeds up according to the latter. Pierre-Louis Maupertuis had already recognized in 1744 that Snel's law would follow if light was speeded up by replacing the speed with its reciprocal in Fermat's principle (the Fermat product $\int(1/v)ds$ becoming what Maupertuis termed the "action," $\int v\,ds$ [Sabra 1981, 157–58]). This suggested to Laplace, and quite likely independently to Malus, that the same procedure had to work also in double refraction.

To understand what is involved in this, consider the effect on a diagram for Snel's law of replacing a velocity with its reciprocal. Start with a figure drawn in Laplace's fashion for least time—in which we draw the Huygens spheres and fix the time interval in order to construct the refracted front from the incident front. The spheres' radii represent distances for the fixed time interval between the two fronts, so they may be thought of as speeds. If we wish to use the same diagram to find the ray paths from least action, we must reinterpret its significance. The radii can no longer represent distances for fixed times; they must now represent times for fixed distances. Similarly, for Huygens's construction the radii of the spheroid must represent times in least action and speeds in least time.

One way to apply least action, then, would have been to calculate the ray velocity from Huygens's construction for a given orientation and then to take its reciprocal. That would give the velocity to be used in the least action principle, and that is essentially what Laplace did, though he did not indicate how he obtained the velocity he used in the principle. Both Laplace and Malus began directly with the principle and used it to obtain two equations in the derivatives of the velocity with respect to angles of refraction and plane of refraction (Frankel 1974, 235 for Laplace and 238–39 for Malus). Laplace then inserted an expression for the velocity, taken a priori, which happened to be just the right one to yield the ray paths determined by Malus's formulas. Malus instead deduced the velocity from the principle given the ray paths. The results are equivalent. With φ the angle between the refracted ray and the optic axis, and the velocity the expression is:

$$v^2 = \frac{1}{a^2} + \left[\frac{1}{b^2} - \frac{1}{a^2}\right] \cos^2\varphi = \frac{1}{b^2} + \frac{1}{a^2} \sin^2\varphi$$

Here a, b are, respectively, the semimajor and semiminor axes of the Huygens spheroid. (This is in fact the correct expression for the reciprocal of the ray velocity on the wave theory.)

Suppose we had no theory at all beyond the requirement that rays take time to travel and that they follow certain paths. Suppose further that we somehow already had formulas for the ray paths in certain crystals. Then to assert that either least time or least action can be used is the same as asserting that n equations in n unknowns can be solved: we will obtain either a velocity or its reciprocal, but we will not at all have shown that either principle governs the phenomenon in any meaningful sense. They are merely analytically compatible with it, as they are with *any formula at all* for a linear ray path. Least time or least action can always be applied after the fact to calculate a ray velocity.

The two principles become meaningful if we begin with the ray velocity rather than with the ray path, because we can empirically test the paths. If we have some independent way of finding the velocity, then we can see whether the path that results proves correct. Laplace gave the impression that he had done just that by providing the velocity and then deducing the path from least action, though he gave no reason for the velocity expression he chose. In the pulse theory, by contrast, we do not have to begin directly with the ray velocity. We begin at a more fundamental level, with the wave surface, from which we can calculate the ray velocity. (We could, if we wished, then use least time to calculate the ray paths.) We do not have to assume it a priori. We must instead assume the wave surface.

This point was not clearly recognized by either Laplace or Malus, nor was it emphasized by Thomas Young, who reviewed Laplace's memoir late in 1809. He remarked that Laplace's assumption of an expression for the velocity reflected "the hasty adoption of a general law, without sufficient evidence; and an inversion of the method of induction equally unwarrantable with any of the paralogisms of the Aristotelian school" (Young 1809, 221). He continued in a different vein: "The law of least action must always be applicable to the motion of light, as determined by the Huygenian theory, supposing only the proportion of the velocities to be simply inverted. Mr. Laplace has therefore given himself much trouble to prove that coincidence in a particular case, which must necessarily be true in all possible cases" (Young 1809, 226).

Young's first point recognizes the "inversion" of Laplace's approach—moving from a seemingly arbitrary velocity through least action to the empirically known laws. But his second point rather obscures the questionable character of Laplace's procedure. Laplace's method did not work because least time becomes least action by inversion of velocities. It worked because it had analytically to work: given the appropriate ray velocity—and it is *always* possible to find one—one can generate from least action (or time) any linear ray path whatever. If, by contrast, Laplace had asserted that he had calculated the ray velocity from the Huygens spheroid, inverted it, and then inserted it into least action, Young's point would have been more cogent (though perhaps Laplace did do just that). Young further argued, in an extension of his first point, that Laplace had not the slightest proof that a velocity that depends upon the angle to an axis is even consistent with the principle of least action, since the latter "is only demonstrable, from mechanical principles, in cases of the operation of attractive forces directed to a certain point" (Young 1809, 227; see Frankel 1974, 243).

Laplace's and Malus's analyses strikingly reveal the limited ability of the emission theory per se to generate empirically testable formulas. With the single exception of the claim that refractive power ($n^2 - 1$) is proportional to density, it did not imply anything that is not independently implied by the assumption that light rays are physical, rather than mathematical, objects. Indeed, I shall argue in the next several chapters that after 1808 the emission theory reduces to this assumption despite Laplace's and Malus's early use of the least action principle (which was never again used by emission theorists). Nevertheless French physicists were apparently deeply convinced by Laplace's and Malus's arguments; for more than a decade afterward the assumption that light rays consist of particles that can be acted on by short-range forces emanating from matter was rarely questioned—even though *no* quantitative work, including Malus's and Biot's, in any way required the hypothesis.

2 The Concept of Polarization

If Malus's only accomplishments had been to confirm Huygens's construction and to discover polarization by reflection, he would still be an important though perhaps not central figure in our story. But he did much more than this; he also provided the basis for a quantitative theory of polarization based on rays. That theory became little more than a vague memory after the triumph of wave concepts in the 1830s, but for about a decade it underpinned the most exciting, innovative area of contemporary research.

The goal of this chapter is to reconstruct Malus's theory of partial reflection. Since he left only a few traces of his reasoning—scattered passages and the odd formula—the much greater power of the wave theory, coupled to the propagandizing efforts of the Cambridge wave men in the 1830s, almost entirely effaced it. At the time, however, it was well known that Malus had made great strides in this regard, and his methods were highly influential. In uncovering Malus's reasoning and the details of his theory, we will be investigating the foundation of optics as it existed during a period of great empirical and, it was felt at the time, theoretical success. It is particularly important to grasp Malus's work because there was much about it that wave theorists—including Fresnel—did not perceive; this has important implications for the full scope of the debates between them and ray theorists.

2.1 Malus's Discoveries

Scattered about Malus's residence in the Rue d'Enfer in the fall of 1808 were the many crystals of Iceland spar that he had been using to establish the accuracy of Huygens's construction. One day he noticed a remarkable thing while gazing through a crystal directly at the sun's image in the windows of the Luxembourg palace.[1] As the crystal turned the two images changed markedly in intensity, though neither ever vanished.

Such a thing had previously been seen only with light that had already passed through a crystal. Yet the light of the sun came directly from the palace windows by reflection. This discovery, together with Malus's nearly unprecedented experimental precision and his meticulous standards of reportage, in-

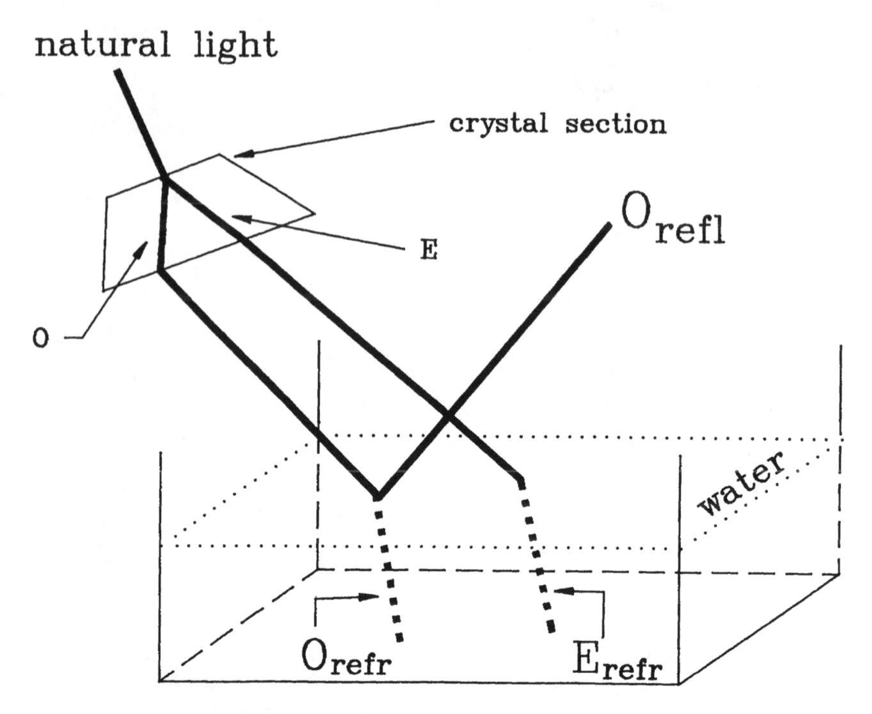

FIG. 2.1 Crystal section parallel to plane of reflection; the *E* ray is completely refracted.

augurated a new era in the history of optics—an era that ironically culminated twenty-odd years later in the widespread abandonment of the most elementary concepts that Malus himself had held and that he used to interpret his discovery. Yet at the time Malus was convinced that, far from requiring major changes, his discovery provided for the first time the possibility of computing how light divides into reflected and refracted parts.

Malus set about determining the conditions necessary to produce the phenomenon. He found at once that the modified beam will be refracted in the ordinary way if the principal section of a crystal receiving it is parallel to the plane of incidence. Further, the angle that produces "this modification" (and that I shall for the present call the "modifying angle") depends upon the substance of the reflector, and it increases with the refractive index (Malus 1809a, 150–51). The modification occurs with internal reflection, and even light reflected from birefringent crystals "acquires the property" (Malus 1809a, 152), as does light reflected by opaque but specular bodies. Metals seem not to produce the effect, but they also seem not to alter the property in light that already possesses it. Above and below the modifying angle, Malus discovered, a "part of the ray is more or less modified, and in a manner

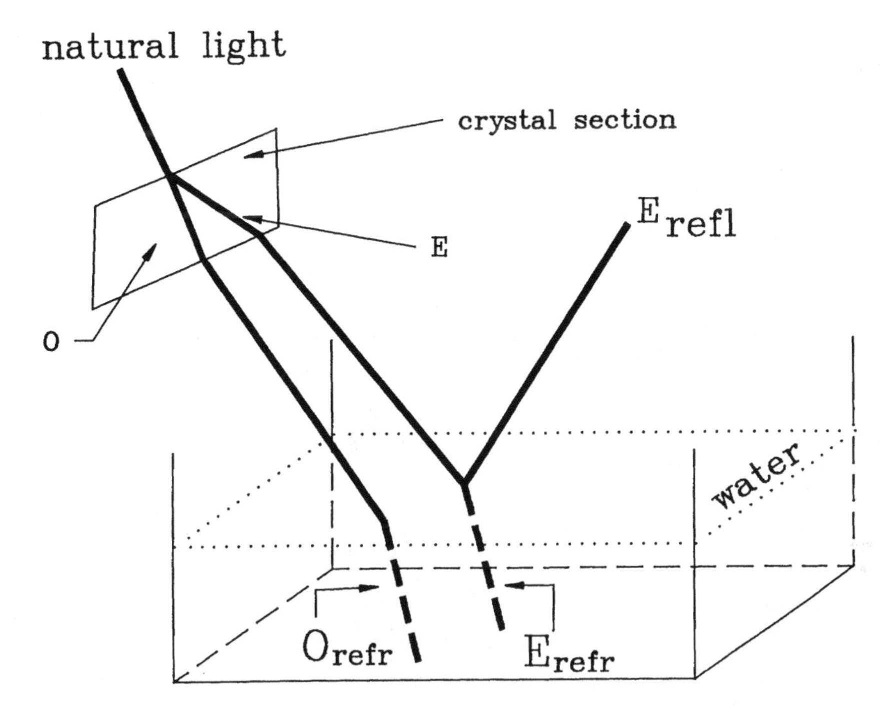

FIG. 2.2 Crystal section perpendicular to plane of reflection; the *O* ray is completely refracted.

analogous to that which occurs between two crystals whose principal sections are neither parallel nor perpendicular" (Malus 1809a, 151), which is to say that a beam so reflected does always divide on entry into a birefringent crystal into two beams, but they are generally of unequal intensity.

None of this could have prepared Malus for the startling observation he soon made (see fig. 2.1). He caused both the ordinary (O) and the extraordinary (E) beams produced by a crystal to be incident on water at the modifying angle. With the crystal's principal section parallel to the plane of reflection on the water:

> [cf. fig 2.1] The ordinary ray, on being refracted [O_{refr}], abandoned part of its molecules to partial reflection [O_{refl}] as would a bundle [*faisceau*] of direct light [i.e., unmodified light], but the extraordinary ray entirely penetrated the liquid [E_{refr}]; none of its molecules escaped refraction. [cf. fig 2.2] Conversely, when the principal section of the crystal was perpendicular to the plane of incidence, only the extraordinary ray produced a partial reflection [E_{refl}], and the ordinary ray was entirely refracted [O_{refr}]. (Malus 1809a, 150)

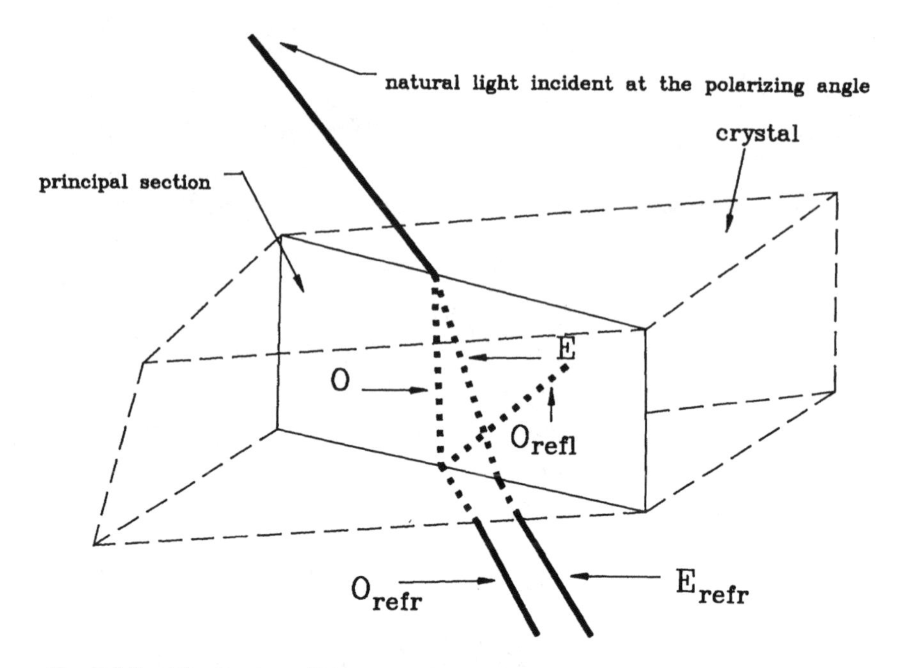

FIG. 2.3 Partial reflection within a crystal.

Beams of light, Malus had discovered, could be made to escape partial reflection—to be entirely refracted—by modifying them and then reflecting them in a plane at 90° to the one in which they were first modified.[2] Observations of internal reflection within crystals provided a cogent illustration of the new phenomenon (see fig. 2.3). Both the ordinary and the extraordinary beams within a birefringent crystal will in general be partly reflected and partly refracted when they strike the crystal's lower surface. But when they hit the surface in the principal section, and the beam of light from which they were originally produced strikes the crystal at the modifying angle, then the extraordinary beam will not be partly reflected at the crystal's lower surface, whereas the ordinary beam will be. Rotate the crystal 90°, and the beams exchange roles.

The ability of a beam to escape partial reflection, and not the bare fact that one could produce the "modification" by reflection as well as by double refraction, was what most excited Malus, for in it he saw the clue to an exact theory of partial reflection. He did not succeed in finishing the theory before he succumbed to the lasting effects of the plague. Nevertheless, he published several results that implicitly contain the theory and that determined the way ray theorists subsequently dealt with "modified" light. From them one can reconstruct the theory in some detail.

In the remainder of this chapter I will, then, be attempting to recover not only a theory that is unknown today, but one that Malus himself did not lay out in a straightforward fashion. Consequently, much that will be said must necessarily remain conjectural in that the intermediate steps that lead to Malus's formulas have disappeared entirely from the historical record. Nevertheless, one can recover his expressions in a way that relies solely upon his fundamental assumption that rays can be gathered into denumerable sets. Since this axiom of ray theory is today so thoroughly unfamiliar, I have provided considerable detail in my reconstruction (particularly in sec. 2.5). This will aid the modern reader in developing a full understanding of how ray theory worked in practice. However, to avoid unwarranted tedium I have relegated the reconstruction of Malus's complete theory of partial reflection to appendix 8. Readers who wish to understand deeply the full power of ray theory should examine this appendix, but one can glean an overall appreciation of the theory from the more limited account in section 2.5.

2.2 The Cosine Law Poses a Problem

Malus's investigations progressed rapidly. He reported them in a series of memoirs, which he gathered together and reprinted, with expansions, in his prizewinning "Théorie de la double réfraction" (Malus 1811b; the prize memoir incorporates many passages drawn literally from the previous articles). There we find the outlines of his theory of partial reflection and the method for counting rays that it assumes.

To build this theory Malus had, we shall see, to employ in a new way a law he had given for determining the amounts into which a modified or, as he named it (Malus 1811a), "polarized" beam divides when it enters a doubly refracting crystal. Malus's law for this division sets the number of rays in each of the resulting ordinary and extraordinary beams respectively proportional to the square of the cosine or the sine of the angle between the plane of polarization of the incident beam and the crystal's principal section. (Malus having defined the *plane of polarization* of a beam as the plane, containing the beam, in which the principal section of a crystal must lie to produce only an ordinary refraction from it.) The constant of proportionality must be the same for both beams, because it represents the total number of rays that escape partial reflection at the crystal's surface. Malus discovered that a similar law holds for the partial reflection of a polarized beam, but with a signal, and puzzling, difference.

To investigate the behavior of polarized beams when they are again reflected at the "polarizing" angle, Malus constructed a simple device, the first polarimeter (fig. 2.4). It consisted of two mirrors pivoted about axes (*aa* and

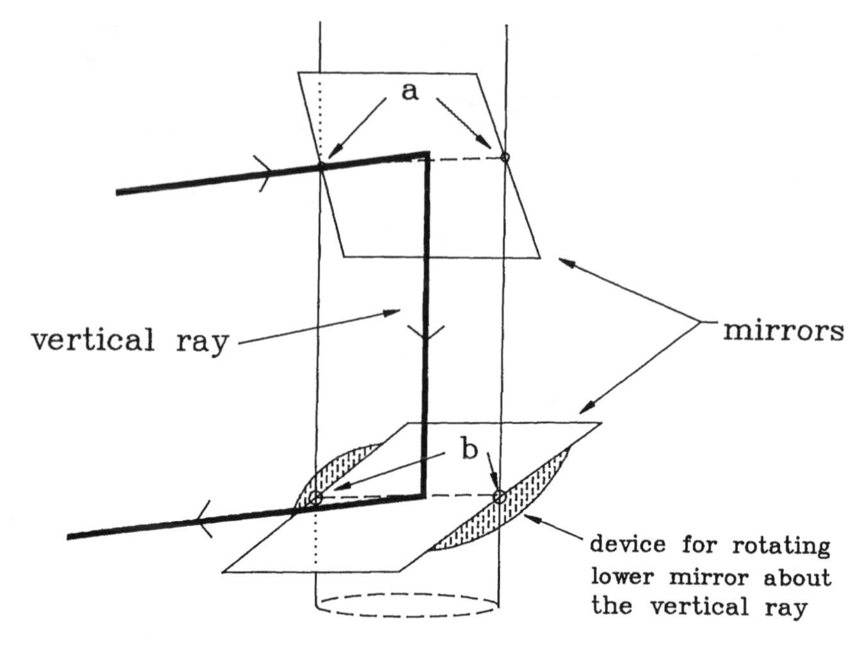

FIG. 2.4 Malus's polarimeter.

bb) in the mirrors' planes, one below the other; these axes always remain parallel to the plane of the horizon. One can rotate the axis of the lower mirror in the horizon plane, however, thereby altering the angle between the planes of reflection on the two mirrors. The incidence on the upper mirror was adjusted to the polarizing angle, and the lower mirror was pivoted about its axis *bb* until the incidence on it was also polarizing. By rotating the lower mirror's pivot axis (*bb*) itself, Malus could vary the angle between planes of reflection (which I shall on occasion refer to as the "planar angle") while holding the angle of incidence proper fixed. He could, that is, vary the angle between the plane of polarization and the plane of reflection for a given incidence.

One can scarcely overestimate the importance of Malus's device or the variations subsequently rung on it. The nature of the images seen in its mirrors powerfully influenced optical theory for almost the next quarter-century. The device directly embodied the two variable factors that Malus determined must be sufficient to determine partial reflection: the angle of incidence and the angle between the planes of reflection. No one had previously suspected that the angle between planes of reflection might have anything at all to do with what occurs, but Malus's device by its very construction constrains the observer to vary the planar angle for a fixed incidence, or the incidence for a fixed planar angle. We shall shortly see that his theory for partial reflection constitutes a method for predicting what can be seen in this kind of device when the planar angle varies.

The root of Malus's direct concern with partial reflection grew from the imperfect analogy I mentioned above between the dependence of reflection on the planar angle and the dependence of a polarized beam's division by a crystal on the crystal's principal section. He soon concluded that the intensity of the light reflected by the lower mirror at polarizing incidence varies as the cosine-squared of the planar angle, much like the intensity of the ordinary beam in a crystal (Malus 1809b, 257). Obviously, because light is conserved, the intensity of the refracted beam within the mirror must then contain a term that vanishes with the sine-squared of the planar angle, and this seems to form a close parallel to birefringence. The reflection plays the role of the ordinary beam, while the refraction plays the role of the extraordinry beam. (And we shall see below that the parallel goes even further in respect to the direction of the polarization itself.)

But there is a major difference, and this forced Malus to construct a theory of partial reflection. Unlike birefringence, where only two beams need be considered, here there are at least three contributions to the light that must be considered, since the refracted light must not vanish even at polarizing incidence. We can easily represent this simple situation in an equation. Let I represent the intensity of the incident beam and, respectively, I_{refl}, I_{refr} the intensities of the reflected and refracted beams for polarizing incidence. And let I_0 represent the amount of light that is refracted when the reflected light is at its maximum intensity. That is, the minimum intensity for the refracted light is represented by I_0. With ε the angle between the plane of reflection and the plane of polarization of the incident light, we have:[3]

$$\left.\begin{aligned} I_{refl} &= m_{refl}\cos^2\varepsilon \\ I_{refr} &= m_{refr}\sin^2\varepsilon + I_0 \end{aligned}\right] \; I = I_{refl} + I_{refr}$$

Now in birefringence the law applies only to the refracted light, so that we may easily determine how well it works without examining the reflected light at all—we may ignore partial reflection. But here the law determines a division between reflected and refracted light, and it does not determine the division completely, since some light always refracts. The analogy between birefringence and the partial reflection of polarized light seems then to be incomplete. In the case of partial reflection, only one of the beams (the reflected one) produced by the process varies with the angle (ε) in the same way that one of the two beams produced by doubly refracting the polarized beam also varies. The other beam—the refracted one—cannot vary in the same fashion as the corresponding beam in double refraction, for if it did, it would have to vanish with the angle ε, which it does not do. Accordingly, if we wish to understand fully the meaning of the cosine law for the reflection of polarized light, we must inquire into the *composition* of the refracted beam, the one that seems to violate the analogy to double refraction. This, we shall soon see, means that Malus had to solve the problem of partial reflection.

2.3 The Asymmetric Ray

In 1809 Malus described a system of coordinates tied directly to the experimental situation that permits its simple description. These coordinates evolved during the next year as Malus sought a solution to the problem of partial reflection. In their final form they embody Malus's and all subsequent ray theorists' understanding of polarization. Introduce with Malus (Malus 1809b) three mutually perpendicular axes *a, b,* and *c*. The first (*a*) lies along the direction of the bundle that is incident on the lower (or first) mirror in Malus's apparatus. The second (*b*) is perpendicular to the plane of incidence on this first mirror. The final axis (*c*) is normal to the plane of incidence on the upper (or second) mirror. Malus's original form of the cosine law for sequential reflection by two mirrors at the polarizing angle used the latter two axes: ε represented the angle between *b* and *c*, since this is the angle between the planes of reflection on the two mirrors (fig. 2.5).

But this specification cannot make distinctions that are useful for solving the problem of partial reflection, because it provides no means of determining how many rays in the incident bundle are reflected and how many are refracted. For that is the problem: *to determine what happens to the rays of light, which are entities that retain their identities as individuals under reflection and refraction.*

In the *Théorie* Malus accordingly showed how to think about partial reflection in terms of rays by assigning a new set of *a, b, c* axes to the rays themselves and then describing partial reflection in terms of its effect on these axes. In the following passage, which I quote at some length, Malus outlined the essential method for analyzing polarization phenomena that subsequent ray theorists took up:

> when a ray is reflected by a mirror at an angle of 54°35′, one recognizes that all of its molecules are arranged in the same way, since, on presenting to the ray a prism of Iceland spar whose axis is in the plane of reflection, all of its molecules are refracted into a single ordinary ray: none of them is refracted extraordinarily. In

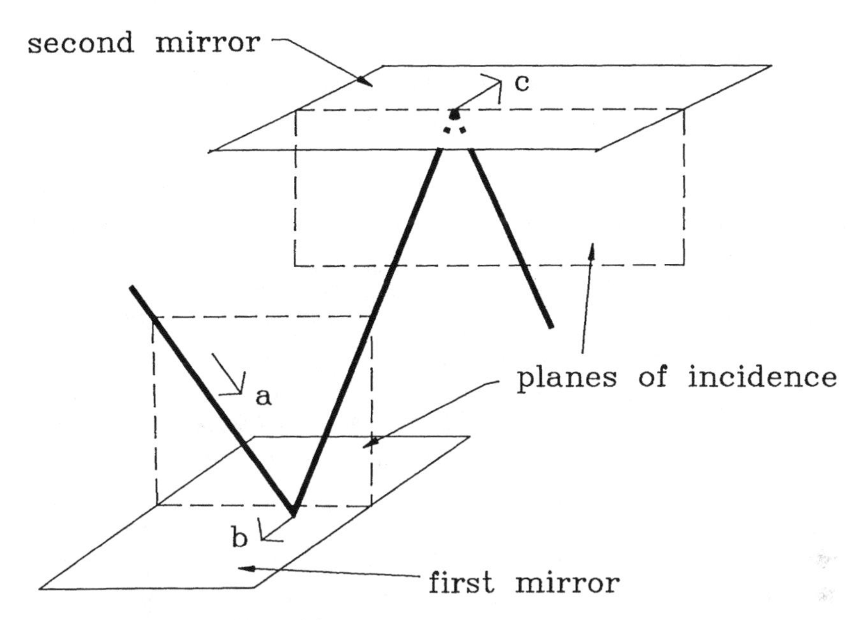

FIG. 2.5 Malus's empirical *a*, *b*, *c* axes.

this case analogous axes of the molecules are all parallel to one another, since they behave in the same way. Call *b* the axis of those molecules that are [*sic*] perpendicular to the plane of reflection [the verb is plural—*se retrouvent*—and belongs grammatically with molecules, but the sense is that *b*—the axis—carries the directionality], none of those whose axis *c* was perpendicular to this plane could be reflected. Therefore if we present to the reflected molecules, and at the same angle [of incidence], a second mirror parallel to their axis *c*, they will be in the situation of those that could not be reflected by the first [mirror]; and we have seen that, in effect, the ray passes entirely into this second mirror. In this case analogous axes of the molecules of the same ray remain parallel after they have experienced the action of the second body. Suppose now that the second mirror, continuing to make the same angle [of incidence] with the reflected ray, is no longer parallel to its axis *c*. Then it must necessarily reflect a part of the luminous molecules. In effect, in order for the axis [*c*] of these molecules to become perpendicular to the new plane of reflection, it will have to describe a certain angle: the quantity of these [molecules] which will come to that position will be proportional to the square of the cosine of that angle, and will again penetrate the second mirror; the other part, which is proportional to the

> square of the sine of the same angle, will turn its axis *c* into the plane of reflection, and the axis *b* perpendicular to this plane, and in virtue of that position will acquire the faculty of being reflected as we saw above. (Malus 1811b:444–45)

These remarks embody many of the assumptions that underlie what I shall call, following Thomas Young,[4] the *selectionist* understanding of polarization: a process in which *groups of rays* are selected and modified. Light consists of spatially discrete physical "rays" (*rayons*): by "physical" I intend solely that the ray possesses a physical identity as an individual object and so can be singled out and counted. Rays may be gathered in sets or "bundles" (*faisceaux*), and the elements of a bundle may share certain properties. For example, in his *Traité d'optique* Malus analyzed bundles that undergo reflection and refraction and that emanate originally from a single point (Chappert 1977, chap. 4). In this study a common origin, linearity, and the laws of reflection and refraction alone determined the relationships among the members of the set in the several situations Malus considered.

Given the prevalence of the emission theory, contemporary terminology can be somewhat ambiguous. Writers in this period often used the word "ray" to refer also to a bundle of luminous objects, perhaps for the following reasons. To an emissionist even the individual physical ray in the sense I defined it above—in the sense that corresponds to the image of a ray track one uses in geometrical optics, with the additional assumption of countability—could itself be thought of as a collection by envisioning it as a succession of discrete parts, since the ray consists by hypothesis of a sequence of particles or "molecules" of light. The *physical ray* in my sense of the term then marks the common trajectory of these parts.

Selectionist counting does not require that the elements that are added together to calculate the light intensity correspond solely to the physical ray proper; they may also include the parts of each ray. However, whether one thinks of a bundle solely as a set of physical rays or whether each physical ray also consists of parts does not in any way affect the essence of selectionist procedure. The count could be taken over space only (assuming solely physical rays) or over both space and time (by constructing a bundle out of physical rays, each of which consists of successive parts). Selectionists apparently did not think the issue was important. They never discussed it, but they often treated the words "ray" and "bundle" as synonyms.

The essential point to grasp is that the intensity of light is measured always and only by an arithmetic count, whether the count involves only physical rays or, additionally, ray parts. Consequently, *selectionists always understood polarization as a property of a group and not of any individual element in the group.*[5] Every ray, no matter what its origin, always possesses a certain directional attribute. Think of a ray as a broomstick with a crosspiece nailed to it at right angles. Like the crosspiece on the broom, the ray's asymmetry moves rigidly with it. Bundles of rays may contain rays with many different asymmetries, since a crosspiece can be found anywhere in a plane at right angles to its ray.

In a bundle of "natural" or "common" light—light that has not undergone specular reflection or refraction—the distribution of asymmetries over the elements (rays) of the set (bundle) may be essentially random according to almost all ray theorists. Each ray is intrinsically asymmetric, but no law governs the variation of asymmetry from ray to ray. (Though we shall see that Brewster much later, and with imperfect knowledge of the wave theory, argued otherwise.)

When the bundle reflects at any angle, a certain portion of its rays acquire a common asymmetry, whereas for the remaining portion either the rays remain unaffected or else other common orientations may be brought about. Such a set, consisting of random asymmetries together with a large number of rays having a common orientation, was later said to be "partially" polarized. If the reflection occurs at the limit or "polarizing" angle, then the rays in the bundle all possess precisely the same asymmetry, and the *bundle*—not its elements (which always possess asymmetries)—is simply "polarized."

Understood in this way, "polarization" is a property not of an individual ray but only of a collection of rays. We must take this quite literally. One could, for example, in principle obtain a beam of light consisting of an unpolarized portion, a portion polarized in one direction, and yet a third portion polarized in an entirely different direction. Polarization selects rays from the original bundle, altering the rays' asymmetries in fixed ways, thereby creating subsets of rays possessing a common asymmetry. The rays in themselves are unaltered physically: like the crosspieces in our broom analogy, the rays' asymmetries can only be rotated about the direction of the ray itself. One can change ray direction and the angle of rotation of the crosspiece about the ray, but nothing else.

With these ideas in mind we can understand Malus's discussion of reflection at the polarizing angle. Implicit in it are two assumptions. First, when a given quantity of light (number of rays) is polarized by reflection, then a "proportional" quantity becomes polarized by the refraction that occurs at the same moment, but in the opposite sense. Second, a ray whose asymmetry (defined

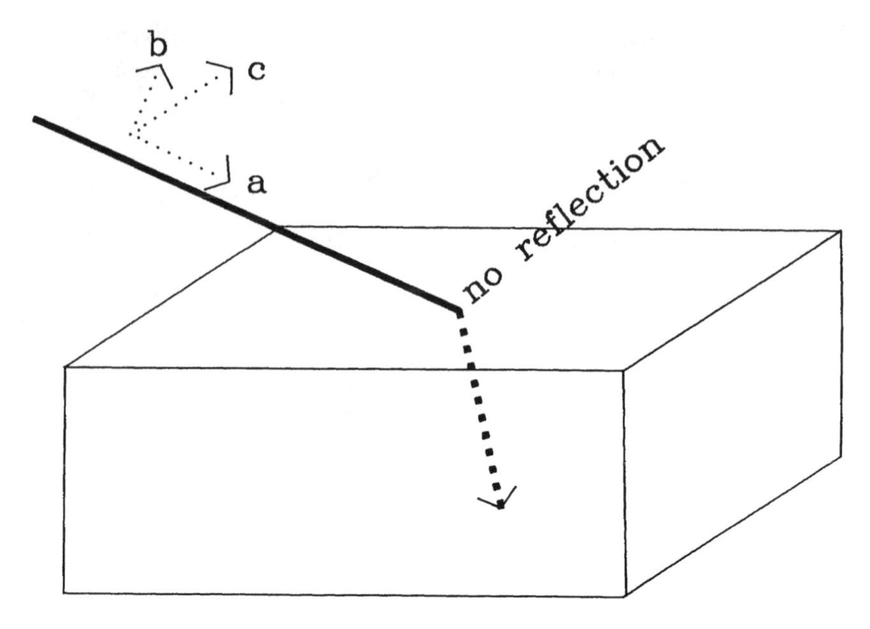

FIG. 2.6 Malus's *a, b, c* axes in the absence of reflection; *c* axis parallel to the mirror.

by the direction of the *c* axis) is normal to the plane of incidence cannot have its *c* axis turned into that plane by reflection.

Then, according to Malus, polarization occurs in the following way. At the first mirror the random asymmetries in the bundle of natural light are transformed in two ways. A portion of the incident bundle has its rays' *c* axes rotated into the plane of incidence—the rays' *b* axes then being normal to that plane. This portion reflects. Second, a portion of the bundle has its rays' *b* axes rotated into the plane of incidence. This portion refracts. Third, the remaining portion of the bundle also refracts, but Malus did not here discuss whether its asymmetries have been affected. Of the rays in the original bundle, those whose *b* axes were already in the plane of incidence could not be reflected. (Remember, we consider reflection at the polarizing angle only.)

With this in mind, consider two orientations for a mirror involved in a second reflection at polarizing incidence. Suppose first that the mirror itself is parallel to the common *c* axes of the rays in the polarized bundle (fig. 2.6). Then the plane of incidence on the mirror contains the common *b* axes of the rays in the polarized bundle, and all of them must be refracted. There will be no reflection at all. Suppose instead that the common *c* axes are not parallel to the mirror (fig. 2.7). Then the *b* axes will not lie in the plane of incidence,

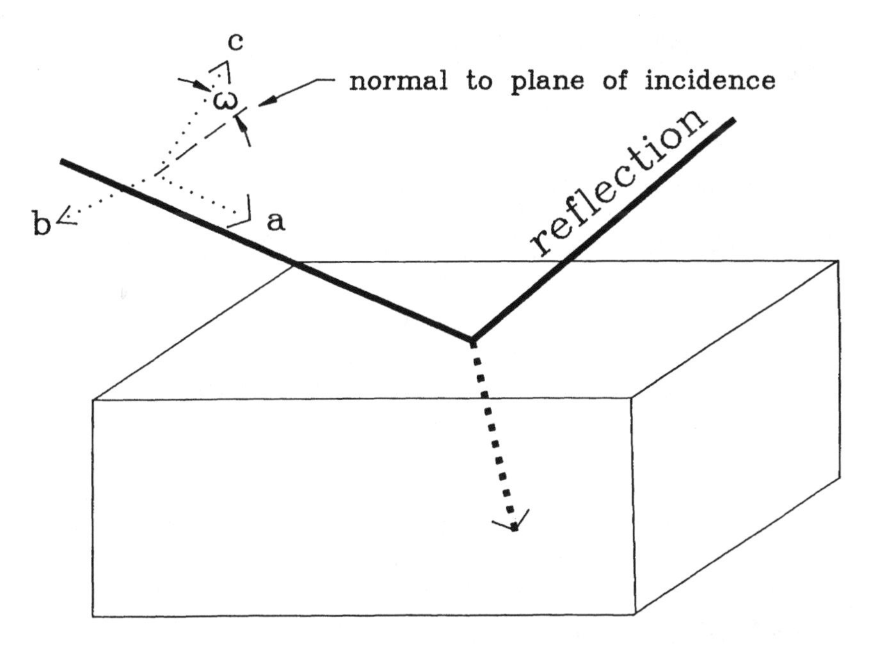

FIG. 2.7 Malus's *a, b, c* axes with partial reflection; *c* axis not parallel to the mirror.

and here we have Malus's must revealing comments. Call ε (ω in fig. 2.7) the angle between the *c* axes and the normal to the plane of incidence on the mirror—thus the complement to the angle between the *c* axes and the plane itself. A portion of the bundle must reflect; the rest refracts. The refraction again consists of two subsets. One subset has had its *b* axes turned into the plane of incidence. The other subset's condition remains for the moment unspecified. The reflected rays have their *b* axes rotated into the normal to the plane of incidence.

According to Malus the number of "molecules" (or "rays" in the sense I have adopted) in the reflected bundle under these last conditions is "proportional" to $\sin^2\varepsilon$. Further, the number of "molecules" whose *c* axes turn normal to the plane of incidence, and that are therefore refracted, is "proportional" to $\cos^2\varepsilon$. Let *N* represent the number of "molecules" in the polarized bundle incident on the mirror. Of these, xN reflect, and yN have their *c* axes rotated normal to the plane of incidence and refract. The remaining rays, $N(1\text{-}x\text{-}y)$ in number, also refract. According to Malus x varies as $\sin^2\varepsilon$, and y as $\cos^2\varepsilon$. Call k_x, k_y the constants of proportionality. Then Malus's law for the count of rays may be written:[6]

Number of rays in the reflected bundle (all by definition have *c* axes in the plane of reflection) $= k_x N \sin^2\varepsilon$

Number of rays in the refracted bundle that must have their c axes in the normal to the plane of incidence $= k_y N \cos^2\varepsilon$

Number of rays in the refracted bundle with unspecified symmetry $= N(1 - k_x \sin^2\varepsilon - k_y \cos^2\varepsilon)$

Malus's task, the task for the ray theory of polarization by reflection, was to determine the values of the factors under the most general conditions, including reflection elsewhere than at the polarizing angle (where another factor becomes necessary). That became Malus's primary goal, one that required careful investigation of partial reflection above and below the polarizing angle.

In this regard—reflection elsewhere than at the polarizing angle—Malus had very early discovered that natural light reflected at any incidence exhibits in some measure the property of bundles reflected at the polarizing angle: their images in a birefringent crystal are unequally intense and vary with the angle between the plane of reflection and the crystal's principal section. This could only mean that the random distribution of asymmetries in natural light is always altered. Most later selectionists thought such a *partially polarized* bundle consisted of a mixture of two subsets of rays, one that has the same symmetry as a bundle reflected at the polarizing angle, whereas the other's elements still have randomly oriented symmetries similar to the randomness in natural light. Partially polarized light is a mixture of polarized and unpolarized light.[7]

2.4 The Added Refraction

The *Théorie* was crowned by the Institute on 2 January 1810 and published in the Academy's *Memoirs* for 1811. Just over a year later Malus presented new results concerning the polarization of refracted light that soon led him to a complete theory of partial reflection [Malus 1811b]. Here he formally introduced the word "polarization." (Though in the *Théorie* he had already written of "polarized" rays and of the "force that polarizes light" [Malus 1811b, 447–48].) He remarked on his choice of terminology:

> These observations lead us to conclude that light acquires in these circumstances properties that are independent of its direction with respect to the reflecting surface and that are the same for the south and north sides [of the ray], and different for the east and west sides. In calling these sides poles, I will call polarization the modification that gives light properties relative to these poles. I waited until now [two and a half years] before admitting this term in the description of the physical phenomena in question; I dared

> not introduce it in the *Mémoires* wherein I published my latest experiments; but the varieties presented by this new phenomenon and the difficulties of describing them force me to admit this new expression, which simply signifies the modification light is subject to on acquiring new properties that are related not to the direction of the ray, but only to its sides taken at right angles and in a plane perpendicular to its direction. (Malus 1811c, 106)

To build a theory of partial reflection Malus needed to examine the polarization of the refracted light, which he did in the following manner (fig. 2.8). In the figure, natural light strikes the plate M_1 at polarizing incidence, producing a reflection (R) and a refraction (E_1) that emerges from the plate. E_1 strikes a mirror (M_2) at polarizing incidence, producing a reflection (E_2).

The beam, E_1, that emerges from the plates (which I shall often refer to as the "emergent beam"), Malus remarked, has "properties analogous to those of" R, the reflection from M_1, but in the perpendicular direction. But the analogy is only a partial one: unlike R, E_2 cannot be extinguished, as we saw above, no matter what the plane of reflection on the mirror M_1 may be. Malus accordingly argued that the emergent beam, E_1, consists of a mixture of two kinds of light. One part is identical in its lack of symmetry to natural light;[8] the other part has complete symmetry, but of the opposite kind to that of R.[9]

Malus went further. He argued that light cannot, however it may be done,

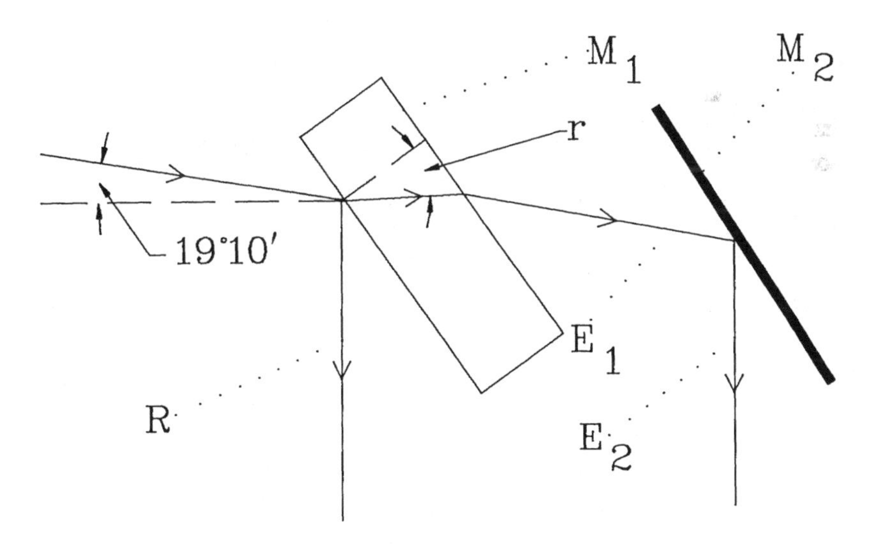

FIG. 2.8 The polarization of the refracted light determined by reflection.

be polarized in one sense without simultaneously polarizing another "proportional" quantity of light in the opposite sense. That claim, which I will call Malus's *law of proportionate polarization,* was the key to a full, quantitative theory of partial reflection, which required carefully investigating the refraction of polarized light, to which he turned in his next (and last) work. Though that article was little read after about 1820, it provides for the first time in optics a theory of partial reflection, a theory that leads to formulas capable of direct comparison with experiment.

Malus began his attempt to obtain a quantitative law with much the same experimental arrangement as before (fig. 2.9). A vertical bundle IO *already polarized in the NS plane* reflects from a plate of glass to the polarizing angle. Begin with IOA, the plane of incidence, set to be the same as the plane of polarization (which contains the NS direction) and examine the emergent beam (B) with a crystal of spar whose principal section remains fixed in the

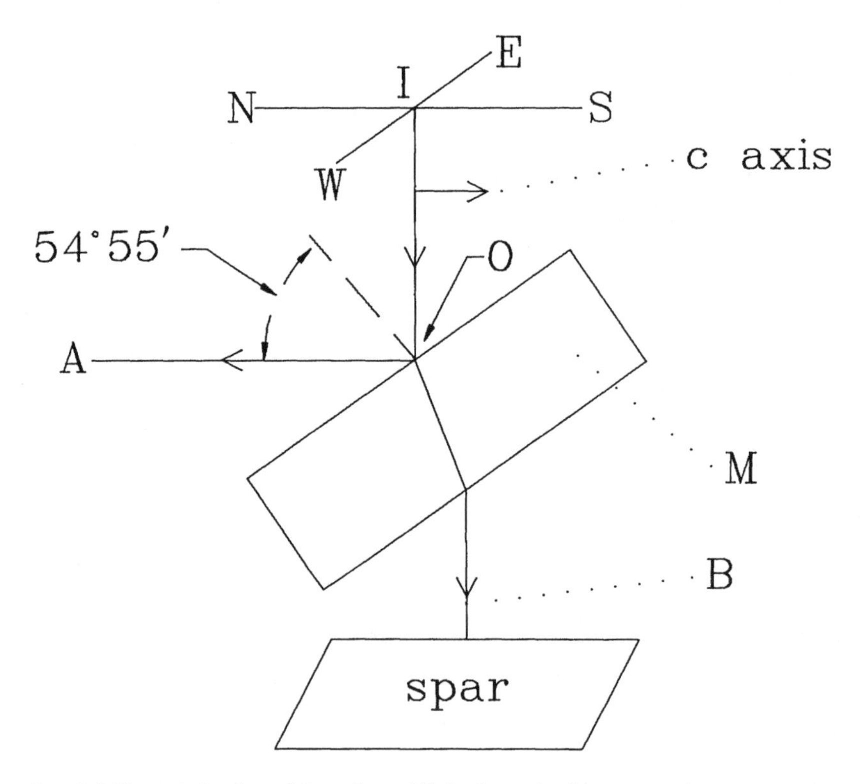

FIG. 2.9 The polarization of the refracted light determined by a crystal.

NS plane. We find that B remains polarized in the same sense as the incident beam IO, since the spar shows only an ordinary refraction.

Next we rotate the reflecting plate of glass about IO as axis until the plane of incidence (IOA) lies along the NW axis, that is, at 45° to the meridian. We find, of course, that the reflected beam decreases in intensity, while the refracted beam increases by an equal amount. However, the light "that is added to that which traversed the glass in its first position" (Malus 1811a, 114) is not polarized in the original plane of incidence, because we now observe an extraordinary beam in the spar. Further, if we alter the plane of incidence from this position, the extraordinary beam decreases in intensity, so that the 45° inclination determines the *maximum* intensity of the added bundle in the refraction.

Malus next concentrated on this additional bundle that arises when the plane of polarization of the light that strikes the glass plate and the plane of reflection from the plate are not parallel to one another. First he set IOA, the plane of reflection, in the NW plane to observe the maximum value of the added light. Then he rotated the spar a bit toward the northeast, whereupon he saw the extraordinary refraction at once diminish and nearly disappear. It appeared again on further rotation. Then, without further explanation, he gave a formula he asserts comes from "theory": "Theory leads to this result, that the light transmitted by the glass in its first position is to the quantity by which it increases after a quarter revolution, as unity to twice the tangent of twice the observed angle" (Malus 1811a, 115).

Let I_0 be the intensity of the emergent beam when the polarization of the incident light lies in the plane of reflection, and let ΔI_0 be the amount by which I_0 increases when the angle between the incident polarization and the plane of reflection is 45°. Then Malus's law can be written in the following way:

Malus's Ratio Law

$$\frac{I_0}{\Delta I_0} = \frac{1}{2\tan(2\omega')}$$

where ω' represents the angle the spar must be rotated through to minimize the extraordinary refraction in it. Malus found that ω' varies solely with the index of refraction, since it depends only on the angle of incidence that is necessary to completely polarize light by reflection.

Take a beam of polarized light, reflect it again at the polarizing angle and compute the intensity of its refraction for a given angle between the plane of original polarization and the plane of reflection using Fresnel's amplitude ratios—which we now know to be extremely accurate. To gain some idea of what actually occurs in an experiment, and therefore how well Malus's ratio law would have seemed to him to work, we may plot the values of log $[I_0/\Delta I_0]$ that are required by the ratio law and by the accurate formulas as a function of the index of refraction as well as their mutual ratio (before taking the logarithm—we must plot the log of the ratios because they can be quite large). The result is striking (fig. 2.10).

The logarithmic curves for the ratios that would be measured in a highly accurate experiment (labeled "Fresnel" in the graph) and for the Malus ratios run nearly parallel to one another. Further, the value required by the former ratio lies between 100% and 120% of the Malus values. Malus's law both follows the general pattern of what reasonably accurate experiments would measure and, if we ignore a factor of two, remains close to it in value.[10]

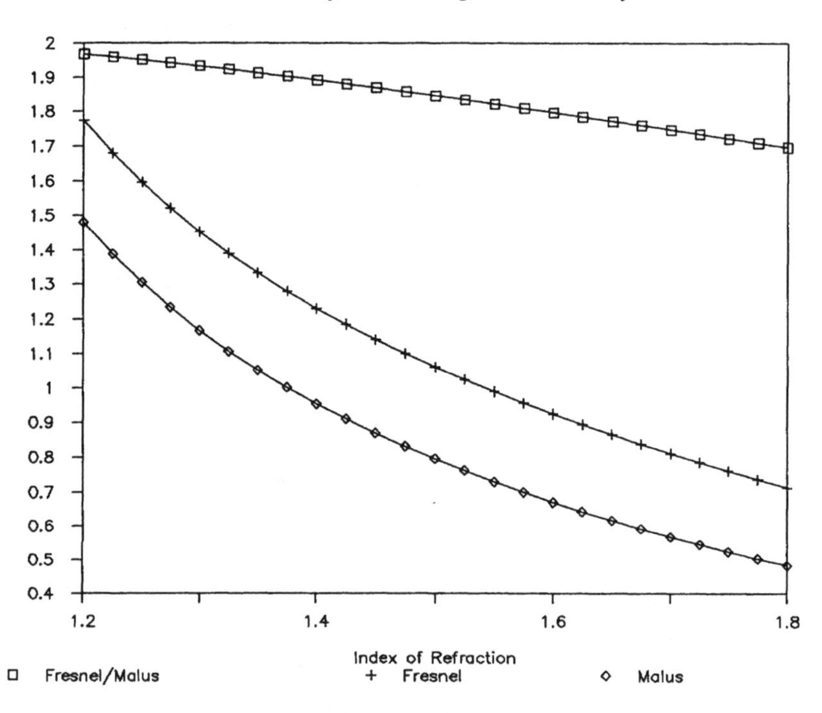

FIG. 2.10 Comparison of Malus's and Fresnel's laws; intensity to change in intensity.

By what "theory" did Malus obtain this law? He did not likely find it by calculations based on particles and forces, since, as an emissionist, he would certainly have said so had he obtained it in this way. Nor did he likely find it simply by correlating observations at different refractive indices; it is too precisely stated for that, and in any case Malus had no way to measure intensity. Though he died too soon to present the full details of this theory, we can nevertheless reconstruct it quite simply. We can see that it provides much more than this single law, which by itself could have only very limited empirical support. It provides a complete account of partial reflection, one that emerges from the very basis of selectionism: the assumption that rays are discrete, identifiable objects.

2.5 Deducing Malus's Ratio Law

Malus never traced the steps to his ratio law, but we can reconstruct them. To do so we must begin with one of his tacit assumptions: that whatever the plane of reflection may be, *the very same portion* I_o of the incident bundle that refracts when the angle between planes of polarization and reflection vanishes continues to refract as the angle changes. This at once means that the constants k_x, k_y must be equal to one another, since the sum $k_x \cos^2\varepsilon + k_y \sin^2\varepsilon$ is equal to I_o. Further, this portion of the refracted bundle always retains the polarization of the incident light. As the plane of reflection rotates away from the plane of polarization, ever more light refracts until, at 90°, all of it does. Again call ΔI_o the extra light refracted at a given angle.

Malus's law of proportionate polarization determines the orientations of the rays in this added bundle. We reflect a portion of the already polarized incident beam into a new plane, thereby altering its plane of polarization by forcing it to the new plane of reflection (since we are always reflecting light at the polarizing angle). This is an instance of the general case covered by the law. Here, then, whatever extra amount ΔI_o of light beyond the usual refracted quantity I_o is refracted must be polarized in the normal to the plane of reflection, since the "proportionate" reflection is polarized in that plane.

In the analyzing spar we will now see both an ordinary beam (from I_o) and an extraordinary beam (from ΔI_o). Malus knew from experiment that the extraordinary intensity reaches a maximum when the new plane of reflection inclines at 45° to the plane of polarization. That necessarily determines the relative proportions of I_o and ΔI_o, and from this we can deduce Malus's ratio law.

Suppose we incline the analyzer's principal section at an angle ω' to the

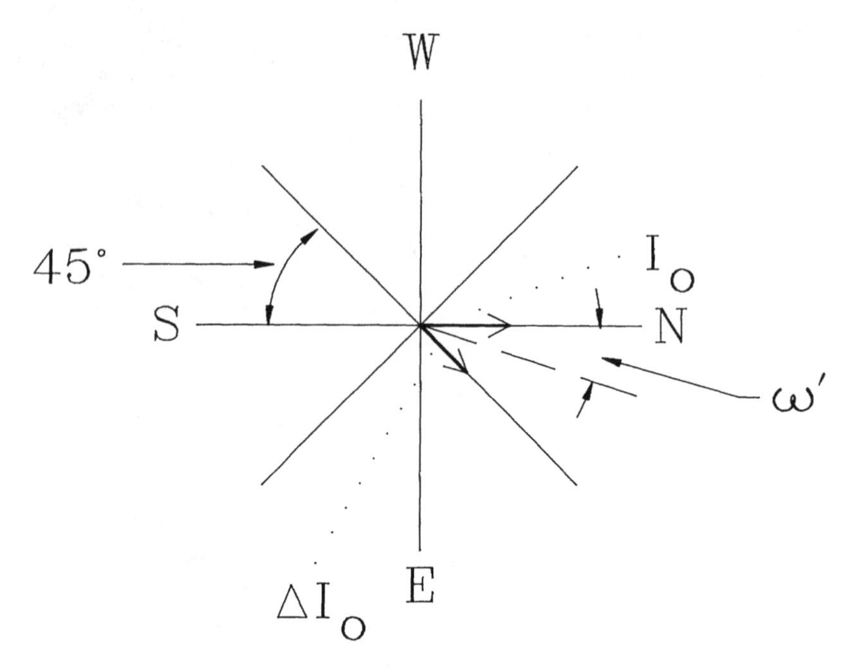

FIG. 2.11 Malus's additional refraction.

plane of polarization of the incident beam when the planes of reflection and polarization are the same. Then, by Malus's cosine law, the ordinary and extraordinary beams in it will have intensities O and E respectively (see fig. 2.11):

$$
\begin{aligned}
\mathrm{O} &= I_o \cos^2\omega' \\
\mathrm{E} &= I_o \sin^2\omega'
\end{aligned}
\qquad (2.5.1)
$$

But when ΔI_o exists—when the plane of reflection is not parallel to the plane of polarization—then it too must contribute to both O and E.

In Malus's experiment the plane of polarization of ΔI_o must be 45° east of north because the plane of reflection was set at 45° west of north, and by Malus's law the extra refracted light is polarized at right angles to the polarization of the reflected light. Consequently the complete O and E beams in an analyzer that is inclined at ω' east of north will be:

$$
\begin{aligned}
\mathrm{O} &= I_o \cos^2\omega' + \Delta I_o \cos^2(45° - \omega') \\
\mathrm{E} &= I_o \sin^2\omega' + \Delta I_o \sin^2(45° - \omega')
\end{aligned}
\qquad (2.5.2)
$$

Because E depends upon ω', at some value of that angle (for given I_o, ΔI_o) E will be a minimum, whereupon $\partial E/\partial\omega'$ will vanish, giving us a condition on the ratio $I_o/\Delta I_o$ as a function of ω':

$$\frac{\partial E}{\partial \omega'} = 2I_o \sin \omega' \cos \omega' + 2\Delta I_o \sin(\omega' - 45°)\cos(\omega' - 45°) = 0$$

Solving, we find:

$$\frac{I_o}{\Delta I_o} = \frac{\cos^2\omega' - \sin^2\omega'}{2 \sin \omega' \cos \omega'} = \frac{\cos 2\omega'}{\sin 2\omega'} = \frac{1}{\tan 2\omega'} \tag{2.5.3}$$

Theory—selectionist theory—does indeed lead to a law that differs from the one Malus gave in print by the very factor of two that causes Malus's law to deviate from the accurate ratio by 100%. Obviously the erroneous factor was very likely a slip of the pen. (Appendix 8 explains where the slip probably occurred.)

Equation (2.5.3) is the first quantitative law for partial refraction: it determines, in a specific situation, the ratio between the amount of light that must always be refracted (and that constitutes all the refracted light when the plane of reflection is parallel to the plane of polarization) and the additional amount of light that is refracted when the angle between the two planes reaches 45°—always at polarizing incidence. The law of proportionate polarization, we shall shortly see, permits us to compute the amount of light that is reflected as well as the amount that is refracted. Malus was filled with enthusiasm by his results. He remarked: "By the simple measure of a single angle [ω'] one may therefore determine the principal element of these phenomena. This quantity once known, one easily deduces, following the theory, the intensity relations of ordinary and extraordinary rays, not only at their maxima, but in all intermediate positions" (Malus 1811a, 115–16).

By measuring the analyzer position under certain conditions, Malus was asserting, one can obtain a value that can be used to solve for polarizing incidence the problem of partial reflection. A problem that now comes to this: to determine for an arbitrary polarization of the incident beam of light what the polarizations and intensities of the various ray bundles that comprise the reflected and refracted light will be. Although Malus did not describe in any detail how one can do this, nevertheless his methods do lead quite directly to a complete solution to the problem. Appendix 8 presents a reconstruction of the formulas Malus would most likely have obtained as well as showing how they can be deduced using the principles of ray theory.

2.6 The Idea of "Depolarization"

Malus at first believed that metals have no polarizing effect at all. On further investigation he changed his mind, and in the process he invented the new, and highly influential, concept of *depolarization*. He wrote:

> Substitute for the mobile mirror, and in the same circumstances, a metallic mirror whose plane of incidence makes a constant 45° angle with that of the meridian; when this mirror is inclined but a few degrees with respect to the horizon, the light it reflects is entirely polarized like the incident light with respect to the plane of the meridian. If the inclination increases, it reflects 1° a certain quantity of light polarized with respect to the plane of the meridian; 2° another quantity of light polarized with respect to the plane of incidence. Above that limit, the light polarized with respect to the meridian begins to reappear, and the light polarized with respect to the plane of incidence diminishes in intensity until the mirror becomes vertical. (Malus 1811a, 117)

This is a theoretical, not an empirical, statement that was based on an adaptation of his selectionist account of partial reflection to metals. He assumed that metals differ from transparent bodies in only one particular, which can best be understood in the case of polarizing incidence. Like transparent bodies, Malus assumed, metals also have a particular incidence at which they produce from the original beam two bundles that are polarized at right angles to one another. However, in the case of metals the bundle that would normally be refracted (and that is polarized perpendicular to the plane of incidence) is instead reflected. Consequently the reflected light cannot be distinguished from natural light. This, Malus felt, explains why metals seem to have no polarizing effect at all on natural light:

> When natural light falls [on a metal at its polarizing angle], the reflected ray contains at once the molecules that are polarized in one sense and those that are polarized in the contrary sense; so that it presents, in its decomposition by a crystal of Iceland spar, the same properties as the natural ray that is reflected at the greatest and least incidences; which makes in this case the proposed limit indeterminable. (Malus 1811a, 118)

However, this means only that one cannot polarize natural light by reflecting it from metals, and that hardly requires, or supports, Malus's assertion. But he had not actually used natural light to investigate metallic properties. He had observed the effect of metallic reflection on an already polarized beam, and that effect does require something like his account, for he found that metallic reflection "depolarizes," in his word, the original beam:

> In submitting an already polarized ray to the reflection of the [metallic] mirror, one avoids this inconvenience [that natural light cannot be observably polarized by reflection from metals]; because instead of observing, as on transparent substance, the angle under which the polarization is complete, one observes on the contrary that for which the depolarization is the most complete. So, for metallic substances, one employs the reflection of an already polarized ray, taking care that the poles of the ray form an angle of 45° with the plane of incidence, and one observes the angle under which the light seems to be completely depolarized like a natural ray. (Malus 1811a, 118–19)[11]

This notion of *depolarization* nicely typifies the selectionist emphasis on the ray structure of a light beam. The loss of polarization, Malus assumed, could be understood in only one way: as the selection of rays from the original bundle and their gathering into new subsets, each with its own asymmetry. We shall soon see that *depolarization* can occur in quite complicated situations, and that this had significant historical results.

The several cases we have thus far examined lead to a general selectionist method for analyzing light, one that, we shall see, was extensively deployed and was still in use as late as the mid-1830s. To determine how a beam of light will behave, one may say that the selectionist had to know three things about it. First, the beam's *ray count,* or the number of rays in each of the beam's constituent bundles; second, the *ray ratio,* or ratio of different kinds of rays in the bundles (when we distinguish among rays on the basis of color); third, the *symmetry* or polarization of the bundles. Given the count, ratio, and symmetry, a selectionist had all the information necessary to construct a theory for any given optical process.

There was no need to delve further into the physics of light to build quantitative theories, as we have seen in detail with Malus's analysis of partial reflection, and as we shall see again in chapter 4 when we examine Biot's theory of *mobile polarization.* Nevertheless Malus, Biot, and most other selectionists before the mid-1820s were also emissionists, and often they did not differentiate between the two. Perhaps it would be better to say that almost no one understood that selectionism, based on the reification of the ray, was itself a theory. It simply did not occur to most scientists that counting rays went beyond the immediate givens of experience. One could grant that the emission theory required hypotheses and yet have vast confidence in it precisely because it seemed to represent so ideally the substantiality of the ray, which never came into doubt. A great deal of later discussion concerning hypotheses in optics entirely missed this point, though confusion on issues deriving from it abounds. Few scientists seem to have been thoroughly aware that part of the argument between emissionists and wave theorists hinged on the issue of whether rays were things at all—or just convenient mathematical constructs. Optics ripened with confusion as the wave theory made inroads in the 1820s. Old concepts that were based on rays no longer fit the new circumstances, but many scientists did not entirely grasp that this was the case. Antoine Parent's easy statement in 1713 that the "rays are nothing other than all the possible perpendiculars in all the points of the waves" (translated in Shapiro 1974, 138) required a great effort of understanding for many in the 1820s and 1830s.

2.7 Ray Physics

In discussing Malus's work I have not examined his many comments on the true physical nature of rays. I have not done so because Malus's quantitative work, as well as much of Arago's, Biot's, and even Brewster's, required only that rays be things, not that they be made of moving particles. This was the essence of Malus's theory of partial reflection. Nevertheless, by the 1810s the close link between the idea of a ray as a thing and the model of a ray as a moving particle amounted, in many people's minds, to an identity. Consequently an empirically successful selectionist analysis was generally assumed to support the emission theory. Later, when Fresnel's work achieved even greater success in the same areas, many selectionists (including Biot and Brewster) did distinguish between the emission theory and ray theory, but even then they did so only implicitly. Many of them never admitted that selectionism was itself a theoretical scheme, which meant they thought, for example, of the quantitative aspects of Biot's account of "mobile polarization" (see below, chap. 4) as being independent of hypothesis, though it epitomized selectionist analysis, as we shall see.

One must, however, never lose sight of the close connection that selectionists inevitably made between ray and emission theory, even though ray theory alone could and did, as we have already seen, yield quantitative laws, whereas the emission theory per se rarely did. Selectionists sometimes worked backward from ray formulas like Malus's for Huygens's construction to speculate about the type and range of the forces that act on optical particles. But they neither succeeded in nor even attempted the reverse, to deduce empirical formulas from forces. At best one could assume a priori an expression for velocity as a function of direction and then apply the principle of least action (see above, chap. 1). But that was an application of emission theory in a very limited sense—that ray speeds increase, rather than decrease, on refraction. This succeeded in yielding formulas for ray directions, but it provided no information about the kind of ray selection that occurs in partial reflection and other phenomena.

A recent biographer of Malus puts the point particularly well. He remarks, referring to Malus's own discussions of forces:

> Malus develops an interpretation [on emission principles of partial reflection] that, in spite of its mathematical apparatus, escapes all quantitative determination. Even if it does not conflict with observational givens, even if it is plausible that a double play of attractive and repulsive forces contributes to the partial reflection of light, nevertheless, because experience is powerless to furnish the least numerical indication of the quantities that are successively introduced, the interpretation advanced here has no hope of experimental verification. (Chappert 1977, 184)

Nevertheless Malus, and after him Biot, Brewster, and others, retained great confidence in the emission theory, in part because they did not admit its empirical poverty since they did not distinguish it until much later from ray theory, which they well knew was no beggar.

The conflict between the wave theory and the emission theory therefore has a considerably wider dimension to it than one might expect were it seen exclusively as a conflict between two well-articulated and widely understood theories, of which only one could generate empirical formulas whereas the other could at best explain most phenomena. (That, at any rate, was the usual view about the time, in the early 1830s, when the emission theory was being increasingly ignored. Lloyd's Report (1834) exemplifies this position, as do several contemporary texts.) For what we have seen in this chapter, and what we shall see again in others, is that the emission theory, as it was then understood, was certainly not powerless to generate quantitative formulas. Precisely because of the very close connection that contemporary scientists drew between the concept of a light ray as a stream of particles and the requirement that a ray must have an individual identity, the quantitative power of the latter was transferred to the former. Certainly one could not—or at least no one ever did—actually use forces and particles to deduce formulas. But one could use ray theory to do so, and this seemed to many scientists a de facto confirmation of the emission theory itself.

Because of this close link between selectionism and the emission theory, it would not be at all historically accurate to separate them until well into the 1830s, when several scientists did begin to make a distinction between them, as we shall see nicely illustrated in the case of Brewster. Before that date any selectionist analysis was inseparably linked to the emission theory, so that a critique of the latter was also a critique of the former. However, on occasion the difference between the two—that selectionism was more general than the emission theory—did surface and lead to problems. We shall see this when we examine the controversy between Biot and Fresnel over chromatic polarization. Here Biot did insist on separating ray theory from the emission theory, but Fresnel would not permit him to do so, provoking in the end an extremely bitter controversy between them.

3 Arago and the Discovery of Chromatic Polarization

3.1 Polarization and Newton's Rings

On 18 February 1811, before Malus's last two memoirs were read, a young member of the Institute described before it the first new discovery concerning polarization since Malus's [Arago 1811a]. Dominique-François Arago was just twenty-three at the time of his election to the astronomy section of the Academy in 1809, and his work on Newton's rings two years later represented the only research for which he could claim sole responsibility. He had reason to be, as he clearly was, jealous and proud of his results. They, and other observations they led to, were quite startling, for they seemed to constitute exceptions to Malus's account of partial reflection.

In this chapter we shall first examine Arago's discovery and the way it fits into the overall selectionist understanding, as well as a criticism of the emission theory that Arago based upon it. Following this, we shall turn to a second discovery he made, one that eventually produced an entirely new area of research in optics—his discovery of *chromatic polarization*. This too was interpreted by Arago in selectionist terms, but we shall also examine how Arago lost control to Biot of the field he had himself founded. Since Arago's loss had important repercussions for Fresnel's interaction with the established Parisian research community, we shall follow his work in some detail.

From the outset Arago fully adopted Malus's terminology and selectionist understanding of polarization (for an example of the extent to which Arago used selectionist concepts, see appendix 9). With these ideas in mind, he decided to examine the polarization of Newton's rings. These rings are formed when light strikes the thin gap between the surfaces of a flat reflecting plate and a lens pressed to the plate. Arago at once discovered that the rings do show partial polarization. He designed a clever apparatus (fig. 3.1) to examine the polarization of both kinds of rings. One portion of a beam of natural light (A) strikes a metal mirror in such a way as to reach from below *at a specified incidence* a combination of two lenses pressed close together. Another portion (B) of the beam strikes the lens combination, *also at this specified incidence,* from above. The two beams that emerge by reflection and by

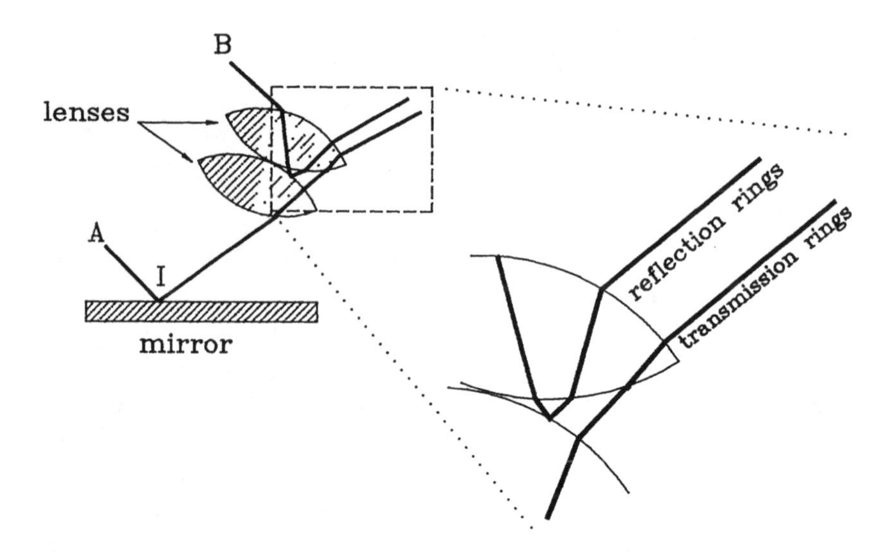

FIG. 3.1 Arago's observation of ring polarization.

transmission from the upper lens then pass through an analyzing crystal. If we block beam B, then only the colored transmission ring appears.[1] If we block beam A (or if we cover the metal mirror with a black cloth), then only the complementary-colored reflection ring appears. If neither A nor B is blocked, both reflected and transmitted light appear together and so reconstitute the color of the original light beam.

Arago found that the ordinary and the extraordinary images formed in the crystal by a reflection ring (deriving from beam B) differ from one another in intensity, and that this difference depends on the incidence of the light on the gap. Indeed, one of the images vanishes altogether when the light strikes at polarizing incidence, showing that the ring is just as fully polarized as would be light reflected at that angle in circumstances where rings do not form. In what follows we shall, following Arago, be examining phenomena that arise from light that strikes at polarizing incidence.

Consequently the polarizing effect of reflection that Malus had discussed does act on the light that forms reflection *rings* as well as on whatever part of the reflected light does not form rings. However, we do not as yet know where the light that forms the rings is polarized—whether it occurs at the partial reflection on the upper surface of the lamina or at the partial reflection on its lower surface—since according to Malus's account both surfaces will polarize light reflected from them in the plane of reflection.

The distinction Arago was implicitly drawing here between two separate parts in the light that is seen when one looks at the lamina from above—one part being involved in the rings, the other not—apparently derives from his reading of Newton's discussion of the phenomenon in the *Opticks.* In this by then standard account, the second surface of the thin lamina of air sends back the light that is seen in the reflection rings, and this light mixes with the light that was reflected from the upper surface in the usual way. According to Newton, both reflections involve the same attribute of rays that occasions rings, namely that every ray has an inherent periodic property or "fit" that disposes it to be reflected or refracted depending on the state of the fit when the ray strikes the surface.

The length of a fit varies with the ray's color. When they are transmitted through the upper surface of the lamina all the rays, whatever their fit lengths, are started out again at the beginning of a fit, so that the rays of a given color reach the second surface in a specific, common fit state. They are then all disposed either to reflect or to refract depending on the distance they have traversed: if the distance is nearer an even than an odd number of fits, the ray is transmitted; otherwise it is reflected. Consequently the light that forms the reflection rings has been produced by the second laminar surface's sending back rays that are all in a particular state of fit. But the first, upper surface of the lamina had also reflected rays *without producing rings* because the light that had struck it contained rays, even of a given color, in every possible fit state. That is, although whether a given ray is reflected or refracted at the lamina's upper surface depends on its fit state, since the rays of a given color are not all in the same state, some of them will be refracted and some reflected, which means that rings do not form from this reflection.

Now Arago had shown that Malus's polarizing property affects the light that forms the reflection rings in the same way it affects the light that does not form rings. In this respect, therefore, polarization by reflection seems to operate in a fashion similar to ring formation in Newton's account: the polarizing action affects light rays that are reflected but that are not involved in forming rings, as well as the light rays that do ultimately form reflection rings—just as Newton's selection of rays for reflection and refraction by their state of fit acts on every ray striking any transparent surface. But what is the state of the light that forms the transmission rings? How has the polarizing action affected it? If the analogy to Newton's "fits" is complete, then presumably the transmitted light that forms rings will be polarized in the same way light usually is polarized by refraction—just as, when they are formed, the transmission rings must be in a "fit" of transmission, whereas the reflection rings must be in a "fit" of reflection. That is, the transmitted light should be polarized perpendicular to the plane of reflection for the case Arago always considers of polarizing incidence.

The rather startling answer to this question was that the transmission rings are *not* polarized in the way Malus's theory seems to require. Again choosing the ring that derives from light incident at the polarizing angle, Arago found that the corresponding transmission ring is in fact polarized in *exactly the same way* as is the reflection ring—in, that is, the plane of reflection. In Arago's words:

> [At polarizing incidence] white light is decomposed into two distinct colored portions [the reflection and transmission rings] that, following different routes, are nevertheless polarized in the same manner, whereas in the general phenomenon discovered by Malus the parts of the incident bundle, transmitted and reflected, are polarized in opposite senses. (Arago 1811a, 16–17)

Malus was himself troubled by Arago's observation of an exception to his laws, and he seems to have tried to undercut the problem by arguing that the equivalence in the polarizations of the reflected and the transmitted rings was only apparent (Malus 1811c).[2] Biot too was troubled by the discrepancy, and he decided it had to mean that Malus's "polarizing force" comes into play only after a certain thickness is traversed. Arago did not accept Malus's rather vague explanation, and he at once set about undercutting Biot's by demonstrating that "unfortunately" he was able to find the usual sort of polarization—obeying Malus's laws—in the refracted light that emerges from laminae that are too thin even to produce Newton's rings. If Biot were correct, this type of polarization should certainly not as yet have occurred.

Arago's own initial reaction to his discovery was to investigate the extent of this rupture between the laws that govern Malus's form of partial reflection and the laws that seem to govern the polarization of Newton's transmission rings. It seemed possible that the kind of partial reflection that occurs in Malus's scheme was different from what occurs in forming Newton's rings. In other words, there might be two unrelated mechanisms that can produce partial reflection: one of them is the usual one involved in Malus's theory, whereas the other one operates independently and determines whether rings form.

But Arago demonstrated in another series of experiments that the mechanisms cannot be entirely distinct, because whenever Malus's sort of partial

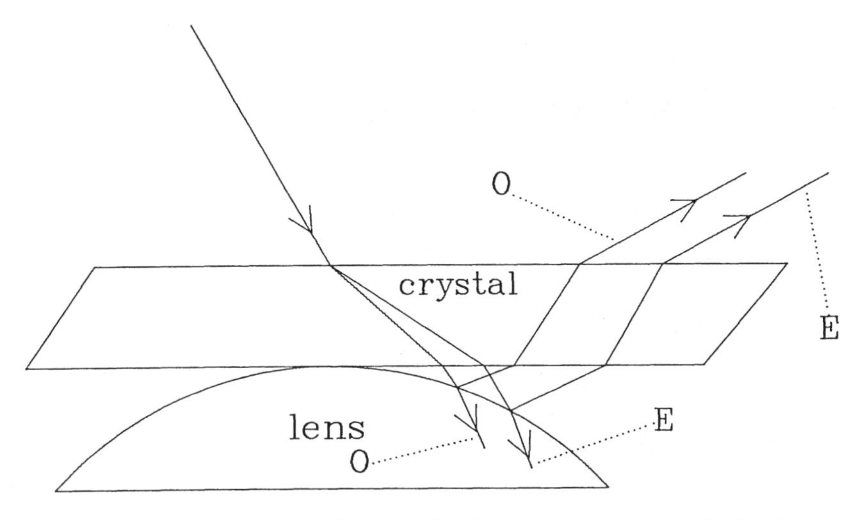

FIG. 3.2 Arago's experiment to show that ring formation requires partial reflection.

reflection disappears then so too do Newton's rings. Pass natural light through a crystal that is pressed against a lens (fig. 3.2). If one rotates the entire apparatus, the angle between the crystal's principal section and the plane of reflection on the lens will change. Here, then, we have a beam of light that, originally unpolarized, becomes polarized by passing through a crystal. After that it reflects at the lower surface of the lamina (on the surface of the lens, that is) and then, having traversed the lamina once more, passes again through the crystal.

Adjust the incidence so that the reflection on the surface of the lens occurs at *polarizing* incidence. Now suppose we turn the crystal so that its principal section is perpendicular to the plane of reflection on the surface of the lens. Then we have polarized light being reflected in a plane perpendicular to its plane of polarization. According to Malus's laws this light will not be reflected at all; it will be completely refracted. If the mechanism responsible for this kind of partial reflection differs from the mechanism responsible for the kind that is involved in Newton's rings, one should still observe them—there should still be reflection and transmission rings. But Arago observed none: when Malus's form of partial reflection ceases, so does the kind involved in forming Newton's rings. From this and other experiments Arago concluded that Malus's sort of partial reflection is a *necessary condition* for the formation of Newton's rings, even though the polarization of the transmission rings violates what usually happens to refracted light.

Arago next sought to determine at which surface of the lamina the light that will ultimately form rings becomes polarized. Does the light that forms a ring acquire its polarization at the upper surface or at the lower surface? To answer the question Arago formed the two surfaces out of different materials by pressing a glass lens to a metal mirror. At an angle corresponding to polarizing incidence for the lens, Arago remarked, the ring is polarized in the plane of incidence. Yet Malus had previously demonstrated that "polarizing" incidence for the metal has a markedly different effect from that for glass. "This would seem to prove," Arago concluded, "contrary to the opinion of Newton and of physicists who, after him, occupied themselves with the same point, that the decomposition of rays takes place at the first surface of the thin lamina, because only it regulates the angle of polarization" (Arago 1811a, 27). To grasp Arago's reasoning here, one must recall that in Malus's view metals polarize by reflection in the same fundamental manner that transparent bodies do, but instead of transmitting the bundle polarized normally to the plane of incidence, they reflect it. The reflected light then consists of two bundles with mutually orthogonal polarizations. If the metallic reflection (at the second surface of the lamina) governed the polarization of the rings, they would presumably seem to be unpolarized. Yet they are not—they show the usual polarization in the plane of reflection. (Recall that we are always examining rings derived from light that strikes at *polarizing* incidence.) Consequently the ring's polarization is evidently governed by the first, not the second, laminar surface.

This result, Arago further concluded, seems to imply that the upper surface of the lamina already acts in the appropriate way to choose the light that will eventually form rings, because it is this surface that determines the *polarization* of the rings—even though the lower surface determines what kind of rings form. Newton and others had assumed the rays that form the rings must be separated from the original set (i.e., the "decomposition" Arago refers to must occur) at the lower surface, because its distance from the upper surface determines what kind of rings will form. The upper surface makes possible the separation that the lower actually produces. But, Arago was apparently suggesting, perhaps this is not correct. Perhaps the separation or "decomposition" somehow already occurs at the upper surface—since it is the upper surface, it seems, that governs the polarization of the light that can form rings. In this case the function of the lower surface would have to be understood in a way very different from Newton's explanation.

Yet a second experiment made the issue acute, because it indicated that the lower surface must be *directly* involved. Arago took the metal mirror and roughened it slightly. Then, he noted, *two* sets of rings appear, one of which (by far the weaker) clearly derives from reflection at the roughened surface of the mirror since the rings vanish when the mirror is smooth. The two sets have "complementary" colors, which, Arago remarks, "suffices to show that they were not formed in the same fashion" (Arago 1811a, 28). Yet both sets are polarized in the plane of reflection when the incidence corresponds to the polarizing angle for the glass lens.[3]

Though the common polarization of the two sets seems to imply that the "decomposition" of the incident bundle occurs at the surface of the lens, that the second set depends upon the state of the metal mirror's surface seems to implicate that surface quite directly. "If one wished," Arago remarked, "rigorously to admit that the first rings are formed at the upper surface of the lamina, one would have to explain why the others, which seem to emanate from the metal mirror, are polarized under the same angle, which, in the emission system, doesn't seem easy" (Arago 1811a, 28).

Apparently Arago understood his observations to constitute a difficulty for the "emission" system. Though unexpressed, his reasoning probably assumed that in the emission theory the rings must be "formed" at one of the two surfaces because the forces that select particles for reflection or for refraction must emanate from these surfaces. He thought the light that may, under the proper conditions, form rings is somehow abstracted from the light that would otherwise be reflected from the upper surface of the lamina: the rings are formed, he remarked, "at the expense of light that would have been partially reflected" (Arago 1812a, 78). It was therefore conceivable that the forces involved in forming the rings were sufficiently long in range that the action of the upper surface was all that was needed to produce the rings. The function of the lower surface would then be to terminate or somehow alter the action of the upper surface in "decomposing" the rays that can form the rings.

In other words, Arago did not think Newton's account, which assigned the formation of the rings to the action of the lower surface in separating from one another the rays that are in different "fit" states, was essential to the emission theory. The forces that, in the emission theory, must somehow govern the separation that leads to rings could in principle emanate from either (or both) surfaces. However, he evidently reasoned, it would be extraordinarily difficult to understand in the emission theory why the polarization of the rings seems always to derive *exclusively* from the action of the first surface, whereas the lower surface can dramatically affect the nature of the rings.

Indeed, the lower surface has the power to obliterate rings altogether when the plane of reflection on it is perpendicular to the plane of polarization of the light striking it.

Arago prefaced his published paper with the following words:

> [In Hooke's work of 1665 one finds] an explanation of the colors of thin laminae having a certain analogy with that M. Thomas Young has given of the same phenomenon, in several very interesting memoirs inserted in the last volumes of the *Philosophical Transactions,* to which I will have occasion to return in what follows. (Arago 1811a, 5)

Arago never did return to Young, at least not explicitly. But we have seen that he thought the "emission" theory was hard pressed to account for the fact that both surfaces of the lamina can influence ring formation. In the version included in his *Oeuvres,* Arago inserted an addendum in which he claimed that through these experiments he arrived "if not at a demonstration of the insufficiency of the [emission theory] at least at showing that it cannot be reconciled with the phenomena except by perpetually adding new properties to those Newton deduced from his experiments" (Arago 1854–58, 10 : 31). He made the same claim in a note included when the article was finally printed in 1817.

In view of these remarks as well as Arago's comments at the time, it seems highly probable that in 1811 he already had at least a vague understanding that Young's new theory required both surfaces of the lamina to be active in order for rings to occur. This seemed in fact to be empirically essential, but the emission theory had trouble dealing with it. Arago certainly did not at this time understand that rays themselves must be considered in a very different way. Rather, we must take literally his reference to the "emission" theory. Most probably Arago could not easily see why, if ring formation is somehow due to forces acting on particles in unusual circumstances, both laminar surfaces can, under the right conditions, influence ring formation yet only the upper one governs ring polarization.[4]

Arago's doubts were certainly vague and ill formed, but they are clearly present in this very early paper. He began the paper with a favorable remark about Young's work on thin films, and in the same note he also claimed he had written a critique of the emission theory that was inadvertently burned at the printer during the long delay in the publication of the third volume of the Arcueil *Mémoires* (which emerged only in 1817).

Arago had therefore written something that, he says, was highly critical of the emission theory at this early date, but he had chosen not to redo after it was accidentally destroyed. In the note he writes that Fresnel's observations of hyperbolic fringe propagation had by then made his own critique of the emission theory unnecessary. No doubt what Fresnel's work in fact did was to make Arago's vague comments seem, even to Arago, hardly to the point.

At some point in his investigations Arago substituted for the air gap a thin sheet of mica (which, though he does not discuss the point, would have the advantage of producing a uniformly tinted field, thereby isolating a given ring—or it would have if it acted merely as a transparent film). The mica sheets, he remarked in August, gave "certain singular phenomena that I did not speak of before [i.e., in February] because they were foreign to the object I had in view, and because it seemed to me very difficult to explain them" (Arago 1811b, 37). These singular phenomena, we shall see, rapidly convinced Arago that he might be able to produce a general theory linking Newton's rings, partial reflection, and birefringence.

3.2 Arago's Discovery of Chromatic Depolarization

On 11 August Arago read a paper in which he described a series of observations that rapidly turned the major concern of contemporary optics into a new direction. He carefully described what he had found:

> On examining, in calm weather, a very thin lamina of mica with the aid of a prism of Iceland spar, I saw that the two images projected on the atmosphere were not tinted with the same colors: one of them was greenish yellow, the second purplish red, while the part where the two images overlapped had the natural color of mica seen with the naked eye. I recognized at the same time that a slight change in the inclination of the lamina with respect to the rays that traversed it caused the color of the two images to vary and that if, leaving the inclination constant and the prism in the same position, one turns the mica lamina in its own plane, one finds four positions at right angles where the prismatic images have the same strength and are perfectly white. Leaving the lamina immobile and turning the prism, one also sees each image successively acquire diverse colors and pass through white after a quarter-revolution. Further, for all positions of the prism and the lamina, whatever the color of one of the beams, the second always presented the complementary color, so that, in those points where the two images were not separated by the double refraction of the crystal, the mixture of these two colors formed white. It is, how-

> ever, well to remark that this last condition is rigorously satisfied only when the lamina has the same thickness throughout. This is only when, in effect, each image has a uniform tint throughout; for in the other cases they both present, even in contiguous parts, very different colors that are the more irregularly disposed the more the mica one uses is unequally [thick]. Nevertheless, the corresponding parts of the images are always tinted with complementary colors. (Arago 1811b, 38–39)

These experiments were performed with skylight on a clear day. Arago soon discovered that the colors varied with the position of the sun and, further, that on an overcast day no colors at all were visible. The only property that could possibly be involved in this was polarization, and so Arago concluded that skylight is generally polarized in directions that depend on the position of the sun. This "naturally" led him, he writes, to try the experiment with light that had been previously reflected. The colors were more vivid, he found, the closer the reflection was to polarizing, which, he wrote, "seems to prove that only polarized rays produce the colors involved here" (Arago 1811b, 40).

Given contemporary principles, there was only one general way to interpret the fact that on passing through a mica plate and then through an analyzer polarized white light yields complementary-colored O and E beams: the common asymmetry of the rays in the incident beam must have been altered in a way that depends in some fashion on the rays' colors and on the orientation of the lamina in its own plane.[5] The incident beam has, in Malus's terms, been "depolarized" in a special way. In Arago's words:

> Mica laminae . . . , placed in a certain way, depolarize the luminous rays that a reflection at the proper angle had already modified. One sees, further, that these laminae seem to act differently on rays of different colors, and that, in consequence, they impress characteristics on the rays of light that distinguish them from both direct light and polarized rays. (Arago 1811b, 42)

Thin plates of mica must therefore disperse the originally common asymmetries of the colored rays, producing as it were a fan with sets of colored asymmetries as spokes. For a given incidence and polarization, there are four orthogonal positions of the mica's principal section at which the dispersion fails to occur. When the beam emerging from the mica is analyzed, the rays of a given color may be thrown into either the O or the E refraction, since Arago does not specify that all the rays of a given color are depolarized in the same way. The spokes of my analogy might occupy several degrees. However this may be, the different colors experience different dispersals, so that the analyzer in a given position will not combine rays in either the O or the E beam in the proper ratio to produce white. Further, by ray conservation whatever color the O beam has, the E beam must have the complementary color. Arago carefully made the point in terms of Malus's "molecular axes":

> If the properties of these . . . rays depend, as has been supposed, on the particular disposition of the axes of the molecules of which they are formed, there will be this difference between the ray that was polarized by reflection or by passage through a doubly refracting crystal and the same ray on its emergence from the mica lamina, that in the first the axes of the molecules of the diverse colors are parallel, whereas in the second there are molecules of diverse tints whose axes have different directions. (Arago 1811b, 42; translated in part by Frankel 1972, 237)

To demonstrate in a graphic way the dispersal of the asymmetries, Arago devised a simple experiment (see fig. 3.3). First reflect a beam at polarizing

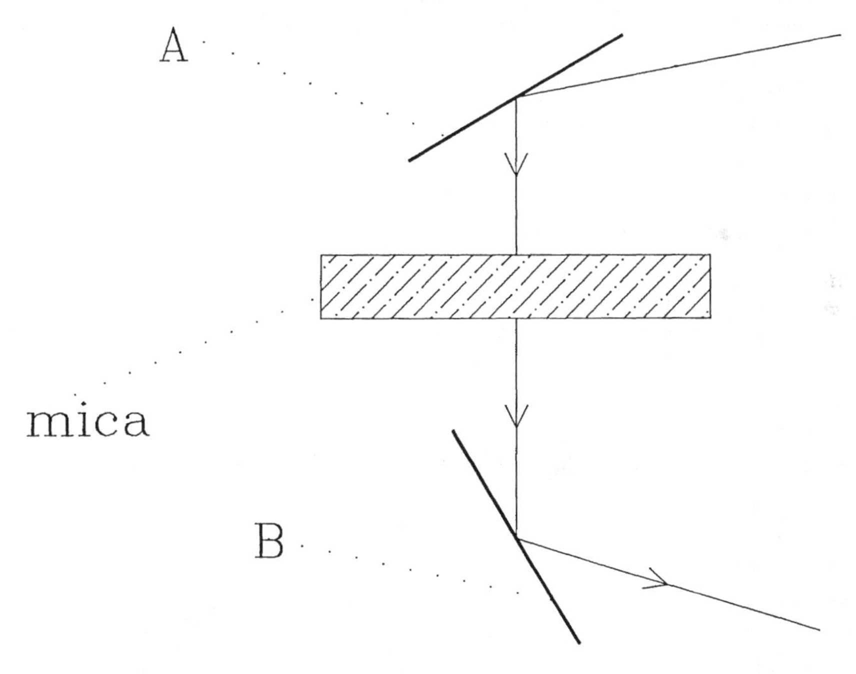

FIG. 3.3 Arago's demonstration of symmetry dispersal.

incidence from mirror A and then pass it through the mica. Introduce a second mirror, B, set again for polarizing incidence, to reflect the light that emerges from the mica. The image of the first mirror seen in the second will be strongly colored, and the tint of the image will change as the mirror rotates about a vertical axis (always retaining polarizing incidence). The tint reflected at a given position of B is found to be complementary to that observed a quarter-revolution before or after. Further, there are four positions of the mica lamina in its own plane at which the image seen in the lower mirror remains white. This device might be called a rotary ray selector: as the bottom mirror turns about the vertical axis, it reflects in a given position some rays from all bundles except those bundles that are polarized in the perpendicular to the plane of incidence upon it. (Whence the complementarity of the tints at quarter-revolutions of the mirror follows at once from ray conservation: whichever rays were not reflected a quarter-revolution before must be reflected now.)

Further experiments indicated that the colors also vary with the thickness of the lamina, but that if it is too thin it loses its "depolarizing" ability (though to a certain point the colors are purer the thinner it is). Arago also sought to find out how thick a crystal could be and still produce this chromatic "depolarization." In this way he was led to yet another discovery. Using a plate of quartz 6 mm thick instead of the thin mica lamina, he found that the quartz did produce colors despite its thickness; that it "gave birth to very remarkable phenomena analogous to those I described" for mica. There was, however, a notable difference between the behavior of the mica and that of the quartz: rotating the quartz in its plane had no effect at all upon the colors, whereas rotating the mica changed them markedly.

Arago developed a theory to explain this discrepancy between quartz and mica that is discussed in appendix 10 as an excellent example of how one can construct explanations for these sorts of phenomena using a principle of ray selection. According to the wave theory, quartz rotates the plane of polarization of light as the light passes through. Arago's theory is of course utterly different from this, but it—like Malus's for partial reflection—could work well empirically if the effect of quartz on light is small. Here one can say that the theory would work if the change in the plane of polarization produced by quartz is very slight. But the change actually amounts to tens of degrees for light that goes through a plate only several millimeters thick. Consequently Arago's theory cannot be reconciled with fairly simple observations that he could himself very easily have made. Obviously he did not make them, and I think it highly probable that it was because he had not realized this implication of his theory. For to grasp it most directly, one should write down equations of the kind we examined above—Malus had certainly done this to reach his reflection laws. But there is no evidence in any of Arago's early work that he ever did any such thing. And this lack of interest in producing equations is, we shall see, symptomatic of Arago's overall failure to understand what his contemporaries demanded of theory and experiment.

3.3 Arago Passed By

Between 11 August and the following spring Arago continued, when he had the time, to pursue his new discoveries, but we do not know precisely what he was doing. He did not have the time to gather his work for a new public reading before Biot read a note on chromatic depolarization that rapidly stripped Arago of his leadership in the new field. Arago at once reacted like a victim of theft. The *Procès-Verbaux* for 20 March 1812 records:

> M. Biot reads a note on some phenomena that depend upon the polarization of light by reflection and refraction.
>
> M. Arago announces that he had in part made the same observations and demands that the Class recognize he had consigned them to the registers of the Observatory. (Arago 1812a, 75)

The Institute appointed Burckhardt and Bouvard to look into this contentious issue. On the following 6 April the *Procès* records that the investigators found "the declaration made by M. Arago [to be] most exactly true" (Arago 1812a, 75), a strong statement indeed. But the notes at issue were published only in Arago's *Oeuvres* forty years later. Furthermore, Arago did not read a third memoir to the Institute until 12 December, eight months after the debacle with Biot over priority. And in the meantime Biot himself read a lengthy memoir on chromatic depolarization, followed six months later by an even longer discussion that had a substantial impact on contemporary optics.

Arago clearly lost control of the field he had himself founded. However, a study of the notes he deposited at the Observatory, and for which he successfully demanded recognition, reveals a rather surprising thing. Unlike Biot's work, which focused from the beginning upon chromatic depolarization, these notes scarcely mention the subject. And Arago's two subsequent memoirs in the general area, as well as his contemporary unpublished "notes" (which were later printed in the *Oeuvres*), though quite directly concerned with crystalline effects, nevertheless have a very different character from Biot's memoirs. Unlike Biot's work, this material is qualitative and nearly devoid of any concern with emission principles, being almost entirely involved with the overall features of what happens to rays rather than with why it happens to them or with representing precisely how it happens.

The notes in fact contain, albeit only in embryonic form, a general theory whose goal was to unite the polarization effects of double refraction and thin crystals with those of reflection. Biot's sudden intrusion into the field forestalled Arago's publication of this theory until both it and Biot's own had been replaced by the wave account. The primary reason for Arago's reticence must wait until I have exposed his theory, but suppose for the present that Arago felt Biot had successfully usurped his position as the exponent of a new field of optics just when he was about to make a particularly impressive contribution.

Arago's professional position made him especially sensitive to an intrusion of this kind. He had been made a member of the astronomy section of the Institute in 1809, at the age of twenty-three and with Biot's support, over Siméon-Denis Poisson, who was five years older and had the support of Laplace. There were probably two major reasons for this (see Home 1983, 248 ff. for a detailed discussion). First, though Poisson had published several very well received articles applying mathematics to physics, the astronomy section was predominantly concerned with observation. Second, together with Biot Arago had made meridian observations in Spain and had returned to Paris as something of a hero after a series of dangerous adventures. "If Poisson and his backers in the geometry section had held hopes of having him elected as an astronomer," remarks Home, "Arago's reappearance in such romantic circumstances put paid to these" (Home 1983, 248). Consequently Arago was appointed over Poisson primarily for reasons of politics and sectional self-image. He had himself done almost nothing of consequence.

Laplace had assiduously and perhaps even improperly supported Poisson, going so far as to attempt to adjourn Arago's nomination until an opening fell vacant for Poisson in the geometry section (since the astronomers would not have him). The bitterness Arago must have felt at the time, bitterness mixed with a sense of inferiority in regard to Poisson, was still present many years later when he wrote his autobiography. He had obtained forty-seven votes to Poisson's paltry four.

> A nomination made with such a majority would at first sight seem unable to have produced serious difficulties; however, that wasn't so. The intervention of M. de Laplace, before the day of scrutiny, was active and incessant to cause my admission to be adjourned until a vacant place in the geometry section would permit the learned assembly to nominate Poisson at the same time as me.

> The author of the *Mécanique céleste* had an attachment without limits for the young geometer, one completely justified furthermore by the beautiful work science already owed to him. M. de Laplace could not stand the idea that an astronomer, more than five years younger than Poisson, that a student, in the presence of his professor at the Ecole Polytechnique, would become an academician before him [Arago having taken classes from Poisson]. He therefore proposed to me that I write to the Academy saying I wished not to be elected until there was a second place to give to Poisson; I replied with a formal refusal. (Arago 1854, 94)

How bitterly humiliated Arago must have been. But after all, what had Arago accomplished by way of scientific research? He lists for us, in his posthumous autobiography, his accomplishments at the time; all of them involve entirely straightforward, though often delicate, measurements. "Laplace," Arago complained, "without denying the importance and utility of these works and researches, didn't see in them more than a promise." To which Arago records Lagrange as having replied:

> "Even you, monsieur Laplace, when you entered the Academy, you didn't have anything striking; you gave only promise. Your great discoveries only came afterwards."
>
> Lagrange was the only man in Europe who could with authority address such an observation to him.
>
> M. de Laplace did not comment on the personal fact; but he added: "I maintain it is useful to hold out to young savants a place as member of the Institute as a recompense in order to excite their zeal." (Arago 1854, 92)

Arago felt a burning need to justify his election. He was apparently convinced he had had success almost in hand when Biot snatched it from him. But what did he have? To understand the depths of Arago's anger and, further, why he fastened so strongly onto Fresnel a few years later, we must understand that more was at stake than an experimental discovery: Arago felt he had had the beginnings of a theory that Biot's subsequent work rendered at best inconsequential—a theory that would have justified the elevated status he had so rapidly achieved.

3.4 Arago's Unification

The "notes" initialed by Burckhardt and Bouvard, which Arago so strongly insisted be recognized, run to only ten pages in volume 10 of his *Oeuvres*. They discuss in essence two broad areas: the colors of striated surfaces and the polarization of Newton's rings. Only one paragraph near the end discusses the colors of mica laminae, and then only to claim that "the colors one perceives on very thin mica laminae are of the kind that form on a lamina of air between two glasses, because, in the two circumstances, the reflected and transmitted colors are exactly complementary" (Arago 1812a, 83). Indeed the major thrust of the notes is to confirm Arago's earlier contention that Newton's transmission rings are formed "at the expense of light that would have been partially reflected" (Arago 1812a, 78) had the rings not formed. But, one might ask, what had this to do with chromatic depolarization?

We find the answer in the last few paragraphs, where Arago advances a major unifying speculation. Take, Arago writes, a very thin lamina of blown glass. Such a lamina, he reiterates (having made the point before) will polarize transmitted light "by reflection"—that is, in the same sense as reflected light—*even if it is too thin to form rings.* If, he continues,

> instead of superposing glass laminae of a certain thickness, one superposes very thin blown fragments, will one not obtain an emergent bundle composed at once of molecules polarized in the two senses? Will not the diverse elements of this new pile, in acting on the light like glass plates of a certain thickness, polarize some rings by refraction, while the rays of the colored rings that traverse are polarized by reflection? Will it not be to this cause that one must attribute the two kinds of polarization that bodies endowed with double refraction impose on a single bundle of direct light? What might give this hypothesis some probability is that all the bodies in which an axis of refraction has been perceived are, in effect, composed of superposed laminae, and further, for certain positions of these laminae with respect to the rays, the effects of double refraction disappear. If this conjecture is verified by direct experiments, that is, if very thin laminae of glass act on light both as thin laminae and as reflecting bodies of a certain thickness [act], one will easily account for the most extraordinary phenomenon of double refraction, on supposing that substances endowed with that property are formed of very thin laminae properly placed and separated from one another. The only thing remaining to explain will be the separation of the images. (Arago 1812a, 83–84)

This passage obviously echoes Newton's opening phrases in the Queries to the *Opticks,* and Arago assuredly intended the two to sound alike. For like Newton in the Queries, Arago was proposing an important unifying speculation. And also like Newton's Queries, Arago's speculations are rather difficult to pin down. The reason doubly refracting bodies polarize light in opposite directions lies, according to this speculation of Arago's, in their structure.

Double refractors are composed of laminae so thin that light passing through them does not actually form Newton's rings. Nevertheless, a part of this light *has* been affected by the very same action that acts to polarize Newton's rings, and so this part of the light is polarized *in* the original plane of incidence of the light that struck the surface of the lamina. In addition to this part, there are also two other parts that compose the light within the lamina—the parts that obey Malus's usual rules for partial refraction. That is, one of these parts retains the polarization of the incident beam, whereas the other part is polarized *perpendicular* to the original plane of incidence. Now as these three parts pass through more and more laminae, the result will eventually be to throw almost all the light into the two parts that are polarized in and perpendicular to the plane of incidence. In other words, a large number of properly oriented thin laminae will take a beam of unpolarized light and produce from it two beams that are polarized at right angles to one another. This, of course, is what happens in a crystal.

The initialed notes say nothing about chromatic depolarization, but Arago already had it in mind. On 14 December he read a paper, which was never published, that extended his theory to cover this phenomenon as well. He assumed that mica consists of laminae that are themselves too thin to produce rings. He further supposed that the laminae are separated from one another by comparatively large gaps. The problem was to explain with this the deviations in the polarization of the several colors he had discovered. Here Arago was far from clear, in part because he never composed a version for publication. However, his idea seems to be this.

The mica laminae polarize the rays in the same way that occurs in doubly refracting crystals. They are themselves too thin to produce Newton's rings, but rings can form in the gaps between them, since the gaps are much thicker than the laminae proper. The light that forms rings in these gaps will be partially polarized. If we suppose that succeeding laminae in mica are oriented in different directions, then this will lead to the formation of colored rings whose partial polarization is different from that of the preceding rings. As the light continues to pass through the mica plate, eventually all of it will become involved in ring formation. Consequently the light that emerges from the plate will consist of rays whose polarizations have been widely dispersed according to their colors.

Arago wrote very little more than this sketch of an explanation, which is in itself purely qualitative and difficult to understand. It also ignores significant aspects of the phenomena—such as the reason for the different paths of the ordinary and extraordinary rays in crystals and why chromatic polarization occurs only in thin plates.[6] However, the important point is Arago's conviction that it was possible to link all polarization processes by paying close attention to the structure of bodies and then modeling their behavior on the formation and polarization of Newton's rings.

Unlike Biot's theory (which we shall examine in the next chapter), Arago's theory was broad in scope and more amenable to qualitative than to quantitative considerations. Arago never attempted to generate formulas from it—he never generated formulas even where it was easy to do so—and his memoirs do not contain numerical, much less tabular, data of any kind. Biot by contrast produced formulas very early on in his work on chromatic effects (though he had nothing like Arago's unifying theory), and his lengthy papers are replete with extensive tables. Of the two, Arago was working in the more traditional, qualitative manner; he was seeking broad principles to encompass several classes of phenomena. Tabular data and formulas do not well fit that kind of endeavor. Biot turned instead to very sharply limited assumptions and pointedly sought to generate formulas from them for specific cases. He made no effort in his early work to link these quantitative results to wider classes of phenomena in any firm way.

Biot's first work in this area therefore follows the pattern established most firmly by Malus. This pattern was rapidly becoming a standard one. It insisted upon the careful tabular presentation of numerical data and the generation of formulas capable of encompassing the material at hand, with little immediate concern to reach out to other, even closely related, phenomena. Biot's early work draws, for example, no firm connection between chromatic depolarization and birefringence: he treats depolarization as a new formal property by which crystals select rays to work upon. In Arago's work both birefringence and chromatic depolarization are supposedly due to hidden partial reflections and ring formation.

The differences between Biot and Arago, then, went beyond questions of priority—What indeed had Arago observed that Biot claimed to have found first?—to issues that hinged upon the changing canons of experimental reporting and investigation, canons that had first appeared in sharp relief in Malus's work on birefringence. It was not merely that, given Biot's work, Arago's might be wrong. It was rather that Arago's work was rendered irrelevant by Biot's. No formulas emerged from it, and perhaps none ever could, nor was it obvious what sorts of numerical data might clarify the issues it raised. Furthermore, we shall also see that Biot rapidly pushed on to emissionist principles, beyond the selectionist ones both he and Arago always deployed. Here also Arago's work would be at best irrelevant, since if it was correct it would presumably follow from emissionist requirements. Caught by Biot on selectionist grounds, where only Biot could produce work in what was rapidly becoming the required style, and outflanked on explanatory grounds by Biot's excursion into emission theory, Arago had every reason to be upset. Biot's behavior, we shall see in the next chapter, gave him even more.

4 Mobile Polarization

4.1 Biot's Intrusion

A few remaining copies of the *Mémoires de l'Institut,* volume 12, contain a note at the bottom of the first page of a memoir read by Biot on 1 June 1812. The note itself is dated 15 February 1813.[1] It reads in translation:

> The original manuscript of this memoir was initialed on every page by M. Delambre, perpetual secretary; it is deposited at the secretariat of the Institute. At the time it was presented to the class, M. Arago had published, of his researches on light, only the extracts printed in nos. 49, 50, 51 of the Bulletin of the Sciences, and in the Monitor of 31 August 1811. Finding myself absent from Paris when he presented his memoir to the Class, I knew only what appeared in the journals just cited, and since then I have had no other communication on them. This is why, *when I wished to occupy myself with this kind of phenomena, I publicly invited M. Arago, in the presence of the Bureau des Longitudes, to have his memoirs initialed by Mm. the secretaries of the Institute, in order to establish invariably the facts or theories he could by then have discovered. This demand seemed only fair, and M. Arago himself seemed to accede to it; but since he neglected to fulfill this formality,* I felt I should recall that the extracts cited above contain all the results that were publicly known as belonging to Arago at the time I read my memoir.

Biot, then, did not hear Arago read his second paper (1811b) on 11 August; he saw only the comments on it in the newspaper (the *Monitor*). He did not say that he had also missed the 18 February reading of Arago's first paper (1811a), but that paper was not published until 1817 and in any case does not concern chromatic depolarization (though we have seen how important the results it contained were for Arago himself).[2]

In the seven months between 11 August 1811 and Monday, 20 March 1812, when Biot read his first "note" and precipitated a direct confrontation with Arago, Arago did not have the time, he later remarked, to pursue his researches. He began to lose control of the new field. The *Procès-Verbaux* for 20 March recall in a few words what must have been a tense and deeply embarrassing moment for Arago:

> M. Biot reads a note on several phenomena that depend upon the polarization of light by reflection and refraction.
>
> M. Arago announces that he had in part made the same observations and demands that the Class recognize he had consigned them to the registers of the Observatory.
>
> Mm. Burckhardt and Bouvard are charged with this examination. (Arago 1812a, 75).

These are the notes discussed in the previous chapter, notes that in fact have very little to do with chromatic depolarization.

A year after these events, just after Biot's first memoir was printed, Arago angrily protested before the First Class in Biot's absence the remarks that Biot had made at the bottom of his memoir's first two pages (Crosland 1967, 234). Laplace tried to make peace between the two, and Biot agreed to retract the note. He and Arago together signed a letter requesting that all copies be returned, and most evidently were (or perhaps not that many had by then been distributed).

Biot had made it impossible for Arago to justify his early appointment to the Institute by carrying forward a new and exciting development. And then, to make matters between the two men even worse, Biot castigated Arago in print for not having taken up the challenge to make all of his work public. This was not the first time Biot had been unpleasant, and not only to Arago. In 1806 he tried to have work he had done with Arago published without listing him as joint author. In the spring of 1811—just before he left Paris—Biot claimed he had already observed partial polarization by refraction when Malus announced it to the First Class; he evidently had done no such thing and knew only what Malus had told Laplace about it in Biot's presence (according to Alexander von Humboldt; see Crosland 1967, 332–33).

Laplace, Humboldt, and presumably other members of the First Class were well aware of, and strongly disapproved, Biot's tendency to avoid acknowledging others' work. Recently published letters from Charles Duchayla, an early graduate of the Ecole Polytechnique, to his friend Louis Poinsot, who was then absent from Paris but who had only a short time before been elected to take Lagrange's place, well capture the emotions of the time. The first letter is dated 15 June, a week before the closed meeting of the First Class at which Arago protested Biot's note (Grattan-Guinness 1981, 676–78). In it Duchayla relates that he had brutally castigated Biot in conversation with the chemist Louis Jacques Thénard, who could say only that Biot's "heart" was not bad despite his being "wrong in this affair." In a second letter after the closed meeting, Duchayla described the events and wrote, "you may easily judge Arago's indignation."

Arago never recovered from these devastating months. He and Biot hardly ever spoke to one another again, despite occasional attempts by their wives and others at effecting a reconciliation. Anything associated with Biot became for Arago at the least distasteful and, given his wounded dignity, perhaps even hateful. And during the next several years Biot not only gained fuller control over the subject that Arago had created, but published long and intricate books and papers linking it to the emission theory. To many people Biot became that theory's primary exponent. Even before these events Arago had expressed doubts about the emission system, or at least about aspects of it. Little wonder that he came to dislike it violently when its most ardent supporter was the man who had thoroughly humiliated him.

No doubt Biot's behavior left much to be desired. Arago was a friend and colleague, and Biot quite likely was aware that Arago desperately wished to justify his appointment. There was no need to issue a public challenge and then to rub salt in Arago's wounds by reminding everyone he had failed. Yet Biot was in fact a much better physicist than Arago. He behaved badly, and perhaps even inexcusably, but he, not Arago, could make good use of the new field. Arago was a capable experimenter and had reasonable, though far from outstanding, mathematical ability. But he lacked Biot's energy and productivity as well as the creative sense to use just the right amount and kind of theory to generate quantitative results.

Arago knew his limitations and could see in others the abilities he himself lacked. When Fresnel came to him a few years later, Arago recognized in him these creative energies, and he used them, at least in part, to take revenge on Biot in a way he himself could never have done. For, we shall see, Arago used Fresnel to humiliate Biot in almost precisely the same way Biot had humiliated him.

4.2 The Failure of Biot's First Selectionist Formulas

When Biot read his first memoir on chromatic depolarization on 1 June 1812, those in the audience who had been following Malus's and Arago's work must at once have understood that he had quantified the newly discovered phenomenon. Biot's introductory paragraphs convey his sense of accomplishment:

> I propose, in this memoir, to make known the law according to which luminous molecules of diverse colors are successively modified by the thin laminae of many crystallized bodies endowed with double refraction. I will deduce from this law how to predict, from only the thickness of the laminae, the color of the rays, be they ordinary or extraordinary, that they polarize by reflection or by refraction, in any given position; finally, I will pull from these results many new and very close analogies between the still unknown causes that produce the ordinary reflection of light and those that polarize it in crystallized bodies. (Biot 1811, 135–36)

Whether or not Biot deduced laws of modification for "luminous molecules," he unquestionably did obtain laws governing the selection of rays by chromatic depolarization. These laws were extremely influential and form the context within which Fresnel's work on polarization was received. It is therefore essential to understand what Biot had done and why he rather than Arago had been successful. I have already mentioned why Arago was not likely to have found or even to have sought formulas; he continued to think, experiment, and analyze qualitatively, whereas from the beginning Biot wished to calculate. This difference between the two had a direct influence on what each looked for in experiments. Arago was concerned only to remark the general dispersal of the asymmetries by mica and to hint at a regularity in the case of quartz that would explain its difference from mica. In empirical practice Arago did not carefully hold certain variables fixed while altering the others according to a regular plan, or at least his written work (both published and unpublished at the time) does not show him to have worked in that way.

Biot, by contrast, clearly wished to find formulas from the beginning, which meant he had to control the variables carefully. In direct consequence, he soon discovered a regularity that had escaped Arago, and in it he found the clue to a quantitative theory of chromatic polarization. Biot, like Arago, used a device consisting of two mirrors that can rotate about a common axis. One could adjust the angles of incidence on the mirrors as well as the angle between the planes of reflection on them (see figs. 4.1 and 4.2).

Biot set the mirrors so that these planes were mutually perpendicular, and he reflected the light from both of them at polarizing incidence. Consequently the lower mirror reflected no light at all when nothing but the mirrors interrupted the beam's path. The purpose of setting up the system in this way was to control precisely the dispersive action on the asymmetries. For then the tint that appears in the lower mirror must consist *entirely* of rays whose asymmetries have been deviated from their original positions—since the rays

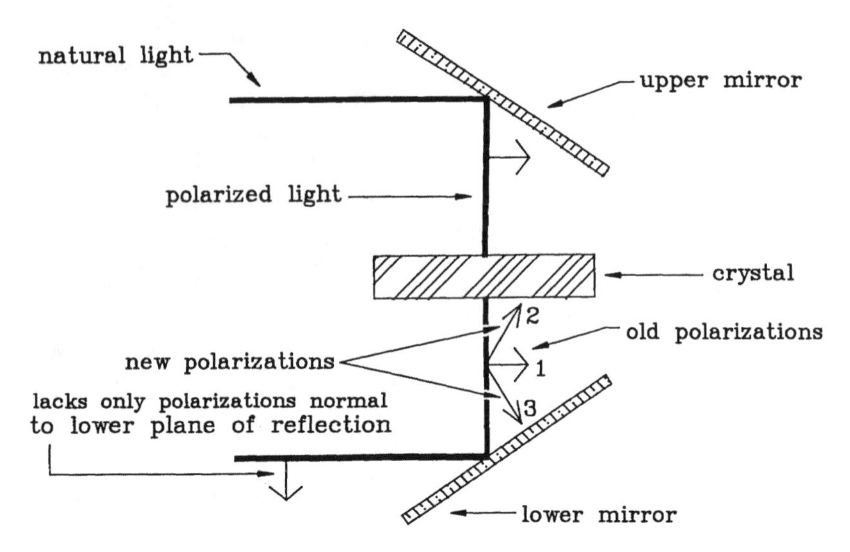

FIG. 4.1 Arago's experiment for chromatic polarization.

whose asymmetries are not affected will continue to pass through the mirror. In Arago's experiments, on the other hand, the plane of reflection on the lower mirror usually remained *parallel* to the light beam's plane of polarization before it entered the lamina. Consequently the lower mirror could reflect any ray except one whose plane of polarization was turned through a right angle. In other words, the lower mirror had to be reflecting many rays that have *not* been deviated as well as rays that have. If the apparatus is used in Arago's fashion, one cannot easily discover which rays have been affected and which rays have not. Using it in Biot's fashion at once shows only those rays that have been acted upon.

Biot's arrangement had another direct advantage. One could rotate the lamina to see how this affects the deviated rays because, again, the lower mirror reflects only these kinds of rays. Biot did so, and almost at once he uncovered a highly suggestive regularity: the tint in the lower mirror did not change at all as the lamina was rotated. That is, the bundle's chromatic composition, or what I called above its *ray ratio,* did not alter as the lamina was rotated. Arago could not have made this discovery, because his mirror contained undeviated as well as deviated rays—and so one could draw no conclusion about the colors present in the deviated set of rays alone.[3] Biot's observation could therefore mean only one thing. The same rays by color must always be deviated by the lamina no matter what its orientation might be.

Biot was particularly surprised that the tint did not change; Arago would not have been even had he experimented in Biot's fashion. Arago had used

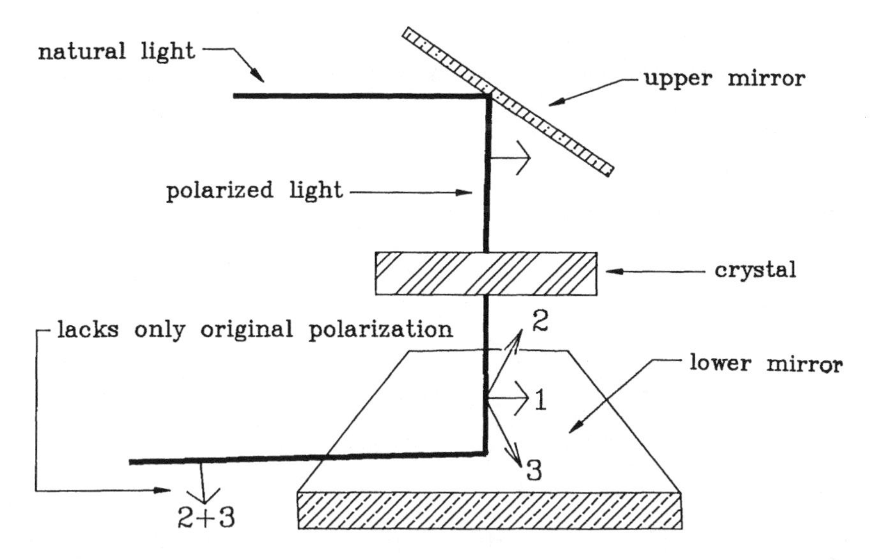

FIG. 4.2 Biot's experiment for chromatic polarization.

mica laminae, and mica had not usually been considered a doubly refracting crystal; since it is transparent only in thin sheets, one cannot easily observe a separation of rays within it. When Arago discovered chromatic depolarization, he accordingly did not think to link it in any way at all with birefringence. But Biot did not use mica. He preferred to try gypsum because unlike mica (which only flakes), it can be sliced fairly easily into laminae of a desired thickness. Gypsum was known to be doubly refracting, and Biot sliced it with the optic axis in the facet, so that the light struck normally to both facet and axis. In these circumstances one would naturally expect the orientation of the lamina to have some effect in Biot's arrangement. But it had none at all.

In a creative application of selectionist principles, Biot hypothesized that double refraction does still occur, but that it operates only upon a certain subset of the incident bundle—the one that his arrangement at once shows to have been deviated. That subset must have a particular ray ratio (tint), which means that it must itself be composed of sets of rays, with the ith set containing x_i rays of the same color such that the ratios x_i/x_{i+1} are fixed. When the lamina rotates it will divide only this particular set into O and E rays, and these two beams will have the same colors because they have the same ray ratios. Consequently the lower mirror will always show the same tint however the lamina may be rotated, since it will reflect at least O or E, and usually parts of both, of the doubly refracted subset and transmit the remaining light.

Biot's hypothesis radically simplified the phenomenon and led at once to testable formulas because one could now apply Malus's cosine law to the doubly refracted portion of the incident beam. They are quite simple to obtain (Frankel 1972, 239–41, discusses them). Begin with two angles (i, α) and two symbols (A, U) for ray counts (intensities). Instead of a lower mirror, use a doubly refracting prism to obtain two expressions, one for each of the two beams in the prism. (We shall in a moment see why only one expression, which we would have if we continued to use a mirror instead of a crystal, would make testing the hypothesis more difficult.) Let (fig. 4.3):

$i \equiv$ the angle between the optic axis of the lamina and the plane of incident polarization;
$\alpha \equiv$ the angle between the principlal section of the analyzing crystal and the plane of incident polarization;
$U \equiv$ the number of rays in the incident beam that are *not* affected by the lamina: call this set S_u.
$A \equiv$ the number of rays in the incident beam that *are* affected by the lamina: call this set S_a.

Since U and A are ray counts, they may alter as the lamina rotates. However, whatever the actual ray count may be, the ray ratio (color) in each of the two sets S_u and S_a must always remain the same.

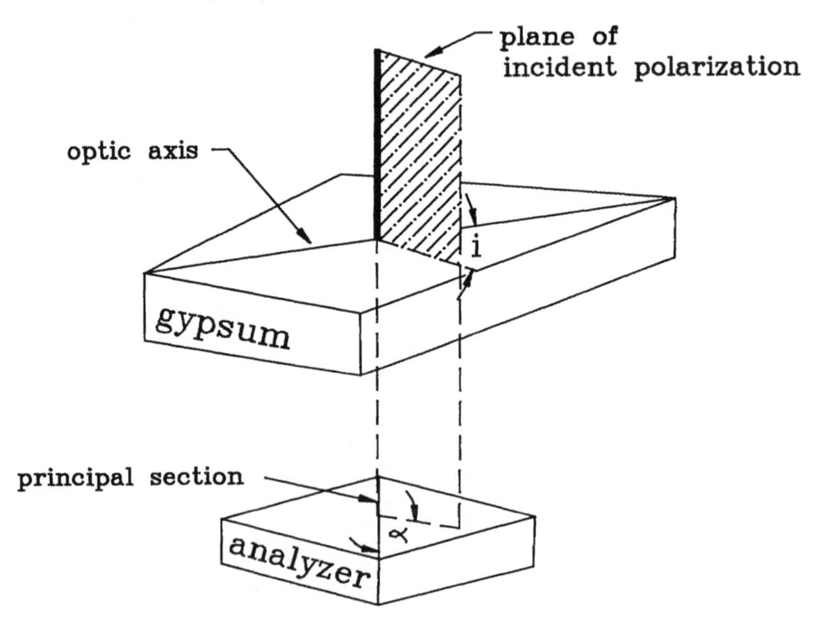

FIG. 4.3 Biot's angles for chromatic polarization.

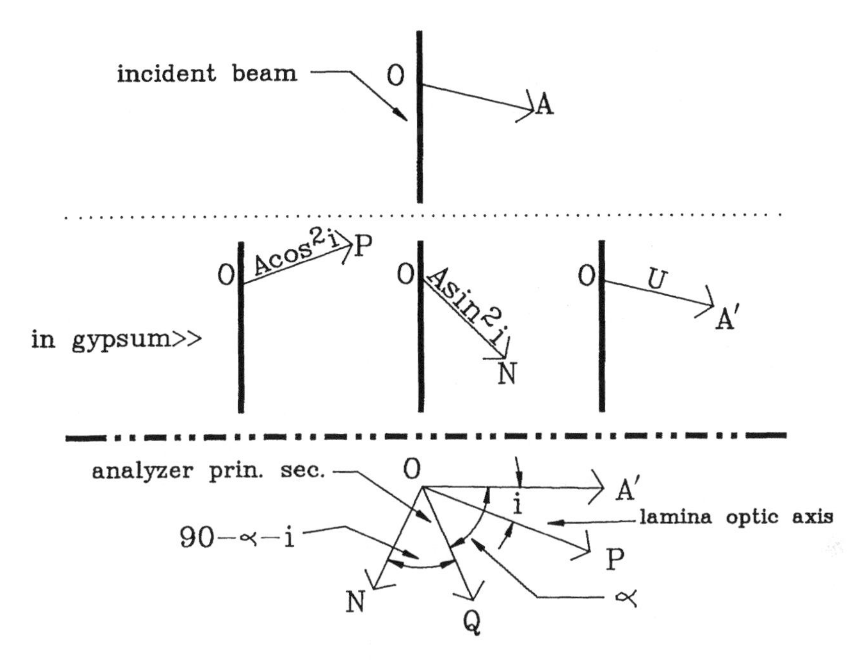

FIG. 4.4 Biot's theory for chromatic polarization.

Biot's formulas follow at once from Malus's cosine law applied first to the lamina and then to the analyzing crystal. The rays in S_a are doubly refracted by the lamina, generating subsets $A\cos^2 i$, $A\sin^2 i$ polarized, respectively, parallel to the optic axis and perpendicular to it. The analyzing crystal therefore receives three distinct bundles: one, U, is polarized at α to the analyzer's principal section; the second, $A\cos^2 i$, is polarized at $\alpha - i$ to the principal section, and the third, $A\sin^2 i$, is polarized at $90° - \alpha - i$. Each of these sets generates both an ordinary (F_o) and an extraordinary (F_e) bundle in the analyzer. Applying Malus's law to each of them and adding the results together, we have (see fig. 4.4):

$$F_o = U\cos^2(\alpha) + A\cos^2(i)\cos^2(\alpha - i) + A\sin^2 i\sin^2(\alpha - i)$$
(4.2.1)
$$F_e = U\sin^2(\alpha) + A\cos^2(i)\sin^2(\alpha - i) + A\sin^2 i\cos^2(\alpha - i)$$

It is essential to grasp properly the significance of U and A according to selectionist principles. They represent the *numbers of rays* in the two subsets formed by the gypsum lamina out of the incident bundle. By ray conservation, their sum must be equal to the number of incident rays minus the number of rays that are reflected and absorbed by the lamina. Furthermore, since the incident bundle is white, U and A are by a trivial implication complementary in color. (Indeed, on selectionist principles the removal of even a *single* ray from a white bundle produces complementary subsets.)

Since the sum $F_o + F_e$ is just $A + U$, then F_o and F_e must also be complementary in color. That is merely a test of ray conservation, as is the fact that F_o and F_e exchange tints when α alters by 90° and could be observed very well simply with a mirror (since at polarizing incidence the mirror would reflect rays that produce only an ordinary refraction in a crystal). However, only with an analyzing crystal can the actual tints in the two beams be simultaneously observed and compared with one another, and only this situation provides, we shall now see, a proper test of the formulas.

To test the hypothesis Biot had to make predictions that go beyond simple ray conservation, since that was not at all an issue. The best way to do this is to find a configuration in which F_o and F_e differ markedly in their ray ratios—in, that is, their tints. Biot did just that. He examined configurations in which the original plane of polarization contains the principal section of the analyzer, whereupon the angle α vanishes, yielding:

$$\begin{aligned} F_o &= U + A(\cos^4 i + \sin^4 i) \\ F_e &= 2A \sin^2 i \cos^2 i \end{aligned} \tag{4.2.2}$$

Here we have an observationally significant result. Whenever an expression for a ray count contains only terms in A or only terms in U, then the set's color must be completely independent of angular changes that do not affect this characteristic, because color depends upon ray ratio, which is fixed in both U and A. In this case, then, F_e must have the same color whatever the orientation of the optic axis, whereas F_o must change color as the axis rotates. This in fact occurs, which must have satisfied Biot greatly when he first succeeded in observing it.

But suppose we examine with Biot the situation in which F_e reaches its maximum, at a 45° angle of the optic axis with respect to the original plane of polarization. Here we have:

$$\begin{aligned} F_o &= U + \frac{1}{2}A = \frac{1}{2}(U + A) + \frac{1}{2}U \\ F_e &= \frac{1}{2}A \end{aligned} \tag{4.2.3}$$

Biot found that equations 4.2.3 are not empirically correct, though precisely how he discovered it is not obvious. He remarked:

> The ordinary image would therefore always contain a portion of white light equal to [$\frac{1}{2}(U + A)$], that is, to half of the total light

> that falls on the lamina [ignoring reflection and absorption], and it would in addition contain a colored portion equal to half of the tint [U]. Now, this is very far from the case: because in choosing appropriate thicknesses of laminae, and placing them in the position we suppose here, one can so attenuate the ray F_o that it becomes completely insensible. . . . For the moment it will suffice to remark that, far from the two rays' blending in the position we are examining, as the formula would have it, they find themselves on the contrary in their greatest separation,[4] in a separation complete, as experience proves on all the laminae one wishes to observe. (Biot 1811, 146–47)

Biot observed two effects that are at variance with the formulas. One of them, the first described, depends entirely upon his selectionist assumptions but requires examining different laminar thicknesses. The second requires only a single thickness but also depends upon Newton's theory of color mixing. He does not say which one first caught his attention, but I think it is likely to have been the latter, since he would first have examined the way the tint changes for a given lamina when it is rotated.

The formulas 4.2.3 hold whatever the laminar thickness may be, though Biot early on discovered that the tints and intensities of U and A depend upon thickness (as in fact Arago had observed before him). That is, the ray ratios in U, A are functions of laminar thickness, as is the absolute ray division between the two subsets. However, when α vanishes and i is 45°,[5] then F_o must always contain *at least* half as many rays as the incident bundle. This, Biot discovered, is "very far" from being true, since one can obtain thicknesses at which F_o sensibly vanishes under otherwise identical conditions. Were 4.2.3 the correct formulas, the implication of this effect would be that rays are not conserved, which is inadmissible.

Biot's remarks at the end of the quotation above concerning color depend upon his assumption that tints can be determined using Newton's color-mixing circle.[6] By "separation" he means the physical distance between tints within the circle. Since we do not know the true tint of either U or A, we cannot compute their true positions within the circle, but we can assign them arbitrary loci and then determine the position of tints formed from them. Given this, one can easily reconstruct what Biot had in mind when he wrote that the "separation" should be a maximum at 45° but is in fact a minimum.[7]

4.3 Success

The inadequacy of the formulas forced Biot to rethink the problem. His first reaction, if one is to judge from his published remarks that this will not work,

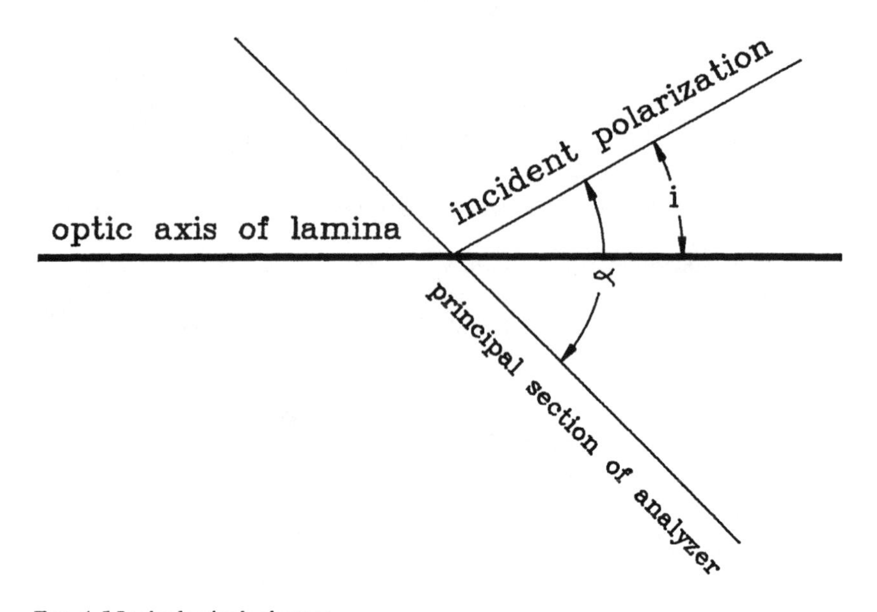

FIG. 4.5 In the lamina's absence.

was to alter the theory as little as possible by expanding it to include Malus's account of partial reflection. Perhaps, Biot apparently reasoned (he only sketched the argument), one should also take account of the light that, in Malus's theory, is refracted with its polarization normal to the plane of incidence; perhaps this light should be added to the light that is affected by double refraction. However, this would add a term of the form

$$A \sin^2 i \left[\begin{matrix} \sin^2 \\ \cos^2 \end{matrix} (\alpha - i) \right],$$

which gives no help. To accommodate the observation that F_o can vanish when i and α are respectively equal to 45° and to 0° requires removing from F_o every term in A, because only if it contains solely terms in U (or, conversely, in A) can it ever vanish under ray conservation.

This procedure—using Malus's theory—was in any case a dubious tack, because Biot experimented at normal incidence, where partial reflection is practically nonexistent (and so the refracted light polarized normally to the plane of incidence must be very small). He soon realized (Biot 1811, 148) that the only way to remove all the A terms from F_o by addition or subtraction was to include terms in the odd powers of the sines and cosines, since these terms could be negative and thereby annul others. Choosing the simplest possible modification required adding $2A \sin(i)\cos(i)\sin(\alpha - i)\cos(\alpha - i)$ to F_o

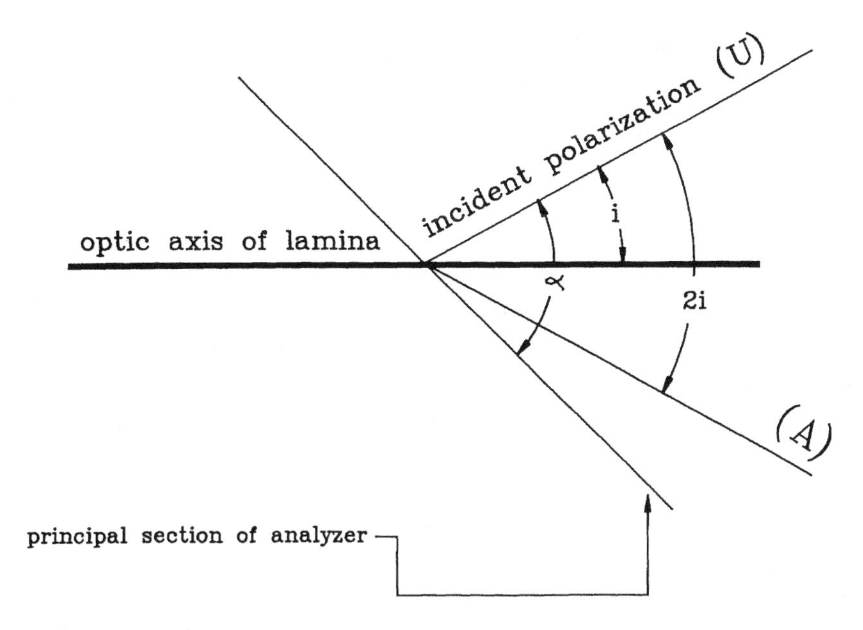

FIG. 4.6 With the lamina in place.

and subtracting it from F_e. Combining terms trigonometrically yielded a particularly simple and, we shall soon see, physically pregnant result:

$$\begin{aligned} F_o &= U\cos^2\alpha + A\cos^2(2i - \alpha) \\ F_e &= U\sin^2\alpha + A\sin^2(2i - \alpha) \end{aligned} \tag{4.3.1}$$

These equations can have only one interpretation in terms of rays. First, they clearly imply, as in Biot's first account, that the incident, white bundle of polarized light divides on entry into two subsets. One of these two subsets (U) is not affected in any way, since it follows Malus's usual law for the division of polarized light that enters a crystal. But the other subset (A) is affected in an unusual way: all the rays in it must have their polarization rotated through an angle $2i$ from one side to the other of the lamina's optic axis (since then their plane of polarization with respect to the principal section of the analyzing crystal will be $2i - \alpha$; see figs. 4.5 and 4.6). These formulas, based directly upon the assumption that rays preserve their identities, are very powerful. With them Biot could solve any problem in which one seeks equal tints and intensities in F_o, F_e, or just equal intensities in the two images whatever their tints may be. Biot could also use them to find under what conditions the images will be white.

One must examine how this works in order to be convinced, as many of Biot's contemporaries were, of the power of his theory. Suppose we wish, in Biot's words, "to find the number of solutions that give equal images, whether in intensity and in tint, or in intensity alone" (Biot 1811, 152). To solve the problem for equal intensities, set F_o equal to F_e, yielding after simplification:

$$U \cos 2\alpha + A \cos(4i - 2\alpha) = 0 \qquad (4.3.2)$$

The ray counts A and U depend upon the thickness of the lamina. We can satisfy 4.3.2 independent of the ratio A/U by requiring both $\cos(2\alpha)$ and $\cos(4i - 2\alpha)$ to vanish. There result four values for α, and, for each α, eight values for i, that satisfy 4.3.2 whatever the actual ratio of intensities (i.e., whatever the laminar thickness) may be:

α	i
45°	0°
135	45
215	90
315	135
	180
	215
	270
	315

These all yield precisely the same expressions for F_o and F_e:

$$F_o = F_e = \frac{1}{2}(U + A)$$

Consequently, both images are white and of equal intensity. That is, only if the images are white can they have the same intensity whatever their intensities may be.[8]

These were powerful and impressive results. But they lack one major thing: they do not determine in any way the ray ratios in A and U. That is, we do not know how to find colors, though Biot knew that they depend upon the thickness of the lamina. The remainder of his lengthy, and very influential, first memoir and most of his subsequent work, including his emissionist explanation of these selectionist formulas, depended on the questions raised by tints.

Indeed, the major part of Biot's memoir concerned the ray ratios, or tints, as they varied with thickness and the angle of incidence. This work was quite influential, since it stimulated David Brewster's and John Herschel's investigations. It depends strongly on Biot's particular experimental arrangements and assumptions, which led him into trouble. He was not able to obtain a formula that could predict the tints at oblique incidences. (See appendix 23 for details.) Many of the internal arguments among selectionists in the late 1810s and early 1820s (e.g., between Biot, Brewster, and Herschel) concerned, we shall briefly see, this very question of how to calculate tints at oblique incidences.

4.4 A New Kind of Oscillation . . .

Biot's second memoir, 371 pages long, consisted of several parts, the first of which was read at the end of November 1812. That the Institute printed an article of such unprecedented length demonstrates how important this work seemed to be. Biot was proposing a major work of explanation and deduction, one that, if achieved, would go far beyond the limitations of purely selectionist principles:

> The work that I have the honor to submit to the class has as its object to determine, by exact and numerous experiments, the manner in which the polarization of light operates in a large number of crystals endowed with double refraction. I hope to show that this phenomenon takes place by a succession of oscillations that the luminous molecules experience about their center of gravity, in virtue of attractive and repulsive forces that act on them. I will deduce from experiment the duration of the oscillations, their velocity, and the law of the forces that produce them; this will give a relation between the size of the luminous particles and the intensity of the forces that act on them, just as the speed of vibration of a pendulum gives a relation between its length and the force of gravity. I will show, by experiment, how one can, at will, accelerate or slow down these oscillations, or even nullify them, or make them go in the opposite direction. Having come to this general cause, I will deduce from it the laws of the phenomena presented by the crystals to which it applies when one reduces them to thin or thick laminae, and when one presents them to a polarized ray; and finally, comparing the consequences of the theory with experiment, one will see the rigorous analogy of these phenomena with colored rings, whence comes the succession of tints that crystals of this kind polarize when reduced to thin laminae; finally, one will see emerge from it a large number of experimental results that I have observed and measured with the greatest care. (Biot 1812, 1–2)

Despite Biot's emphasis on the oscillations of light particles, he rarely recurs directly to "molecules" but concerns himself for the most part with effects on rays and their asymmetries. As we shall see below, however, it would be entirely wrong to divorce selectionist from emissionist concepts at this stage in Biot's thought, because he certainly saw little distinction between them.

The first part of the memoir reiterates previous discoveries and mentions an important new one: that two *thick* plates set on one another with their axes at right angles (the axes being as usual in the interface) behave in chromatic depolarization like a single thin plate equal in thickness to their difference. The memoir's second part, read on 7 December, begins as usual for Biot with a repetition and summary, as well as a statement that he intends to propose no "hypothesis" but will rest with the "facts":

> These are the facts that will serve as my base. I do not propose at all to seek a hypothesis that explains them, I wish only to bring them together and to reduce them by mathematical considerations to a single general fact that will be their abbreviated expression, and from which one will then be able to draw, by calculation, not only the facts I just recalled, but also all the compound phenomena that can result from their combinations. (Biot 1812, 60)

We must take Biot seriously here, despite his very frequent discussions of light particles and forces, and despite his unbreakable devotion to emissionist concepts. For this second part of the long memoir relies almost exclusively on the analogy Biot had proposed between chromatic polarization and Newton's rings.

Biot was certain that Newton's table of the colors corresponding to various air gaps results physically from the separate arithmetic laws that govern the homogeneous colors that compose white light. By analogy, chromatic polarization must have beneath it a similar regularity for homogeneous light, which Biot illustrated with an example:

> Accordingly we can specify what would happen, for example, if, after having polarized a homogeneous violet ray, taken at the last extremity of the spectrum, one has it traverse perpendicularly one of our gypsum laminae; one need only follow the laws for violet [Newton] rings. If we designate the average thickness at which the extreme violet begins to be polarized for the first time by the lamina by $2e$, it will continue to be polarized thusly at thicknesses $2e'$; $6e'$; $10e'$; $14e' \ldots (4n - 2)e'$, following the progression of the odd numbers [this being the modification of Newton's arithmetic law]; and on the contrary, it will conserve its primitive po-

> larization at the intermediate thicknesses 0; $4e'$; $8e'$; $12e'$ . . . $(4n - 4)e'$, following the progression of the even numbers; and this without end or limits, since we have proved by experiment that the same accesses endure across masses as thick as four centimeters and more, in which, to judge by the primitive thickness e', these alternations must have occurred many hundreds of times, many thousands of times. (Biot 1812, 65–66)

If the analogy is apt, then a homogeneous beam of light must be alternately polarized in one of only two azimuths—in $2i$ or 0°—as the laminar thickness changes arithmetically. These two polarizations correspond analogically to transmission and reflection. If the analogy to Newton's rings is truly rigorous, then at intermediate thicknesses the homogeneous beam, at least if it were to leave the lamina at that point, must consist of two subsets. One of these must be polarized at 0°, whereas the other must be polarized at $2i$—since at intermediate thicknesses in Newton's rings some light reflects and some is transmitted.[9] Biot takes great care to emphasize this point, and for good reason: it permits his theory to work well despite its generally incorrect estimation of the polarization of the light that emerges from the lamina.

For to this point Biot's theory does indeed work rather well. It is correct that homogeneous light when it emerges from the lamina will be polarized along the $2i$ azimuth at arithmetically spaced thicknesses when (but only when) the optic axis lies in the facet (Mascart 1891, 2:12–13). Furthermore, at intermediate thicknesses the light oscillation will have components both along 0° and along $2i$, which is qualitatively consistent with Biot's account[10] because his theory does not generally require the emergent homogeneous beam to be polarized entirely along only one of the two possible azimuths. It is important to emphasize this point, since Biot was later unable to convince his critics (Fresnel in particular) that he could never have supposed otherwise.

Biot was quite convinced that his theory was essentially impregnable, that it was in fact entirely independent of hypothesis—independent, that is, of the actual physical nature of light. He remarked:

> Physicists' researches on colored rings may discover many more phenomena unknown to Newton; they may even lead one day to a knowledge of their cause, but they will not lessen the existence proper of the fits and their periodicity; now, these are the only properties that served Newton as a basis for constructing the table of thicknesses that reflect or transmit the diverse tints of the compound rings, and therefore the consequences I deduce from comparison of this table with the phenomena of crystallized laminae

> can also not be destroyed by new considerations, as long as I limit myself to linking these two classes of facts by the laws of periodicity that are common to them, without pretending to conclude any relation between the different or similar causes by which they are produced. (Biot 1812, 70)

Nevertheless, Biot's comparison of his account with Newton's fails at several points. First, Biot never did examine homogeneous light to see whether there really was a periodic law for individual colors. He took it for granted that a successful use of the table for compound colors had to mean that the homogeneous phenomena were periodic. In fact the homogeneous phenomena will be periodic in Biot's sense, but only if the optic axis lies in the interface,[11] which leads to the next difficulty with the comparison.

Even if the optic axis does lie in the interface, nevertheless there is a major difference between chromatic polarization and Newton's rings with respect to what can possibly happen to a ray. In the case of the rings, any given ray must certainly be either transmitted or reflected; no other result may occur (without absorption), whether rings form or not. If rings do form, then the ratio of reflected to transmitted light for a given color repeats arithmetically. But in chromatic polarization a ray need not *in principle* be polarized in one of only two azimuths—other polarizations are certainly conceivable. Reflection and transmission therefore correspond only imperfectly to Biot's alternate polarizations, because the former two are the only conceivable cases, whereas the latter two are not.

In Biot's understanding, a ray's polarization in fact does oscillate within the lamina between the two extremes of 0° and $2i$. The period of the oscillation is analogous to the fit length in Newton's rings. However, the ray on emerging from the lamina can be polarized along only one of these two azimuths, even though within the lamina it necessarily ran through every intermediate polarization as well. If it happens to be the case that the lamina's thickness is such that the ray has oscillated, say, an integral number of times from the 0° azimuth, then it emerges with this polarization; if it has oscillated an integral number of times from the $2i$ azimuth, it emerges polarized in that way. If the thickness is intermediate between these two extremes, then the ray may emerge in either one of these polarizations, though Biot had no rule to predict how many rays in a set with a given color will emerge polarized along 0° and how many will emerge polarized along $2i$, at a given intermediate thickness—just as one could not predict how much light reflects, and how much refracts, between fits.

On the whole, then, Biot's account was to this point without serious conceptual or empirical difficulty, despite Fresnel's later criticism of it, as long as we do not inquire too deeply into precise causes. There are, however, two related phenomena that did raise questions, and Biot was well aware of them. First, why, in Biot's language, does "partial" (i.e., chromatic) become "total" (the same for all colors) polarization when the laminar thickness becomes large enough? Second, why do thick laminae with crossed axes act like thin laminae equal in thickness to their difference?

The first problem was always a difficult one for Biot, or indeed for any selectionist account. He tried again to invoke a parallel with Newton's rings. As the thickness increases, he argued, the colors will "mix more and more at the two extremes of the oscillation," yielding, finally, two white images—just as, in thick films, the colored rings ultimately "mix." In Newton's rings, however, the "mixing" results from the decrease of ring width with thickness, making it impossible to distinguish the individual rings from one another. Nothing at all similar can occur in Biot's laminae because, in his experiments, the colors in U and in A are already superposed in each image.

He was more successful with the second question. In fact he answered it with complete success, though Fresnel later missed this point. (See appendix 12 for Biot's solution.) And Biot was also successful in providing explanations and related formulas for optical rotation, which he soon turned into a major tool for investigating the chemical structure of liquids that exhibit the phenomenon.

4.5 Biot as an Emissionist

Biot was utterly convinced that his "theory of oscillations," as he called his account in these early papers, or "mobile polarization," as he later termed it (and as it became known), represented the action of forces upon particles of light. One sees this in four ways. First, in several places Biot explicitly examined an equation of motion for the particles to see how his selectionist theory delimited the force laws. Second, he frequently referred to repulsive forces associated with a crystal's optic axis. Third, he considered that the arithmetic law applied to polarization represents only the extremes of a continuous oscillation that a ray (particle) undergoes within the lamina; that is, the ray is not, within the lamina, polarized only at 0° or at $2i$, as it must be on emergence, but oscillates between these extremes with a frequency unique to each homogeneous color. Finally, Biot usually wrote about "molecules" and not "rays" of light. (Though one must be careful to read this word in context, since it is often used only to refer to the least elements in the beam.)

Yet a careful study of Biot's memoirs does not provide a single instance in which Biot actually obtained an empirically testable formula by starting with particles and forces, despite his claim at the very beginning of his 1812 paper to have done just that. Referring to his formulas for F_o, F_e, he remarked:

> Although these formulas are only the empirical representation of facts and their abbreviated expressions, nevertheless the manner in which I found them, and so to speak molded them upon experiment, left me in no doubt that they contained the true laws; and this appears even more so today, that I can derive them directly from a consideration of the very forces that act in these circumstances upon the molecules of light. (Biot 1812, 3)

In fact Biot never did any such thing. Rather, he constructed a particulate model that was consistent with what, he felt, he already knew to hold true. Every passage here and elsewhere in Biot's work that considers particle dynamics provides at most a demonstration of consistency with some phenomenon, and usually a qualitative demonstration.

Biot's great confidence in the construction of consistency proofs devolves from his complete belief in selectionist principles. For him, as for almost every one of his contemporaries until the mid-1820s, light could be thought of in only one way, as a set of individually identifiable and manipulable rays. His accounts did, in fact, make extensive and productive use of this basic selectionist assumption in actually deducing formulas. Although these formulas do not depend in any direct way upon forces and on light particles, nevertheless by this date, in France at least, no one could think of any way to envision a substantial ray unless it was thought of as a sequence of particles. Biot's complete confidence in the emission theory was therefore enormously reinforced by his ability to deduce workable formulas from principles that were based on the physical reality of rays.

Yet one need not think that way at all. In the eighteenth century a number of writers employed a fluid, spread like an ether throughout space, that generated light rays by its rectilinear motion rather than by vibration (Cantor 1983, chap. 4, discusses these "fluid" theories in detail). Though these several theories were almost entirely qualitative, and usually concerned with issues beyond the physics of light, nevertheless their existence demonstrates that the substantiality of the ray—in them identified with a given quantity of moving fluid—need not presuppose particles and forces. These last are merely one way of making the ray concrete. Of course, in France during the 1810s this was also the only acceptable way to do just that.

As Biot's work became more detailed, expansive, and elaborate, so did the connections that he drew between it and the emission theory become increasingly tight. And the formulas that he (and Malus) achieved were born out of emission principles. The idea that optical phenomena are produced by forces acting on particles of light was, if not fruitfully quantifiable, immensely suggestive.

Consider Malus's theory of partial reflection. As we have seen in detail, it depends upon the idea that light rays have asymmetries that can be turned about in various ways during reflection and refraction. The reason Malus thought at all in this way was his belief in the emission theory, for the properties of light particles are as fixed as the properties of material particles. Consequently they can only be reorganized into different groups; their essential nature cannot be changed. Suppose, by contrast, that light consisted of physical rays whose properties are not necessarily fixed. Then one might think that polarization alters the nature of the individual ray itself, in which case ray-counting procedures would hardly seem likely to generate results. Precisely because the emission theory forbids one to play with the nature of the rays—because it *limits* what can be done—it had considerable suggestive power.

Seen in this way, then, the connection between what I have called selectionism and the emission theory becomes close indeed. When Biot conceived of "mobile polarization," he was not thinking of abstract physical rays. Molecules of light, deviated by a complex array of forces, were foremost in his mind. This suggestive model made it possible for him to obtain quantitative results—not because one could deduce workable formulas from it by analysis, but because the model sharply limited the kinds of effects that could possibly occur. We shall see below that the mechanical structure of the ether had a very similar function in Fresnel's work—it did not by itself generate quantitative results, but it limited the possibilities.

Over time Biot developed an elaborate scheme of particles and forces that imbued nearly all his discussions. Biot's own remarks are detailed, prolix, and scattered throughout his various publications. Perhaps the clearest, and most succinct, description of these ideas was written in 1834 by Humphrey Lloyd. Although Lloyd was presenting a brief for the wave theory, nevertheless he carefully presented Biot's views. Lloyd remarked:

> When a ray of light enters a crystal, the component molecules are supposed, in the theory of M. Biot, to receive different motions round their centres of gravity, dependent on the nature of the forces exerted upon them by the particles of the body. Sometimes the molecules of the ray are turned by the operation of these

> forces, so as to have certain lines in each, denominated axes of polarization, all in the same direction; and this arrangement of the molecules is maintained throughout the whole of their future progress. There are other cases, however, according to this author, in which the molecules *oscillate* round their centres of gravity in certain periods, during their entire progress through the crystal; while in others, finally, they receive a motion of *continued* rotation. . . .
>
> The phenomena of *fixed* polarization are ascribed by M. Biot to the operation of certain forces, which he denominates polarizing forces. In the case of uniaxial crystals these forces are supposed to act in the planes containing the two rays and the axis of the crystal,—the ordinary polarizing force tending to arrange the axes of the molecules in the plane containing the ray and the axis, while the extraordinary polarizing force draws them towards the perpendicular plane. If the molecules were similarly circumstanced in every respect, they would necessarily obey the stronger of these forces, and there would be but one plane of polarization. This, however, is supposed not to be the case. Owing to the different phases of their fits, at their incidence upon the crystal, the molecules are disposed to yield more readily to one or other of these forces; so that when a polarized ray meets a double refracting medium, some of the molecules fall under the influence of the ordinary polarizing force, and have their axes of polarization turned into the plane containing the ray and the axis of the crystal, while others are actuated by the extraordinary force, and have their axes arranged in the perpendicular plane. The number of molecules that yield to one or other of these forces, or the intensity of the two polarized rays, is supposed to depend on the angle that the plane of primitive polarization makes with the two planes just mentioned. When the plane of polarization coincides with the former, the extraordinary force has no effect, and the ray receives only the ordinary polarization; the converse takes place when the plane of polarization coincides with the perpendicular plane. Similar suppositions were made to account for the phenomena of polarization in biaxal crystals. (Lloyd 1834, 381–82)

Lloyd's brief summary nicely encapsulates the most important elements in Biot's physical theory of *mobile polarization.* Molecules of light rotate about their centers of gravity and are acted upon by matter in ways that depend upon their orientation. It would be incorrect to think of Biot's scheme as a *predictive* model for light, because it cannot be used to calculate anything.

But as an *explanatory* model it provides a way to understand why the rays of light behave in the ways that Biot's formulas require. Indeed, one might say that Biot's model is a translation into the physics of forces and particles of his theory for the behavior of rays. Molecules of light replace the discrete rays of the formulas, and instead of simply specifying what happens to the rays, one says instead that "forces" produce whatever changes occur. Precisely because of the intimate connection that Biot drew between rays and light molecules, he and many of his contemporaries were convinced that he had, as he claimed, deduced laws of modification for "luminous molecules."

PART 2

Fresnel, Diffraction, and Polarization

5 Fresnel's Ray Theory of Diffraction

> Fresnel was a great engineer, a man who, faced with a concrete problem, knows how to find the best solution, the one that leads to the best result with the minimum of effort and time.
>
> C. Fabry, *Oeuvres choisies*

5.1 Augustin Jean Fresnel

In 1784 the architect Jacques Fresnel, from the small town of Mathieu near Caen in northwest France, was employed by the Duc de Broglie to direct the repair of his chateau. There he met, and soon married, Augustine Mérimée, whose father, a noted lawyer and author of a treatise on feudal law, directed the Broglie estate. They had four sons: Louis, Augustin Jean (born 10 May 1788), Léonor, and Fulgence. In the confusing and dangerous years immediately following the Revolution in 1789, the Fresnel family retired to Mathieu, where Jacques died in 1805 (Fabry 1938).

Augustin spent his childhood in the countryside and apparently did not even learn to read until he was nearly eight. He was, however, an excellent little engineer, who built rather dangerous devices that had to be proscribed (Arago 1830, 478). When he reached thirteen Augustin and his brother Louis were sent to the Ecole Centrale in Caen, where Augustin proved a mediocre student in the classics. He evidently detested Latin, and "someone who knew him well" remarked that his "memory seemed to refuse to learn words and his spirit, avid from childhood for positive knowledge, seemed to rebel against classical learning" (Fabry 1938, 640). But the Ecole Centrale had an excellent mathematics teacher, and Augustin rapidly distinguished himself in the subject, rising here at least to the top of his class. He spent only four years in Caen; at age seventeen he followed his brother Louis to the Ecole Polytechnique in Paris, where he ranked seventeenth out of 130 entrants.

At the Ecole Polytechnique Fresnel followed what was by then the standard curriculum, which involved two intensive years or "divisions" of study (Langins 1980). The first division covered algebraic analysis, the differential and integral calculus, mechanics, descriptive geometry, "elements" of chemistry, and "general" physics.[1] The second division extended the subjects of the first or treated them at a higher level, adding about two hundred lessons in the detailed applications of descriptive geometry and practical chemistry. By the end of the second division, then, the neophyte polytechnician had become well versed in mathematics, mechanics, and statics. He knew something of hydrostatics and hydrodynamics and had been introduced to the concepts of Lavoisieran chemistry as well as to practical chemical techniques.

The "general" physics lessons included ten distinct topics (Langins 1980; Ecole Polytechnique 1799).[2] These lessons seem to have lacked an explicit discussion of microscopic structure (though, no doubt, the sessions on capillarity and on general material properties raised the subject). Nevertheless they certainly introduced Fresnel to the major topics of the day in optics. By contrast, the chemistry lectures, which were given by Antoine-François Fourcroy and Louis-Bernard Guyton de Morveau (who took them over from Claude-Louis Berthollet in 1805), were certainly permeated with detailed discussions of caloric and its role in combustion and chemical change. These questions had long been of interest to Fourcroy, who had accepted the general Lavoisieran chemical theory by 1786. What is of special significance is the account of heat and light that Fourcroy gave in his 1800 *Système des connaissances chimiques*. Smeaton (1962, 105) nicely describes the concepts:

> [Fourcroy] presented and obviously favored Monge's hypothesis, that caloric and light were two states or modifications of the same substance, fire itself: less dense and moving more slowly in caloric, more dense and moving more rapidly in light. According to this theory, light could be converted into caloric, and vice versa. Fourcroy now distinguished two main types of combustion, fast and slow. In a fast combustion, the caloric of the oxygen gas was released rapidly, with the formation of a flame, while in a slow combustion no light was evolved, but only heat, sometimes so slowly that it was scarcely perceptible. . . . The same theory was presented in Fourcroy's last book, the third edition of *Philosophie chimique* (1806). He agreed that the identity of heat and light had not been proved, but thought that all phenomena could be explained by assuming that they were the same substance with different modes of motion, and he repeated his theory of fast and slow combustion, as given in 1800.

Compare this with the first letter written by Fresnel to his brother Léonor (5 July 1814) that discusses physical topics. It begins:

> I permit myself several doubts concerning the theory of caloric and of light. I recall that at the Ecole Polytechnique I could not well conceive how so much light and heat could be disengaged by the combustion of charcoal: its resultant being a gas, there cannot be great approach of the molecules. I later saw, in the Annales de Chimie, that Berzelius cited this example in particular, and many others, as objections to the French theory. (Fresnel, *Oeuvres,* 2:820)

Comparing Fresnel's remarks with Fourcroy's theory, it seems evident that his interests, or worries, derived initially from ideas that were presented to him in the chemistry, rather than the physics, sessions at the Ecole—much as Malus before him was concerned primarily with chemical questions. The caloric theory of the day ascribed to molecules of matter affinities for the substance of caloric, which was considered to be highly subtle and self-repulsive; caloric was responsible for all thermal, and perhaps luminous, effects. During Fresnel's years at the Ecole caloric theory had, however, not yet been extensively applied to purely physical problems, and consequently Fresnel's understanding of it was primarily chemical.

The chemists, Berthollet in particular, did not develop elaborate accounts of caloric structure and its link to matter but preferred rather general descriptions that permitted them to use the hypothesis without detailing the molecular processes involved. They developed a delicate balance between demonstration and presumption in order to preserve the usefulness of caloric theory; they learned, that is, to ignore certain kinds of questions because everyone knew they led nowhere. Malus, in his first solitary thoughts at Lesbieh, had apparently not yet learned to avoid questions of that kind, but he examined them in an effort to clarify the theory, not to destroy it. In later years he learned, by concentrating on the behavior of rays rather than on their physical structure, generally to put aside these kinds of questions in optics.

Fresnel's letters clearly reveal that he was extremely eager to make a discovery of almost any kind. After leaving the Ecole in 1806 he joined the Bridges and Roadworks (Ponts et Chausées) as an engineer for three years. He was then attached to roadworks near Napoléon-Vendée (1809). In 1810 he thought of a new type of hydraulic ram and, the next year, of a new process for producing soda cheaply (Fresnel, *Oeuvres,* 2:810–19). Fresnel's maternal uncle, Léonor Mérimée—once teacher of design at the Ecole Polytechnique, by then permanent secretary of the Ecole des Beaux Arts, sometime chemist and intimate of Thénard and Joseph-Louis Gay-Lussac (and father of the writer Prosper Mérimée)—in August 1811 sent Fresnel's description of the soda process to the chemist Louis-Nicolas Vauquelin, who promised to repeat Fresnel's experiments. He was busy with something else for the next few months, however. On 31 October Mérimée wrote Fresnel a very encouraging letter, urging him to repeat his experiments on a larger scale and promising that, after Vauquelin had finally criticized the work, they would arrange for its publication in the *Annales de Chimie,* a promise that must have greatly excited Fresnel's hopes (Fresnel, *Oeuvres,* 2:813).

By the fall of 1811, then, Fresnel's scientific interests seemed to have been firmly set in practical chemistry. But Vauquelin seems never to have kept his promise, and so the following March Mérimée went to see Gay-Lussac and Thénard about Fresnel's process. In April he wrote Fresnel a crushing letter:

> My dear friend. Thénard made a little trip to take his ill mother to a skillful doctor, and I was obliged to await his return before giving him your letter. I went to find him yesterday, and here's what his reply to me was: Your process is good, but it seemed to him to be more costly and more difficult to execute than the one in use. . . . The other process you dreamed of is already employed at the Gare by a certain M. Huskin. (Fresnel, *Oeuvres,* 2:816)

This seems to have rather strongly dampened Fresnel's enthusiasm for practical chemistry, and two years later we find him asking his brother Léonor to subscribe for him to the *Annales de Chimie* and buy the latest edition of Haüy's *Physique,* and mentioning that he had recently seen a piece on something called the "polarization of light": "I read in the Monitor some months ago that Biot read at the Institute a very interesting paper on the 'polarization of light.' I've about broken my head over it, but I can't divine what this is" (Fresnel, *Oeuvres,* 2:819). By the spring of 1814—nearly a decade after leaving the Ecole—Fresnel's interests had begun to turn to physical rather than chemical questions, but his earliest considerations were nevertheless rebellious reactions to the chemical concepts he had learned at the Ecole.

The brief passage I quoted above from Fresnel's letter to Léonor of 5 July 1814 concerns combustion and caloric theory and clearly reflects ideas he had picked up in the chemistry lessons at the Ecole. However, it also shows quite well that Fresnel did not pick up the agnosticism that was essential to the fruitful use of caloric theory. One can see this illustrated in his objection to the theory: How can it be, he wonders, that so much heat and light are released in the combustion of charcoal when the product is a gas whose molecules are no more closely packed together than are the molecules of the combusting oxygen? Moreover, he continues, since the "integrant" molecules of the product have greater masses than the oxygen molecules, they should retain more heat—heat, that is, should be absorbed, not released.

Both of Fresnel's objections, and others he offered later, are answerable, but only if one takes the position, as Berthollet did, that the affinity of matter for caloric should be treated as a fact without providing a detailed mechanism to explain it. So, in this case of the combustion of charcoal, one assumes that the affinity of the gaseous product for caloric is much less than that of oxygen for caloric, which requires caloric to be released on combustion. To Fresnel this release seems odd because he wants a mechanism for it: he would like to make it a general rule that caloric is released in chemical reactions between gases or between a gas and a liquid or a solid only when the gaseous product occupies less volume than the combining substances. He perhaps had in mind here an analogy to adiabatic compression, in which the usual reasoning of the time ascribed the temperature increase to the transformation of caloric from a latent, or bound, state to a more or less "free" state as a result of forcing molecules closer together. Fresnel would also like to make the affinity of a gas proportional to its molecular mass. What distinguished Fresnel's views, then, from those of many of his contemporaries was what one might call a demand for structure: he was unwilling to accept the affinity approach to caloric without at least sketching a mechanism for it.

Other problems with received theories also caught Fresnel's attention at this time; he wondered, for example, whether the emission theory can explain stellar aberration, and whether optical particles differ in the speeds with which they are launched according to their degree of refrangibility. In the case of aberration, Fresnel soon learned, his doubts were based on his inadequate knowledge of what had been written on the subject. As for the problem of particle speeds, here the difficulties could be answered in several ways, which Fresnel would not at first have been aware of because he was unfamiliar with the bulk of contemporary literature in the emission theory (Fresnel, *Oeuvres,* 2:824–29).

Taken together, Fresnel's objections to caloric theory and to the emission theory of light are far from being conclusive, or perhaps even very compelling to a contemporary; all of them could be answered in one way or another. Some of Fresnel's objections stemmed from ignorance; others demanded more than contemporary theory intended to provide. Should we then assume that these objections were themselves the direct sources of Fresnel's turn to the wave theory? Or do they rather reflect Fresnel's intention to unearth something to quarrel with in contemporary theory in order to reply with a new discovery? For it seems quite certain, given the evidence of his letters to his uncle and brother, that Fresnel was extremely eager to discover something, that he almost succeeded with the chemical process for soda sulfate, and that, having failed there, he turned to questions of theory with the same goal in mind. Writing to his brother on 3 November about a certain N** he had been corresponding with, Fresnel reveals his deep admiration, one might almost say obsession, for "discovery": "N** is wrong, in my opinion, to prefer metaphysics to physics, [to prefer] a subject of eternal dispute to the science that does the greatest honor to the human spirit, and in which discoveries are made every day" (Fresnel, *Oeuvres,* 2:829–30).

None of this explains why Fresnel turned to the wave theory of light in particular as a vehicle for discovery, but that question may not be answerable. Certainly by 1814 he saw in it—in its assimilation of both light and heat to vibrations in an ether—a solution to the more damaging, in his view, objections to contemporary theory:

> I tell you I am strongly tempted to believe in the vibrations of a particular fluid for the transmission of light and heat. One would explain the uniformity of the speed of light as one explains that of sound; and one might perhaps see, in the derangement of the fluid's equilibrium, the cause of electric phenomena. One would easily conceive why a body loses much heat without losing weight, why the sun has for so long shined upon us without diminishing its volume, etc. (Fresnel, *Oeuvres,* 2:821–22)

Given the extensive discussion of acoustic laws at the Ecole, we have at least a partial source for Fresnel's comprehension of wave propagation, but only a partial one, because understanding of the kind of interference—the formation of a stable spatial pattern—that must be assumed in the wave theory of light was not at all well developed during the eighteenth century. It has recently been argued (Kipnis 1984), in fact, that the major purpose of eighteenth-century considerations of wave interactions was to explain why waves, after interacting with one another, can thereafter continue on as though the interaction had not occurred at all. Thomas Young and, independently, Fresnel, analyzed instead what went on during the interaction, though neither apparently realized the novelty of doing so for any kind of wave.

Fresnel evidently wrote a paper, now lost, that contains his earliest ideas on the wave theory, which he called his "rêveries." He sent the piece to André-Marie Ampère in the fall of 1814—initiating a long and friendly association between them—but, his uncle wrote him on 20 December, Ampère had not by then written anything about it. However, his uncle continued, Ampère agreed with many of Fresnel's arguments, nevertheless remarking that much of the piece had already been anticipated by Arago.[3] On the evening of 19 December Mérimée dined with Arago and Ampère, and he asked Arago to look at Fresnel's paper, which he promised to do. The documentation for this period ceases here, beginning again about six months later, so we do not know what went on during that momentous dinner when Arago first became aware of Fresnel.

On Napoleon's return from Elba the following March, Fresnel, having greeted the event by joining the Duc d'Angoulême's resistance, was suspended from the corps and put under surveillance at Nyons.[4] He was fortunately able to obtain permission from the local police prefect in July to join his mother at Mathieu. On the way to Mathieu he stopped in Paris to visit Arago, and it was at that time that he evidently first began to consider the wave theory, and in particular how one might use it to explain diffraction. From Mathieu he wrote to Arago requesting help, and Arago replied on 12 July with a list of authors, including Grimaldi, Newton, Jordan, Brougham (!), and Young. Indeed, though we cannot know for certain, quite possibly Arago had actually suggested diffraction to Fresnel as an interesting subject when they met in Paris—perhaps one that Arago even then hoped would prove a difficult problem for the emission theory, which he by this time disliked as much as he disliked Biot, its most prominent exponent.

On 20 September, still in the countryside at Mathieu, Fresnel again wrote Arago, announcing that he had discovered an explanation of the colored fringes produced by diffraction. He pleaded that his observations were not as accurate as he would have liked because he had had to fabricate his own instruments. Moreover, he had not been able to obtain the books and papers Arago had mentioned; in any case he could not read English. It is indeed fortunate that Fresnel did not read English, for otherwise he would have seen Young's use of the principle of interference and very likely have been discouraged from further pursuit of the wave theory, much as he had turned from practical chemistry to physics.

Between 1815 and 1818 Fresnel wrote six major papers on diffraction as well as a number of "fragments" and "notes" on the subject. Few of these were published until the 1860s, when his brother Léonor and Emile Verdet collected them in the *Oeuvres*. The documentation preserved in the *Oeuvres* is sufficiently complete to permit following Fresnel's work very closely during these critical years for the wave theory. I shall begin by examining in some detail his first paper on diffraction, the "Premier mémoire" of the fall of 1815, together with a "Complément" that he appended to it. Neither was published at the time. They were followed in the spring of 1816 by a "Deuxième mémoire," which was published and first gave him a name, together with a "Supplément," which was not published but in which he modifies in an important way his first theory. Then, in the spring of 1818, Fresnel submitted a note on diffraction to the Academy that contains the essence of his final theory, including the Fresnel integrals. That theory, together with extensive criticisms of the competing emission view, was ultimately published as the famed "Mémoire couronné" that won Fresnel an Academy prize.

In this part I will, then, be concentrating almost entirely on diffraction, which did occupy most of Fresnel's time during the crucial two and a half years between the fall of 1815 and the spring of 1818, though, we shall see below, he was also working on polarization during this period. Fresnel's work on polarization was much more controversial than his analysis of diffraction, and it was also considerably more difficult to understand. Many of the early reactions to Fresnel's wave theory, and his receiving the Academy prize, reflect a fairly widespread understanding of how the principle of interference can be applied successfully, if not to wave fronts, then at least to rays of light taken two at a time (a point made by Kipnis 1984, chap 7). That position reflects the emergence of a break between selectionist and emissionist principles—for here physicists concentrated on interactions between rays rather than upon the physical principles that are necessary to explain diffraction in the emission theory (though the latter too could be done—see Kipnis 1984, 261–63). This permitted those few people who were able to appreciate Fresnel's quantitative work not to accept, or even fully to understand, the concept of the wave front as Fresnel used it while nevertheless adopting a principle of interference for rays.

5.2 The First Experiments

One of the major characteristics of Malus's and Biot's work, as we have seen in some detail, was their devotion to precise measurements and to the full reporting of their results. This contrasts markedly with most work in optics before this time, including Thomas Young's. Young rarely gave detailed measurements or discussed the precise structure of his experiments, and for some phenomena that he described he gave no measurements at all—such as for single-slit diffraction. Yet this is not to say that Young's work lacked either experimental precision or quantitative detail. It had both, but it was not always clear from his articles either how he had obtained his formulas, how he had actually tested them, or how precise his experiments were. In fact Young's experiments, for example, for the fringes that are produced outside the geometric shadow by a narrow diffractor, were very likely as precise as Fresnel's. It has recently been remarked that Young chose to compare the wavelengths calculated from observations of fringes of different orders. Fresnel, we shall see, compared the measured with the theoretical positions of the fringe using an independently calculated wavelength. This has the effect of apparently increasing the accuracy of Fresnel's experiments over Young's—though in fact Fresnel's are about as accurate as Young's if one computes from them the wavelengths implied by his observations.[5]

Fresnel tells us that he began his observations with the fringes formed outside the geometric shadow of a narrow object, but that he discovered the principle of interference while examining the much fainter fringes formed within its shadow:

> For a long time I stopped at the external fringes, which are the easiest to observe, without bothering about the internal fringes. These latter are the ones that finally led me to an explanation of the phenomenon. I had already many times glued a small square of black paper to one side of an iron wire that I used in my experiments, and I had always seen the fringes inside the shadow disappear opposite this paper; but I was seeking only its influence on the external fringes, and I shut my eyes to the remarkable consequence that this phenomenon was leading me to. It struck me as soon as I occupied myself with the internal fringes, and I at once had the following thought: since intercepting the light from one side of the wire makes the internal fringes disappear, the concurrence of the rays that arrive from both sides is therefore necessary to produce them. (Fresnel 1815a, 16–17)

Young had also observed these internal fringes, but for him they were apparently a further demonstration of what he already believed to be the case, whereas for Fresnel they were instrumental in suggesting the principle of interference to him in the first place.[6] Fresnel had, then, been first concerned with the fringes that appear outside the geometric shadow. The principle of interference occurred to him when, seeking only the effect of light from the far edge of the diffractor on these fringes, he saw the pattern inside the shadow vanish when this light was completely blocked. If one may judge by the order in which he presented his results in the account he sent to Arago, he first measured the loci of the external fringes and compared his results with the positions that the principle of interference requires. He then turned again to the internal fringes, which he also measured carefully.

Fresnel's first two attempts to obtain formulas for diffraction were, like Young's before him, based directly upon the interactions of rays and not upon the behavior of wave fronts. The problem he set himself was to construct a configuration of light source, obstacle, screen, and rays that permits computing the loci of total constructive or total destructive interference. Fresnel considered that only two rays had to be taken into account: one directly from the source and the other, he apparently assumed from the outset, from the near edge of the diffractor. In this Fresnel differed from Young, who had at first considered that the second ray might come from points near, as well as at, the diffractor's edge.[7]

The basic structure of Fresnel's theory is extremely simple. The interference pattern is governed at each point by the interaction of two rays that ultimately derive from a single source. If the path lengths of the rays are integral multiples of a wavelength, then they should produce a bright fringe; if the lengths are half-integral multiples of a wavelength, they should produce a dark fringe (see fig. 5.1). Consider a source S situated directly above the midpoint K of an opaque obstacle AB, the geometric shadow of AB being IJ with center K′. The point P_e outside the geometric shadow receives light directly from S by ray SP_e and, Fresnel assumes, indirectly from edge A by the broken ray SAP_e. Since the path lengths of these rays generally differ, interference occurs. Point P_i within the geometric shadow receives light, Fresnel further assumes, by two inflections, one from edge A (ray AP_i) and the other from edge B (ray BP_i). Again the path lengths generally differ.

One obvious objection to this purely binary scheme—a scheme in which *only two* rays determine the interference pattern for a given wavelength at a given point—is that points outside the geometric shadow should receive rays from edge B as well as from edge A and the source S. Indeed, Fresnel did suppose that ternary combinations of this kind do occur, and we shall see that he considered the effect of the rays from edge B to be significant in the region between I, the edge of the shadow, and the first major maximum outside the shadow. However, he also assumed that, if the diffractor is sufficiently large (in practice 1 mm or more)—as he took care to ensure in his experiments—then the external effects of the rays from B will not be significant. It is essential that the B rays be ignorable outside the shadow, since otherwise Fresnel would have had to contend with a ternary combination. And he would not

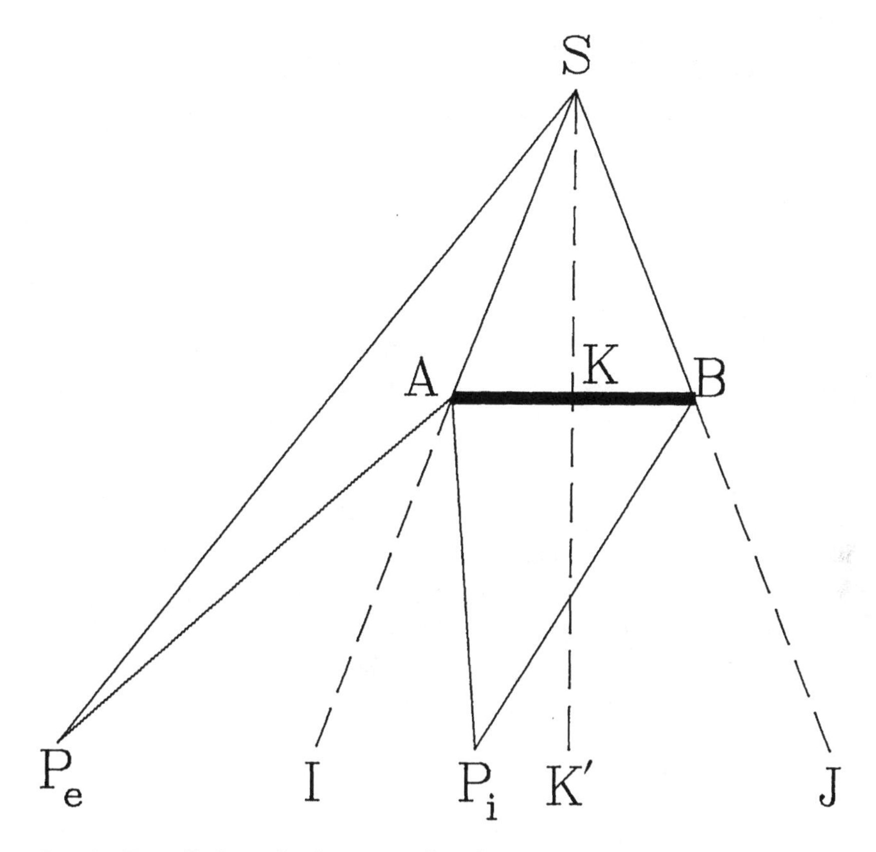

FIG. 5.1 Fresnel's theory for the narrow obstacle.

then have been able to calculate interference loci because he did not have a general method for combining the effects of more than two rays.

The "fringe" of a given order corresponds to a given path difference, d, where d is $(SA + AP_e) - SP_e$. Since SA is fixed, requiring d to be a given number determines a hyperbola with foci S and A. Fresnel obtained an approximation to this curve by computing the intersection of two circles, one centered on S, the other on A. These circles have direct physical significance, since they represent the two interfering waves. For the distance P_eI $(= y)$ from P_e to the edge of the geometric shadow he found the approximate expression:

$$y \approx \sqrt{\frac{2db(a + b)}{a}} \tag{5.2.1}$$

Here a is the source and b is the screen distance to the diffractor.

Since Fresnel was living in the country, he was forced to invent his own apparatus to test equation 5.2.1. His first experiments were accordingly simple in structure but difficult in execution. Placing a light source S behind the wire, Fresnel attempted to measure the widths of the dark fringes outside the wire's geometric shadow. Numerous difficulties impede accurate measurement. First, one needs a steady, pointlike source of fairly high intensity. The only intense source available was the sun (though Fresnel later became a sufficiently acute observer to see the diffraction pattern produced by starlight passing the edge of a branch). The sun's light passed through a sheet of tin punched with a small hole; to ensure sufficient intensity, Fresnel used a large lens to focus light on the hole, which then shone into a darkened room beyond. The major problem with this arrangement, he found, is that solar motion rapidly displaces the image.

Fresnel accordingly included a very convex "lens" in the light's path—one with a focal length of "six lines," or about 13.5 mm. This artifice substantially slows the solar motion, but one pays the price of ill-defined fringes at points closer than 500 mm to the focal point (probably because the path differences from different points on the surface of the highly convex lens become significant fractions of a wavelength). Since Fresnel did not actually have a lens so convex, he cleverly deposited a drop of honey on a hole made in a copper sheet. With this arrangement he could estimate the focal point to about 0.5 mm.

Fresnel early discovered that he could observe the fringe pattern directly through a lens without disturbing it—making unnecessary an opaque or translucent screen, which previous investigators had used, and thereby considerably improving the accuracy with which the fringe widths (the distances between the left- and right-hand fringes of a given order) can be measured. This procedure requires a micrometer, which Fresnel did not at first possess. He accordingly turned to traditional methods: he used a white piece of paper as a screen and then applied a ruler directly to it. Knowing the distance to the paper, the distance to the source, and the wire's diameter, Fresnel could then compute the width of the geometric shadow. The difference between the fringe and the shadow widths gives $2y$ in equation 5.2.1.

There are, as Fresnel understood, several sources of inaccuracy in such an experiment. First the source to wire (a) and the wire to screen (b) distances must be well known. Still, inaccuracy here will not systematically affect a series of experiments with different a, b of the kind Fresnel performed. Second, the wire diameter (c) must be known to a very high degree of accuracy, since it enters as a constant parameter in every computation. Moreover, one must ensure that the wire's effective diameter remains the same whatever a may be—which will not be the case if the wire is cylindrical. Fresnel insisted on such a high degree of accuracy that he chamfered the wire's edges.[8]

Perhaps the most vexing observational difficulty is due to the inhomogeneity of sunlight, which produces a spectral diffraction pattern. To locate, for example, a green minimum in such a pattern, one must actually find the apparent boundary between red and violet. This is not always simple, and it is especially difficult if, as Fresnel first had to do, one uses a screen instead of a lens to observe the pattern. Finally, one must independently compute the wavelength, λ, for spectral green. To do so Fresnel employed Newton's computations in the *Opticks* (1704, book 2, part 2) for the sizes of the air gap in a Newton's rings apparatus. At a given spectral maximum in the transmitted ring pattern the total path traversed between two successive reflections in the gap must obviously be an integral number of wavelengths. Newton's table does not give the gap thickness for the transmitted green, but it does give the thickness for reflected red of the first order and reflected violet of the second order. Transmission green occurs between these two, so all Fresnel had to do was to add the air thicknesses for the two bounding reflection rings and convert to millimeters, yielding a wavelength of 5176 Å.

Since Fresnel's observations involved the first minimum outside the geometric shadow, he should have obtained correct results by substituting $\frac{1}{2}\lambda$ for d in equation 5.2.1, but he did not, as Young had not before him. In fact, Fresnel found that one must use λ for the *minimum*, whereas $\frac{1}{2}\lambda$ gives the locus of the maximum. This can only mean, he remarked later in his first paper (Fresnel 1815a, 26), that the sources on the diffractor's edges must oscillate half a wavelength out of phase with the ray that comes directly from the source S. Inflection, therefore, somehow shifts phase 180°.[9]

Like Malus and Biot, Fresnel generated tables of observations. His first table (Fresnel 1815a, 15) has fifteen entries, and from them I find a mean difference between his formula for the first external minimum and his observations of only 0.13 mm, which is about 3% of the mean predicted position. This is deceptively good. Using highly accurate formulas, I calculate a mean difference between Fresnel's observations and the true loci of 0.19 mm (5%) and a mean difference between the latter and his theory of 0.25 mm (6%). Clearly the observational accuracy of 5% easily encompasses the difference of 3% between theory and experiment. Why, though, did Fresnel's formula fit his observations better than the apparent accuracy seems to permit?

The answer reveals factors that are inherent in narrow diffractor experiments of this kind. Unlike the semi-infinite plane, the narrow diffractor produces an intensity curve that contains many subsidiary maxima and minima (see appendix 3). For the most part these irregularities are not observationally distinguishable even though they may be as much as 0.5 mm apart. In fact, we can use them to determine the sensitivity of Fresnel's eye to intensity changes. In the case *a, b* respectively 17 mm, 1033 mm, Fresnel located the minimum at 39 mm from the shadow's center. There are actually two minima near this point (see fig. 5.2). The first is at 38.31 mm, the second at 38.73 mm. The intervening maximum is at 38.54 mm. Clearly Fresnel could not here distinguish a difference in the logarithm of the intensity of about 0.028, or about 1%. If he could have seen a 0.5% difference, then he would have located the minimum at 38.3 mm instead of at the 39 mm he actually saw. In cases like this one of multiple subsidiary minima, Fresnel usually located the minimum somewhat past the intervening maximum, and this tends to improve its apparent accuracy by shifting the observed locus closer to the one predicted by his formula, since the formula always gives too large a value.

The agreement between theory and experiment, though very good (3%),

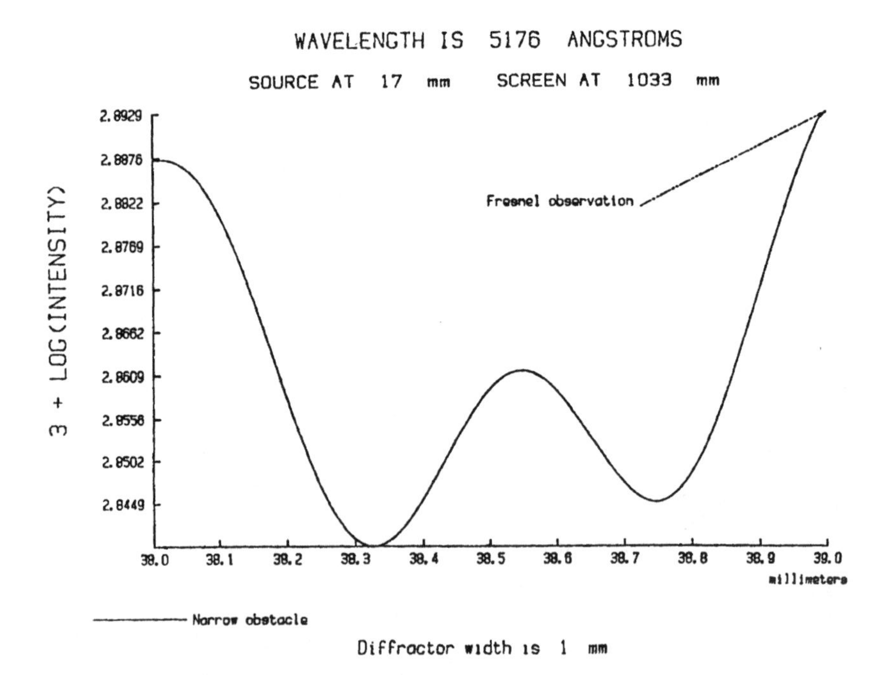

FIG. 5.2 The sensitivity of Fresnel's eye.

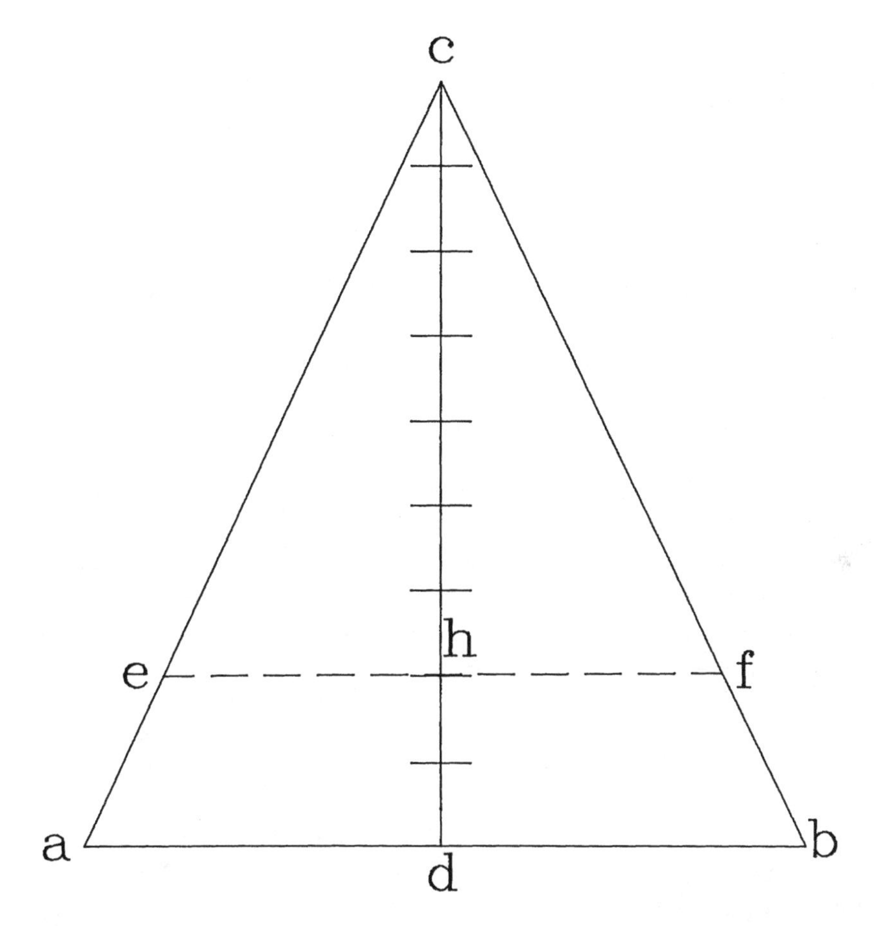

FIG. 5.3 Fresnel's micrometer.

was perhaps not entirely satisfying for a theory that does deep violence to contemporary views. Fresnel felt that one implication of the theory required particularly careful confirmation—that the fringes trace out hyperbolic paths with distance from the diffractor: "One very remarkable consequence of this theory of diffraction is that a given fringe propagates not in a straight line, but along a hyperbola, whose foci are the luminous point and one of the edges of the wire, for the external fringes" (Fresnel 1815a, 19).[10] To confirm this startling implication of the theory Fresnel devised a crude but effective micrometer that allowed him to observe the fringes directly (see fig. 5.3).

We have encountered the principle of Fresnel's micrometer before. Malus had used a similar device to measure the amplitude of the aberration in double refraction (see chap. 1). Fresnel simply fastened at their ends (c in fig. 5.3) two narrow strings of equal lengths, and he then constructed a scale to judge distance from the juncture. The scale (cd), marked in millimeters, was 218 mm long; length ab was 5 mm. Using this device as, in effect, a screen observed through a lens, Fresnel moved it laterally until ca, cb respectively intersected the first-order red/blue boundaries on either side of the center of the shadow, say at e and at f. He then noted the distance ch. Since ef is equal to ch(ab/cd) and he could measure ch to the nearest millimeter, Fresnel had the fringe width ef to the nearest 0.025 mm. (The main disadvantage of the device is that Fresnel could not observe widths greater than 5 mm.)

To confirm the hyperbolic propagation of the fringes Fresnel made six measurements, involving two values for a and six values for b. Here his observations reach an accuracy of 0.1 mm (from 0.19 mm), while the difference between his formula and the true loci also improves, to 0.11 mm (from 0.25 mm). The difference between the formula and observation improves to an astonishing 0.02 mm (from 0.13 mm), though only five observations with the micrometer are listed. Clearly most of the apparent improvement in the agreement between Fresnel's formula and his observations is fictitious: the actual advantage afforded by the new apparatus improves accuracy by about $2\times$, certainly not by $6.5\times$. Much of the improvement is due to the sensitivity of Fresnel's eye, since he again tends to shift the true minimum away from the shadow and so toward the locus required by the formula.[11] But no one at the time could have been aware of this limitation, so that Fresnel's new observations sufficed very well to demonstrate that the fringes do follow some curve that is concave toward the diffractor: the differences between the observations and the loci implied by linear fringe propagation reach 0.2 mm, which Fresnel knew to be considerably greater than his experimental accuracy.

Having examined the fringes outside the shadow with successful results for his formula, Fresnel turned to the fringes within it—the ones that had suggested the principle of interference to him in the first place. It is quite simple to obtain a formula for these fringes, because the sources are then on a line parallel to the screen. We take (fig. 5.1) y as the distance P_iK' and x as KK'; then if P_i is a minimum locus it lies at the intersection of two circles, one centered on A, the other on B, whose radii differ by $\frac{1}{2}n$ where n is an odd integer. (We may ignore the phase change on reflection, since it is the same at both points.)

$$\left[y - \frac{1}{2}c\right]^2 + x^2 = r^2$$

$$\left[y + \frac{1}{2}c\right]^2 + x^2 = \left[r + \frac{1}{2}n\lambda\right]^2$$

Eliminating x yields:

$$y = \frac{1}{2}\frac{n}{c}\left[r\lambda + \frac{1}{4}\lambda^2\right]$$

Fresnel dropped λ^2 as small and approximated r by b,[12] obtaining:

$$y = \frac{1}{2}\frac{b\lambda n}{c}$$

Fresnel listed four observations for the internal fringes: using the first two and the last, I find that his mean observational error here is 0.01 mm, that his theory errs by about the same amount, and that the differences between theory and observation run 0 mm, 0.02 mm, and 0.07 mm. Fresnel remarked that the observations are always greater than or equal to the theoretical values, which he thought pointed to a common inaccuracy in his value of 1 mm for the wire's diameter, an inaccuracy due to the closeness of the screen or else to problems in separating the colors.

The most interesting of Fresnel's four observations is the third, where a is 1490 mm and b is 592 mm and where Fresnel claimed to have observed the third-order internal fringe. His formula places this fringe at 0.765 mm; he observed it at 0.805 mm. This raises a question in that both these loci are considerably *outside* the geometric shadow, which is at 0.699 mm (fig. 5.4). Evidently Fresnel actually observed the pronounced minimum that occurs just within (0.696 mm) the shadow, since the difference between the shadow and the observation is only 0.07 mm. What is particularly noteworthy is that Fresnel's formula places the third "internal" minimum unambiguously outside the geometric shadow, where heretofore we applied the "external" formula 5.2.1.

There are two likely explanations for this apparent inconsistency. First, perhaps Fresnel did not compute the position of the shadow, in which case he would have been unaware of the problem. This is possible but hardly probable considering the important role the shadow edge plays in the binary theory as a theoretical boundary. Second, Fresnel may have been well aware that the third minimum, both observationally and theoretically, was outside the shadow, yet still considered it an "internal" fringe. This last is almost certainly what he had in mind.

We need only to recognize that, as Fresnel saw it, the distinction between *internal* and *external* fringes does not refer directly to the geometric shadow. Rather, it concerns the comparative significance of the three optical sources: the prime emitter and the two edges of the diffractor. Within the shadow, it is clear, where rays from the prime emitter cannot penetrate, only the diffrac-

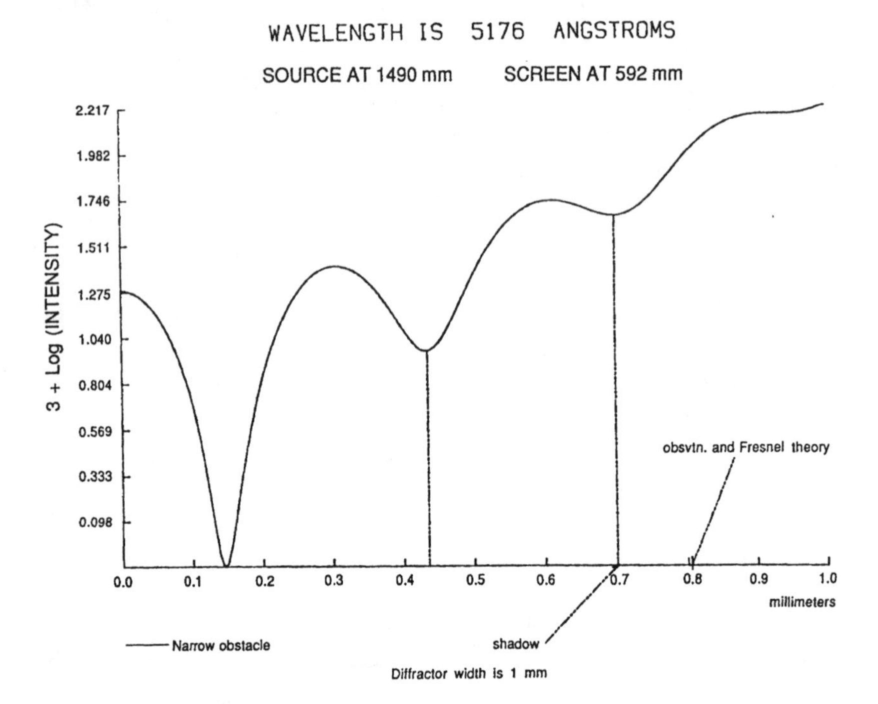

FIG. 5.4 Minima outside the shadow.

tor's edges are sources, and the "internal" formula is uniquely applicable. But outside the shadow all three sources send rays to all points, though with unequal intensities. One employs the "external" formula only for fringes that are sufficiently far away that the intensity of the ray from the far edge of the diffractor is small in comparison with the light from the near edge. What, in Fresnel's opinion, were the limits? If we assume that Fresnel knew his third observation was outside the shadow, then it seems to follow that the "internal" formula remains applicable at least up to the first major "external" maximum—which in this case occurs well over 0.4 mm from the shadow edge. In this nearly half-millimeter wide region the "internal" formula continues to hold.

I emphasize what might seem to be a niggling point because it is precisely this kind of situation—though not this one in particular—that ultimately betrayed the inadequacy of this first binary ray theory to Fresnel. I shall discuss this below, but here it is important to understand the precise circumstances in which observable difficulties with this binary theory may arise.

The region between the shadow edge and the first "external" minimum is peculiar in one major respect: the interference pattern is assumed to be governed primarily by two rays, one of which (AP) lies entirely outside the shadow, whereas the other (BP) is partially inside and partially outside it (fig. 5.5). In a proper wave theory the edge of the geometric shadow is not of direct physical significance: the edge of the diffractor delimits the wave and, we shall see, may also have dynamic significance. But in Fresnel's binary theory the shadow edge is very important because it demarcates a region in

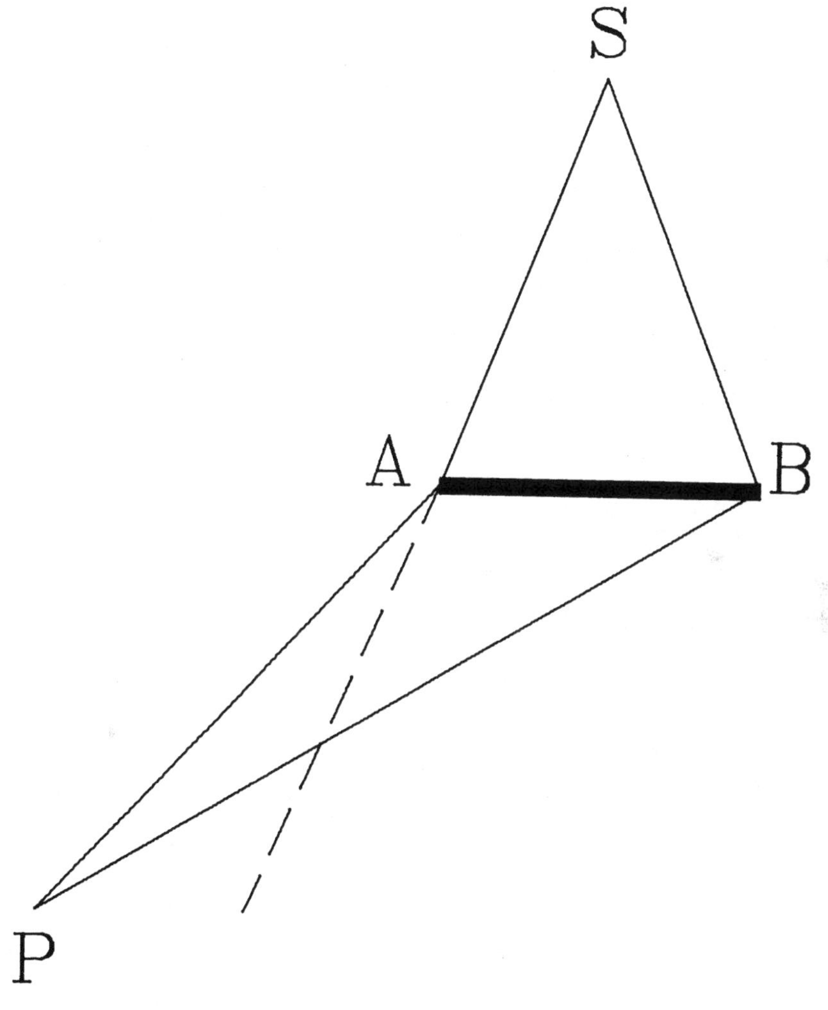

FIG. 5.5 The significance of the shadow edge.

which only two rays interfere with one another from a region in which three rays may interfere. The binary theory deals with only a single pair of rays, so Fresnel had to divide this second region again in two: in both parts one considers the interference of only two rays. However, in the part nearest the shadow one of the two interfering rays must emerge from within the shadow, a region where only two rays exist, into a region where three rays actually exist, though only two are taken into account at any place.

Until Fresnel fully adopted Huygens's principle, he thought of the shadow edge as a physically distinguished boundary: because some interfering rays must cross from inside the shadow to outside it whereas some do not, Fresnel could, and did, use this difference to explain empirical anomalies. This was possible only because he had not entirely abandoned rays for waves: he still thought in terms of relationships between rays rather than between propagating luminous surfaces. At this stage he used waves primarily as computational devices rather than as intrinsically significant physical things. Fresnel can be said to have begun abandoning rays in a deep sense only when the shadow edge lost this type of importance for him, and this probably occurred simultaneously with his adoption of Huygens's principle. Before we examine how this change took place, I must prepare the ground by considering a special form of Huygens's principle that Fresnel seems to have held very early and that sat somewhat uneasily with his account of interference.

5.3 Fresnel's Principle

One of Fresnel's major characteristics was his continuing concern with the physics of light: his first considerations in the wave theory had been deeply involved with such matters as the relationships between heat, light, and chemical transformations. These concerns did not lapse with his discovery of the principle of interference and his early experiments. For example, at the end of an addendum to his first paper he wrote:

> Analogy leads me to believe that heat, as well as light, is produced by vibrations and not by the emission of caloric. . . .
>
> It is much more natural to think that heat and light are uniquely due to the vibrations of caloric; because one sees that heat and light are always produced when a lively chemical action imparts large motions to the molecules of bodies, whether these molecules approach or move away from one another.
>
> Whatever system one adopts for the production of heat and light, one cannot doubt the continual vibrations of caloric and of bodies' particles: the force and the nature of these vibrations

> must have a great influence on all the phenomena embraced by physics and chemistry, and it seems to me that one has hitherto ignored them too much in studying these two sciences. (Fresnel 1815b, 59)

Fresnel's continuing concern with the underlying physical structure of the wave theory involved a complex mix of ideas that both impeded and furthered his work in different ways at different times. This is particularly well illustrated by his adoption of what I shall call *Fresnel's principle.* This is a principle that bears a certain resemblance to Huygens's but that is deeply restricted and strongly linked to dynamic considerations, whereas the import of Huygens's principle is almost entirely kinematic. On the one hand this principle enabled Fresnel to construct theories of reflection and refraction; on the other, the physical understanding of wave dynamics that underlies the principle prohibited, or at least strongly discouraged, him for some time from employing Huygens's principle.

The principle Fresnel developed asserts that a wave impinging on a material body sets its particles in vibration, with the result that each such particle becomes a source of secondary radiation that propagates both in the original medium (reflection) and in the refracting medium at appropriate speeds. This material secondary radiation, therefore, generates the reflected and refracted fronts. What this principle does not assert is that waves propagating in homogeneous ether (or pure "caloric," as Fresnel envisioned it) are themselves constructed out of secondaries; the fronts proper are not yet for Fresnel the loci of secondary emission as in Huygens's principle. This limitation was an essential one that derived from Fresnel's physical understanding.

In its fullest sense Fresnel's principle is implicit in several of his discussions of the law of reflection, which he presented in a deceptively simple fashion. I quote his discussion of it in full (fig. 5.6).

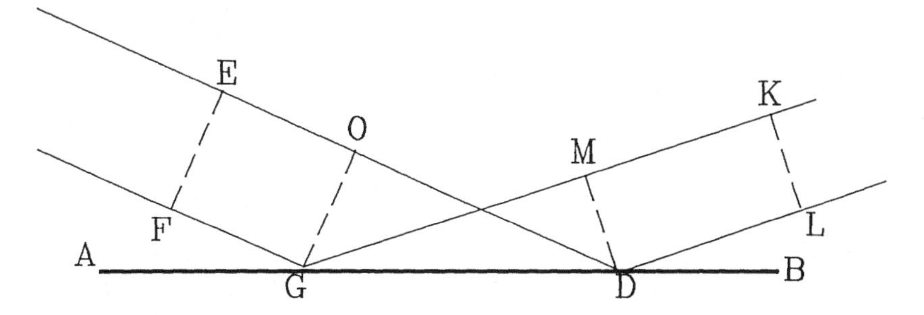

FIG. 5.6 "Fresnel's principle" for reflection.

> Let AB be the surface of a polished body; ED and FG, two neighboring rays; GK and DL, the same rays reflected. I suppose that the points F, E, K, and L are, in the two rays, the corresponding loci of the same vibrations, so that one has ED + DL = FG + GK. [Since] the two incident rays FG and ED vibrate synchronously, the two points F and E are on the same perpendicular. When the angles KGB and BDL are equal to the angles AGF and EDA, the points K and L are also on the same perpendicular to the reflected rays. But when the angle of incidence is no longer equal to the angle of reflection, the corresponding points K and L are no longer on the same perpendicular to the reflected rays, and their vibrations are opposed: now one can always find two incident rays at such a distance from one another that the discord is complete for the reflected rays—that is, a half-wavelength—and since they have equal force, their vibrations destroy one another. (Fresnel 1815a, 29)

Fresnel's demonstration shows that he fully grasped the concept of a wave front as a locus of constant phase. And it clearly presumes that the points of the interface AB that are successively struck by the incident front then become sources of secondary radiation from which rays can be drawn in all directions—that is, they themselves radiate at least into the upper hemisphere. (The theory of refraction presumes radiation into the lower hemisphere as well.) The proof proceeds essentially by choosing special pairs of these secondary rays at a given angle of reflection. They are so chosen that there exists a common normal to the two rays such that the points of the rays that lie on this normal differ from one another by a half-wavelength in the distances that light has traveled to reach them—resulting in total destructive interference between these points. This can be done for all ray pairs except those that are reflected at angles equal to their corresponding angles of incidence, whereupon points on the reflected rays and on a common normal to these rays are uniquely in phase with one another.[13] (In this demonstration, though Fresnel did not specifically so state, the claim is limited to plane fronts—as the accompanying diagram in his paper clearly shows—and he wrote of "the" angle of incidence and reflection. One can easily generalize his proof to spherical fronts, a fact of which he was no doubt well aware, preferring, however, to keep the textual demonstration as simple and as direct as possible.)[14]

Fresnel's brief demonstration depends directly upon the principle of interference coupled to radiation from material secondaries in the interface. Note that he (tacitly) assumed the interference takes place only between points that lie on the common normal to a given pair of reflected rays. As it stands, this

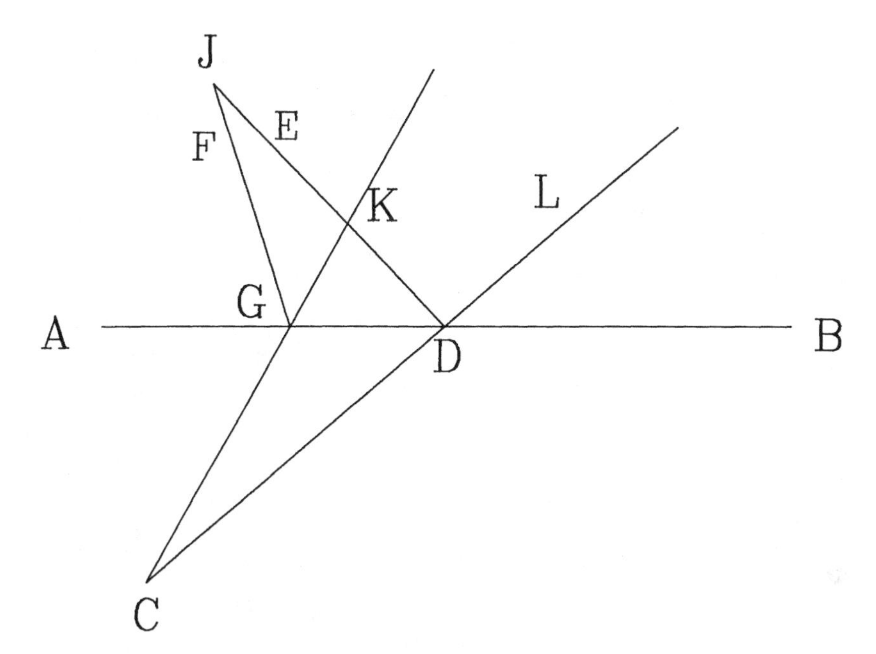

FIG. 5.7

contrasts rather strikingly with Fresnel's previous application of the interference principle solely to rays that intersect one another rather than to parallel neighboring rays. However, Fresnel no doubt had in mind an approximation: take a secondary ray that is not reflected at the angle of incidence and then find another secondary, emitted from a different surface point, that is nearly parallel to it, so that the rays meet at some point far (in comparison to a wavelength) from the reflecting surface where they differ in path by a half-wavelength. This can be done for any far point in the reflected field and for any such ray through that point—the unique ray through the point that is reflected at the incident angle is reinforced by neighboring rays that meet it at the point and that are nearly parallel to it. (I thank Joseph Keller for pointing out to me the omissions in Fresnel's own argument.)

There is no hint in Fresnel's argument—or in the similar one he gave for refraction (Fresnel 1815a, 30–31)—that secondary emission takes place anywhere but at the interface. Indeed, in his "second" memoir, which was for the most part a reworking with corrections of the first one for publication, Fresnel made this limitation quite clear:

> The most natural hypothesis [to explain diffraction] is that the molecules of the body set in vibration by the incident light be-

> come the centers of new undulations. Analogy leads one to suppose that in reflection the molecules that compose the surface of the reflecting body also become centers of new luminous undulations. How is it that these undulations are sensibly propagated only in a direction that makes with this surface an angle equal to the angle of incidence? That is easy to explain when one sees that, in every other direction, the vibrations of the reflected rays are mutually contrary and destroy one another. (Fresnel 1816a, 118)

In comparing reflection with inflection and using secondary radiation in both cases, Fresnel was emphasizing that this kind of radiation occurs uniquely when light impinges upon matter and does not occur in the free ether. At this stage in his thought he would have found Huygens's principle dynamically unnatural: since the purpose of secondary radiation in reflection theory is precisely to produce retrograde radiation, to use the concept for a freely propagating wave would have made little sense, as we can readily see from remarks he made a year later (1816) when he abandoned the original binary theory:

> When nothing troubles the regularity of the undulatory movement produced by a luminous point, it is clear that all the waves must be perfectly spherical and have the luminous point as their center. It is true that at each point of space where the ether is condensed it presses and tends to dilate in all directions; but this dilatation can take place only in a direction perpendicular to the spherical surface to which this point belongs, because a similar presssure makes itself felt at the same instant throughout [the surface's] extent. (Fresnel 1816b, 161)

Fresnel, then, thought that light propagates in right lines, not because Huygens's secondary waves conspire with one another to form succeeding fronts, but because a given element of a given front is prevented dynamically by its contiguous neighbors from dilating in any direction but the front normal. Not only does this dynamic understanding of rectilinear propagation not suggest Huygens's principle, it denies the very existence of secondary radiation in homogeneous media: "This dilatation takes place *only* in a direction perpendicular to the spherical surface." We must take this limitation quite seriously: for many months Fresnel simply did not think that a front sends out radiation except along the normals to the front. I shall show below how this dynamic image was modified only slightly as a result of the demands of experiment in order to yield Fresnel's first departure from the binary theory—his concept of the "efficacious" ray—in a way that remained essentially at odds with the requirements of Huygens's principle. To reach the latter—and hence to abandon rays entirely as basic elements—Fresnel had to abandon the adverb "only" in the quotation above, an abandonment that caused him some difficulty.

5.4 Successes of the Binary Theory

Fresnel sent his first memoir, which contained the binary ray theory, his explanations of reflection and refraction by means of the principle of interference, and the tables of experimental results, to the Academy via his uncle Léonor on 15 October 1815, having written Arago on 23 September. Arago replied on 8 November; he had in fact obtained the job of examining Fresnel's paper for the Academy (Fresnel, *Oeuvres,* 1:35–39). He at once remarked to Fresnel that many of the experiments had already been done by Young, "who, in general, envisages the phenomena in a manner quite analogous to that which you have adopted." However, Arago did not know Young's work well enough to realize that he too had discovered the law of hyperbolic propagation, and he fastened on this law as a new proof of the wave theory and as a strong argument against the "système à la mode." He accordingly asked Fresnel to make a new series of observations capable of showing the curvilinear path very clearly. (Recall that Fresnel had thus far made only a few measurements.)

To this Fresnel replied on 12 November. His letter turns at once to Young. One can imagine Fresnel's reaction to Arago's statement that Young had anticipated him, given his unfortunate experience with the abortive process for soda and his great desire to make a new discovery. Only the original French fully conveys the mix of trepidation and belligerence in Fresnel's reaction:

> Ce que vous me dîtes du docteur Young me fait désirer de connaître plus précisément en quoi je me suis rencontré avec lui. Vous concevez quelles peuvent être à ce sujet les petites inquiétudes de mon amour-propre. Je voudrais bien savoir s'il s'explique nettement sur la manière dont il conçoit l'influence que les rayons lumineux exercent les uns sur les autres. (Fresnel, *Oeuvres,* 1:61–63)

In loose translation:

> What you tell me of Doctor Young makes me want to know more precisely where I have collided with him. You may imagine the

> uneasiness of my self-esteem on this point. I would like to know if he cleanly explains the manner in which he conceives the influence of the luminous rays on one another.

On 20 November Fresnel again wrote to Arago, sending him new observations on hyperbolic propagation (Fresnel, *Oeuvres,* 1:64–69). His experimental method remained essentially the same, but he increased the length of the micrometer to measure fringe widths as large as a centimeter. He reported eight measurements, three taken at a source distance of 1988 mm and five at 3 m. Here, I find, his observational accuracy reaches a mean of 0.05 mm, and the difference between his formula and his observations is about the same (on the order of 1.5%). The extraordinary accuracy of the binary formula is readily explained: Fresnel very carefully kept the source distance above 1.5 m and the screen distance above about 3 m. These, I find, are precisely the limits within which the formula for the external fringes in narrow obstacle diffraction works well.[15]

During the month following the completion of his first paper, Fresnel had also been working on a "Complément" to it, which he sent in on 10 November. The "Complément" is particularly interesting because in it Fresnel examined the limitations placed on the accuracy of his formula by such things as the finite sizes of the source and the focusing lens; he also developed a theory of the diffraction grating, and he considered Newton's rings.

We need not closely consider the first set, except to remark that Fresnel's familiarity with the demands made by wave fronts was clearly growing rapidly. For example, he was able to demonstrate that (at least for a plane front) the maximum phase difference at the focus due to the sequential refraction of different parts of a front by a lens is insignificant at small angles of incidence—though it can become significant.[16]

Fresnel's account of the diffraction grating was essentially a modification of his theory of reflection. He envisioned a set of extremely thin parallel striations in an otherwise smooth reflecting surface. These striations are not in the same plane as the remainder of the surface and so may be considered independently of it. Each one is so thin that it constitutes in effect a linear source of secondary radiation, emitting rays in all directions. Since the striations are finite in number, their rays interfere constructively wherever, considering any given pair of striations, the path difference is an integral number of wavelengths. (See fig. 5.6: G and D now represent striations, and OG is the incident front.) If, Fresnel found, d is the distance between contiguous striations, i the angle of incidence (again assuming a plane front), and r the angle of reflection, then complete constructive interference ensues wherever $\sin(r) = \sin(i) - n(\lambda/d)$, where n is a positive integer. Fresnel considered the colored image so formed to superpose upon the usual uncolored reflection (which is not itself greatly affected unless there are many striations in a small area). Indeed, he noted that the separation of the spectral images varies reciprocally with d, though he had as yet not tested the formula.[17]

On 26 February 1816 Arago, having been "charged" with Poinsot to examine Fresnel's paper, described his own attempts "as much as the state of the sky permitted" to "verify" "the laws to which this talented physicist was led, and that seem to me destined to constitute an epoch in science" (Fresnel, *Oeuvres,* 1:75–77). In so doing Arago had tried to use a thick transparent glass instead of an opaque plane as an obstacle and had seen the internal fringes disappear in consequence. On using thin laminae of mica instead, at Fresnel's suggestion, Arago saw these fringes reappear but in displaced positions, as Fresnel had told him would happen. These several effects follow from the theory's requirement that light slows on refraction (since then the "internal" fringes are shifted toward the shadow in amounts depending on the thickness and index of the refractor).

Arago was by this time particularly interested in anything that cast doubt on the emission theory (or, perhaps better put, that cast doubt on the theory whose major champion was Biot). This was why he had strongly pushed Fresnel to make more accurate observations of hyperbolic propagation, and he was perhaps also instrumental in leading Fresnel to perform an experiment in which interference takes place without the intervention of an obstacle (Arago 1816). Arago was well aware that emission theorists could always argue, however weakly in his opinion, that diffraction results from the forces material objects exert on light particles. But then one presumably needs to have such an object whenever interference occurs. It seems probable that Arago pushed Fresnel to think of a situation in which interference would occur without material edges nearby. The result was Fresnel's mirror experiment, which he introduced with the following remarks:

> In order to eliminate any idea that the action of the edges of the body, the screen, or the small holes [might be involved] in the formation and the disappearance of the interior fringes, I sought to produce similar ones by crossing rays reflected from two mirrors, and I managed to do so after several difficulties. I would mention in passing that only the vibration theory could give one the idea of this experiment, and that it is sufficiently hard to per-

> form that it would be nearly impossible to discover it by accident. (Fresnel 1816b, 150)

This difficult and delicate experiment marks the high point of Fresnel's first binary theory and became almost canonical in the texts on the wave theory that began to appear in the early 1830s. (Fresnel discusses it in section 32 of his 1822 "De la lumière"; appendix 13 below provides details.) Arago's discussion of the mirror experiment appeared almost simultaneously with his and Poinsot's highly favorable report, dated 25 March, on Fresnel's paper. (Arago almost certainly wrote the report, and I doubt it contains much of Poinsot's.) The concluding paragraph of the report strongly urged the First Class of the Institute to encourage Fresnel to apply his theory to other phenomena and to "clarify several points that are still a bit obscure" (the points are not specified; Arago and Poinsot 1816, 87). By this time Fresnel was busy rewriting his paper for publication in the *Annales,* a task he found rather burdensome—the result was the "Deuxième mémoire," which was published in the *Annales* for March.

Fresnel moved temporarily to Paris at Arago's instigation—he was technically granted a vacation specifically to repeat his experiments. At first he was enthusiastic: he had at last begun to achieve the kind of recognition he had so long desired; he wrote to his brother on 16 February that Arago "attached the greatest importance to my discovery" (Fresnel, *Oeuvres,* 2:833). By early March, burdened with the rewriting, his enthusiasm had lapsed a bit, and his morale had fallen quite far by the end of the summer. As a result of a visit Arago paid to England in the company of Gay-Lussac, Fresnel realized that many people regarded his work as no different from Young's. On 25 September he wrote his brother a bitter and despairing note:

> [Because of my health and because I am too old to begin a teaching career] I have decided to remain a modest engineer of bridges and roads, and even to abandon physics, if circumstances require it. I would resolve to do so the more easily that I now see it's a stupid plan troubling oneself to acquire a small bit of glory, that they'll always quarrel with you about it. M. Arago, who returned from England, told me that they saw my memoir as a commentary on doctor Young, and that they found useless and pretty insignificant the new proofs I added to his. M. Arago hotly defended me and announced to doctor Young that he would insert a small memoir in the *Annales de Physique* where he would take our side. But all this does not satisfy me. Fie on contested glory! (Fresnel, *Oeuvres,* 2:836)

Here we see starkly revealed Fresnel's tremendous desire for recognition, his despair at Young's having anticipated him, and his anger at the English failure to see anything novel in his own work. But mid-October sees a revival of his determination; he will prove his worth by pressing far beyond Young, by making new discoveries:

> I've taken fairly philosophically the unpleasant news that came to me from England. It left me somewhat bitter, but it could not make me lose my taste for physics, and I felt that I had to respond to the charges of plagiarism with new discoveries. It seems to me that my last two memoirs offer fairly interesting ones. (Fresnel, *Oeuvres,* 2:837)

5.5 The Efficacious Ray

During the late spring and early summer of 1816 Fresnel worked hard to extend the range of his theory in a determined effort to make "new discoveries." At some point he set up an experiment with a single slit instead of an obstacle and with a fairly large screen distance (fig. 5.8). Within the geometrically illuminated region HSI, any point P_0 receives radiation from the edges A and B as well as directly from the source S. A and B are mutually coherent and have the same phase, while both are coherent with S, whose phase is 180° different from theirs. This region, then, must be optically equivalent to the area, in the case of narrow obstacle diffraction, between the shadow edges

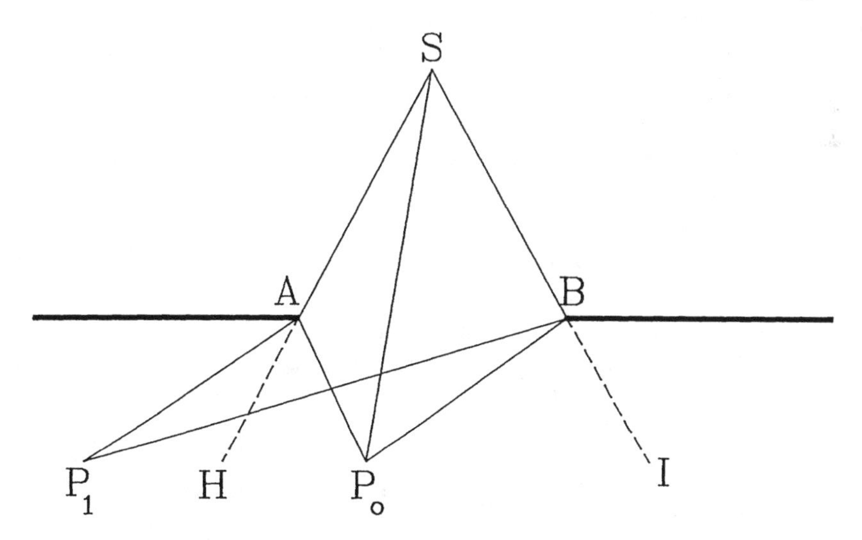

FIG. 5.8 Fresnel's rays for single-slit diffraction.

and the first major maximum outside the shadow. Outside region HSI a point P_1 receives radiation only from the edges, so here we have purely "internal" fringes in the narrow diffractor sense; that is, this area is optically equivalent to points within the geometric shadow of the narrow diffractor.

The fringe pattern within HSI, then, cannot be calculated using the binary theory because it deals only with a single pair of rays at each point in the illuminated field. And so Fresnel did not at this time attempt to compute fringe positions in this region. However, careful observation near the shadow edges might enable one to distinguish the pattern due solely to interference between the rays from the edges, just as one can in the similar region for the narrow diffractor, wherein the internal fringes should be observable up to the first major external maximum. Although the transition between a region where purely internal fringes occur to a mixed fringe is abrupt, occurring always at the edge of the geometric shadow, nevertheless the internal pattern continues into the mixed region—here, however, it is superimposed upon an external pattern. The latter may either fully dominate after some distance or else may remain of more or less equal significance with the internal pattern (Fresnel 1816b, 155–56).

Since the fringes outside region HSI are internal, they are governed by the following formula:

$$y_n = \frac{b\lambda(2n - 1)}{2c} \tag{5.5.1}$$

Here y_n is the distance to the center of HSI, b is the distance between the planes of the slit and screen, and c is the slit width. So the distance between consecutive minima is a constant—they are equally spaced—and this distance should be twice the distance from shadow center to the first fringe (or, equivalently, it should be equal to the width of the first fringe):

$$y_{n+1} - y_n = \frac{b\lambda}{c} = 2y_1 \tag{5.5.2}$$

At this point in his "Supplément" Fresnel does not provide much detail concerning the sequence of his observations—indeed, he here gives almost no detail at all. This perhaps reflects a certain disorder in his researches that followed his discovery that formula 5.5.1 is in error by about 100%. Although Fresnel did not provide numerical data on the problem, it is so striking that precise data are hardly necessary. Since Fresnel's discovery of this anomaly was purely empirical—it was one he did not expect—I shall attempt to recreate the situation in which he found himself by providing experimentally significant numbers based upon values for the parameters that are typical of the ones Fresnel used in his other experiments. In this way we can gain a sense of what Fresnel actually saw, though one hardly needs such an elaborate calculation to understand in retrospect *why* he saw it.

Consider the kind of situation that Fresnel was examining: a single slit, say 0.5 mm in width, together with a fairly distant screen, say 2250 mm. Set the source at 250 mm. According to formula 5.5.1, the first three minima outside the illuminated region should be located at 3.51 mm, 5.86 mm, and 8.19 mm. The intervals between consecutive minima should therefore be 2.34 mm. But Fresnel saw something quite different, as one can see from figure 5.9. Here we find that the first three minima outside the illuminated region are situated at 4.68 mm, 7.02 mm, and 9.36 mm. Clearly their actual positions differ from those required by Fresnel's formula by a nearly constant 1.17 mm—which is almost exactly the locus of the very first fringe according to the formula. Note, though, that the fringe spacing is precisely as required by the formula. The difficulty, therefore, is that the fringes seem to be displaced away from the center by half the fringe spacing, which is the same as the locus of the first fringe according to the formula.

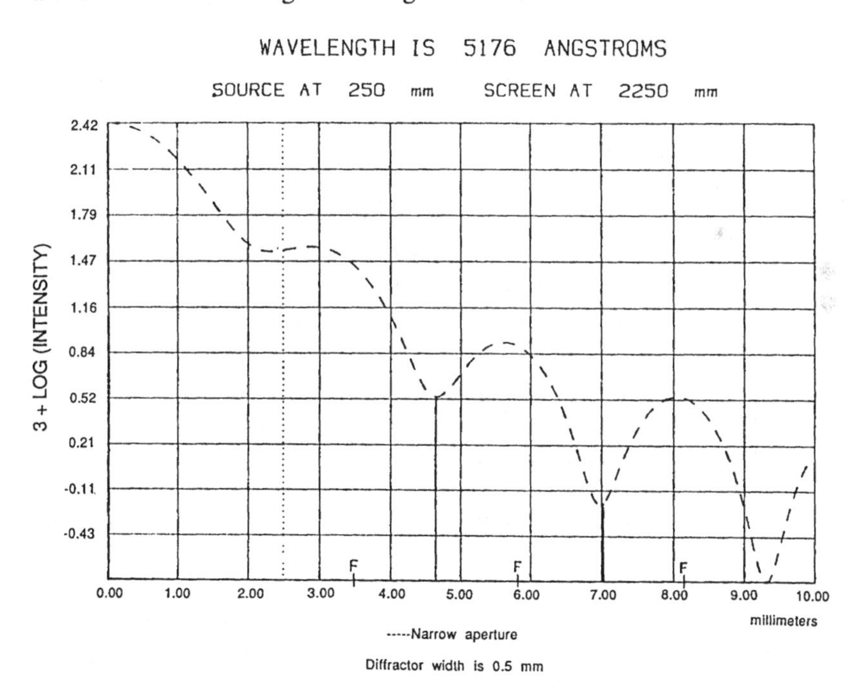

FIG. 5.9 What Fresnel saw in single-slit diffraction.

However, as Fresnel remarked, observational difficulties make it very hard indeed to locate these fringes precisely. Nevertheless, it is possible to determine whether the fringe widths outside the geometrically illuminated region are closer to being odd than even multiples of the fringe interval or vice versa: according to formulas 5.5.1 and 5.5.2 the fringe widths, $2y_n$, should always be odd multiples of the fringe interval, $y_{n+1} - y_n$. However, the width nearest the edge of the illuminated region, 9.36 mm, is four, not three or five times the spacing, and the differences are well over 2 mm, which is vastly greater than Fresnel's observational accuracy:

$$\begin{aligned} \delta &= 2.34 \text{ mm (fringe spacing)} \\ 3\delta &= 5.02 \text{ mm} \\ 4\delta &= 9.36 \text{ mm (true width} = 9.36 \text{ mm)} \\ 5\delta &= 11.7 \text{ mm} \end{aligned}$$

This is similar to what Fresnel very probably saw. Consequently formula 5.5.1, which follows directly from his first binary theory, fails dramatically for single-slit diffraction. For what Fresnel had in effect discovered was that the positions of the bright and the dark fringes outside the illuminated region are interchanged from what his formula requires.[18]

Fresnel's first thought on discovering this anomaly was to see whether the failure could be produced in other ways. Noting that the problem seems to arise when one of the rays just begins to lie entirely within the geometric shadow while the other lies only partially within it, Fresnel reasoned that in similar circumstances in obstacle diffraction the formula should also fail.

Recall that the failure concerns only internal fringes. This type of situation occurs outside the geometrically illuminated region in single-slit diffraction. In obstacle diffraction internal fringes occur within the geometric shadow. However, the two situations are not precisely analogous to one another. In the case of the internal fringes within the shadow of the obstacle, we do not, as we do in the similar case for the single slit, have one ray that lies entirely within the shadow, with the other ray lying partly within it and partly outside. To obtain this latter situation for the narrow obstacle, one of the rays must emerge from within the shadow into the illuminated region where the direct ray from the source becomes involved, and where the external formula dominates after a certain distance. But the first external bright fringe is preceded by a region that contains a barely visible pattern wherein the internal fringes continue to dominate. Here, then, Fresnel recognizes (1816b, 158) that we have a situation akin to the one that generates the anomaly in the single-slit case.

This is precisely the situation for that peculiar observation in Fresnel's first memoir that I discussed above. There the shadow locus was about 0.7 mm, and Fresnel's observations placed a minimum at 0.805 mm. The nearest true minimum is within the shadow. However, Fresnel's internal formula implies a third-order minimum at 0.766 mm, or only 0.04 mm from the observation. This difference was of no significance in Fresnel's early experiments. But if, as Fresnel now did, one uses homogeneous (red) light with a wavelength of 6230 Å and large screen distances, then marked discrepancies arise. For example, with a 1 mm obstacle, a source at 1 m, and a screen at 6 m, there is a true minimum at 4.48 mm. The nearest minimum implied by the internal formula is second order and situated at 5.61 mm, for an easily observable difference between theory and experiment of 1.13 mm. Again the internal formula unquestionably fails.

Fresnel was faced with unequivocal conflicts between theory and experiment for both single-slit and obstacle diffraction. To solve the problem and to salvage the method, if not the early form, of the binary theory, Fresnel decided to displace the loci of inflection away from the edges of the diffractor. He remarked:

> In seeking the cause of this difference between the results of calculation and those of experiment, some new reflections and observations made me doubt the exactness of a hypothesis from which I had calculated my formulas: that the center of undulation of the inflected light was always at the edge proper of the opaque body or, what is the same, that inflected light could come only from rays that have touched its surface. (Fresnel 1816b, 158–59)

Still thinking essentially in terms of binary ray combinations, Fresnel thought the difficulty could be removed simply by altering the phase difference between the interfering rays—because the difference between theory and experiment seems to involve only a constant displacement. Since the problem is that the bright and dark fringes exchange positions, then an extra half-wavelength path difference was called for. This at once raises a question that the first binary theory had managed to avoid completely: the only way to effect the change is to displace the loci of the inflected rays away from the diffractor edges, but then one is faced squarely with a physical question: These new loci must lie on the front, but why does oblique radiation occur from these points? For we saw above that a cardinal assumption of Fresnel's physical theory at this time was that each point on the front acts *only* along the normal to the front; there was no notion at all of oblique radiation of any kind.

To make his new assumption—that the rays are inflected slightly away from the edges, and not at them—physically reasonable, Fresnel considered what happens to the tangential pressures in the front when part of it is blocked by an obstacle. He argued that the edge will have the effect of removing the tangential action on the part of the front that borders it. Since radiation in directions oblique to the front's normal is, according to Fresnel's understanding, precluded by these very pressures, we now have a justification for the shift: near the edge, *and only near the edge,* oblique radiation can occur because the pressure balance is upset. At distances even slightly farther away the situation remains essentially unaltered, so that Fresnel could, he felt, concentrate entirely on the region very near the diffractor. In his words:

> When nothing troubles the regularity of the undulatory movement produced by a luminous point, it is clear that the waves must be perfectly spherical and have the luminous point as center. It is true that at every point of space where the ether is condensed, it presses and tends to dilate in all directions; but this dilation can occur only in a direction perpendicular to the spherical surface to which this point belongs, because a similar pressure makes itself felt at the same instant throughout its extent. It is no longer the same when the vibratory movement is intercepted in a point of space; and one conceives that the extremities of the waves can give rise to new waves; but these become sensible only in those directions where they mutually reinforce, and cannot propagate in those where they are contrary. (Fresnel 1816b, 160–61)

Note Fresnel's wording here—"one conceives that the *extremities* of the waves can give rise to new waves." He has very carefully limited the effect of the diffractor to points near the edge. Only there does he allow "new waves"—oblique radiation—to arise. Despite this limitation, Fresnel's shifting of the loci of inflection off the material object and onto the near portion of the front was a radical one, because it fundamentally altered the role of the diffractor from an active one (as a secondary emitter) to a passive one (as an interrupter of the front).

But Fresnel's understanding was not entirely dynamic. Or better put, he did not at this point draw a clear distinction between dynamic effects and the effects of interference. To see this, consider with him a diffractor AB, and let AC″ be the interrupted front (fig. 5.10). From screen point F, and with lengths AF + $\frac{1}{2}n\lambda$, where n is a positive integer, draw FA, FC, and so on to the front. Each successive distance from A, C, C′ . . . to F is half a wavelength greater than its predecessor. Then, Fresnel remarked:

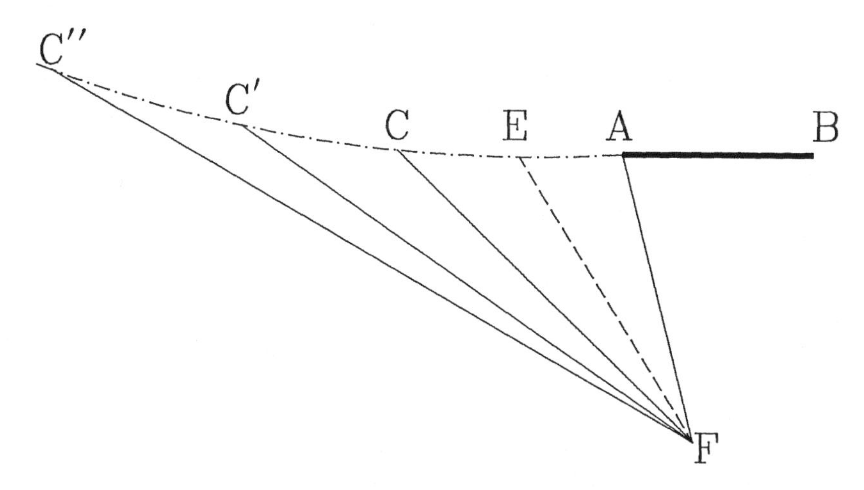

FIG. 5.10 Fresnel's efficacious ray.

> Now all the vibrations that leave are CC′ in this oblique direction [toward F] are in complete discord with the vibrations leaving from the corresponding points of AC. But all those that are born on CC′ are already greatly enfeebled by those of the succeeding arc C′C″ and can probably produce a diminution of no greater than a half in the undulatory motion emanating from AC: excepting the extreme arc, each part of the direct wave finds itself comprised between two others that destroy the oblique rays it tends to produce. (Fresnel 1816b, 161)

Fresnel's argument begins purely kinematically by noting that the radiation from each point of CC′ is in complete discord at F with a corresponding radiation from some point of AC. However, he continued, the emanations from CC′ are "already greatly *enfeebled* by those of the succeeding arc C′C″." Why are they "enfeebled"? If interference alone were responsible, then the oblique radiation from C′C″ should almost completely obliterate that from CC′, because the two have nearly the same distances and orientations with respect to the point where their rays meet. Instead, in Fresnel's view their effect is weakened by only about half. In light of his remarks on the "pressure" in the front, we must therefore recur to *dynamics* as well as to interference to understand what is going on.

In terms of dynamics any arc that is situated between two others simply does not radiate obliquely, because dynamic pressures forbid it. In terms of interference each arc *may be considered* to produce oblique radiation that by itself can destroy only half the effect of one of its neighbors—the second half of the neighbor's effect is destroyed by the other arc that it (this neighbor) touches. The arc AC therefore has a net effect in oblique directions because it has only one neighbor: from a dynamic point of view half the tangential action that is necessary to stop its oblique radiation is missing, while from the standpoint of the interference principle one has only half the light that is necessary to produce total destruction. These two points of view were evidently complementary ones for Fresnel at this time.

Now since each point of the terminal arc produces rays, Fresnel chose its center B′ as the "principal" point of emission—the point from which one may for computational purposes assume all the rays on the arc to originate (supposing that the obliquity B′F to AB is sufficiently large that the curvature of AC can be ignored). And this produces what Fresnel was seeking, because the line B′F, that becomes the new inflected ray, is a quarter-wavelength longer than AF (and so, as we shall in a moment see, two such rays in a single-slit experiment will have the extra half-wavelength difference that is necessary to accommodate the results of experiment). Fresnel named it, for obvious reasons, the "efficacious ray."

The principle of interference therefore becomes a significant factor only when the front points are near the diffracting edge. *Everywhere else it is simply superfluous,* because dynamic factors—tangential pressures—preclude oblique radiation. Indeed, the entire thrust of Fresnel's remarks—the whole purpose of considering either dynamics or interference for arcs on the front that are far from the edge—was to explain why the oblique radiation from these arcs can always be ignored. I think it highly probable, then, that Fresnel did not at this time thoroughly distinguish the interference principle for oblique radiation from the dynamics of pressure in the front, but that neither dynamics nor interference for this kind of radiation ever had to be brought into a calculation. Fresnel's goal, in other words, was *solely* to provide a convincing argument for shifting the locus of radiation off the edge of the diffractor and onto the front at a point a quarter-wavelength from the edge.

The efficacious ray readily explains the anomalies Fresnel had observed. Recall that, in the case of the single slit, the problem was that the first minimum is twice as far from the center as the binary formula required; that is, it was at $b\lambda/c$ instead of at $\frac{1}{2}(b\lambda/c)$. Fresnel had derived his internal formula by taking the slit edges as the radiators. Thus if δ is the difference between the radii of fronts from the edges of the slit, then:

$$\left(y - \frac{1}{2}c\right)^2 + x^2 = b^2$$

$$\left(y + \frac{1}{2}c\right)^2 + x^2 = (b + \delta)^2$$

Setting δ equal to $\frac{1}{2}\lambda$ in the previous theory gave the locus of the first minimum as $b\lambda/2c$. Now, however, the "efficacious" rays emanate from C and D, not from A and B, and CF, DF are, respectively, $\frac{1}{4}\lambda$ greater and smaller than AF, BF (see fig. 5.11). Previously we had AF, BF respectively equal to b, $b + \delta$. Whence CF, DF are now equal to $b + \frac{1}{4}\lambda$, $b + \delta - \frac{1}{4}\lambda$. This gives a first minimum at DF $-$ CF $= \delta - \frac{1}{2}\lambda = \frac{1}{2}\lambda$, whence $\delta = \lambda$. This gives $b\lambda/c$ for y instead of $\frac{1}{2}(b\lambda/c)$, and we have removed the anomaly for slit diffraction—the bright and dark fringes exchange positions.

Despite this shift of the radiating points off the diffractor and onto the front, the geometric shadow continues to play a distinguished physical role in the theory. In consequence one has not passed to a theory founded principally on waveforms rather than on rays—one has not as yet fully abandoned a selectionist outlook. The new efficacious ray theory rejects the previous account's reason for emphasizing the shadow edge, but it introduces a new one of its own. The earlier theory, which I shall continue for convenience to call the binary theory (even though the new one also employs only binary ray combinations), used the shadow to distinguish internal from external interference: the former involves rays that, aside from path differences, have the same phase as the source; the latter involves rays that, in addition to the path differences, differ in phase by 180°. In the new theory no phase differences enter except those due to path lengths, so the shadow loses its former function in this regard.

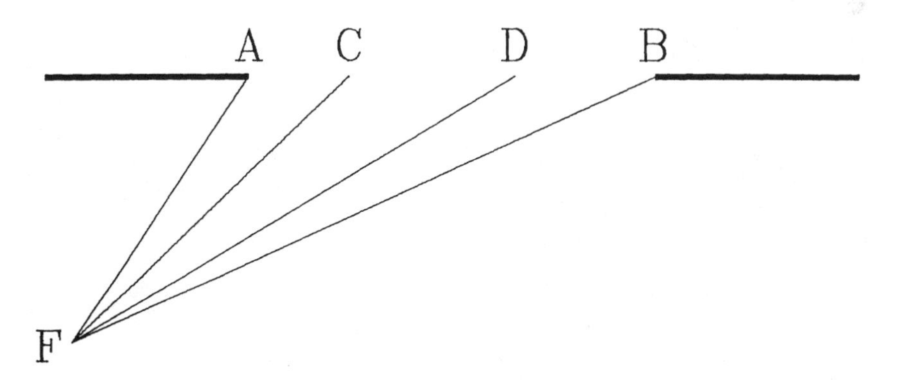

FIG. 5.11 Efficacious rays in single-slit diffraction.

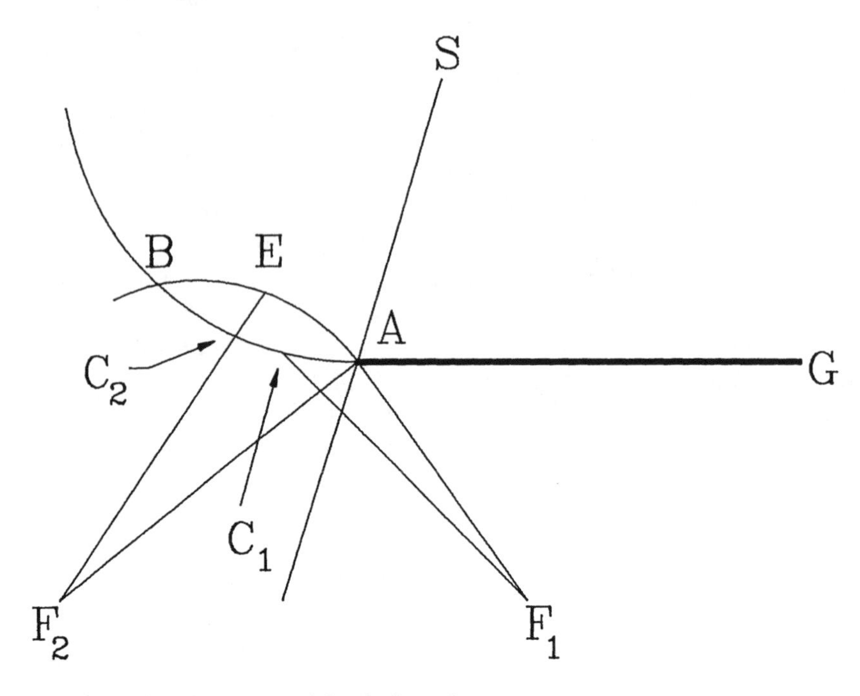

FIG. 5.12 Efficacious rays and the shadow edge.

However, the shadow still has a privileged role to play because efficacious rays that penetrate the shadow region are a quarter-wavelength longer than rays to the same points from diffracting edges, whereas efficacious rays outside the shadow are a quarter-wavelength smaller than a ray from the edge. In figure 5.12, F_2 is a screen point outside the geometric shadow, and F_1 is a point within. AC is the front from S intercepted by the obstacle AG. Consider F_1. The efficacious ray, C_1F_1, is greater than AF_1. In the case of F_2, first describe a circle of radius AF_2 that intersects front AC in B, forming the arc AEB. Clearly, rays drawn from F_2 to AC are smaller than AF_2 as long as these rays lie on AC between A and B. That is, if the greatest distance between the two arcs AEB and AC_1B is greater than or equal to a quarter-wavelength, then it is possible to draw an efficacious ray F_2C_2 such that the difference $AF_2 - F_2C_2$ is $\frac{1}{4}\lambda$. Since Fresnel's dynamic argument for the existence of the efficacious ray depends upon a disturbance in equilibrium created near, and limited to the vicinity of, the obstacle edge, we must choose F_2C_2 on AC between A and B, that is, the efficacious ray outside the shadow is smaller than a ray from the edge.[19]

The continued, albeit altered, significance of the shadow has decided advantages. Consider, for example, the internal fringes that are produced by a narrow diffractor within the shadow region. Here the efficacious rays are each a quarter of a wavelength greater than the corresponding rays from the edges. If we redo our computation for the slit with CF, DF now respectively equal to $b + \frac{1}{4}\lambda$, $b + \delta + \frac{1}{4}\lambda$, we obtain δ equal to $\frac{1}{2}\lambda$ and so $\frac{1}{2}(b\lambda/c)$ for the minimum of first order—the very same formula given by the binary theory, and one that here works very well. Of course, Fresnel had found a discrepancy for internal fringes just outside the shadow of the narrow diffractor, but the new theory also explains this problem: since here the near efficacious ray is $\frac{1}{4}\lambda$ greater than the ray from the edge nearest to it, whereas the ray from the other edge is $\frac{1}{4}\lambda$ greater than the corresponding efficacious ray, we retrieve the same situation we had with the internal fringes of the single slit; that is, we obtain the empirically necessary swap between bright and dark fringes.

Evidently the new theory, which is still based on binary combinations that grant the shadow locus a privileged physical significance, does explain the anomalies Fresnel had observed for internal fringes in both single-slit and obstacle diffraction. But can it also deal with the external fringes produced by the obstacle? Can it, that is, also retrieve the empirically accurate original formula? Here, we shall now see, Fresnel encountered a serious difficulty that may have served as a stimulus to his subsequent consideration of oblique radiation from points on the front that are quite far from the edge.

Consider for simplicity the semi-infinite plane (fig. 5.13). The efficacious ray BF, which is shorter than AF by a quarter-wavelength, here interferes with the direct ray SRF. Set *y, a, b* respectively equal to FG, FA, AG. Then *d,* the path difference (SB + BF) − SF, is simply SA + AF − SF − $\frac{1}{4}\lambda$, since BF equals AF − $\frac{1}{4}\lambda$. This gives to first order in the path difference and wavelength:

$$y = \sqrt{2\left(d + \frac{1}{4}\lambda\right)b\,\frac{a+b}{a}}$$

At the first minimum d must be $\lambda/2$, since we now have no phase change due to inflection at the edge itself, whence we obtain Fresnel's result (Fresnel 1816b, 169):

$$y = \sqrt{3\lambda b\,\frac{a+b}{2a}}$$

The problem is that the previous formula, which worked to within a tenth of a millimeter, has a factor of 2 under the root, not $\frac{3}{2}$. This difference is much too large to be compatible with observations. (For the distances used in Fres-

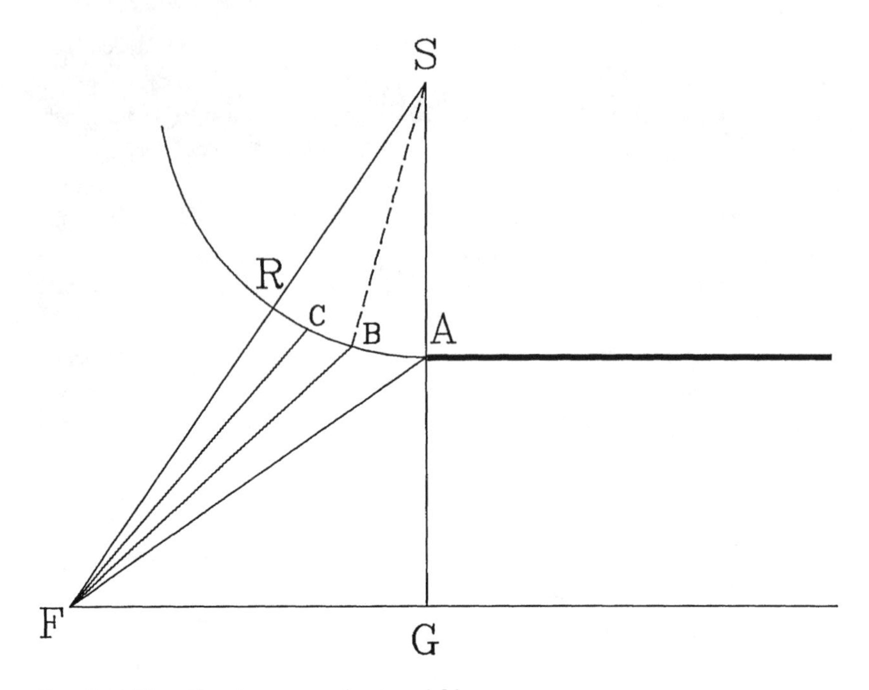

FIG. 5.13 The efficacious ray and external fringes.

nel's first experiments the new formula differs from the old by up to half a millimeter.) Fresnel was well aware of the situation, for which he presented the following argument:

> Perhaps the direct waves experience a slight change in curvature toward their extremities, in the part that concurs in fringe formation, in such a manner as to remove farther from the shadow their point of intersection with the undulations of the efficacious rays. But the laws governing the efficacious rays are still not well enough known for the hypothesis to be considered a necessary consequence of the phenomenon. (Fresnel 1816b, 169)

This failure on Fresnel's part to retrieve the original formula for external diffraction fringes shows almost conclusively that he did not permit oblique radiation to occur except from points very near the edge of the diffractor. For if he had, he would have been able very simply to solve this problem. Suppose Fresnel did assume that all points on the front radiate obliquely. Then take a field point P outside the geometric shadow of the diffractor, and from it draw a line (SP) to the source and a line (EP) to the edge of the diffractor (fig. 5.14). Consider the front EDI at the moment it passes the edge E of the diffractor. The part DI of the front that extends from SP out to infinity must always have the same net effect and so need not be considered. But the portion DE of the front between SP and the edge E varies in size, and so it governs the interference pattern. About P draw a circle with radius DP, which intersects EP in L. If D is not too far from E, we may consider DEL to form a right triangle. Join the midpoint M of DE to P. Suppose EL is half a wavelength greater than MF. Then the oblique radiation from M and that from the point of the front very near E destroy one another. One can continue pairing rays along DE with half-wavelength differences (taking one ray from DM and the other from ME and thus, in later parlance, constructing half-wave "zones" on the front), so that P must be the locus of a dark fringe.[20] However, the difference between EP and DP is one wavelength, which means that we retrieve the original formula for external diffraction, since it was based on a 180° phase change at the edge E.

If, therefore, Fresnel had at this time used oblique radiation, he would have been able to retrieve the original, empirically correct formulas for both exter-

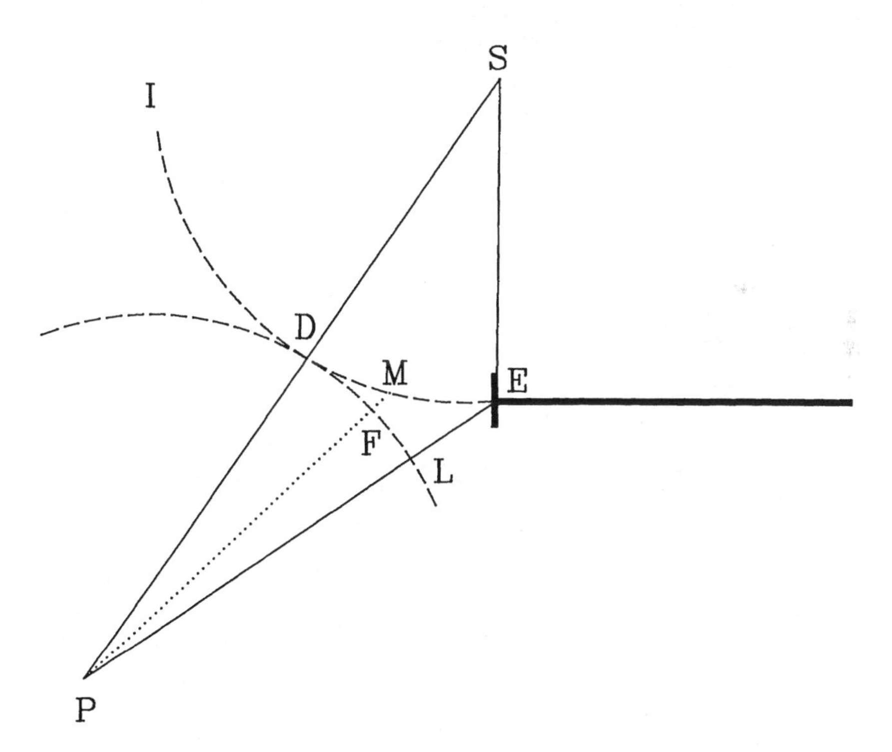

FIG. 5.14 How to retrieve Fresnel's original formula for external fringes.

nal and internal fringes. This would have meant that the efficacious ray could be used only for calculating internal fringes: it would have had no intrinsic physical significance, since it would become a purely computational device that can be used only where the interference of oblique radiation yields the same result. Instead, the structure of Fresnel's argument for the efficacious ray, together with his attempt to use it where it leads to an empirically incorrect result, shows conclusively that he had not as yet altered his understanding that radiation oblique to the front is inconsequential except near an edge.

One other experiment Fresnel introduced to support his introduction of the efficacious ray is particularly interesting in this regard, because he was later to claim that it proved the existence of oblique radiation from all points of the front. He took a metal plate and cut it, placing it 4 m from the source (fig. 5.15). Here we have in effect an obstacle ABCD and two narrow slits C′CEE′, D′DFF′. Observing close to the plate, Fresnel found that the fringes

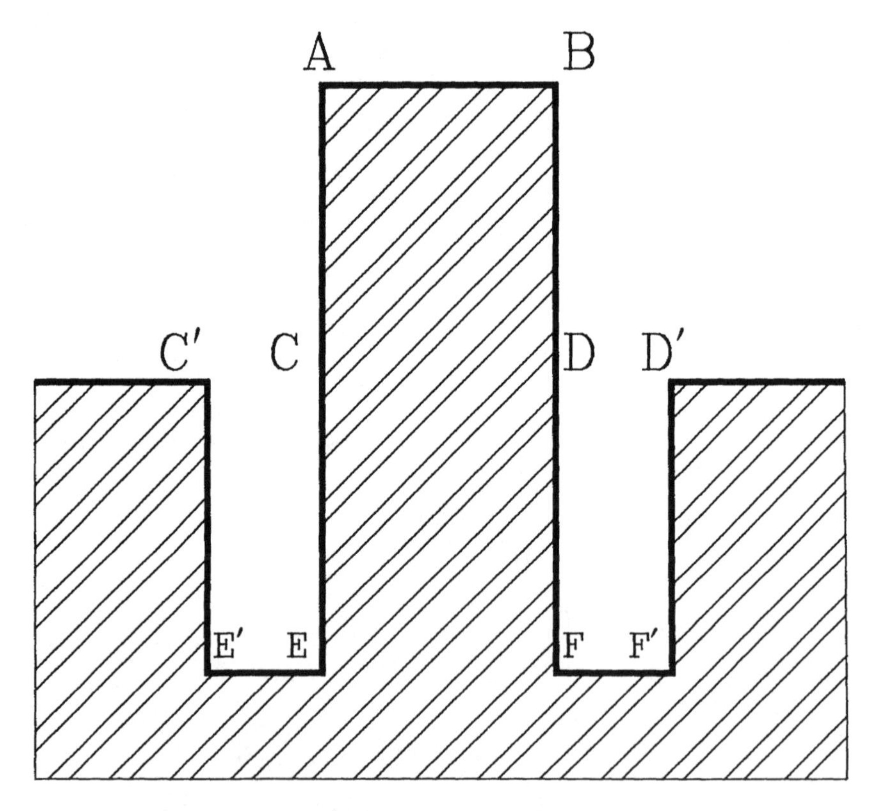

FIG. 5.15 Demonstration of the efficacious ray.

produced outside the geometric shadow of CDFE by light from the slits are much purer in color and greater in intensity than the fringes inside the shadow of ABCD. This should not be so, on the original binary theory, because according to it the former fringes are due to rays that are inflected from the edges C′E′, CE, DF, D′F′, whereas the latter fringes are due to the two inflections AC, BD—and the purity and intensity of a fringe is the more pronounced the fewer rays from different loci concur in producing it: here an extremum for C′E′, DF will not coincide with one for C′E′, DF, and this should impair fringe formation.

But the efficacious ray can explain the observation. In Fresnel's words:

> When the illuminating arc . . . is equal to the aperture of one of the slits, then it sends twice as much light into the shadow CDFE as into [the shadow] of ABDC, because in the upper part the oblique vibrations of the portion near the direct shadow destroy half that of the illuminating arc; whereas in the shadow CDEF, where it is intercepted, it can no longer produce the same effect. (Fresnel 1816b, 166–67)

The argument is quite simple. When one observes from sufficiently close to the plate, the zone near the edges from which the efficacious ray emanates will be of the same order of magnitude as the widths C′C, D′D of the slits. In the shadow of CDFE, therefore, two efficacious rays—one from the center of C′CE′E, the other from the center of D′DFF′—interfere with one another. The arcs from which these rays emanate are terminated on *both sides* by an edge. Now in the shadow of ABCD one also has two efficacious rays from the two arcs near the edges ACE, BDF. However, these arcs are terminated by an edge on only one side; on their other sides they are pressured by the contiguous portions of the front, and this "destroys by a half" the light they can send into the shadow. Hence the weakness of the fringe pattern they produce. Nowhere in this argument, as in the others we have examined, did Fresnel suggest that radiation from points on the front that are sensibly removed from the edge has any effect at all.

The early summer of 1816, then, was a time of only partial success for Fresnel. He had obtained a group of formulas that were accurate to better than a tenth of a millimeter, but he could not deduce them all from a single theory. Although he had now begun to consider rays emanating from points on the front proper, nevertheless his dynamic understanding of wave propagation made it difficult for him to conceive of an arbitrary point on the front as an oblique radiator. Finally, even if Fresnel had seen how to use Huygens's principle at this time he would not in any case have been able generally to calculate with it—except through a zone count—because he did not as yet know how to superpose wave forms with phase differences other than 0° or 180°. Within a year and a half he had apparently understood that the efficacious ray must be replaced by oblique radiation from all points of the front in order to achieve a successful set of formulas. And he had also discovered the equations governing superposition, which made mathematically possible his adoption of Huygens's principle by the early spring of 1818.

6 Huygens's Principle and the Wave Theory

6.1 The Superposition Equations

The supplement to the second memoir, containing the new theory of the efficacious ray, was presented to the Academy on 15 July 1816. Fresnel sent nothing further on diffraction until the spring of 1818, when, in a short note, he introduced integral methods based on Huygens's principle combined with the principle of interference. During the two intervening years Fresnel was far from idle, as we shall see in detail below. Working with Arago, he developed laws for the interference of polarized light, and most important here, he discovered how to superpose wave forms algebraically.

In a memoir presented on 19 January 1818 (1818c) as a supplement to one on the reflection of polarized light (1817b), Fresnel introduced for the first time trigonometric wave expressions. His goal in doing so was not to improve diffraction theory but to develop a general theory of chromatic polarization, which was not possible without a method for computing wave resultants. He therefore posed, and solved, the following general question: Given the intensities [amplitudes] of an arbitrary number of systems of [coherent] luminous waves and their respective positions, or their different degrees of accords and discords [phase differences], to determine the intensity of the total light (Fresnel 1818c, 488).

Eighteenth-century work on the wave equation for the string, and even for the membrane, could not help here. That work was predominantly concerned with the kinds of solutions that evolve from a given initial configuration rather than with the resultant of a set of distinct, arbitrarily specified waveforms. Furthermore, the eighteenth-century controversy over the generality of trigonometric solutions to the wave equation introduced complications that Fresnel, who was looking for the quickest route to his goal of superposing waves, would have sedulously avoided. He chose to do so, in effect, by constructing an ordinary differential equation instead of the partial differential wave equation (whose three-dimensional generalization was not as yet developed in any case) and then introducing propagation as a physical hypothesis.[1]

To obtain the solution he needed, Fresnel essentially assumed that the ether particles oscillate harmonically, which quickly gave a $\sin(2\pi t)$ for the speed, v, of the particle—taking a period as the unit of time. Fresnel then simply argued that the speed of a given particle at time t must be equal to the speed of the emitting particle (the source of the wave) at the time of emission, $t - (x/\lambda)$, where λ is the wavelength and x is the distance from the source to the field point. Whence:

$$v = a \sin\left[2\pi\left[t - \frac{x}{\lambda}\right]\right] \tag{6.1.1}$$

The next problem, as Fresnel saw it, was to consider the meaning of an arbitrary constant addition i to the argument of the sine function in equation 6.1.1:

$$v = a \sin\left[2\pi\left[t - \frac{x}{\lambda}\right] - i\right] \tag{6.1.2}$$

Algebraically decomposing 6.1.2, Fresnel obtained:

$$R = a \cos(i)\sin\left[2\pi\left[t - \frac{x}{\lambda}\right]\right] - a \sin(i)\sin\left[2\pi\left[t - \frac{1}{\lambda}\left[x + \frac{1}{4}\lambda\right]\right]\right]$$

$$= [a \cos(i)]\sin\left[2\pi\left[t - \frac{x}{\lambda}\right]\right] - [a \sin(i)]\cos\left[2\pi\left[t - \frac{x}{\lambda}\right]\right]$$

That is, a single waveform 6.1.2 with an arbitrary phase i can always be considered to arise from the *interference* of two other waves with amplitudes $a \cos(i)$, $a \sin(i)$ that differ in phase by 90°. This, Fresnel at once pointed out, involves the same kind of process as the composition of two mutually perpendicular forces of magnitudes equal to the component amplitudes. Whence the magnitude of the resultant of two waves that differ in phase by 90° is the square root of the sum of the squares of their amplitudes, and the phase of the resultant is the arctangent of the ratio of the component amplitudes. Fresnel generalized the procedure to compute in a similar way the resultant of two waves that differ in phase by an arbitrary amount.[2]

Fresnel's realization that a wave with an arbitrary phase can always be decomposed into two others that differ in phase by 90°, though mathematically simple, was nevertheless of great conceptual importance. Previously he had never considered the resultant at an arbitrary point of even two waves, and he had not considered the resultant at any point of more than two waves.

He could now examine both situations. For the first, but not for the second, his algebraic decomposition for an arbitrary phase difference sufficed. To compute the resultant of more than two waves anywhere, it is essential to construct a reference phase from which the several interfering waves differ by their respective amounts in order to apply the quarter-wave decomposition. This requirement, I shall now show, determined the structure of Fresnel's new mathematics for diffraction on the basis of Huygens's principle.

6.2 The Fresnel Integrals

6.2.1 The Pole

Although Fresnel was now capable of examining analytically the multiple interferences that arise from secondary radiation across the entire wave front, he had to alter his previous stringent requirement, based primarily on dynamics, that effective oblique radiation arises only near the edge of an obstacle. By the spring of 1818 he had done so:

> [Since] the impulsion that was communicated to all the parts of the primitive wave was directed along the normal, it is clear that the motion they tend to create in the ether must be more intense in this direction than in all others, and that the rays that would emanate from them, if they acted alone, would be more feeble in proportion as they diverged from this direction. But [since] the effects produced by the rays that emanate from the primitive wave destroy one another nearly completely when the [rays] are sensibly inclined to the normal, the rays that appreciably influence the quantity of light received by each point P can be regarded as of equal intensity. In extending the integration to infinity, I suppose, for purposes of calculation, that this holds also for the other rays, inasmuch as the inexactitude of this hypothesis should not bring a sensible error in the results. (Fresnel 1818b, 174–75)

This is a viewpoint considerably different from the one Fresnel had held in 1816. He has altered his previous understanding that, *except in the immediate vicinity of an edge,* points on the primary front simply do not produce oblique radiation because of the dynamic balance in the front or because they completely obliterate one another through interference. This, he now thinks, was not a thoroughly accurate way to understand what takes place. Instead, he now begins with the oblique radiation that is sent out from all points on the front, whether or not they are close to an edge, and he substantially ignores dynamic factors. Instead, he introduces a limitation that preserves the physical spirit of the dynamics while removing its purpose of obliterating the oblique radiation: the intensity of radiation from a point on the front, he now assumes, decreases rapidly and continuously according to some unknown function of its inclination to the normal. (Hereafter I shall call this function the *inclination factor.*) The inclination factor, that is, continues Fresnel's emphasis on the dynamic aspects of wave propagation by insisting that the radiation is indeed most intense—but only most intense—along the normals to the front. However, Fresnel does not use it to explain physically why waves propagate in straight lines except near an edge, as he had used his previous dynamic argument.

The whole thrust of Fresnel's previous argument for the efficacious ray had been to show why effective oblique radiation can occur *only* near an edge, regardless of the actual configuration of the objects that block the light. This could be understood in what Fresnel had taken to be two complementary ways: one may say that the dynamics of pressure in the front simply preclude oblique radiation unless matter upsets the balance, or as a complementary way of thinking one may say that, except near an edge, oblique radiations always obliterate one another through interference. To this way of thinking the actual variation in the intensity of radiation with obliquity is irrelevant, since except near the edge the oblique radiation always has no effect at all. But if that radiation can have an effect even when it comes from points that are not close to an edge, then it becomes important to understand the variation of intensity with obliquity. In other words, Fresnel has introduced the inclination factor for two quite different reasons: first, to preserve the dynamic demand that the radiation must be most intense along the normals to the front; but second, to permit oblique radiation to be dealt with as though it were very nearly as effective, in the right circumstances, as direct radiation. These circumstances, we shall now see, concern the inclination of the radiation to the line that joins the field point to the source, rather than the contiguity of the radiation to the edge (see fig. 6.1).

Fresnel's new way of thinking shifts the physically important point—the point that determines a region that can produce effective oblique radiation—from the edge of the diffractor (and so from the locus of the geometric shadow) to the line SP that joins the source and the field point. Hereafter I shall refer to the physically distinguished point R on the front as the "pole" with respect to the field point P. The core of Fresnel's new understanding is that the oblique rays that are sensibly inclined to SP annul one another (also see Fresnel 1819b, 297). Thus, for a given point P (fig. 6.1), construct

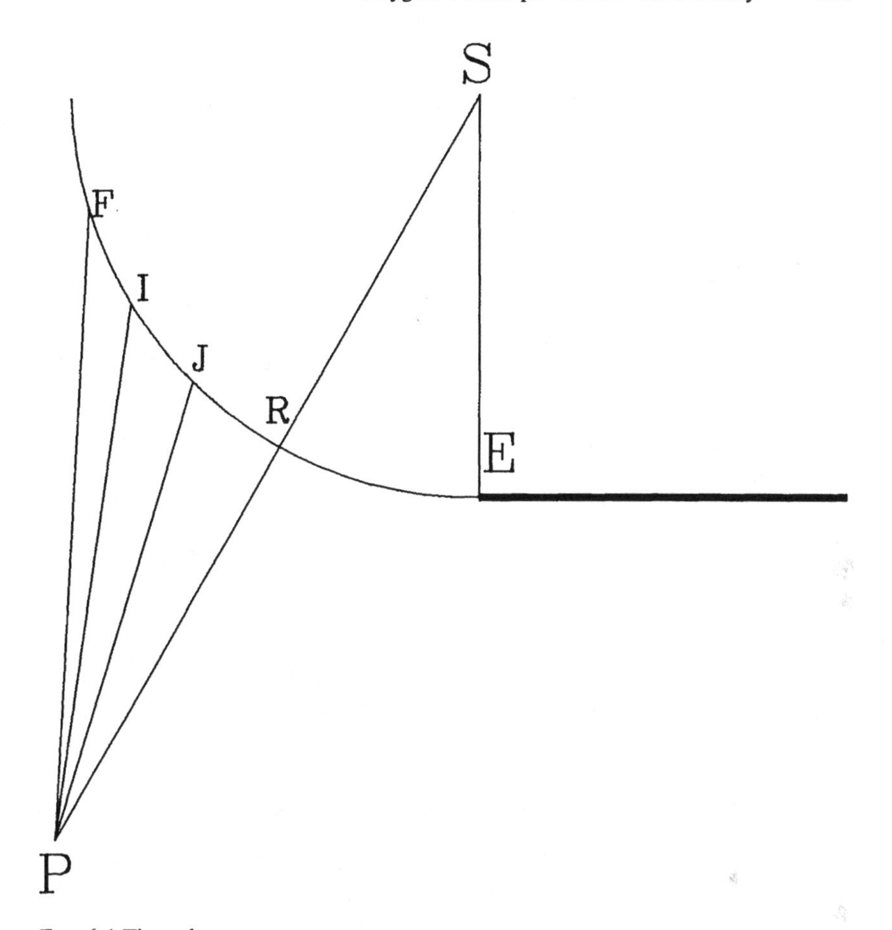

FIG. 6.1 The pole.

successive lines JP, IP, FP, . . . that differ from one another by a half-wavelength. Unless J, I, F, . . . are near the pole R, the arcs IJ, FI, . . . will be nearly the same length. And though they are sensibly inclined to SP, they nevertheless do radiate obliquely, albeit much more weakly than points closer to the pole. But since Fresnel presumed that the effect of an arc is proportional to its length, it follows that FI and IJ will destroy one another's radiation at P. Note the essential requirement that they must be far from the pole, not from the edge, in order to destroy one another. For unless they are far from the pole, these arcs will not have nearly equal lengths, and so they will not annul one another. In this argument, then, the purpose of the inclination factor is to make reasonable the assumption that oblique rays from points near the pole are nearly as effective as the ray from the pole itself. Its purpose here is to explain why oblique rays remain strong, not why they become weak at great obliquities.

The significance of the edge E of the obstacle now derives entirely from its proximity to the pole. If the edge is close to the pole, it blocks some of the rays on its side of the pole that are not sensibly inclined to the front normals—some of the radiation that actually produces the light at the field point—and so a fringe results. In other words, whereas the function of the edge had previously been to *permit* oblique radiation in its vicinity, now its function is to *obliterate* many of the effective oblique rays that would otherwise reach the field point. And so the farther away from the pole the edge is, the less effect there will be on the light at the field point, because it will block fewer effective rays (rays that are not sensibly inclined to the normal). Instead of determining a single point from which the only significant oblique radiation that occurs must be estimated, the edge now specifies a point beyond which radiation that would otherwise produce an effect no longer acts.

It is important to understand how completely Fresnel's understanding of this matter has changed, for with this new way of thinking he has thoroughly linked the principle of interference to Huygens's principle: the fringe pattern, he now thinks, is governed by the interference of rays that are emitted from every point of the front. Fresnel had indeed previously asserted that one can consider oblique radiation to be eliminated through the mutual interference of rays at all points of the front, but in his earlier understanding the interference was considered to occur between an arc on the front and its contiguous neighbors. Any interference between the arc and regions farther away on the front was irrelevant. Consequently, the whole purpose of the edge was to eliminate one of the contiguous neighbors and so to permit the arc to radiate obliquely. Except for this single arc at the edge of the obstacle, all the others on the front remained completely ineffective in producing oblique radiation. Indeed, the *only* other arc on the front he had thought optically important was the one that radiated directly to a field point. Consequently, as the field point moved about it received radiation from only two regions on the front: one remained fixed near the edge of the obstacle, whereas the other moved with the line joining the field point to the source. *Every other part of the front remained optically irrelevant, including the parts between the edge and the pole*—thereby making Huygens's principle itself irrelevant. In Fresnel's new understanding this has completely changed. The arc about the pole remains optically significant, but so now are the parts of the front between the pole and the edge. The edge therefore produces its effect by eliminating an entire section of the front instead of by eliminating a very small region, *for every point in this region is important in determining the light that reaches the field.* Far from being irrelevant, Huygens's principle is now the very essence of the phenomenon.

The only evidence from the period between Fresnel's invention of the efficacious ray and his replacement of the edge by the pole is the single note on diffraction (1818b) that already introduces wave front integration. Consequently we cannot know the precise course of Fresnel's thoughts. Nevertheless it seems quite probable that he continued to puzzle over the fact that the efficacious ray could not be used to retrieve the empirically successful formula for the external fringes produced by a narrow obstacle. This meant that one had, in effect, to change the locus of emission of the oblique ray that governs the interference pattern depending on whether one was calculating external or internal fringes. There was clearly no way to use a single indirect ray for both cases, so Fresnel may have begun to consider using more than a pair of rays. He would then have had to weaken his previous restriction that effective oblique radiation did not occur anywhere but near the edge. This may have led him to the idea of an inclination factor, which necessarily shifts attention from the edge to the pole. His task would then have been to rederive all his previous, successful results from this new scheme, which he does accomplish in the "Mémoire couronné." This must all have been done between July 1816 and, at the latest, early spring 1818.

6.2.2 Integrating over the Wave Front

In the next section I shall discuss whether Fresnel is likely to have examined methods of counting zones on the front in order to retrieve his original results before he turned to the much more complicated integral methods, with their higher accuracy. However the 1818 note itself turns at once to the more intricate theory, which posed a number of analytical difficulties. One of the major problems Fresnel faced involved choosing a usable coordinate system for integrating the radiation across an entire unobstructed front. It is not feasible to set a fixed system—one that is independent of the position of the field point—because, as we shall presently see, the integrands cannot then be reduced to readily manipulable forms. For example, one might naturally think first to set a fixed origin at one edge of an obstacle. But if one does so, then the integrands become extremely complicated functions for which Fresnel's method of wave decomposition has no analytical benefits. If we try to build a mathematical structure on the basis of Fresnel's original dynamic understanding that the obstacle edge is very significant, we reach an analytical dead end. Here, then, Fresnel's new focus upon the importance of the pole also had a direct mathematical benefit.

Suppose we choose the pole as the origin—an origin that will then vary with the position of the field point even though the optical configuration remains the same. Further, let us choose the phase of the front at the pole as a reference for employing Fresnel's wave decomposition. Then we have at once the solution to the problem of interference, because the path differences of the radiation from the front to the field point P had, in effect, already been computed by Fresnel in his very first theory, as we shall shortly see. That early result yields an expression for the difference between the distance of a point on the front to P and the distance of the pole to P; the expression is proportional to the square of the distance z along the front from the pole to the point in question. One can then immediately employ Fresnel's decomposition to obtain a set of sine terms and a set of cosine terms, the elements of each set differing among one another only in amplitude, not in phase. If, by contrast, we fixed the origin at an edge and computed path differences in this coordinate system, we would not obtain expressions that differ solely by quadratic terms in the distance to the origin, and then Fresnel's decomposition would lead to an impossibly complicated integrand.

I shall for a moment step aside from Fresnel's own very simple derivation (see below) to consider a more intricate but revealing expression than the one he himself obtained. Consider a source S, a diffractor edge E, and a field point P (see fig. 6.2). The front (which I take to be plane for simplicity) is EQC. Take the origin of coordinates at an arbitrary point O′. To find the resultant at P we must consider the various paths $r + s$ that pass through the point Q on the front (Born and Wolf 1975, 382–83). By construction:

$$\left.\begin{aligned} r^2 &= r'^2 + \xi^2 - 2x_0\xi \\ s^2 &= s'^2 + \xi^2 - 2x\xi \end{aligned}\right] \quad \begin{aligned} \mathrm{Q} &= \mathrm{Q}\,(\xi,0) \\ \mathrm{S} &= \mathrm{S}\,(x_0,z) \\ \mathrm{P} &= \mathrm{P}\,(x,z) \end{aligned}$$

Expanding in powers of ξ/r' and ξ/s' we have:

$$r \approx r' - \frac{x_0\xi}{r'} + \frac{\xi^2}{2r'} - \frac{x_0^2\xi^2}{2r'^3}$$

$$s \approx s' - \frac{x\xi}{s'} + \frac{\xi^2}{2s'} - \frac{x^2\xi^2}{2s'^3}$$

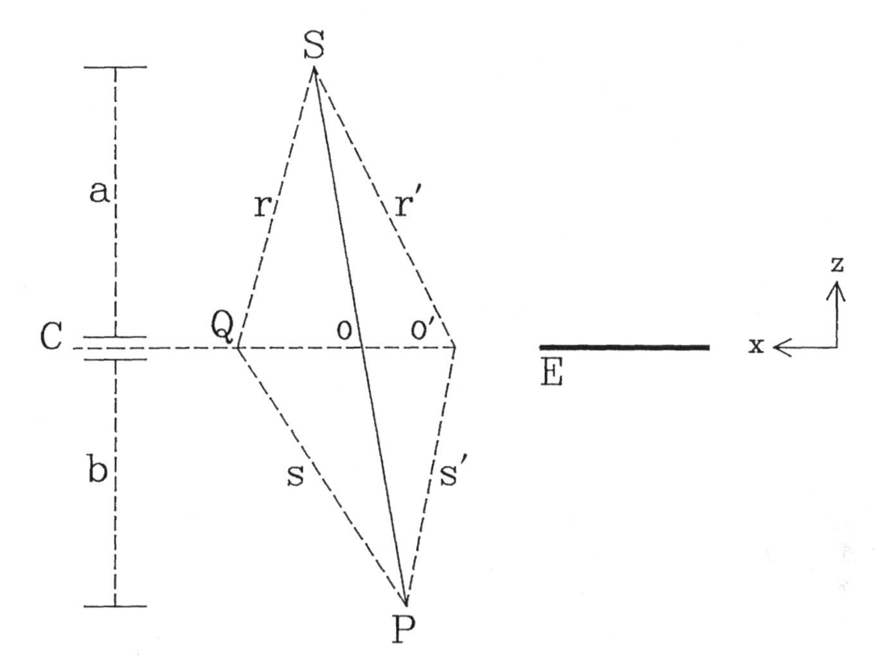

FIG. 6.2 Ray paths and the pole.

Whence if we set $-x_0/r'$, x/s' respectively equal to l_0, l we have:

(6.2.1)
$$(r + s) - (r' + s') = (l_0 - l)\xi + \frac{1}{2}\left[\left[\frac{1}{r'} + \frac{1}{s'}\right]\xi^2 - l_0^2\frac{\xi^2}{r'} - l^2\frac{\xi^2}{s'}\right]$$

Now the magnitudes of l_0 and l are, respectively, the magnitudes of the cosines of the angles made by SO′ and PO′ with respect to the x axis (here the line EC). If the origin, O′, is arbitrarily located—or even placed at the edge E—then the path lengths $r + s$ differ among one another by the right-hand side of 6.2.1, which contains terms that are linear as well as quadratic in ξ. This makes even numerical integration impossibly difficult. Clearly the linear terms must be removed, and they will be if the origin is set at the pole, here O, because l_0 and l are then equal to one another, yielding:

$$(r + s) - (r' + s') = \frac{1}{2}\xi^2\left[\frac{1}{r'} + \frac{1}{s'}\right](1 - l^2)$$

But l^2 is also the square of the sine of the angle between SP and the z axis, so that $1 - l^2$ is the square of the cosine of this angle. If we limit ourselves to field points near the normal, as Fresnel did, then this factor is nearly unity, and we finally obtain:

$$\delta = (r + s) - (r' + s') = \xi^2 \frac{r' + s'}{2r's'} \tag{6.2.2}$$

This is precisely Fresnel's expression for the path differences, though he did not obtain it in this way—if he had, he would also have obtained the inclination factor in the form $1 - l^2$.

Fresnel obtained equation 6.2.2 almost at once from an expression he had derived in his very first theory once he realized the necessity of referring the scheme to the pole and not to the edge. In figure 6.3 we have a source C, a field point P, a semi-infinite screen AG, and a primary front AmMm′. If we take the origin at M, then the path differences m′s′ on either side of M and referred to MP as a standard length are the same functions of distance z along the front from M. From Fresnel's deduction of his first formula (5.2.1) for the external fringes in obstacle diffraction we have:[3]

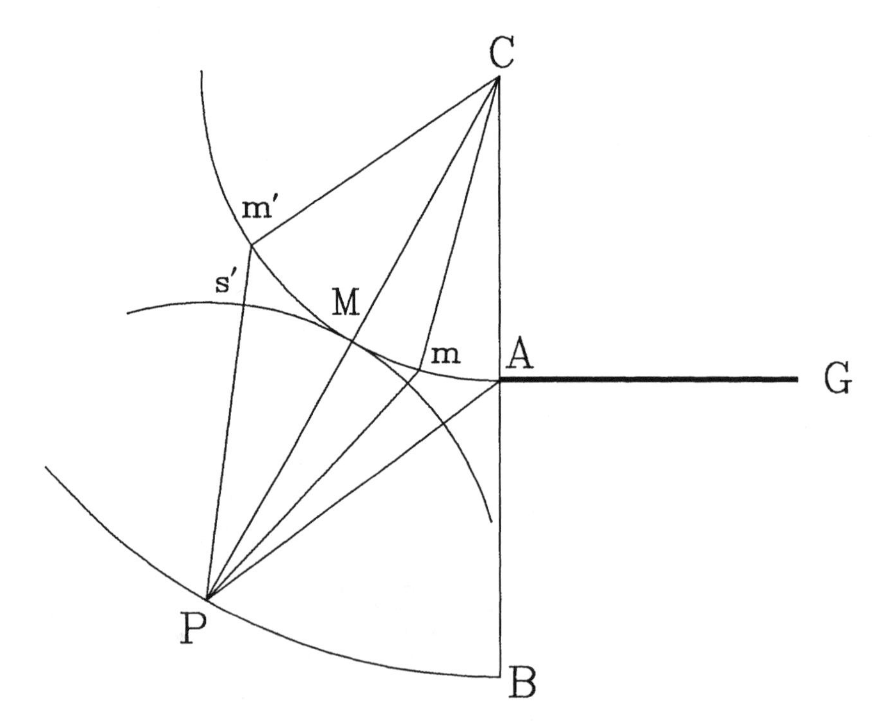

FIG. 6.3 Fresnel's path differences.

(6.2.3)
$$m's' \approx \frac{z^2(a + b)}{2ab}$$

This is just 6.2.2: a is the source and b is the screen distance. (In the April note Fresnel did not give the proportionality factor in 6.2.3, though he knew it.)

Given equation 6.2.3, Fresnel could at once deploy his quarter-wave decomposition. Making explicit what he left implicit in the note, consider the general expression for the amplitude at P due to a secondary from m′ (ignoring a common phase $-2\pi(MP/\lambda)$):

$$\sin\left[2\pi\left[t - \frac{CM + m's'}{\lambda}\right]\right]$$

$$= \sin\left[2\pi\left[t - \frac{CM + m's'}{\lambda}\right] - \pi z^2 \frac{a + b}{ab\lambda}\right]$$

Decompose this into two waves that differ in phase by 90°, but one of which has the same phase as the direct wave from M:

(6.2.4)
$$\overbrace{\cos\left[\pi z^2 \frac{a + b}{ab\lambda}\right]}^{\text{amplitude}} \underbrace{\sin\left[2\pi\left[t - \frac{CM}{\lambda}\right]\right]}_{\text{same phase}}$$
$$+ \overbrace{\sin\left[\pi z^2 \frac{a + b}{ab\lambda}\right]}^{\text{amplitude}} \underbrace{\sin\left[2\pi\left[t - \frac{CM}{\lambda}\right] - \frac{\pi}{2}\right]}_{90^\circ \text{ phase dif.}}$$

We now have an infinity of such pairs to consider; to count them we assume with Fresnel that the length dz of an element of the front measures the number of wavelets in it. With this the problem of diffraction has been essentially solved, since the square of the resultant amplitude is just the sum of the squares of the quarter-wave components:

$$\left[\int dz \sin\left[\pi z^2 \frac{a + b}{ab\lambda}\right]\right]^2 + \left[\int dz \cos\left[\pi z^2 \frac{a + b}{ab\lambda}\right]\right]^2$$

The limits of integration differ for each screen point P, since the origin M moves with the screen point. This has the unfortunate but unavoidable result, as Fresnel rapidly learned, that it is impossible to solve a diffraction problem generally unless there is only one finite limit involved. This is true for the semi-infinite plane, which Fresnel was therefore able to solve generally.

Here the limits run from the edge of the diffractor to the pole M, and from the pole to infinity. Fresnel knew that $\int_0^\infty {\sin \atop \cos}(\frac{1}{2}\pi z^2)$ are both $\frac{1}{2}$.[4] If the edge is at u relative to M, then the square of the amplitude at P is proportional to:

$$\left[\frac{1}{2} + \int_0^u dz \sin\left(\frac{1}{2}\pi z^2\right)\right]^2 + \left[\frac{1}{2} + \int_0^u dz \cos\left(\frac{1}{2}\pi z^2\right)\right]^2 \tag{6.2.5}$$

Even though u depends upon the position of P, Fresnel could nevertheless obtain general expressions for arbitrary values of u and then substitute the appropriate values to obtain the correct results for a given configuration. Were there two such limits—as there are in single-slit and in narrow obstacle diffraction—this would not be possible.

The difficulty now was to evaluate these integrals, since eponymously known as the "Fresnel integrals." Fresnel noted that one could employ either series or "partial" (i.e., numerical) integration methods. Using the latter (a form of the series—see appendix 3—was subsequently obtained by Cauchy), Fresnel tabulated the integrals for values of u from 0 to 5.1 in steps of 0.1. He did not describe his method here, but it was no doubt the same one he shortly employed in his prizewinning memoir (Fresnel 1819b, 317). In essence Fresnel considered the integrals

$$\int_0^{a+2p} {\sin \atop \cos}(gv^2)dv$$

for small p, with g a constant. He replaced v with $u + p + a$ to change the limits of integration (now taken over u) to $-p$, $+p$, expanded the trigonometric expressions by identities, neglected a term in p^2, and then integrated. This method—which Emile Verdet discovered in Fresnel's papers—leads in the end to the following formulas for numerical integration, which are accurate up to $\frac{1}{4}t^2$, where t is small:

$$\int_i^{i+t} dv \cos(gv^2)$$
$$= \frac{\sin\left[g\left(i + \frac{1}{2}t\right)\left(i + \frac{3}{2}t\right)\right] - \sin\left[g\left(i + \frac{1}{2}t\right)\left(i - \frac{1}{2}t\right)\right]}{g\left(i + \frac{1}{2}t\right)}$$

$$\int_i^{i+t} dv \sin(gv^2)$$
$$= \frac{-\sin\left[g\left(i + \frac{1}{2}t\right)\left(i + \frac{3}{2}t\right)\right] + \cos\left[g\left(i + \frac{1}{2}t\right)\left(i - \frac{1}{2}t\right)\right]}{g\left(i + \frac{1}{2}t\right)}$$

With this we encounter Fresnel's prodigious computational ability. Using these formulas, I have recalculated Fresnel's tables with a machine, and I find that his error amounts to a mean of only 0.0003. (The integrals have values between 0 and 1.) The difference between Fresnel's values and more accurate ones computed using Cauchy's series is about twice as large, but still only 0.0006, an accuracy that, we shall see, permits experiments to be calculated to a very high degree of exactitude indeed.

To compare theory with experiment, Fresnel had to find the maxima and the minima of the intensity curve plotted against the locus of the field point P with respect to the center of the geometric shadow or some other fixed point. These can be found for the semi-infinite plane rather simply by using Fresnel's tables of integrals, and he accordingly did so. First examine the table to find the extrema for the entries in it, which gives them to an accuracy of 0.1 in the variable u (equation 6.2.5). This is not good enough. To obtain higher accuracy, Fresnel added to u a small arc t and extremised the resulting integral (Fresnel 1819b, 321). If, he found, A and B are the respective cosine and sine integrals for an extremum at the tabular value u, then the more accurate extremum is situated at $u + t$ where:

$$\sin\left(\frac{1}{2}(u^2 + 2ut)\right)$$
$$= \frac{\pi u A - \sin\left(\frac{1}{2}\pi u^2\right)}{\sqrt{\left[\pi u - \sin^2\left(\frac{1}{2}\pi u^2\right)\right]^2 + \left[\pi u B + \cos^2\left(\frac{1}{2}\pi u^2\right)\right]^2}}$$

Fresnel computed the first four extrema for the 1818 note. Again using his formulas, I find his accuracy in computation to be (with the exception of the first minimum, which he corrected in his final diffraction memoir) within 0.0002 in the value of u, which is about the same as his accuracy in computing the table itself.

We can now determine the error in millimeters that Fresnel's computational error of about 0.0003 in u leads to. The distance of P from the shadow center is:

$$u\sqrt{\frac{b\lambda(a+b)}{2a}} \tag{6.2.6}$$

Using typical values of a, b and from Fresnel's early experiments, I find that an error of 0.0003 in u leads to an error in the locus of only 0.0004 mm—vastly smaller than Fresnel's observations ever reached. Consequently his computational accuracy was, as he knew, unquestionably well adapted to his observations.

In the note Fresnel used equation 6.2.6 to show why his old formula for the external fringes, derived from the binary theory, had been reasonably accurate. That formula gave for the first minimum $\sqrt{[2b\lambda(a+b)/2a]}$. Fresnel's integral computation gives 1.899 for u in equation 6.2.6 at the first minimum, so that the old formula was accurate to within 0.101 in the value of u. This yields in general an error of about 0.1 mm, which was just about the observational accuracy of the first experiments.

So we see that the combination of Huygens's principle with the principle of interference solves the major problem that had been posed by the efficacious ray—namely, to retrieve, within the limits of observational accuracy, the original formula for the external fringes in obstacle diffraction. But thus far Fresnel had not considered a wire or a slit, and the internal fringes produced by them under certain conditions were the original reasons for his having introduced the efficacious ray. In the note Fresnel mentioned that he had been able to deal successfully with the slit, but he gave few details there. They were provided in his prize memoir, and they constitute the major empirical triumph of the new theory.

6.3 The Prize Memoir

6.3.1 The Announcement

Scarcely three months elapsed between the deposition of the curt note at the Academy on 20 April and Fresnel's submission of a major manuscript on 29 July that, with the addition of two notes he later appended, was published in 1826. This memoir won Fresnel an Academy prize and contains a full working out of his integral diffraction theory, though it was essentially complete in the April note. Considering the vast scope and the detail of this "Mémoire couronné," Fresnel must already have worked out most of it, and have performed many of the experiments it describes, before the spring of 1816. Recall that he had first presented his quarter-wave decomposition, which is crucial to the integral theory, on 19 January of this same year. Most likely, then, he did much of the work on the integral theory in the fall of 1817 and the winter of 1817–18. The immediate stimulus to this intense, concentrated work was the prize competition itself.

On 17 March 1817 the Academy publicly announced that it had decided to offer a prize for a memoir on diffraction, the closing date for the competition being 1 August 1818 (Fresnel, *Oeuvres,* 1:xxxv–xxxvii [Verdet's introduction]). There is little doubt that the hope, especially on the parts of two members of the committee—Biot and Laplace—was for an empirically successful theory of diffraction that avoided wave principles (Frankel 1976, 159–62). A week before the official announcement (6 March) Fresnel's uncle Léonor wrote him a letter that nicely captures the tone of the discussions that led to the formulation of the prize question:

> You should have received at least fifteen days ago a letter from your defender Arago, who encountered me as he came from enduring a hard battle with the *émissionaires,* who found it appropriate to put the *diffraction of light* back into question, and [who] have proposed a prize for whoever would best explain it according to the doctrine they have adopted. Arago, caught up short, took the enemy attack head on, called his [supporters], and managed to stop the invasion; that is to say, he obtained that your memoir would be mentioned in the program.
>
> He at first thought that you should not descend into the arena, but publish in the *Annales* everything you find to be new, so that in the report on the prize one can say: None of the contestants solved the problem. . . .
>
> Yesterday I saw Ampère, who asked me news of you and strongly told me to write you to put yourself in the ranks and to send your memoir to the contest, with the new observations that you have made and that you may yet make. "He will easily win the prize," he said to me; "for him and for the cause he must compete."
>
> I made some objections, founded on the partiality of the commissioners if they were chosen from the section of *Biotistes.*—Ampère replied that there was nothing to fear, that General Arago would not miss, when the commissioners are nominated, making known the impropriety of nominating party men, and that what will happen is what always happens whenever one warns the Republic that citizen Laplace wishes to dominate. (Fresnel, *Oeuvres,* 2:841–82)

Arago did succeed in having the contest question mention Fresnel's work, and he also succeeded in having both himself and Gay-Lussac placed on the board. But he did not succeed in having the question formulated in a way suited to the kind of answer Fresnel could submit. It is worth reproducing in translation the full question that the Academy posed to capture the commission's intensions:

> The phenomena of diffraction, discovered by Grimaldi, and subsequently studied by Hooke and Newton, have lately been the object of research by many physicists, notably Young, Fresnel, Arago, Pouillet, Biot, etc. They observed the diffracted bands that form and propagate outside the shadow of bodies, those [bands] that appear within the shadow proper, when the rays pass simultaneously the two sides of a narrow body, and those [bands] that form by reflection on the surfaces of a limited region, when the incident light and reflected light pass very near their edges. But as yet no one has sufficiently determined the motions of the rays near the bodies proper, where their inflection occurs. The nature of these motions therefore today provides that aspect of diffraction that is of the most consequence to deepen, because it contains the secret of the physical mode by which the rays are inflected and separated into diverse bands of unequal directions and intensities. This is what determined the Academy to propose this research as a prize subject, setting it forth in the following manner:
>
> 1. To determine by precise experiments all the effects of the diffraction of direct and reflected luminous rays, when they pass separately or simultaneously near the extremities of one or more bodies, limited or indefinite in extent, having regard to the distances between these bodies as well as to the distance of the luminous focus from which the rays emanate.
> 2. To conclude from the experiments, by mathematical inductions, the motions of the rays in their passage near bodies.
>
> The prize will be awarded in the public session of 1819, but the contest will be closed 1 August 1818, and so the memoirs must be submitted before this date, so that the experiments they will contain may be verified. (Fresnel, *Oeuvres,* 1:xxxvi–xxxvii [Verdet's introduction])

If Fresnel had taken these questions literally he could not have submitted a memoir, because the questions are phrased in a way that presupposes that rays retain their identities as they "pass" by obstacles. Even in Fresnel's original, binary theory this did not occur, because the inflected rays were generated at the diffracting edge proper—they did not pass by it. The efficacious rays were supposed by Fresnel to arise only near the diffracting edge as a result of its having upset the dynamic equilibrium in the front; so here too the committee's way of phrasing the question did not fit. It fit Fresnel's integral theory even less well, because in it rays are defined exclusively with reference to the wave front—there is no remaining implication or assumption that rays are useful physical entities in their own right. The committee's questions therefore went beyond selectionism, in which rays are the elements of analysis, to impose the extra condition that the rays must all come from the source, that they are not subsequently generated. That might easily (though not necessarily) be true if rays consisted of particles, so the questions were implicitly asking for an account that would support the emission theory and not just a theory based on rays.

Fresnel was pressed by Arago and Ampère to compete (Fresnel, *Oeuvres,* 1:xxxvii)—and they may have had to press him to do so because the very form of the committee's questions did seem to prejudice the issue. To avoid any possible preemption this time, Fresnel submitted to the committee the brief note that contained the essence of the integral theory. The committee consisted of Laplace, Biot, Poisson, Arago, and Gay-Lussac, and it received only one entry besides Fresnel's, whose author was never revealed.[5]

6.3.2 Criticisms of the Emission Theory

The "Mémoire couronné" (hereafter referred to as M.C.) contains little new theoretical material. However, it brings together Fresnel's several researches on diffraction and presents new, accurate, and detailed experimental results in confirmation of the integral formulation, results that reached an observational accuracy of 0.001 mm. The M.C. also details several important computational methods that permit the integral formulation to be applied to narrow obstacle and to aperture diffraction.

Fresnel began the M.C. with an attempt to convince the committee, the majority of whom were, in the slang of the period, *émissionaires* (partisans of the emission theory), that their theory could not be correct (Fresnel 1819b, 247–61). The thrust of his criticism rested on the grounds that the emission theory does not have sufficient latitude to encompass the phenomena without Draconian measures:

> In the emission system [unlike the wave theory] the progress of every luminous molecule being independent of that of the others,

> the number of diverse modifications to which they are susceptible seems to be extremely limited. One can add a motion of rotation to that of transmission; but that's all. As for oscillatory movements, their existence is conceivable only within media that would maintain them by an unequal action of their parts on the different sides of the luminous molecules, supposing [these sides] endowed with different properties. As soon as this action ceases, the oscillations must also cease or change into motions of [continuous] rotation. So the motion of rotation and the diversity of faces of a given luminous molecule are the only mechanical resources that the emission theory has to represent all of the permanent modifications of light. They seem very insufficient, if one pays attention to the multitude of phenomena that optics offers. One will be even more convinced of this on reading Biot's treatise on experimental and mathematical physics, in which the principal consequences of Newton's system are developed with much detail and clarity. We will see that, in order to explain the phenomena, it is necessary to accumulate a very large number of diverse modifications on each luminous particle, modifications that are often very difficult to reconcile with one another. (Fresnel 1819b, 250–51)

His brief for the wave theory proper rested on several foundations, beginning with what he claimed was its conceptual simplicity. In this he did not include mathematical simplicity, since the emission theory, being founded on particle mechanics, is in principle much simpler analytically than the wave theory. However, Fresnel argued, "In the choice of a system one must consider only the simplicity of the hypothesis; that of computation can have no weight in the balance of probabilities" (Fresnel 1819b, 248). Many phenomena, he continued, that in the emission theory require multiple, intricate hypotheses are very simply explained in the wave theory.

Neither these nor any other of Fresnel's several arguments for the wave theory and against the emission theory were taken to be conclusive by many, perhaps by most, *émissionaires*. It was, as Fresnel well understood, usually possible to produce emission explanations for most, and perhaps all, phenomena then known. In Fresnel's eyes these explanations—of the sort presented by Biot in his *Traité*—were complicated, improbable, and inadequate to completely explain the phenomena. Nevertheless explanations were available, so that Fresnel's brief for the wave theory had in the end to rest on the grounds of economy and fecundity: that only the wave theory could accommodate the phenomena without introducing many gratuitous hypotheses, and that only it gave rise to new predictions and to quantitative formulations. These argumentative sections of the M.C. were not instrumental in winning Fresnel the prize. That was achieved by his massive theoretical analysis buttressed by precise experimental detail.

6.3.3 Zones of Equal Length

The bulk of the M.C. is concerned directly with the wave theory and proceeds almost pedagogically, no doubt because Fresnel assumed, probably correctly, that most of the committee would be unfamiliar with wave methods. First Fresnel introduced the principle of interference, linking it to the old binary ray theory, which he then applied to the narrow diffractor, where he pointed out the necessity in this approach of assuming a 180° phase change on inflection. Here he remarked the close agreement between theory and experiment concerning the general features of the phenomenon (hyperbolic propagation and dependence of the pattern upon the distance of the source—which posed a major problem for the emission theory).

He then demonstrated the inadequacy of the binary theory by discussing the experiment he had performed in 1816 with the slit metal plate (see sec. 5.5, fig. 5.15). Recall that in 1816 Fresnel had used this experiment to show the superiority of the efficacious ray to the binary theory. Here, in the M.C., he repeated nearly literally his discussion of the observations and his explanation of why they show the failure of the binary theory. However, instead of introducing the efficacious ray as he had in 1816, which he had at that time argued could alone explain the facts, he concluded instead that:

> It follows from the experiments I just reported that we cannot attribute the phenomena of diffraction only to the rays that touch the edges of bodies, and that one must admit that an infinity of other rays separated from these bodies by sensible intervals nevertheless find themselves diverted from their original directions and also concur in the formation of the fringes. (Fresnel 1819b, 277)

Here, then, an experiment that two years before had been used only to displace the sole ray involved in producing a fringe from the edge of the diffractor to a point near it was now invoked in support of there being an "infinity" of rays involved.

Having demonstrated the empirical inadequacy of, he argued, both the emission theory and the old binary ray theory, Fresnel turned in the second section to Huygens's principle, which he prefaced with an exposition of his method for compounding coherent waves of different amplitudes and phases.

He introduced Huygens's principle with the following words: "The vibrations of a luminous wave in each of its points may be regarded as the sum of the elementary motions that would be simultaneously sent there by each of the parts of that wave, acting in isolation, considered in any of its anterior positions" (Fresnel 1819b, 293). Fresnel seems by this time to have regarded Huygens's principle as a method of decomposition for waves that is closely akin to what one does in projecting static forces along various axes. Beginning with an unobstructed wave, which always remains evenly illuminated, Fresnel envisioned decomposing it into secondary wavelets according to the principle. As in a decomposition of equilibrated static forces, the resulting system is fully equivalent to its undecomposed progenitor. This was, to Fresnel, more than an analogy, because it is precisely when equilibrium does not subsist that Huygens's principle is directly useful—when, in diffraction, an obstacle destroys the balance that subsisted in the unobstructed wave. In statics removing a support destroys the equilibrium and so permits forces that were previously in balance to produce motion. In optics the obstruction of a wave also destroys an equilibrium and thereby permits interference to produce fringes (see footnote * to Fresnel 1819b, 295).

Again Fresnel faced the question of the unknown inclination factor, which in the end he had to justify removing from the computations. His argument in the M.C. differs little from his statement in the April note. However, in a footnote added in press (Fresnel 1819b, 296–97), Fresnel effectively assumed that the factor varies as the cosine of the angle to the normal. Thus at 90° there is no radiation at all, whatever the dynamic balance, or lack of it, along the front may be. To forbid retrograde radiation he recurred again to dynamics, as we shall see below when we discuss his controversy with Poisson over Huygens's principle.[6]

In the thirteen sections that follow the enunciation of Huygens's principle but precede his turning to integrals, Fresnel discussed his old concept of the efficacious ray and applied it to the internal fringes of the narrow obstacle, where it works well, though no better than the binary theory, since here they give the same result (Fresnel 1819b, 299–300). These pages may provide some clues to the actual route Fresnel had followed in moving from the efficacious ray to Huygens's principle.

Fresnel knew that the efficacious ray does not work for external fringes. But perhaps surprisingly, he does not explain why it fails here. In fact he does not even apply the efficacious ray to the internal fringes of a narrow slit, which is why he had invented it in the first place. In other words, Fresnel did not recapitulate his own history in the M.C. except to demonstrate the insufficiency of the old binary theory—a theory first developed by Thomas Young. It is simple enough to understand why Fresnel did not discuss the failure of his second theory: Why emphasize for a prize competition something that he alone had invented and that had not worked well? He of course knew why the efficacious ray cannot work for external fringes, since using it alone eliminates from consideration a great deal of light that strongly influences the diffraction pattern (viz., the remaining light between the edge and the pole). But what is rather surprising is that Fresnel did not deploy his method of dividing the wave front into half-wavelength zones in order to generate formulas for external fringes.

And yet Fresnel did here introduce true zone computations for the first time (Fresnel 1819b, 302–5). He had not discussed this "simple geometric" method, as he termed it, in the April note—though he certainly must have had it well before then. The method is in principle quite simple, and it can be universally applied. One determines whether a field point P is a locus of a dark fringe or of a bright fringe by counting whether, from P, an odd or an even number of half-wave zones are visible on the front.

Fresnel first applied the method to explain the pattern of internal fringes produced by a very narrow slit—the pattern that had originally led him to the efficacious ray. As I remarked above (chap. 5), one can readily understand what occurs in this situation by means of the following argument, which Fresnel gives for the first time in the M.C. (1819b, 300–301). In figure 6.4 S is the source, P is the screen point, and AG is a very narrow slit. Fresnel writes:

> I suppose [AG] first to be sufficiently narrow that the dark bands of the first order are within the interior of the geometric shadow of the screen and sufficiently far from the edges B and D. Let P be the darkest point of one of these bands; it is easy to see that it must correspond to a difference of a wavelength between the two extreme rays AP and PG. In effect, if we conceive another ray PI taken in such a way that its length is halfway between that of the two others, in consequence of their pronounced obliquity to the arc AIG, the point I will be nearly in its middle. This arc will therefore find itself composed of two others, whose corresponding elements are sensibly equal, and will send to the point P contrary vibrations, which will in consequence destroy one another. (Fresnel 1819b, 300–301)

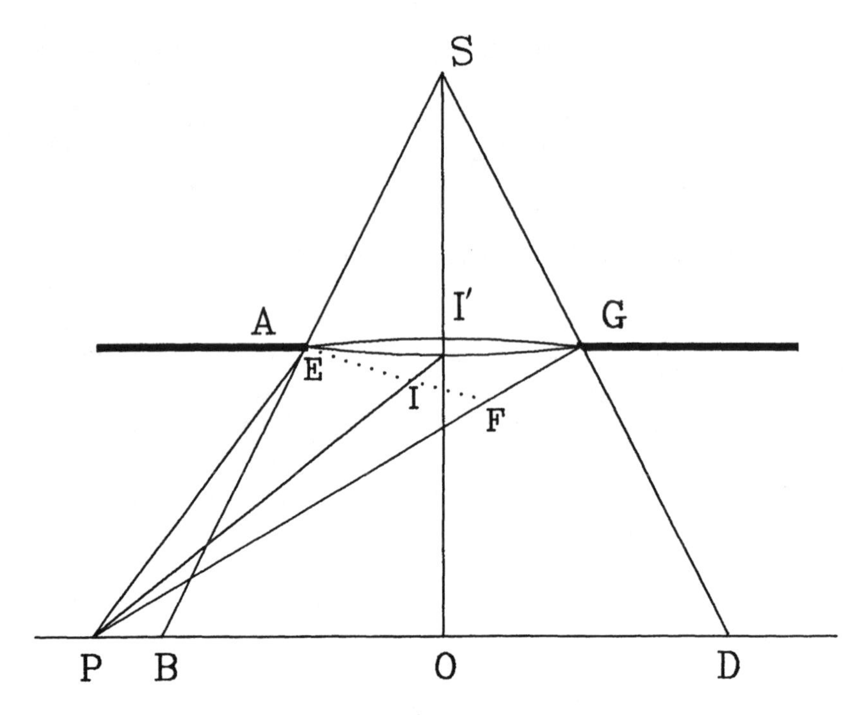

FIG. 6.4 Fresnel's zones and the single slit.

But Fresnel did not go on to apply the method to external fringes. Moreover, he argued that it does not work even here if the field point is so close to the edge of the geometric shadow that the rays from the front are not strongly inclined to the front AIG. It no longer works because in these circumstances "the corresponding elements of the arcs into which we have supposed [AG] to be divided can no longer be considered to be equal to one another, but are sensibly larger on the side that is closer to the [fringe]." In Fresnel's understanding, then, one can add zones together *only* if they have very nearly the same lengths. If they do not then they cannot compensate one another.

His views are very nicely illustrated by what he does next. Admitting that the method cannot be used when, for example, the slit is rather large, Fresnel conceives the idea of using a lens to transform the front in such a fashion that the zones on the transformed front will all be very nearly the same length. In Fresnel's figure (fig. 6.5) the front is transformed by a lens into AI′G, which has center O. I′ is the intersection of SO with the transformed front. From A and with P as center, Fresnel drew arc AEL. Since, he noted, AEL and AI′G are both circles that are concave toward PO, then the path differences

I′P−AP, GP−I′P will be equal to one another *only* if the corresponding arcs AI′, I′G are also mutually equal. If we set the path differences all equal to a half-wavelength, then we have a succession of half-wave zones on the front *that are equal in length.* Whence we can, Fresnel reasoned, simply add them up in pairs to determine the effect at P.

The reason Fresnel had to transform the front by refraction in order to apply his method of zone counts to a large slit was that, in his treatment, only a zone's *length* determines its effect. As a result his method seems to be extremely limited in its application: it requires equal-length zones, and this means that in most circumstances the front must be transformed in some fashion. Now, instead of following Fresnel, assume for a moment that the amplitude a zone produces at a distance from it is indeed directly proportional to its length, but that it is also inversely proportional to its distance. Then the method of zone counting becomes fully general. Suppose that the boundaries of some zone are at distances r and $r + \delta$ from the field point; then the length

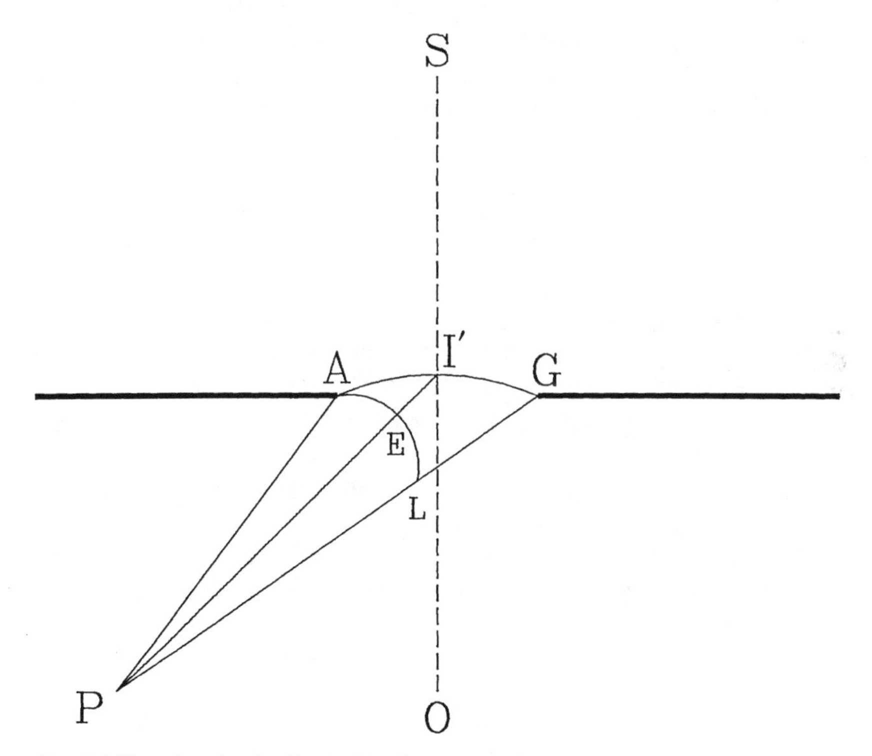

FIG. 6.5 Transforming the front with a lens.

of this small arc on the front will be nearly $r\delta$. Consequently the amplitude due to the zone at the field point will vary as $r\delta/r$: it will therefore be independent of r. All zones with the same δ produce the same effect, and so we can at once proceed to determine the net effect by a simple zone count.

One might be surprised that Fresnel did not follow what seems to be an obvious procedure: he had only to know that the amplitude falls off reciprocally with distance from the source. And he assuredly did know this, since he was well aware that what we term the wave's intensity, and that he assimilated to its vis viva, or kinetic energy, falls off reciprocally with the square of the distance. However, it is far from obvious that Huygens's wavelets behave precisely like the wave fronts they conspire together to produce. The amplitude in any direction about the front is the same at a given distance from the source; it varies with inclination to the front's normal for a wavelet. Fresnel may at this time simply have decided not to consider the dependence of amplitude on distance for a secondary because he was not fully confident of the propriety of treating wavelets like fronts in this respect.

As a consequence Fresnel was *not* able to provide a "geometric" account of the internal fringes that are formed without a lens outside, but near to, the illuminated region produced by a comparatively large slit. Nor was he able to produce such an account for the external fringes produced either by a slit (within the geometrically illuminated region) or by an obstacle (outside the geometric shadow). He simply did not have a method for doing so, because he demanded that zones can be compounded only if they have equal lengths, which they will certainly not have whenever external fringes are produced. Instead of simply abandoning this situation, Fresnel chose to pose a different question: Given a pair of different slits, or a pair of different semi-infinite screens, how must the dimensions of the configurations be related among the members of a given pair in order to produce a fringe of a given order? This problem could be solved "geometrically" without counting zones simply by requiring that the distances between the rays that interfere in each configuration must be the same in both (Fresnel 1819b, 306–12).

Fresnel, then, did not consider his zone method fully general (which we shall again see below), and in particular he did not use it to derive the original binary formula for the obstacle—the one he knew worked reasonably well. In view of this we must, I think, conclude that Fresnel probably turned to integral methods almost from the moment he conceived the idea that permitting radiation from all points on the front might produce a consistent theory. For he could not otherwise have known that this hypothesis would produce an empirically accurate formula for the external fringes.

6.3.4 Applying the Integrals to the Single Slit

We need not examine Fresnel's integral analysis in the M.C. of the semi-infinite plane, since it is essentially the same as in the April note. Using (red) light with a wavelength of 6380Å, Fresnel tabulated the differences between theory and experiment for minima of various orders, obtaining a mean difference between the two, I find, of only 0.007 mm (Fresnel 1819b, 332–35). Since I also find his computations here to be correct to about 0.0002 mm, his observational accuracy evidently reached an astonishing 0.0066 mm, which at once permitted him to show that the old binary formula for the external fringes was unacceptable because it deviates from experiment by about a tenth of a millimeter (a fact that even a zone calculation could not have shown, since it leads to the old formula).

The semi-infinite screen posed a comparatively simple problem because it involves only one limit. The narrow obstacle and single slit, on the other hand, involve two limits. For these cases Fresnel accordingly invented a method of computation that is simple in principle but requires a vast amount of calculation. Consider the solution for the single slit (fig. 6.6). The geometric shadow at S occurs at a distance $c(a + b)/2a$ from O, the center of the geometrically illuminated region. Then with the origin at the pole M, Fresnel's solution is:

$$\psi^2 = [-C_A + C_B]^2 + [-S_A + S_B]^2$$

$$C_{A,B} = \int_0^{A,B} \cos\left(\frac{\pi(a + b)z^2}{ab\lambda}\right) dz \qquad (6.3.1)$$

$$S_{A,B} = \int_0^{A,B} \sin\left(\frac{\pi(a + b)z^2}{ab\lambda}\right) dz$$

Introducing the change of variable v equal to $z\sqrt{[2(a + b)/ab\lambda]}$, we have the new limits v_A, v_B instead of A, B:

$$C_{A,B} = \int_0^{v_{A,B}} \cos\left(\frac{1}{2}\pi v^2\right) dv \qquad \left[\begin{array}{l} v_A = \overline{MA}\sqrt{\dfrac{2(a + b)}{ab\lambda}} \\ v_B = \overline{MB}\sqrt{\dfrac{2(a + b)}{ab\lambda}} \end{array}\right.$$

and similarly for $S_{A,B}$. The computational problem is that the difference $v_B - v_A$ depends upon a, b, c, and λ, so no general solution is possible.

Nevertheless, since for any given set of a, b, c, and λ the difference $v_B - v_A$ (hereafter δ) is always $c\sqrt{[2(a + b)/ab\lambda]}$, we could first set v_B and

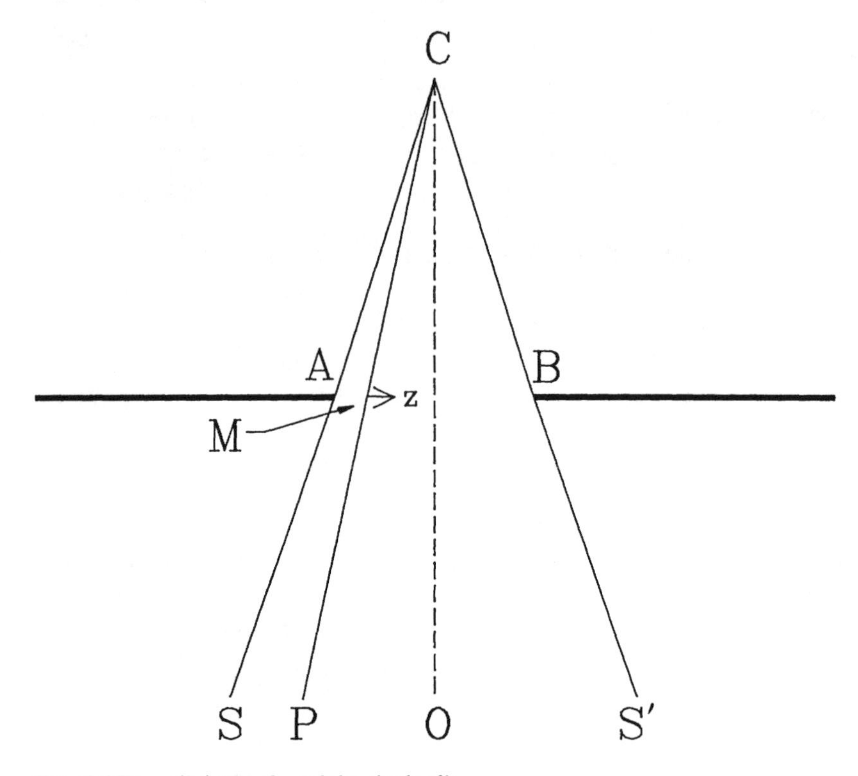

FIG. 6.6 Fresnel's integrals and the single slit.

at once obtain v_A. This was apparently Fresnel's idea, though his discussion is cryptic. Consider points within the geometrically illuminated region, where v_A, v_B have opposite signs. Begin with v_B at the right aperture edge, where it vanishes. This locates screen point P at the right geometric shadow S′. Then compute v_A as $v_B - \delta$. Fresnel's tables give the integrals at intervals of 0.1 in v, so we need only look up the nearest v'_A, v'_B to our computed v_A, v_B. Then we correct these values for the differences $v'_A - v_A$, $v'_B - v_B$ using the approximations ($v' = v + t$):

$$C(v') - C(v) \approx \frac{1}{\pi v}\left[\sin\left(\frac{1}{2}\pi v(v + 2t)\right) - \sin\left(\frac{1}{2}\pi v^2\right)\right]$$

$$S(v') - S(v) \approx \frac{1}{\pi v}\left[-\cos\left(\frac{1}{2}\pi v(v + 2t)\right) + \cos\left(\frac{1}{2}\pi v^2\right)\right]$$

Using the corrected values of C and S we then find ψ^2. Next increase v_B by

0.1 and repeat the procedure until v_B is 5.5, which is the last entry in the table (or until you have as many extrema as you wish).

We now have a set of intensities that we may examine for extrema. Having found them, we assume that each extremum, together with its neighboring intensities, lies on a curve of the second degree in the values of the corresponding v'_A. If, for example, we have a triplet (v'_{A_2}, ψ_2^2), (v'_{A_E}, ψ_E^2), (v'_{A_1}, ψ_1^2) wherein ψ_E^2 is the extremum, then we suppose that for all ψ^2 between ψ_1^2 and ψ_2^2:

$$\psi^2 = dv'^2 + ev' + f \qquad (6.3.2)$$

Defining:

$$v' \equiv v'_{A_2} - v'_{A_1}$$

$$v'' \equiv v'_{A_2} - v'_{A_E}$$

$$p' \equiv \psi_2^2 - \psi_1^2$$

$$p'' \equiv \psi_2^2 - \psi_E^2$$

then substituting the triplets into equation 6.3.2 yields for the constants d and e:

$$d = \frac{p'v'' - p''v'}{v'v''[v'_{A_2} - v'_{A_E}]}$$

$$e = \frac{p'}{v'} - d[v'_{A_2} + v'_{A_1}]$$

Setting $d\psi^2/dv'$ equal to zero to obtain an extremum v yields Fresnel's formula 6.3.3:

$$v_M - v_{A_2} = \frac{p''v'^2 - p'v''^2}{2(p'v'' - p''v')} \qquad (6.3.3)$$

This method is extremely good, since the integral correction is itself accurate to five decimal places. I find, using integrals computed by machine, that Fresnel's calculated values for the loci of the extrema are correct to better than 0.01 mm.

It is worthwhile taking an example to see how intricate and demanding these computations can be. Consider Fresnel's third observation, wherein a, b, c are, respectively, 2.01 mm, 401 mm, 1 mm. To find the minima we begin with v_B equal to 1.5 and proceed upward to 4.8 in steps of 0.1. Each step requires four integral corrections, for a total of 136. Using a machine, I find from Fresnel's procedure that the first minimum v'_{A_E} occurs in the triplet:

$$v'_{A_2} = -1.262$$
$$v' = -1.162$$
$$v'_{A_1} = -1.100.$$

These give for the true locus, using equation 6.3.3, -1.175. Fresnel has -1.181, but a variation in v of 0.01 yields a variation of only 0.01 mm in the screen locus, and, indeed, I find with Fresnel that the locus is 0.14 mm—which is also the one obtained through calculation using Fresnel integrals accurate to better than five places.

Fresnel tabulated seventeen extrema with various values for b and c. If he had proceeded as I just did, this would have required hundreds of calculations. What he more likely did was to find the approximate loci of the extrema from the old binary formula and then compute in that neighborhood. This would require tabulating at most four or five values for v_B, reducing the number of computations for each extremum to about twenty or so—still large, but less than a fifth of what would otherwise be required.

Fresnel carried out the same sort of computation for the narrow diffractor, which differs from the slit only in that the limits run from each edge to infinity, and here too he found a difference between theory and experiment of less than 0.01 mm. This concluding section of the memoir was a computational and experimental tour de force.

Despite the preponderance of *émissionaires* on the prize committee, only willful blindness could possibly have prevented their granting Fresnel the prize. The accuracy of his observations, their close agreement with his formulas—better than 1.5% in almost every case—and the clarity of the mathematical analysis, if not the physical suppositions, were simply too obvious for even determined opponents of the wave theory to have shut their eyes to—and Fresnel had an ally in Arago and perhaps a friend in Gay-Lussac. But to accept a formula's approximate validity is not necessarily to accept the theory that implies it—or even to accept the formula's complete generality. Certainly Laplace, Biot, and Poisson were not converted to the wave theory by the success of Fresnel's formulas for diffraction. Indeed, I shall presently show that Poisson, for one, never accepted Huygens's principle, on which Fresnel's integral theory is based, in part (but only in part) because he was not able entirely to cease thinking in terms of rays. But the power of the formulas was indisputable, especially when Arago demonstrated that an unusual implication Poisson himself drew from them in fact occurs (Arnold 1978, 182–88, discusses the deduction; see appendix 14 for details).

6.4 Objections to Huygens's Principle

6.4.1 Arago's Report and Young's Reaction

Arago wrote the commission's report (Arago 1819), and it was published in the *Annales de Chimie* in May 1819. The report begins by describing the very high accuracy of Fresnel's measuring technique for placing fringes. Arago then discusses the establishment through such exact experiments of hyperbolic fringe propagation outside the geometric shadow. Turning next to the internal fringes, he briefly discusses Young's experiment with two pinholes followed by the much more vivid two-mirror experiment described in the submitted memoir (whose author is not identified in the report, though no one on the commission could fail to have realized it was Fresnel). To this point—nearly halfway through the report—Arago has not mentioned the principle of interference in any form. In the next paragraph he gives it *as a formula* for binary ray combinations only, which are all that need be considered in the mirror experiment. He presents the principle as a method of calculating the mutual effects of two rays without at this point mentioning that the constant (the wavelength) in the formula that determines the fringes has any physical significance other than what is contained in the formula itself.

Arago goes on to remark that Fresnel's highly accurate experiments suffice to show that one must, in diffraction, consider a fringe to result from the deflection to its locus of many rays passing the obstacle's edges. How, Arago asks rhetorically, can this occur? The "limits" of the report do not permit him to "follow the author" on this point, but instead he presents what one might call a selectionist account of Fresnel's integral method—one that avoids any mention whatever of Fresnel's "elementary waves." It is worthwhile quoting Arago in full on this point, because his statement represents the only way Fresnel's integral theory could be understood by people who continued to think in terms of rays (not that there were many who understood even this much):

> The author envisions on the borders of an opaque body a portion of a sphere whose center would be the radiant focus, and he supposes that, from each point of this surface, *elementary luminous rays* depart in all directions and with sensibly equal intensities as long as they do not deviate far from the normal; he does not take account of the *rays* that are very inclined [to the normal, since

> these], in his hypothesis, destroy one another; he determines finally the intensity of the light resulting from the concurrence and the reciprocal influences of all the *rays* that are slightly inclined to the normal, by assimilating [these influences] to forces that make angles among one another proportional to the differences between the paths traversed, the difference *d*, which we have already spoken of, corresponding to a complete circumference. From that, the intensity of the light in all the points of space situated behind the body relative to the radiant focus are found to be represented by an integral formula that embraces each special case. (Arago 1819, 235; emphasis added)

In the M.C. Fresnel had discussed Huygens's principle in detail and had deduced his integral formulas from it. He had not written of "elementary rays." Quite the contrary; he had insisted throughout on discussing "elementary waves." I do not think Arago's replacement of "waves" with "rays" in his report to the commission was prompted entirely by the fact that the other members would certainly have objected to the word, though this was no doubt an important factor. Rather, I think it more than likely that Arago himself had difficulty accepting what was in truth the core of Fresnel's theory of diffraction, the aspect that removed it entirely from selectionist principles: its reliance on Huygens's principle.

Arago could have phrased it differently had he not himself continued to think essentially in terms of rays: he could, for example, have discussed, not "elementary luminous rays" (an obvious replacement of Fresnel's "elementary waves" with ray terms) but only the method of calculating path differences from different points on the "sphere whose center is the radiant focus." Of course it is entirely possible to think of Fresnel's integral theory in this way—in terms of rays—since the formulas encapsulate everything that can be known empirically about obstacle diffraction, and since in diffraction ray counts are not involved, inasmuch as each point of the front emits in all directions. By contrast, in partial reflection the assumption that rays exist as individuals leads to results entirely different from anything that can be obtained from the wave theory, because ray counts and ratios enter directly into the formulas. Consequently diffraction could not, and indeed did not, pose an impossibly difficult problem for selectionist understanding, though it raised some nasty questions for the emission theory.

Even Thomas Young found Fresnel's use of Huygens's principle difficult. In a letter to Arago sent on 4 August he remarked:

> You will imagine how greatly I have been interested with the two principal papers in the *Annales de chimie* for May [Arago's report: the second section of the M.C., containing the integral theory, was published later in 1819 in the *Annales* and was sent by Fresnel to Young in September]. Perhaps, indeed, you will suspect that I am not a little provoked to think that so immediate a consequence of the Huygenian system as that which Mr. Fresnel has very ingeniously deduced should have escaped myself, when I was endeavoring to apply it to the phenomena in question: but in fact I am still at a loss to understand the possibility of the thing; for if light has at all times so great a tendency to diverge into the path of the neighboring *rays* and to interfere with them as Huygens supposes, I do not see how it escapes being totally extinguished in a very short space, even in the most transparent medium, as I have observed in my first paper on the subject. I cannot, however deny the utility of Mr. Fresnel's calculations. (Fresnel, *Oeuvres,* 2:746–47; emphasis added)

Young, then, had to this point seen only Arago's report, but its account, based on rays, of the integral formulas was enough for him to generate at once a quick theory of diffraction by a circular opening, despite his doubts about Huygens's principle (though Young seems to have incorrectly estimated the path difference).[7]

What did Young mean when he wrote that, were Huygens's principle true, then the light would rapidly be totally extinguished? In the Bakerian lecture to which he refers, Young had remarked:

> [Huygens] supposes every particle of the medium to propagate a distinct undulation in all directions; and that the general effect is only perceptible where a portion of each undulation conspires in direction at the same instant; and it is easy to show that such a general undulation would in all cases proceed rectilinearly, with proportionate force; but, upon this supposition, it seems to follow, that a greater quantity of force must be lost by the divergence of the partial undulations, than appears to be consistent with the propagation of the effect to any considerable distance. (Young 1802, 24)

From this it seems that Young did grasp the principle but that he could not see why the resultant even at the common tangent to all the secondaries would not over time become extremely small, inasmuch as the secondaries would carry energy away in all directions. Obviously, he concluded, "some such limitation" to prevent oblique radiation "must naturally be expected to take place" to avoid this difficulty. Young assumes, in fact, that almost no radiation will occur in oblique directions. This is essentially the same assumption that Fresnel had originally made, though Fresnel, unlike Young, seems not in his early work to have considered Huygens's principle at all. The reason Fresnel did finally introduce the principle, whereas Young did not, was that Fresnel undertook a series of exact and detailed experiments, in particular for the single slit, whose results could be accommodated only with two inconsistent formulas in the absence of integral methods.

A month and a half after Young sent his letter to Arago, Fresnel wrote to Young directly for the second time: the first letter, in 1816, had been to tell Young of his work and to emphasize Fresnel's accurate experimental techniques—thereby distinguishing his work from Young's as much as he could at the time (Fresnel, *Oeuvres,* 2:737–40). Arago had let Fresnel see Young's letter of 4 August; on 19 September Fresnel sent Young two copies of his M.C. (as printed in the *Annales*) and took the opportunity to reply to Young's criticism of Huygens's principle (and to correct his erroneous estimate of path difference for the circular opening).

Fresnel had of course never read any of Young's memoirs because he could not read English. And it did not occur to him that Young's objection to Huygens's principle was essentially the same one that had for so long blocked Fresnel's own use of integral methods. Fresnel thought that he was referring to the total energy of the wave, that perhaps he felt interference would destroy the conservation of vis viva—though Young was in fact referring not to the extinction, but rather to the dispersal of the wave's energy. Accordingly Fresnel demonstrated very simply, using his decomposition for a pair of coherent waves with arbitrary phase differences, that the total vis viva is not changed by interference (Fresnel, *Oeuvres,* 2:749–50). This was a point that many scientists seem not to have appreciated for quite some time.[8]

But the most interesting part of Fresnel's letter to Young is its direct defense of Huygens's principle, for he insists that the principle is a "rigorous consequence of the coexistence of small motions in the vibrations of fluids." He writes:

> Huygens's principle seems to me, just as much as that of interference, to be a rigorous consequence of the coexistence of small motions in the vibrations of fluids. A derived wave may be considered to be the assemblage of an infinity of simultaneous disturbances; one may therefore say, according to the principle of the

> coexistence of small motions, that the vibrations excited by this wave in an arbitrary point of the fluid situated beyond it are the sum of all the agitations that each of the disturbing centers would there give rise to acting in isolation. In truth, by the nature of derived waves, these centers of disturbance cannot produce retrograde motion, and the elementary waves that emanate from it would not have the same intensity in directions oblique to the primitive impulsion as along the normal to the generating wave. But it is evident that the decrease of intensity must follow a law of continuity and may be considered to be insensible within a small angular interval. (Fresnel, *Oeuvres,* 2:748)

In the M.C. as well, Fresnel had partially conflated Huygens's principle with the principle of the composition of small motions. But they are not in fact one and the same, nor does Huygens's principle as used by Fresnel follow from the principle of composition. The latter asserts that the resultant amplitude at a given point may be calculated by adding together all the waves that would, considered individually, reach that point at a given time were the others not present. It is analytically equivalent to the assertion that the equation of propagation is linear, that the simple sum of any number of solutions to the equation is itself a solution. Huygens's principle does require the principle of composition, but it asserts something more.

The principle of composition concerns the resultant at any given point of the set of waves that arrive at that point at a given moment. It does not have anything to say about the individual waveforms themselves. Huygens's principle as used by Fresnel, on the other hand, takes a given waveform and dissects it into secondary elements. These elements must individually generate waves whose amplitudes vary in highly restricted ways with distance from the elements and with the angle to the original waveform's normal. One then constructs the primary wave at future times by applying the principle of composition to these elements. The principle of composition, that is, shows us how to combine these elements at a given point in space at any subsequent time. But it does not in itself permit us to perform the original spatial dissection into elements that have precisely the right kind of amplitudes to satisfy the demands of Fresnel's integral theory. Fresnel had not attempted before 1823 to explain how the necessary dependence of amplitude on inclination comes about (indeed, we have seen that he went to some lengths to avoid considering it), but in that year Poisson argued that Fresnel's integrals implicitly require certain dependencies, and that these are not justifiable if light obeys the same kinds of laws that govern the propagation of waves.

Before we turn to Poisson's objections, it is instructive to understand that Fresnel's version of Huygens's principle can be thoroughly justified only by an intricate analysis of the three-dimensional scalar wave equation (Baker and Copson 1939)—and in the absence of this kind of a justification arguments of Poisson's kind could not be readily dismissed. In essence Huygens's principle expresses the fact that the solution to the wave equation outside a surface S beyond which the wave function has no singularities involves an integral over S. The elements of the integral are the same functions of time and distance as the primary wave and are, in effect, Huygens's secondary waves. This general integral was not obtained until the last quarter of the nineteenth century by Gustav Kirchhoff, but particular versions of it were found by Sir George Stokes in 1849 and by Hermann von Helmholtz ten years later. Understood in this way, Huygens's principle is a result of analysis rather than of dynamics (as Poisson insisted it must be) or of kinematics (as Fresnel insisted, since he identified it with the composition of small motions). Huygens's principle can, *given the integral theorem,* be granted dynamic significance, but even then only with much effort (Baker and Copson 1939, 28–32). In the absence of this general solution, many people who were sympathetic to the wave theory found it extremely hard to accept or even to grasp the full significance of Huygens's principle.

6.4.2 Poisson's Critique

In early September of 1818, several months after submitting his memoir to the prize commission, Fresnel visited Laplace in the company of Arago. He wrote his brother a short account of what happened:

> M. Becquey had repeated [to Laplace] a conversation I had had with him on the subject of systems of physics, and in which I let slip that *nature does not dread difficulties of analysis,* and that those the theory of undulations presents are not at all a probability against it.—Apparently M. Becquey changed some of my expressions a bit, because M. de Laplace concluded from this that I did not believe in the utility of analysis. I replied to him that on the contrary I felt strongly that it was indispensable for giving physical theories mathematical rigor; but that it seemed to me that difficulties of calculation should never enter into the balance of probabilities when it concerns choosing between two systems. He said to me that in this regard he wasn't of my opinion, and sought to quarrel with me about Huygens's principle, which serves as the basis of my new theory of diffraction, and that he did not con-

> ceive, I think, in the same fashion as I do. (Fresnel, *Oeuvres,* 2:848–49)

Later in the letter Fresnel remarks that Laplace had not as yet read his memoir, but that he had heard enough about it to know that it relied on Huygens's principle in some way. Laplace himself understood the principle quite well, at least as a method for constructing rays, since he had used it, as I remarked above (chap. 1), to generalize Huygens's own demonstration that the wave theory implies the principle of least time for rays. Moreover, Laplace did not tell Fresnel that he objected only to the use of the principle in optics; he evidently objected to using it for any kind of wave motion at all. In this he was hardly alone, because his closest colleague, Poisson, though he told Fresnel at this time that "the multiplicity of hypotheses required by the Newtonian theory much diminishes his confidence in it" (Fresnel, *Oeuvres,* 2:849), nevertheless found Huygens's principle impossible to accept.

It was four years before Poisson fully expressed his objections. At a meeting of the Institute in the spring of 1823 Poisson apparently criticized Fresnel's account of refraction. On 5 March Fresnel replied that, as he understood Poisson, the objection was that the disturbance at a given point in the refracting medium will not be the same function of the time as in the first medium because a series of waves must strike the boundary (Fresnel, *Oeuvres,* 2:183–85). That, Fresnel replied, makes no difference, because the theory presumes that the incident wave train is regular and indefinitely long. But this was just a shot across the bows. The next day (6 March) Poisson presented Fresnel in writing with a series of new objections that struck at the heart of the integral theory (Fresnel, *Oeuvres,* 2:186–89).

Poisson's principal objection was to the use Fresnel thought he could make of the "principle of the coexistence of small oscillations." The principle as Poisson understood Fresnel's use of it asserts that "if one has a system of material points executing very small vibrations, one determines the motions of the system, after any interval of time, by combining all the motions that would occur, if each of its points vibrated in isolation throughout the entire interval of time." But this, Poisson remarked, is an *extension* of the principle as used by Daniel Bernoulli and needs to be justified. In substituting the motions of the parts for the motion of the entire wave "you increase the difficulty of the question, because you must now know the motion each particle would have, if it were alone, and the motion it would send throughout the entire system, which is much harder than knowing the motion of whole waves."[9]

The major problem in this regard, as Poisson saw it, was Fresnel's assumption that the secondary oscillations have measurable effects in directions other than those in which they themselves vibrate, namely, along the normals to the primary wave. Poisson, that is, felt that the inclination factor had to be an exceedingly rapid function of the angle, which would prohibit the kind of lateral radiation Fresnel needed. Poisson had, however, no analytical demonstration of the point, either for an isolated oscillator or for an oscillator that forms a part of the primary wave. But he believed that the radiation must occur almost entirely along the normals to the front because "it is only in this way that one may conceive, in the theory of undulations, the propagation of an isolated, thin streak of light, which the adversaries of that theory deny the possibility of." This remark reveals how deeply Poisson was still thinking, and continued to think, in selectionist terms. I shall return to it in a moment when I examine Fresnel's response, but first we must consider Poisson's public critique.

Pressed by Fresnel to make known his objections, Poisson did so in a letter printed in the *Annales de Chimie* for March. That letter again concentrates on Fresnel's assumptions concerning secondary waves. It amplifies and makes more precise several of the objections he had made in his personal letter to Fresnel. It does not explicitly mention the "isolated streak," but it does insist that an oscillator can produce an effect only very near its line of vibration (Poisson 1823b). This letter was preceded in the same issue of the *Annales* by an "Extract" in which Poisson amplified his objections, founded on his understanding of fluid dynamics, to Fresnel's account of reflection and refraction (Poisson 1823a).

The major thrust of Poisson's public letter again concerned Fresnel's use of Huygens's principle. Poisson quoted Fresnel's statement of the principle from the M.C., and then he proceeded to gloss it, remarking that by it Fresnel meant that

> if one decomposes at any instant the portion of fluid in motion into an infinity of infinitely small parts; takes one of these parts with its true speed and condensation, and seeks the motion that it would produce in the fluid, if it alone were disturbed; that one does the same for all the other similar parts, and after an arbitrary time compounds, for each point of the fluid, the speeds and condensations that this point would have received from all these elementary motions, supposed mutually independent, then one will effectively have the speed and condensation that will exist at that instant and at that point.

Put this way, Poisson admits, the principle is certainly true because in this form it is essentially the same as the principle of the coexistence of small motions: since each part of the fluid has by presumption its "true speed and condensation," we have said nothing more than that the disturbances produced by a set of oscillators with given velocity and displacement at a given time can be compounded linearly to produce the single disturbance at a later time due to the set taken as a whole.

But Poisson in effect says, So what? One must know what the "true speed and condensation" is, and this Fresnel cannot know a priori; it can be found only by analysis. Yet Fresnel makes certain assumptions in his theory that amount to a specification of this kind. Poisson proceeds to draw them out. He takes the square root of Fresnel's diffraction integrals, multiplies it by a constant p, and notes that this must, on Fresnel's own account, represent the (maximal) speed u at a given point of a fluid particle. In the absence of an obstacle the integral limits run from $-\infty$ to $+\infty$. Performing the integration gives for u:

$$\frac{pab\lambda}{a + b}$$

where a, b, and λ have essentially their usual significance.

Now, Poisson continued, for a "simple wave"—that is, for a spherical primary wave—u must vary reciprocally with the distance from the source of the disturbance. Suppose that at a given point on the primary wave (which is at a distance a from the source) the particle has a speed v. Then at the distance $a + b$ from the source the particle speed u must satisfy the equation

$$\frac{v}{u} = \frac{a}{a + b}$$

Whence the constant p must be equal to $v/(b\lambda)$. But why, Poisson objects, should the speed of a particle in the medium depend upon the reciprocal of the wavelength?

Turning again to Fresnel's integrals, Poisson remarks that they imply, as they stand, the existence of retrograde radiation, which does not occur in optics. But it can occur in fluids when one is dealing with an isolated pulse. In general, Poisson had shown, a spherically symmetric region over which there is a distribution of velocity and condensation will generate both forward and retrograde pulses unless certain restrictions are placed upon the relationship between the initial velocity and condensation over the pulse.[10] Poisson's point was therefore twofold. First, Fresnel had not shown that the appropriate conditions of velocity and condensation over a front can be satisfied consistently with his other assumptions concerning the parts of the front. Second, no oscillator can in any case produce much oblique radiation, so that the entire theoretical foundation of Fresnel's integral theory falls, whatever conditions he might assume. Neither of these criticisms, Poisson carefully remarked, at all affect the *empirical* validity of the diffraction integrals: "Observe that I attack here only your demonstration, and not at all the laws of diffraction that you have found, and whose exactitude you have established by experiments more precise than any hitherto made in optics."

Fresnel was not assuaged by Poisson's praise for his formulas and experimental acumen. "In a word," he remarked in public reply, "according to you I have arrived at a correct result by false reasoning" (Fresnel 1823i). Fresnel was of course not "convinced" by Poisson's objections, to which he replied at some length. Concerning Poisson's deduction of the inverse dependence of the speed of a particle of the medium, Fresnel remarked that Poisson "must prove the falseness of this consequence," not that Fresnel had to prove it correct. In any case Fresnel offered a qualitative argument for it that I (like Poisson) find impossible to follow. To Fresnel, Poisson's deduction of this relationship was a natural and not at all unreasonable consequence of the diffraction integrals.

Turning to the stickier problem of retrograde radiation, Fresnel first remarked that Poisson's objection will not hold if a secondary simply does not radiate except in directions between the tangent and the normal to the front. Note that Fresnel *did not*—either here or in the M.C.—attribute the absence of retrograde radiation to interference: he instead assumed that it *does not occur at all.*[11] Nevertheless Fresnel was not immune to the force of Poisson's basic objection, that he had ignored the relationship between condensation and velocity. At this point in his reply Fresnel more or less pushed the problem to the side, but he returned to it at the end after attempting to deduce a formula for the inclination factor.

That, indeed, was the most significant of Poisson's points—that there can be no substantial radiation except along the normals to the front, so that the inclination factor had to be a very rapid function of the angle. For if this were true, then the foundation of the integral theory was undermined. In his published letter Poisson had not linked this claim to the existence of a "thin streak" or beam of light, but Fresnel saw at once that this was what had truly prompted Poisson's claim, which Fresnel took seriously indeed: "I arrive finally at the major and direct objection by which, if it is well founded, you overturn the base of all my calculations . . . [that] the absolute speeds are sensible only in the direction of oscillation of the center of disturbance" (Fresnel 1823i, 220). Fresnel attacked the problem in two ways. First, he reiterated that there is no such thing as a "beam" in the sense Poisson means, so that the primary reason Poisson had given in his private letter for limiting the disturbance entirely to the line of oscillation does not stand up. Second, Fresnel attempted to deduce the inclination factor.

Fresnel's first response to Poisson again highlights a major, perhaps the major, difference between selectionism and the wave theory: that selectionism presumes the existence of physical rays whereas the wave theory reduces them to mathematical constructs. Despite his facility in applying Fresnel's integrals, Poisson seems never to have grasped this fundamental point. Verdet remarks:

> Poisson was preoccupied to the end of his life with the difficulty that, according to him, the propagation of a thin streak of light opposes to the theory of undulations, and from a note added to his memoir on the equations of motion of crystallized bodies it follows that he believed he had found the solution, but the sufferings of his final illness didn't permit him to write it down. This circumstance does not seem to indicate that he ever accorded sufficient attention to the experimental laws of diffraction (Fresnel, *Oeuvres*, 2:196)

Poisson never fully appreciated that the wave theory replaces rays as physical objects with wave fronts. He continued to feel that any theory of optics must somehow yield rays that have a physical basis. The only way to do this in the wave theory as he understood it was to limit the possible spatial extent of a disturbance to a well-defined, nearly linear region. Groups of these regions would constitute bundles of rays.

Fresnel's counterargument went entirely past Poisson. Fresnel well understood that anything empirically similar to what used to be meant by a "beam" of light can be produced only by passing a wave through an opening that is large in comparison with a wavelength. In the wave theory a "beam" is an approximation to the true facts. To a selectionist like Poisson it is a fundamental thing, and so any theory must accommodate it as a basic reality, not as something that requires artificial conditions for its production. To Fresnel a "beam" does not exist unless a front passes through a large enough opening; to Poisson "beams" are de facto created by light sources.

Obviously Fresnel could not easily have convinced Poisson on this most elementary point, though he twice emphasized it. If Poisson continued to insist that "beams" must exist in a way that the wave theory forbids, then very little could convince him otherwise until he could fully grasp that there are no empirical facts requiring physical beams. What Fresnel could do, however, was to attempt to show that a linearly oscillating fluid element will generate detectable off-axis radiation. For if it cannot, then Fresnel's integral theory fails entirely. Were he successful in this, it is true, Poisson could still reply that this merely shows the complete bankruptcy of the wave theory, since it could then not accommodate a physical "beam," but that would at any rate leave the argument at a point where the differences almost entirely escape the powers of logical persuasion. Either Poisson would eventually understand that physical rays are not empirically necessary—making the wave theory tenable on this score—or he would not. According to Verdet, he never did grasp this fact.

Fresnel constructed figure 6.7 in an effort to deduce the inclination factor. AC marks the direction of a linear oscillation of a given point in the fluid. Fresnel next constructs two lines, AB and AD, at arbitrary angles to AC. Along AC the line segments Ac, AR respectively represent the magnitudes of the fluid point's displacement and speed at a given moment.

Referring to the "parallelogram of forces," Fresnel then constructs points d, Q on AD and points b, P on AB such that bc and PR are parallel to AD

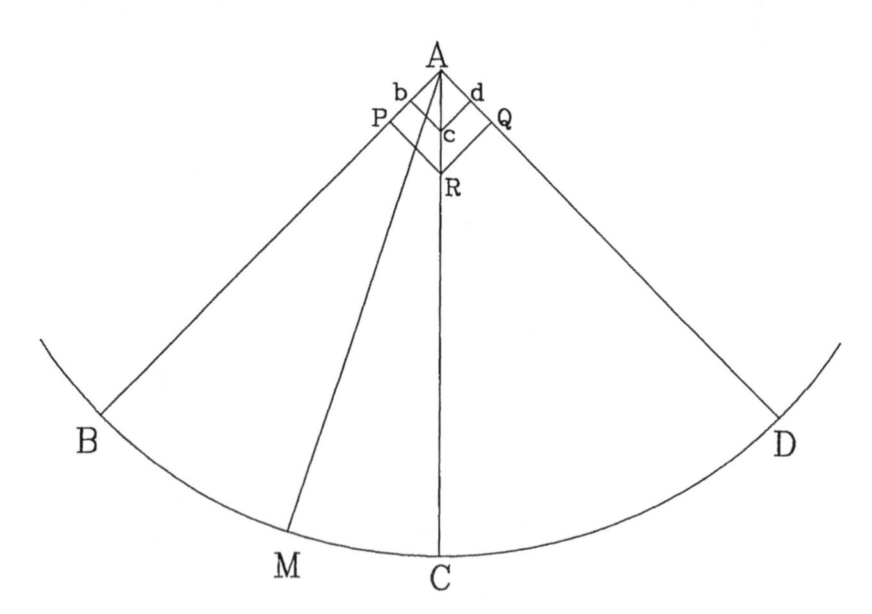

FIG. 6.7 Fresnel's diagram for the inclination factor.

while dc and QR are parallel to AB. Then, he continues, "in virtue of the principle of small motions" the lengths "Ab and Ad are those by which one must suppose the molecule A to have been displaced in order that the motions that would be produced in the fluid for each of these disturbances considered separately reproduce, by their reunion, the motions resulting from the single displacement Ac." In a note to Fresnel's letter Poisson remarks that he finds Fresnel's use here of the "parallelogram of forces" vicious (*vicieux*) in the sense of being circular, that he doesn't "understand it well."

Fresnel's argument is indeed peculiar. What he assumes is that a displacement Ab together with a *distinct* displacement Ad will produce the same effect in the fluid as a single displacement Ac that is directed along the diagonal of the parallelogram Abcd. If, that is, we consider A to be the locus of two distinct fluid points, one of which moves to b while the other simultaneously moves to d, then the effect in the fluid of these two displacements will be the same as if a single point went in this same time to c. This is not simply an application of the parallelogram of forces but in fact amounts to a physical assumption. In Fresnel's words, "The lengths Ab and Ad are those in which the molecule A would have to be supposed displaced in the directions AB and AD so that the motions that would be produced in the fluid by each of these disturbances considered separately reproduce, by their reunion, the motions resulting from the single displacement Ac." That Ab and Ad may be considered "separately," though they do not in fact exist separately (being projections of Ac), is an assumption. In Poisson's view the assumption is not legitimate—in fact he claims not to understand Fresnel's reasoning—simply because Ac cannot produce effects except along its line, or Ab except along AB, or Ad except along AD.

Fresnel next considered a front BCD that is centered on A and that is due to the linear oscillation of the point A along AC. He assumed that the particle displacements or speeds (he considers either indifferently) on BCD vary symmetrically on either side of AC. He further assumed that the same law of variation also applies to the displacements Ab, Ad considered as individuals. Finally, he also assumed that the speeds or displacements in BCD along a given line of oscillation are proportional to the speeds or displacements of the corresponding oscillation at A and in the same line; for example, that the displacement in B is proportional to Ab, that in C to Ac, and that in D to Ad. This carries to its logical conclusion Fresnel's presumption that the decomposition of a motion is no different in effect from considering two distinct linear oscillations with appropriate speeds or displacements along the directions of the decomposition.

Fresnel next takes a special decomposition in which ∠BAC is equal to ∠CAD; we shall represent the single angle by α. Then, by the parallelogram construction, AP sin (2α) must be equal to AR sin α, so that

$$\frac{AP}{AR} = \frac{\sin \alpha}{\sin(2\alpha)} = \frac{1}{2 \cos \alpha}$$

The crux of the argument follows immediately. Consider a point M on BCD, with x representing ∠MAC. The disturbance along AR will produce at M a disturbance equal to some function $\psi(x)$ times the magnitude of the disturbance produced by it at point C. Fresnel now replaces the effect at M that is due to Ac with the combined effects at M of Ab and Ad, each considered separately, using his assumption that ψ applies to these disturbances as well as to their single equivalent Ac. To obtain their effects, then, we must divide the effect due to Ac by 2 cos α and multiply it by the appropriate function ψ, which gives:

$$\frac{\psi(\alpha - x)}{2 \cos \alpha} + \frac{\psi(\alpha + x)}{2 \cos \alpha}$$

This, Fresnel continues, must hold even if M is situated at C, where x vanishes and $\psi(x)$ is one, which implies that

$$\psi(\alpha) = \cos(\alpha)$$

So cos α is the inclination factor, as Fresnel had asserted in the M.C.

But if it is the correct factor, why is there no retrograde radiation, for which cos α will be -1, not zero? To explain why there will not be retrograde radiation, Fresnel takes advantage of the distinction between displacement (in effect condensation) and velocity that he had built into figure 6.7. He writes:

> Consider, for example, an element of a wave at the moment when its molecules are pushed forward, that is, in the direction of propagation of the derived wave: one knows that this forward motion is accompanied by a condensation, that is, by a mutual approach of molecules; if the molecules were only displaced and therefore were without velocity at the same instant, an expansive force would result from this mutual approach that would push the fluid in the rear as in the front and would thereby produce a retrograde wave similar to the one it would excite in the forward direction, but in which the absolute speeds [speeds of the molecules] would have the opposite sign; if, on the other hand, the molecules were in their equilibrium positions [undisplaced] and

> received at that instant only the speeds that push them forward, a wave toward the rear would also result, as one forward, since the molecules would be followed by those behind, and so on; the retrograde wave would have the same intensity as the wave propagating forward, and it would displace the molecules of the fluids in the same direction. These two motions therefore counter one another in the retrograde waves owing to the condensation and to the molecules' velocities, whereas they reinforce one another in the two waves that propagate forward.

This difficult qualitative argument (a version of which appears also in the M.C., 1819b, 296, note 1) in no way depends upon the principle of interference, for it applies to an isolated pulse. It amounts to the assertion that from any given *front* four waves propagate in pairs: two forward and two retrograde. Each duo consists of a velocity wave paired with a wave of condensation.[12] In the forward-propagating duo these two waves reinforce one another; in the retrograde pair they cancel one another. Fresnel's physical argument therefore forbids retrograde radiation by requiring the velocity and condensation over the front to satisfy a certain relationship with one another. Poisson knew that one could avoid retrograde radiation from pulses in this way because the result follows from his own analysis. But the result follows only for complete fronts. That is, for a given distribution of velocity and condensation over some closed front, there will result four waves, two condensational and two of velocity, with one of each contracting toward the center and the other propagating outward. If a certain condition on the condensation and velocity over the initial front is satisfied (see my note 11), then the two retrograde waves will annul one another.

The problem is that this result was not derived for the elementary disturbances into which Fresnel supposes the front can be subdivided. They might as individuals radiate backward, and Poisson found it questionable whether, in view of this, the secondaries will have no net effect in the reverse direction. He emphasized this point in a note dated 29 June that Verdet found and appended to the publication of Fresnel's letter in the *Oeuvres* (Fresnel 1823i, 226–27). Poisson in effect insisted that Fresnel must use his "cosine law" (that the inclination factor is cos) to deduce "after a time t" the total effect at an arbitrary point K of all the secondaries when each of them has a specified velocity and condensation (viz., one that satisfies the condition that the front as a whole will not produce retrograde radiation). This will test the validity of substituting secondaries for the front as a whole. "That's the question I spoke of to M. Fresnel," Poisson concluded, "who has, in his theorem of the lateral speeds expressed by the cosine, all that's necessary to resolve it." Poisson, then, wanted Fresnel to calculate explicitly the net effect of the secondaries using his inclination factor and not to use an analysis appropriate only to a complete front to dispose of the retrograde wave.

We should not reject Poisson's demands out of hand. For Fresnel himself, as we have just seen, accepted the terms of the argument—that retrograde radiation must be eliminated by choosing appropriate conditions for velocity and condensation. At no point in his argument with Poisson—or in the M.C. for that matter—did Fresnel suggest that the principle of interference should be used to explain rectilinear propagation. The reason he did not suggest it was that Fresnel, like Poisson, apparently insisted that rectilinear propagation must be guaranteed even for a single pulse, whereas interference requires a long wave train. In this context Poisson's criticisms were perhaps justified, since it is a difficult matter to formulate Huygens's principle at the outset in a way that will guarantee the rectilinear propagation of a single pulse.[13]

6.5 Understanding of Interference in the Early 1820s

Poisson was not alone in regarding Huygens's principle with suspicion or in continuing to treat the ray as a physical entity. Indeed, throughout the 1820s the ray remained the basic unit of analysis even in diffraction, with most accounts considering Fresnel's and Young's original binary theory to have been adequate (Kipnis 1984, 258).

The task of examining in detail the penetration of Fresnel's ideas into the texts of the 1820s and 1830s is a difficult one, and I shall not attempt it. Instead I shall contrast two texts printed in France in 1823 and 1824 to see how they dealt with diffraction (see also Kipnis's remarks in 1984, 255–57). I have chosen these two texts because one of them represents the very common qualitative cast of physics texts of the time, whereas the other takes a somewhat more quantitative approach, and the contrast between the two in their attitudes to the two theories of light is striking. The 1824 text, by F. S. Beudant, who was a professor of mineralogy and member of the faculty of sciences at the Paris Academy, was actually a third edition. It is on the whole devoid of mathematics and was aimed at giving some idea of the "exact sciences" to young people "on terminating their education and at the moment of entering the world."

The sixth book of Beudant's *Essai d'un cours élémentaire,* 116 pages long, is concerned entirely with light. Only one and a half pages are devoted to diffraction, in a chapter entitled "Dispersion of Light into Colored Rays." These pages, as well as everything else in the book, are fully selectionist in approach. Beudant continues to believe that diffraction involves the separation from one another by inflection of rays at substantial distances from the diffractor's edge: "All the phenomena of diffraction agree in demonstrating that the luminous rays that pass near bodies are not inflected only at their surfaces, but also at very sensible distances." But he also refers to Fresnel, remarks the hyperbolic propagation of fringes, and notes that "if we cause two *rays* of light to meet in any way, their encounter gives rise to a series of dark and light fringes" (Beudant 1824, 556; emphasis added). He has, then, understood that a Newtonian separation of rays is not sufficient to explain diffraction, and that the separated rays must meet one another to form fringes. But for Beudant the bare meeting of a pair of rays suffices to give rise to fringes. There is no evidence that he understood to any substantial degree even the limitations on the origin of the rays that a binary theory requires. Nor does he seem to know how to compute fringe loci, though, since his text is substantially devoid of mathematics, it may be that he simply regarded the point as too technical.

He mentions the wave theory briefly in an earlier chapter, remarking that, according to it, "all the phenomena of light are explained as in the theory of sound" (Beudant 1824, 381). But since Beudant had only a rudimentary account of sound waves, this was little more than words. Of greater interest than this quick comparison is Beudant's recognition that diffraction casts substantial doubt on the "emanation" (emission) theory:

> There is really at this point only one objection, in truth quite powerful, against the emanation hypothesis adapted to the phenomena of light. This objection is furnished by the experiments that M. Fresnel made on the diffraction of light, and that are today completely inexplicable in the emanation hypothesis; they are explained, on the contrary, with the greatest facility in the hypothesis of vibrations.

But he has chosen nevertheless, he says, to adopt in this text the "emanation" theory:

> It would without doubt be useful, in the interests of science, to present here the two theories, in applying the one and the other to diverse phenomena; but in the necessity to choose in order not to make my work too long, I had now to limit myself, as in the first edition, to adopting only one. *I chose that of emanation, not because I believe it to be better founded, but because in putting, in a way, more materiality in the phenomena it is easier to grasp.* (Beudant 1824, 383; emphasis added)

In fact Beudant has hardly a word to say about the "emanation" hypothesis. What he does insist on is the physical reality of the ray, which is essentially what he meant by saying that the "emanation" hypothesis puts more "materiality in the phenomena." This was a statement of selectionist principle, and I doubt that Beudant could possibly have given a coherent account of even a binary ray theory, much less of one based on Huygens's principle. He was too strongly convinced that the ray is the unit of optical analysis.

The contrast of Beudant's text with one published the year before and written by Eugène Péclet is striking. Péclet, who was then professor of "physical sciences" at the royal college in Marseilles, also avoided mathematics, but he included extensive diagrams. Furthermore his text, unlike Beudant's, has a definite quantitative cast, beginning with Péclet's insistence on exact measurement and the estimate of errors:

> One must never lose sight, in these researches, that the imperfection of our organs and of our instruments does not allow us to make absolutely exact observations; they will never rigorously satisfy the laws that govern them; one must require only that the differences be smaller than the probable errors in the instruments. Further, the series of observations must be very extensive; for one would otherwise risk obtaining, not a general law, but a law that would apply only within the period observed. (Péclet 1823,v)

Péclet rather thoroughly adopted Fresnel's final theory of diffraction, including its use of Huygens's principle (to the extent that he, like Fresnel, regarded it as "a consequence of the general principle of the coexistence of small motions" [Péclet 1823, 492]), and he even mentioned the inclination factor. He rejected the "emission system" entirely and gave a very brief (and entirely inadequate) zone construction for diffraction by a semi-infinite plane. He grasped the physics of the theory to the extent that he, very unusually for the time (see Kipnis, 1984, 265–66), explained the necessity of a point source.[14]

Beudant, it seems, remained a selectionist, whereas Péclet had apparently abandoned selectionism. Neither of them approached optics in quantitative detail, but of the two Péclet had by far the more quantitative an outlook. Consider finally a German text, also of 1824, that sits somewhat uneasily between rays and waves: Andreas Baumgartner's *Die Naturlehre*. Fourteen chapters in the second volume were on light (for a total of 148 pages), and Baumgartner took a more quantitative approach than either Beudant or Péclet. He felt that the "vibration" theory must be taken quite seriously, since

> it explains most optical phenomena from the mere nature of a vibrating motion and has gaps only where the artifices of mathe-

> matical analysis known till now are insufficient to represent the laws of vibrational motion, while in the emanation hypothesis many experiments can be explained only by coercion and additional hypotheses that violate all analogy and others cannot be explained at all even with additional hypotheses. (Baumgartner 1824, 2:6)

Despite his sympathy for the "vibration" theory, Baumgartner was at this time incompletely emancipated from rays. He continued to think of luminous bodies as emitters of "rays," which he thought could be assumed whatever the ultimate nature of light may be (Baumgartner 1824, 2:9), and this led him into difficulties when he considered diffraction, for which he gave an interestingly confused account.

He examined obstacle diffraction, for which he drew an interrupted wave with rays emanating from points on the front to a point on the screen (Baumgartner 1824, 2:112–13). He then used a zone construction. For the internal fringes produced by an obstacle he cut off half-wavelength zones, ending up essentially with Fresnel's theory of the efficacious ray. (Two rays slightly displaced from either edge of the diffractor meet at the screen point.) His argument is extremely vague. What is worse, for the semi-infinite plane, again using a zone construction, he ends up with Fresnel's original binary theory, in which a ray emanates from the edge proper, though he gave almost no details of the argument.

The problem was that, though he recognized the need for an integral formulation (Baumgartner 1824, 2:114), he was still thinking essentially in terms of combinations of rays, and he combined them inconsistently in his zone count. This, we shall see below, is even more apparent in his discussion of polarization, which scarcely deviated from selectionist language. Though Baumgartner clearly supported the "vibration" over the "emanation" hypothesis, he had not as yet fully understood that rays must be entirely abandoned except under the limiting conditions that obtain in geometrical optics (and these conditions were themselves obscure to most people for many years).

One of the difficulties people had at this time in forming a clear conception of the wave theory was where to read about it. A major source seems to have been Fresnel's "De la lumière" of 1822, which covered most of his work to that point and was printed as an appendix to the French translation of Thomas Thomson's *Chemistry* (Fresnel 1822c). His account here was not deeply quantitative and, in its experimental discussions, considered for the most part situations that can be analyzed in terms of binary ray combinations (viz., the inclined mirrors experiment). However, both Beudant and Baumgartner refer explicitly to volume 11 of the *Annales de Chimie,* which contains the second section of the M.C. itself, where Fresnel developed and applied the integral theory. Much of this, as well as its less quantitative analogue in the "De la lumière," was incomprehensible to Beudant and certainly very difficult for Baumgartner. Péclet, on the other hand, seems to have understood it quite well, though we shall see that even he had difficulty understanding Fresnel's work on polarization,[15] which is discussed in "De la lumière," and to which I now turn.

7 The Puzzle of Polarization

7.1 The Effect of Polarization on Interference

Several weeks after Fresnel finished the supplement to his second memoir on diffraction in mid-July 1816 (in which he introduced the efficacious ray), he tried, apparently at Arago's suggestion (according to Fresnel 1819a), to examine the fringes produced by the two beams polarized at right angles to one another that are always generated by passing an unpolarized beam through a crystal. He writes:

> I tried in vain to produce fringes with the two images of a luminous point in front of which I had placed a rhomb of calcspar, despite the care I took to have the extraordinary ray traverse a plate of glass, whose thickness was determined in such a fashion as to nearly compensate the difference between the numbers of undulations formed in the crystal by the ordinary and the extraordinary rays, so that on gently inclining the crystal I could establish an exact compensation. But the space within which I hoped to perceive the fringes being so narrow, and being furthermore occupied in part by the bands projected by the edge of the glass plate, I had recourse to another method, which did not have these disadvantages: I received the rays that had traversed the rhomb of gypsum on a small unsilvered glass, whose thickness was so calculated that the difference between the numbers of vibrations of the rays reflected by its first and second surface should be a bit larger than that resulting from the double refraction, so that, by an easy procedure, one could find an inclination such that the differences became equal; and yet this second try had no greater success than the first. (Fresnel 1816c, 387)

Fresnel accordingly began to suspect that "the two systems of waves produced by light in crystals endowed with double refraction have no influence on one another" (Fresnel 1816c, 387). On thinking it over, he realized that this must of course be so, because one can obtain slices of gypsum so thin that the path difference in them between the ordinary and the extraordinary refractions is only two or three wavelengths, yet they show no color. Since, Fresnel further reasoned, the only difference between the ordinary and the extraordinary light seems to be their polarizations, it seemed likely to him that perpendicularly polarized rays of light will not interfere with one another.[1]

To test his conclusion, Fresnel developed a new experiment:

> From a very clear gypsum crystal I carefully detached a lamina nearly a millimeter thick, and I cut it into two parts that I fixed on each of the slits in the copper plate [through which light from an effective point source entered], setting their axes in perpendicular directions. Then, observing the shadow of this apparatus with a lens, I saw two systems of fringes separated by a fairly considerable white interval, as the theory foresaw. They of course derived from the action of the left ordinary rays on the right extraordinary rays [the crystal sections being perpendicular], and [the action] of the right ordinary rays on the left extraordinary rays. One again sees from this experiment that rays polarized in opposite senses cannot produce fringes, because those in the middle [which appear when the crystal sections are not perpendicular] had disappeared. (Fresnel 1816c, 392–93)

Arago apparently felt that this was at best an indirect test of the lack of interference between perpendicularly polarized rays, because what Fresnel observed was in fact an interference pattern, though it differed from what it would have been if perpendicularly polarized rays did interfere with one another. Accordingly Arago insisted on a direct test. To obtain only two perpendicularly polarized beams, Arago thought to use a pile of plates polarizer, which he himself had extensively examined in past years. Two piles were built, each containing fifteen leaves. Unpolarized light from a very small source strikes a copper plate in which two narrow parallel slits have been cut. Each slit illuminates one of the two piles at about 30° incidence, thereby highly polarizing the light that emerges from each pile. Fresnel found that when the planes of incidence on the two piles are mutually perpendicular, "no trace of fringes" can be found. But when the sections are parallel to one another then fringes, albeit very irregular ones, do appear.

It naturally occurred to Arago that one could restore the fringes by manipulating the light after it had passed through the piles. Suppose that the planes of reflection from the piles are mutually perpendicular, so that no fringes occur. Put a crystal behind one of the two piles in such a fashion that the beam emerging from that pile will be split into two others, and such that these two will not be polarized at right angles to the beam from the other pile. Then presumably fringes should reappear. But they do not (Fresnel 1816c, 389). This led Fresnel to the important realization that interference requires more than a common point source and nonorthogonal planes of polarization: for beams to interfere they must also have *originally* been polarized in the same plane.[2] This fact was exceedingly difficult to explain at the time, and Fresnel did not attempt to do so. Instead, he and Arago summarized their results in five laws (Fresnel 1819a, 521–22):

1. "In the same circumstances that two rays of ordinary light seem to destroy one another, two rays *polarized in contrary directions* exert no appreciable action on one another."

2. "Rays of light polarized in a single direction act on one another like natural rays: so that, in these two kinds of light, the phenomena of interference are absolutely the same."

3. "Two rays *primitively polarized in contrary directions*" may thereafter be brought to the same plane of polarization, without *nevertheless acquiring by this the faculty of influencing one another.*"

4. *"Two rays polarized in contrary directions, and thereafter brought to analogous polarizations, influence one another like natural rays, if they derive from a beam primitively polarized in a single direction."*

5. "In the phenomena of interference produced by rays that have undergone double refraction, the place of the fringes is not determined uniquely by the difference of paths and by [the difference] of speeds; and in some circumstances, which we have indicated, it is necessary to take account, in addition, of a difference equal to a half-wavelength."

The first textual variant printed in the *Oeuvres* (dated 30 August) (on which see note 1) contains a note that introduces the concept of a transverse oscillation in an effort to explain at least point 1—the lack of interference between perpendicularly polarized waves. This passage, I shall argue, is extraordinarily significant for grasping what Fresnel did, and what he did not, understand by transversality at this time. It reads in full:

> Two systems of waves in which the progressive [longitudinal] motion of the molecules of the fluid are modified by a transverse motion of going-and-coming, which would be perpendicular [to one another] and equal in intensity, could exert no action each on the other, when the discordance of the transverse motions corresponds to the accord of the progressive motion, or reciprocally, because then the resultants of these two forces in each system would be in perpendicular directions. There is also another hy-

> pothesis that could explain the absence of fringes in circumstances otherwise favorable to their production: that would be transverse vibrations that would offer simultaneously condensed and dilated nodes on the same spherical surface, from which points of accord and discord would result [that are] so closely spaced that the eye, unable to distinguish them, would have the sensation of continuous light. On the surface of water one often sees waves rippled in this way in the direction of their length. But I have tried in vain [*inutilement*] to explain the phenomena with these hypotheses, of which the first was indicated to me by M. Ampère. That one would furthermore not be sufficient, for one would also have to explain how light finds itself modified in this way by reflection or by double refraction. (Fresnel 1816c, 394).

This is a complicated statement and is not present in the second variant, dated two months later. From it we see that Fresnel understood Ampère to have suggested that the usual oscillation, presumed to be normal to the wave surface, is "modified" by a transverse motion on reflection or double refraction—not that it consists exclusively of such a transverse motion. That was indeed a simple, and obvious, suggestion, because polarization required some sort of asymmetry about a ray, which in the wave theory is normal to the front.

But Fresnel's remarks here, and later, on the addition of a transverse motion to the usual one are vague and difficult to make complete sense of. The difficulty derives from Fresnel's insistence, to which I shall return in a moment, that one can only add a transverse component to the usual longitudinal oscillation. One cannot entirely replace the latter. Indeed, Fresnel (and probably Ampère) continued to think that unpolarized light must be purely longitudinal. This makes it very difficult to understand why there should be no fringes at all even when beams are polarized at right angles to one another, since some longitudinal motion is always present. The first part of the passage quoted above constitutes an attempt to provide conditions on the phase relationships between the transverse and longitudinal motions in polarized light without doing away entirely with the latter, which will accommodate this fact.

Fresnel has provided two possible explanations (see fig. 7.1). In the first, if the longitudinal components of the two interfering waves have no phase difference, then their transverse components must differ in phase by 180°. For then the resultant (R_1) of the components of the first wave will be perpendicular to the resultant (R_2) of the components of the second wave. Alternatively (and rather obscurely) Fresnel hypothesizes that the transverse components in

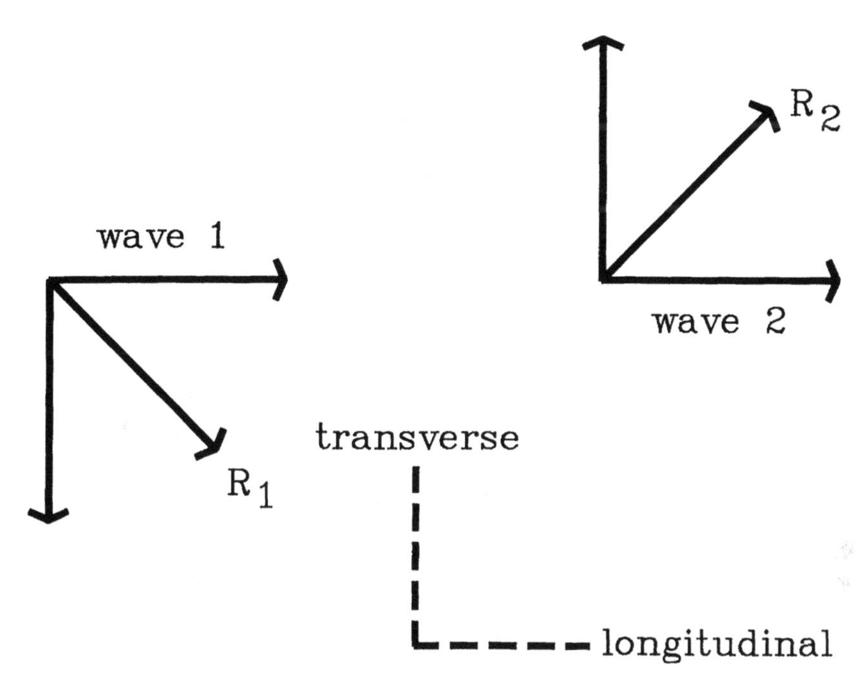

FIG. 7.1 Fresnel's first explanation for the lack of interference between perpendicularly polarized waves.

the two waves in effect produce a secondary interference pattern. This pattern superposes on the pattern that is due to the longitudinal components. The result, Fresnel seems to argue, will be to obliterate all visible fringes, because those that do occur will be very closely spaced.

Of course there would be no need to invent such complicated and obscure explanations for the absence of interference if polarized waves were entirely transverse, but at this point Fresnel did not believe this was a reasonable possibility. To judge from his later remarks, which I will discuss below, Fresnel had two reasons for continuing to insist on longitudinal oscillations. First, he could not see on dynamic grounds how to uncouple longitudinal from transverse waves. Second, longitudinal waves also continued to have a major *kinematic* function in that Fresnel still thought of unpolarized light as light that completely lacks the asymmetry that is present in polarization.

These two points—the dynamic and the kinematic—were strongly linked. There were two ways to understand the nature of *unpolarized* light, according to Fresnel (Fresnel 1818e, 526–27). Either the oscillation in an unpolarized wave is normal to the front, or else it contains simultaneously all possible asymmetries. This last possibility can have little meaning in the wave theory, though it is precisely the selectionist position. Therefore if *polarized* light were to be entirely transverse, then the purely longitudinal oscillation in the unpolarized beam must somehow be converted into a purely transverse motion. This seemed to Fresnel to be very difficult to understand dynamically. The only alternative was to assume that the action of polarization only adds a transverse oscillation to the longitudinal one. However, this conception makes it difficult to explain why perpendicularly polarized waves do not interfere with one another—as Fresnel's intricate and obscure remarks above show well.

Fresnel's reluctance at this point to assume that polarized light can be entirely transverse has, then, two related sources. First of all, the very idea of a transverse oscillation seemed difficult to understand dynamically. The ether would have to be most unlike, for example, air or even water, either of which can support (it was thought) only longitudinal waves. Nevertheless, both Fresnel and Ampère did think that, at the least, a transverse component could be added to the usual longitudinal one—and they offered this possibility without specifying a dynamic mechanism that could account for it. Furthermore, we shall see in a moment that, according to Fresnel's later testimony, he and Ampère had had no difficulty in thinking of polarized light as a purely transverse oscillation. But, they had wondered, how could the act of polarization destroy the longitudinal oscillations in the unpolarized light, and why do subsequent reflection and refraction not restore these oscillations? In other words—the second source of Fresnel's reluctance to adopt complete transversality—Fresnel and Ampère did not think that *unpolarized* light could itself be transverse: it had to be longitudinal because it never shows any asymmetrical behavior. Even if Fresnel had at this point constructed (as he later did) a medium that could support *only* transverse oscillations, he would still have faced this thorny problem of explaining the natures of unpolarized and partially polarized light.

What Fresnel was missing at this point was what one might call a *kinetic* conception of polarization rather than a static one. He, like his contemporaries, still thought of polarization in terms of an intrinsically stable directionality. Polarization, he thought, is temporally fixed. It changes only as a result of an external action, and it is not the usual condition for light. Selectionists, of course, also thought this way.[3] However, for them a change in the polarization of a beam required the alteration of ray asymmetries. Fresnel naturally could not accept a conception based on collections of rays. But at this point in his work he did not have a thoroughly different position to offer because he had not, as it were, set polarization in motion.[4] The difficulty he faced was how to retain what he thought was essential in the old concepts of polarization while adapting them to the demands of the wave theory. This was extremely hard to do because even the terminology surrounding "polarization" was imbued with concepts that do not fit the wave theory well, but that Fresnel initially tried to adapt.

Abstract for a moment from the fact that each of the three possible states for a light beam in the selectionist understanding of polarization derives from the properties of collections of rays. Fresnel never adopted any such notion. However, he did tacitly use an idea that is implicit in the *definitions* of "polarized," "unpolarized," "partially polarized," and even "depolarized" used by the selectionists: that the *polarized* and the *unpolarized* are two extreme conditions, with *partial polarization* being a mixture of the two. We can see this by recognizing first that Fresnel did not assume that the transverse and longitudinal oscillations could ever have phase differences other than 0° or 180°. That is, the transverse component might reach a maximum at the same moment as the longitudinal component, or it might do so a half-period later, but those were the only two possibilities. This means that their resultant must always point in the same direction unless the relative amplitudes of its transverse and longitudinal components are changed—which is precisely what occurs during polarization or during depolarization. For Fresnel, then, the light vector is always temporally fixed in direction.[5]

According to this way of thinking *polarized* light either is light that entirely lacks a longitudinal oscillation or, more probably, is light in which that oscillation is the least possible. *Unpolarized* light, at the other extreme, is light in which the transverse oscillation either is the least possible or, perhaps, is entirely lacking. Finally, *partially polarized* light is a *mixture* of these two extreme conditions in the sense that two significant oscillations, one transverse and the other longitudinal, might exist in a given wave simultaneously and independently of one another.

Fresnel later extended the idea of a mixture by considering it possible to define a kind of *depolarized* light as a mix of two perpendicularly polarized, independent waves. That is, Fresnel in effect conceived of two kinds of mixtures. One is a mixture of conditions in a single wave, the other is a mixture of two waves in different conditions. In both cases there are two physically distinct effects to consider. In the first case there are the transverse and longitudinal oscillations, united, it is true, in a single wave, but to be considered independently of one another. In the second case there are two physically distinct, though coherent, waves that are united in a single beam.

In both cases—a mixture of conditions or a mixture of waves polarized at right angles to one another—we can if we insist consider the set of conditions or the set of waves each to form a single physical oscillation at any given instant and point. If we do this then the waves as well as the conditions might themselves be thought of as projections of their resultant (along the wave front or its normal) at that instant and point. To see what I mean by this, take first the case of mixed *conditions* in a single wave. We might think of the wave's general state of oscillation as one that points in some direction in a plane normal to the wave front. Thus, in figure 7.2 we might specify a state of polarization in the following way. In an *unpolarized* wave the angle (∠RCL) between the resultant CR of the transverse and longitudinal oscillations and the direction CL of the ray is the least possible. In a *polarized* wave ∠RCL is the greatest possible, and in a *partially polarized* wave it lies somewhere between the two extremes. In this way of thinking one associates the state of polarization with the direction of the resultant with respect to the direction of the ray.

Take next the case of mixing two coherent waves that are polarized perpendicularly to one another. Let us use Fresnel's first hypothesis—that such waves contain both transverse and longitudinal oscillations, but that the resultant of these in the one wave is perpendicular to their resultant in the other wave (see figs. 7.3 and 7.4). In figure 7.3, C_1R_1 is the resultant of the two oscillations in wave 1. C_2R_2 is the resultant in wave 2. When we compound the two waves, the total resultant will always be CR, which does not vanish—

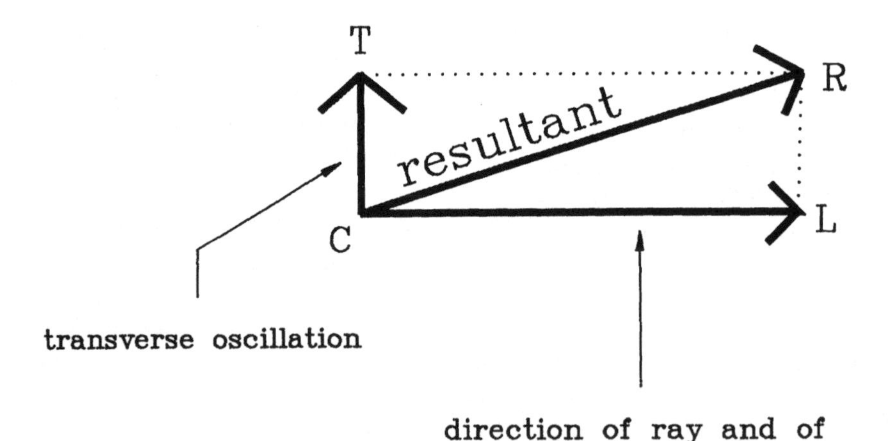

FIG. 7.2 Mixed conditions in a single wave.

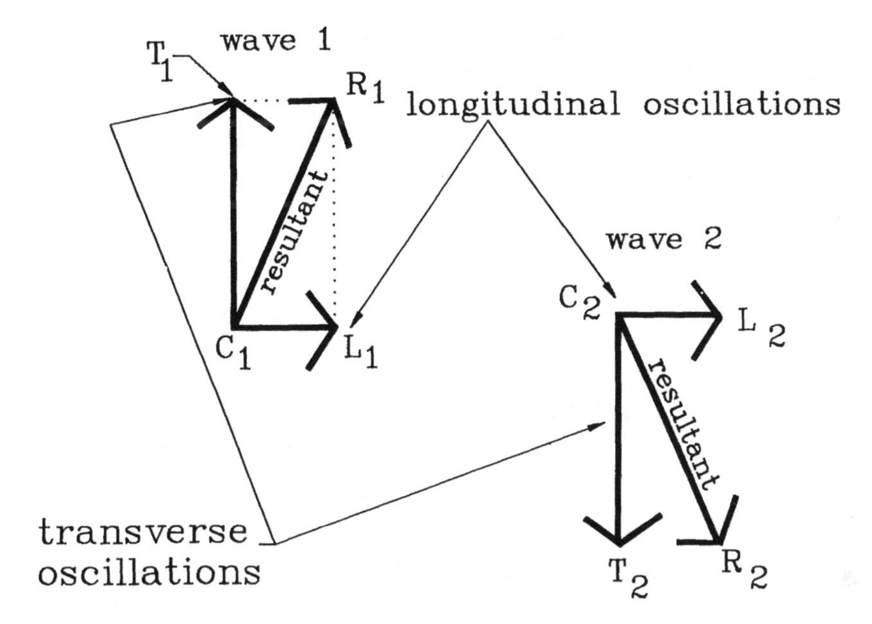

FIG. 7.3 Perpendicularly polarized waves.

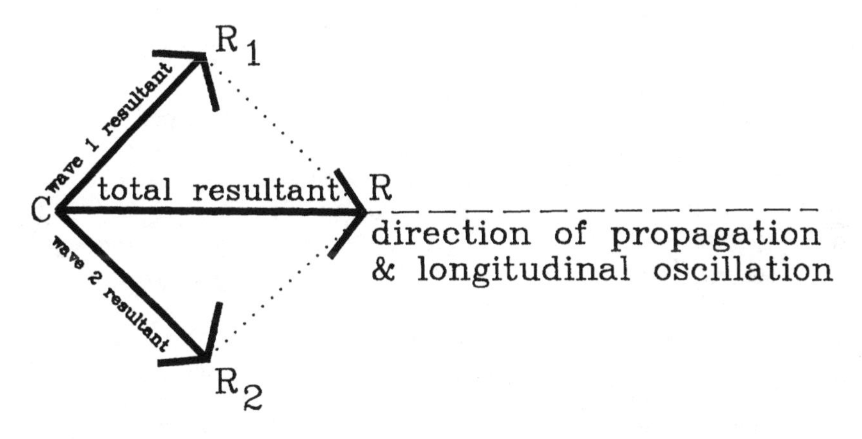

FIG. 7.4 Compounding wave 1 with wave 2.

so that the waves do not interfere with one another. But what is particularly striking here is that CR, in this hypothesis, *lies along the direction of the ray.* In other words, the wave that results from compounding two perpendicularly polarized waves is itself *unpolarized*—it is purely longitudinal. We can, as it were, imitate the lack of polarization of a single wave by compounding two polarized waves.

It is very important to understand (as this example shows) that for several years Fresnel did not associate phase differences with anything beyond the actual production of interference patterns. He had, to be precise, no notion that phase has anything at all to do with the state of polarization of a wave. That is determined entirely by the ratio between its longitudinal and transverse components; whatever phase difference the components may have is not important for the wave's polarization per se. However, it *is* important for explaining why waves polarized at right angles to one another never interfere.

Fresnel's way of thinking was intimately associated with his belief that polarized light is light in which the longitudinal oscillation has been reduced to a minimum but perhaps not eliminated altogether. In 1821—after he had fundamentally altered his ideas—Fresnel publicly discussed his earlier refusal to assume that polarized light can contain *only* a transverse oscillation. He remarked:

> While I was occupied with drafting my first memoir on the coloration of crystallized laminae (in September 1816), I remarked that polarized luminous waves act on one another like forces perpendicular to the rays, which would be directed in their planes of polarization, because they neither enfeeble nor fortify one another when their planes are at right angles, and that two systems of waves show opposite signs independently of their path difference when their planes of polarization, at first united, separate and then return to a common plane, in setting themselves along one another's prolongations. M. Ampère, to whom I communicated these experimental results, had the same thought about the opposition of sign resulting from the course of the planes of polarization. We both felt that these phenomena would be explained with the greatest of simplicity if the oscillatory motions of polarized waves took place only in the plane of the waves. But what happens to the longitudinal oscillations along the rays? How are they destroyed by the act of polarization, and how do they not reappear when polarized light is reflected or refracted obliquely by a glass plate? (Fresnel 1821c, 629–30)

Here again we see that Fresnel had at first entirely rejected the possibility that *unpolarized* light might itself consist solely of a transverse oscillation. Indeed, it almost certainly simply did not occur to him for several years that such a thing might be possible. He in fact found the issue so confusing that when, in 1819, he attempted to develop formulas for partial reflection he at first ignored transversality altogether, as we shall see in the next chapter.

Because the issue here is a profound and confusing one, it is worthwhile pushing it a bit further. Fresnel's recollection in this quotation might seem to imply that the major, even the sole, difficulty he had had was in conceiving how to uncouple transverse from longitudinal motions. That is, the problem might seem to be exclusively a dynamic one, a question of how to build a structure for the ether that would have the correct properties. This was without doubt a central difficulty for Fresnel, as were other dynamic issues that arose in diffraction theory in connection with Huygens's principle. But like those problems, this one probably would not have blocked him for nearly five years had he thought that purely transverse waves would enable him to solve every difficulty presented by polarization—as he eventually did decide. When, for example, Fresnel had begun to think it was essential to shift the radiating point in diffraction theory from the diffractor edge to a point on the wave surface near it, he adapted his dynamic understanding to necessity. He could have done the same here and later provided a justification for it in the ether's structure (which is, in fact, what he finally did do).

It therefore seems probable that something in addition to the structure of the ether troubled him or, rather, shaped his understanding of polarization. And that, I argued above, was the continuing hold on him of the idea that the general state of polarization involves the mixture of two temporally stable extreme states: one state, which represents unpolarized light, involves predominantly longitudinal oscillations, whereas the other extreme involves predominantly, but probably not exclusively, transverse oscillations.

Fresnel's problem was therefore not that he had ever had any insuperable objections to permitting the ether to support transverse as well as longitudinal oscillations. Far from it, though Fresnel and Ampère did not at first even think to entirely replace longitudinal with transverse oscillations for polarized light. But they did think it was entirely possible to add a transverse component to explain how *unpolarized* light can acquire an asymmetric property that it in itself does not possess. But, Fresnel soon found, this combination cannot easily be adapted to the empirical demands of *polarized* light. Accordingly it occurred to both him and Ampère (to judge from the 1821 remarks) that the properties of polarized light could best be explained if it were entirely transverse. However, this in turn raised the problem of how, in effect, to transform a longitudinal into a transverse oscillation by reflection. And there had to be a longitudinal oscillation before the light was polarized, because that was the only way to understand how unpolarized light can show no asymmetry and can always interfere with itself. The sense of an entire vocabulary that had been constructed to talk about "polarization" had to be changed in profound though subtle ways in order for Fresnel to move away from this *spatial* distinction between the polarized and the unpolarized.

Despite Fresnel's difficulties in developing a clear understanding of polarization, he did not need much more than the basic requirement that rays polarized at right angles do not interfere with one another in order to generate an alternative to Biot's account of chromatic polarization. This new account led in 1821 to a major controversy with Biot, one that marks the first public encounter between a selectionist and a wave theorist. We must examine this explanation to grasp the development of Fresnel's understanding of polarization.

7.2. Fresnel's First Explanation of Chromatic Polarization

Fresnel, like Young before him, attributed chromatic polarization to interference (Young 1814). However, Young had assumed that the interference takes place between the ordinary and the extraordinary waves in the crystalline lamina, which Fresnel now knew to be impossible because they are polarized at right angles to one another. He accordingly went on to deploy this very lack of interference to produce a new theory that explains the necessary role of the analyzer and accommodates Biot's observations. This theory is most important for understanding how Fresnel thought about interference between polarized beams even though he had not as yet adopted the hypothesis of transversality.

The core of the theory was simple. Assume that the lamina acts precisely like a birefringent crystal. It forms two beams according to Malus's law from the original, polarized beam. Each of these beams, on emerging from the lamina and entering the analyzer, generally produces two other beams. Interference occurs *in the analyzer* in pairs of the ordinary beams, on the one hand, and the extraordinary, on the other.

Retaining Biot's symbols and assuming normal incidence Fresnel supposed F to represent the intensity of the incident, polarized beam. Then in the lamina two beams appear (see fig. 7.5).

$i \equiv$ the angle between the plane of polarization of the ordinary light and the optic axis of the lamina.
$\alpha \equiv$ the angle between the plane of polarization of the ordinary light and the principal section of the analyzer.

ordinary $F_o = F \cos^2 i$ polarized along i° {bd°}

extraordinary $F_e = F \sin^2 i$ polarized along $i^\circ + 90^\circ$ {bc}

In the analyzer, whose principal section lies at α to the original plane of polarization, each of these divides again to produce:

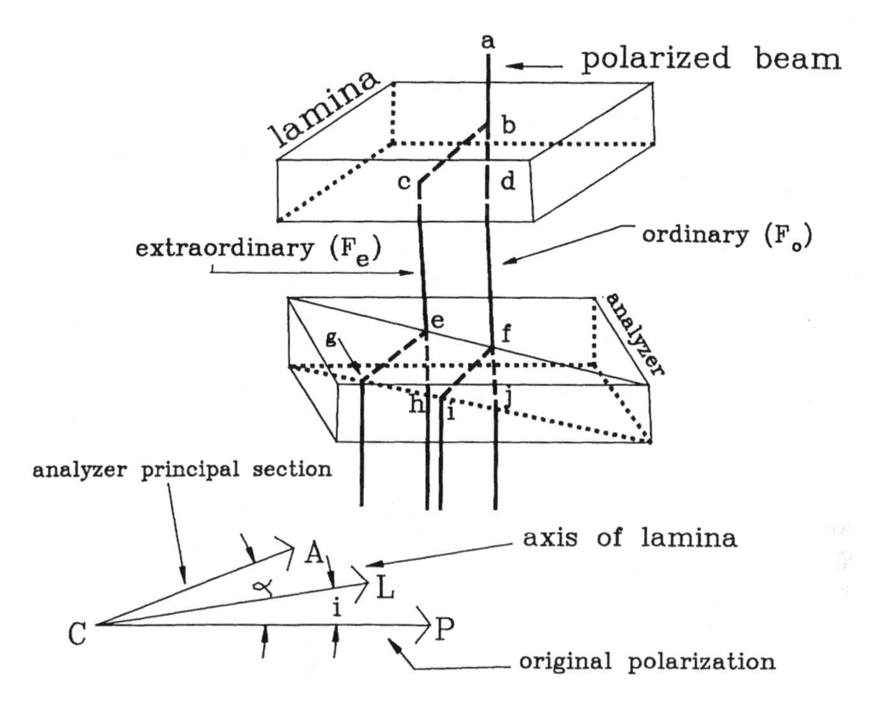

FIG. 7.5 Fresnel's interfering rays in chromatic polarization.

$$\text{ordinary} \quad \begin{cases} F_{oo} = F\cos^2 i\cos^2(\alpha - i)\ \{\text{fj}\} \\ F_{eo} = F\sin^2 i\sin^2(\alpha - i)\ \{\text{fi}\} \end{cases}$$

$$\text{extraordinary} \quad \begin{cases} F_{oe} = F\cos^2 i\sin^2(\alpha - i)\ \{\text{eh}\} \\ F_{ee} = F\sin^2 i\cos^2(\alpha - i)\ \{\text{eg}\} \end{cases}$$

Here F_{oo} and F_{oe} derive from F_o, while F_{eo} and F_{ee} derive from F_e. Since F_o and F_e travel at different speeds, there is a phase difference between them. As it stands the ordinary pair of rays in the analyzer (F_{oo} and F_{eo}) have the same phase difference between them as the extraordinary pair (F_{oe} and F_{ee}) have between themselves. Consequently the two images—the ordinary and the extraordinary—should have the same color. But they are complementary. To accommodate the complementarity of the colors, I shall for the present simply assume with Fresnel that the two waves in one of the images have an extra half-wavelength phase difference.[6]

These, then, are Fresnel's first alternatives to Biot's selectionist formulas. As they stand the formulas cannot in any way be reduced to Biot's, nor his to

Fresnel's, because Fresnel's are not actually formulas at all; one cannot simply add F_{oo} to F_{eo}, or F_{oe} to F_{ee}, because the expressions take no direct account of the phase differences due to the different paths the rays have traversed—nor could they, because he did not as yet know how to find the resultant of a pair of waves with an arbitrary phase difference.

Nevertheless, even these expressions could be applied qualitatively to reach many of the same conclusions Biot had obtained from his more quantitative ones. Appendix 16 provides the details of this work, including a discussion of how Fresnel was able to deal with the fact that the colors in each image were complementary to what the theory first leads one to expect—a fact that would later receive its fullest explanation in the hypothesis of transverse waves.

In this form Fresnel's account was not a marked improvement over Biot's because it could do nothing that Biot's account could not also do. Furthermore, Biot could in principle compute chromatic effects using Newton's color circle, since he could determine the loci in the circle of the two parts into which the lamina divides a white beam. Fresnel could do nothing similar. In comparing his theory with Biot's Fresnel could not therefore point to its greater computational power, since it had none. But he did claim it was superior to Biot's because it does not require thin laminae to behave differently from thick ones.[7] Further, Fresnel regarded the necessity of having to assume that the polarizations of Biot's *U* and *A* beams become fixed only when they leave the lamina to be "difficult to admit," since "It does not seem that the surfaces of crystals have a polarizing action on light different from that of transparent bodies" (Fresnel 1816c, 438)—and the surfaces of transparent bodies do not act in this fashion.

These themes were to be repeated later in the polemic with Biot instigated by Arago. But what Fresnel never remarked was that the quantitative structure of Biot's theory did not depend upon emission principles but only upon the individual reality of rays. So Biot did not consider Fresnel's criticisms here to be fatal ones. Nevertheless Fresnel had certainly put his finger on the weakest point in Biot's theory: its difficulty in explaining why thick laminae behave differently from thin ones. For on this point Biot could not provide even a criterion for distinguishing the thick from the thin, whereas in Fresnel's theory the distinction is precisely the same as the distinction between thick and thin gaps in Newton's rings.

Because this inability of Biot's theory was one of Fresnel's most telling criticisms of it, I shall spend a few moments drawing it out a bit further. In the wave theory a large distance obliterates the effects of interference, other things being equal. Fresnel never discussed this point in any detail, evidently thinking it was obvious. And it is fairly simple. Suppose we have a plate whose thickness is $n\lambda$, where λ is some wavelength, and the integer n must be very large. Consider another wavelength λ' that is so close to λ that $\lambda' - \lambda$ is extremely small—so small that it requires the very large number n times $\lambda' - \lambda$ to produce $\frac{1}{2}\lambda'$. Then we have:

$$n(\lambda' - \lambda) = \frac{1}{2}\lambda' \rightarrow n\lambda = \left(n + \frac{1}{2}\right)\lambda'$$

That is, where constructive interference occurs for λ, there will be destructive interference for λ'. However, by assumption $\lambda' - \lambda$ is extremely small, which means that the colors that correspond to λ and λ' may be indistinguishable. Consequently, and despite interference for each wavelength, the visual field will show no apparent interference pattern. If the plate is thin then λ' cannot be close to λ because n will not then be very large. Consequently the pattern becomes visible. This explains the absence of interference fringes for white light striking a thick plate or film.

And even if homogeneous light is used, fringes will not appear with thick plates. The reason is much the same even though here we have only a single wavelength. With a variably thick plate (as in a Newton's rings apparatus) the successive rings decrease in width until they are finally too thin to distinguish from one another, so that in either homogeneous or inhomogeneous light the field becomes sensibly uniform at large distances from the center. With a uniformly thick plate the distance between points on the plate surface that is necessary to change constructive to destructive interference becomes smaller as the plate thickness increases, so again overlapping occurs in both homogeneous and inhomogeneous light. And all of these explanations for the effects of thickness can be carried over almost without change to the case of chromatic polarization.

Ironically, then, Biot's own argument that the difference between thick and thin laminae in chromatic polarization is the same as the difference between thick and thin films in Newton's rings holds in the wave theory, but it does not easily hold in his own theory. For unlike Fresnel, Biot could not explain what there is about chromatic polarization in thick laminae that corresponds to the decreasing widths of Newton's rings with thick air gaps. Here then was the true weakness of Biot's theory in comparison with Fresnel's. But it was hardly a critical weakness in view of the fact that Fresnel could not calculate colors whereas Biot could. In any case the problem was not that Biot's theory contradicted something that was known to be true, but rather that it could not easily explain the thing.

7.3 The Reflection of Polarized Light

Within a year Fresnel had begun to examine how polarization behaves in reflection, where many of the same kinds of issues I have just discussed also arise. This had been Malus's major concern, and it had led him to create selectionism. In a remarkable introductory paragraph to a memoir that was not printed (November 1817), Fresnel rejected the most basic of Malus's selectionist claims, a claim no one previously had questioned:

> A very simple experiment, and one in which I scarcely expected to obtain a new result, led me to the discovery of the singular phenomena that are the object of the memoir I have the honor to present to the Academy. In receiving on an unsilvered glass a beam of light divided in two by the action of a rhomboid of calcspar, and observing the double image of the illuminating aperture with another, I noticed that rotating the second rhomboid always made each of the four images successively disappear, whatever the angle of incidence and the azimuth of the plane of reflection with respect to the plane of primitive polarization. Many liquids I substituted for the glass having produced the same result, I concluded that *completely polarized light still conserved this property after its reflection by transparent bodies.*
>
> It seems that this observation has escaped scientists who have concerned themselves with polarization, because M. Biot doesn't mention it in the chapter of his treatise on physics where he speaks of the influence of reflection on polarized light.
>
> I also consulted Malus's memoir concerning the same object, read to the Institute on 27 May 1811 [Malus 1811a]. But the manner in which he exposes the results of his experiments makes me suppose that he regarded the light as partially depolarized by its reflection on the glass, when it took place in another plane than that of the primitive polarization, and at an inclination larger or smaller than 35°25′; because he says that the reflected light contains: (1) a portion of light polarized with respect to the primitive plane of polarization; (2) another portion polarized with respect to the plane of incidence. Furthermore, if he had noticed that the reflected light conserves all the appearances of complete polarization he would no doubt not have neglected to mention it.

> Having trouble believing that this observation could have escaped such an able physicist, I was on the point of suspecting, despite analogy, that possibly light polarized by reflection behaves differently, in this circumstance, than light polarized by the action of a crystal, and that it was the first Malus had used. I even repeated these experiments with a more convenient apparatus, which M. Arago was good enough to lend me, and I always saw each image entirely disappear during a rotation of the rhomboid, when the incident light had been very completely polarized by the preliminary reflection. (Fresnel 1817b, 441–42)

Fresnel insisted, then, that polarized light remains polarized when it is reflected from transparent bodies. For him this was not a necessary fact: it had always been entirely possible that light remained polarized after reflection or that it was partially depolarized. Since in Fresnel's understanding at least through 1820 the general state of polarization involved a mixture of two extreme conditions (longitudinal and transverse oscillations), it was a question of fact to determine what the actual mixture was after reflection. However, though Fresnel did himself think in terms of mixtures, nevertheless his concept of a mixture was radically different from that of his selectionist contemporaries, since his concept applied to a single ray whereas theirs applied to collections of rays.

Because of this difference Fresnel did not appreciate that Malus could be correct in arguing that the reflection might be partially depolarized despite Fresnel's own observations. He did not see that, if the extra polarized beam that selectionists used to explain the effect of reflection is sufficiently small, then Malus could be correct in claiming that reflection partially depolarizes light.[8] Fresnel missed this point because he did not think in terms of rays, and it is the ability to manipulate rays in groups that makes it possible for Malus's theory to work well. Consequently Fresnel felt that Malus must have made an almost inconceivable error in his experiments. Indeed, he found it so hard to believe that Malus could have missed the observation that he at first thought there might be something different about light polarized by a crystal and light polarized by reflection. That is, perhaps light polarized by double refraction, which Malus had used in his work, behaves differently when reflected than does light polarized by reflection. But Fresnel found no difference at all.

This apparently new discovery was of capital importance, and Fresnel made it a "general principle" that polarized light "does not at all lose this property in its reflection by transparent bodies." The only thing that happens is "a change in the azimuth of polarization." This set a plan of experiment:

to observe the deviation of the plane of polarization by reflection. Fresnel first established a criterion to use for judging whether the reflecting surface has in fact produced an effect. He remarked that, if there are no actions except for pure reflection, then the light's "plane of polarization ought not experience any other change of direction than that which results from the reflection itself, all of the motions of the fluid being symmetrically reproduced in the reflected beam; *so in this hypothesis the plane of polarization of the reflected ray would be the image of the plane of polarization of the incident ray*" (Fresnel 1817b, 444; emphasis added).

But which plane of symmetry produces the "image" Fresnel refers to here? According to him, *the symmetry plane must be the plane of reflection itself* and not (as one might expect) the plane normal to it. This means that the position of the plane of polarization after reflection will lie on the *opposite* side of the plane of incidence from its locus before reflection if there is no other action than the reflection itself. Fresnel apparently had some sort of a model in mind based on what happens when a moving fluid element is reflected. But it is hard to see what sorts of motions would behave in this way.

In any event, Fresnel polarized a light beam and then reflected it at various angles, holding its original plane of polarization fixed at 45°. He found that at low incidences there is little effect in his terms. That is, the plane of polarization merely flips across the plane of reflection. However, as the incidence approaches the polarizing angle, the plane of polarization of the reflected light gradually moves toward the plane of incidence. Above the angle of polarizing incidence, the plane remains on the same side of the plane of incidence as before reflection. Consequently, in Fresnel's terms it has been deviated by some action of the surface from where it would have been were there no effect. Finally, at grazing incidence the plane apparently remains nearly the same as before reflection (see fig. 7.6). Furthermore, Fresnel found that the deviation from the symmetric position depends in some way on the initial direction of polarization. He hoped to uncover the law of the deviation by independently altering the angle of incidence and the angle of polarization, but to this point he had, he admitted, not succeeded.

Fresnel pursued his investigations further to examine what happens to polarized light when it undergoes total internal reflection. Although he went to some effort to explain the effects this kind of light produces, as with ordinary polarized light he was blocked by his inability to compute the resultant of two waves with an arbitrary path difference. (Appendix 15 discusses in detail this important work, work that, like his account of chromatic polarization, re-

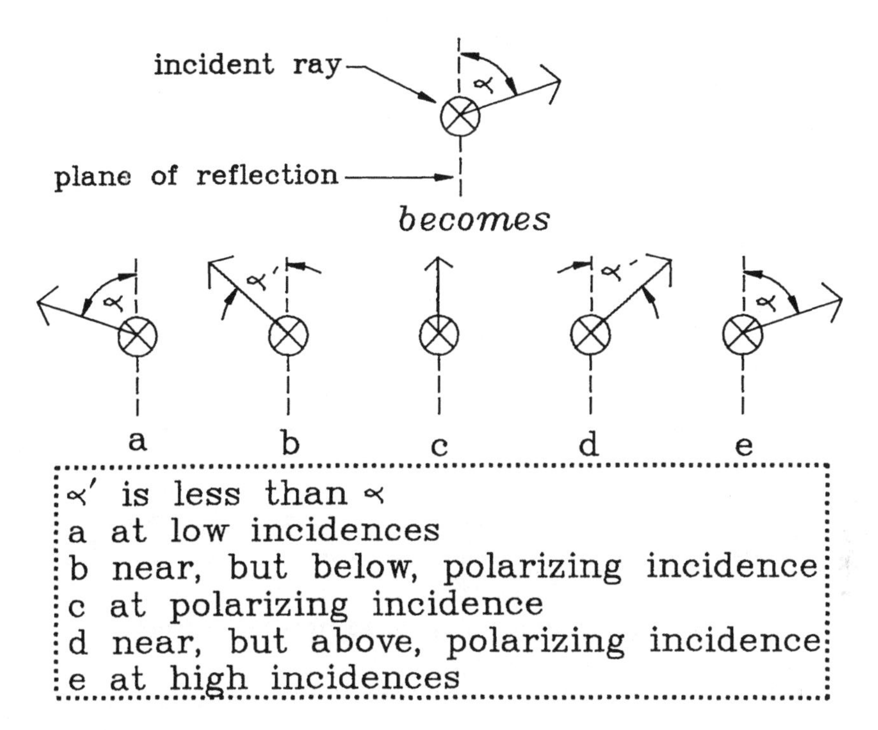

FIG. 7.6 The effect of reflection on polarization; looking along the ray from behind it.

ferred effects to the influence of the analyzer in causing polarized beams to interfere with one another by bringing their polarizations to the same plane.) Within two months he had solved that problem, as we have already seen. And with that he was able for the first time to compete with Biot's chromatic calculations for ordinary polarized light[9] and to provide as well a quantitative analysis for the chromatic effects produced by polarized light that has undergone total internal reflection. Yet here too there is not the least indication that Fresnel had made any substantial changes in his understanding of polarization. Or perhaps it would be more accurate to say that he had extended his initial understanding along lines that are essentially consistent with it.

8 Transverse Waves

8.1 The Effect of Interference on Polarization

Fresnel's memoirs on polarized light before 1821, most of which were not published until the *Oeuvres*, often generate a peculiar feeling in the modern reader, a sense that one is reading familiar words but that there is something odd about the way they are put together. Fresnel may, for example, fail to draw a conclusion that to the modern eye seems almost explicit in his discussion; or perhaps he does not mention what we today consider essential in a given case. These seeming incongruities are all due to the way Fresnel for a long time thought about the principle of interference, a way that, in the case of polarized light, was intimately linked to the experimental apparatus—the analyzing crystal—that he invariably used to examine the light.

Precisely because Fresnel had demonstrated that waves polarized at right angles to one another never form fringes, he remained for some time convinced that such waves must always be treated entirely independently of one another. Accordingly, the purpose of the analyzer was to make it possible for the principle of interference to be used by instrumentally forcing the two waves to interact with one another in the only meaningful sense Fresnel could at this time envision—fringe formation. One may say that the analyzer in effect substituted for a theory that specifies the resultant of the oscillations. Since it physically unites two waves that do not otherwise form fringes with one another, and since every form of experiment with polarized light used either an analyzer or (equivantly) a reflecting mirror, Fresnel did not have to know what such waves do to one another before the apparatus acts upon them in order to calculate what occurs in the analyzer itself. He was therefore able to develop a group of theories for particular phenomena based on the hypothesis that the waves are split into independent parts that the analyzer later reunites. These theories lead to formulas that in fact agree with the results we would obtain today, and that Fresnel himself obtained in and after 1821.

Where, to put it somewhat too precisely, we today (and Fresnel after about 1820) see the effects of phase differences on the projection of the oscillation into two mutually perpendicular waves, Fresnel saw only the enforced superposition, in a fixed direction determined by the experimental device, of independent (though coherent) waves with fixed polarizations. Where we today see the components of a single, directed physical quantity, Fresnel saw two waves that have no necessary intrinsic relation to one another.

The clearest example of how this works in practice can be found in Fresnel's discussion of the chromatic effects produced by quartz and by certain liquids. A salient characteristic of these colors is that they do not change when the quartz plate is rotated, holding the original plane of polarization and the analyzer's principal section fixed in position. The production of colors by polarized light, according to Fresnel, here involves the enforced superposition of sets of independent waves that are polarized at right angles to one another and that have traveled different distances from a common source. Consequently the colors that are produced by quartz crystals (cut normal to their optic axes), and by certain liquids, must literally involve a species of *double refraction*.

Already in January Fresnel had developed a way to simulate the behavior of quartz: the gross phenomena can be imitated by a process of two pairs of total internal reflections separated by a double refraction (see appendix 4.2). That process begins with a polarized beam and then generates from it by total internal reflections within a glass rhomb two waves that are polarized at right angles to one another and that have a path difference of a quarter-wavelength. Each of these waves then passes through a thin crystal lamina and so generates within the lamina an ordinary and an extraordinary beam, yielding four waves (and new phase differences) in all. Finally, from each of these four two new waves, polarized again at right angles and with a quarter-wave phase difference, are generated by a last pair of internal reflections. These eight waves—four at each of the two orthogonal polarizations—differ from each other in phase in such a fashion that, when brought to a single plane by an analyzer, they produce by interference a color pattern that is independent of the orientation of the device between the original direction of polarization and the analyzer.

Fresnel did not believe that quartz consisted of microscopic birefringent sandwiches. Rather, the purpose of his model was to provide a physical embodiment of the splitting and recombination of waves that had to occur to produce the right kinds of chromatic effects. Whatever effects these independent waves have upon one another before they are united in the analyzer was not, at this point, of any interest to Fresnel. He was concerned only with the production of colors by interference, and interference occurs when, and only when, the several waves are joined together by the analyzer.

Whatever the actual mechanism that produces the separation and recombination, any substance that behaved chromatically like quartz cut normal to its axis had to be birefringent in the limited sense that it must always generate from any given beam two others that are orthogonally polarized and that travel at different speeds. Biot had discovered that certain liquids, turpentine among them, do behave chromatically like quartz, and these had to be birefringent in Fresnel's sense. In principle one could indirectly detect this "double refraction," and Fresnel was actually able to do so scarcely a month after achieving his equation for combining waves with different phases (Fresnel 1818d).[1]

In all his discussions to this point Fresnel had maintained a strict physical separation between the waves that in the end are forced to interfere with one another. This enforced distinction is particularly striking here, in the case of quartz, because of what the net effect in the analyzer will be. Fresnel had found that the ordinary image will be proportional to

$$\sqrt{\frac{1}{2} + \frac{1}{2}\cos[2i + 2\pi(o - e)]}$$

where i is the angle between the initial plane of polarization and the principal section of the analyzer and $o - e$ represents the path difference (in wavelengths) that arises within the crystal lamina in Fresnel's model (see appendix 4.2). In January he left the expression in this form. But in March (1818d, 669) he noted that it can be reduced to:

$$\cos[i - \pi(e - o)]$$

And, he further noted, the extraordinary image will therefore be:

$$\sin[i - \pi(e - o)]$$

But he goes no further. Though he was well aware of it, he does not discuss the obvious implication that this is equivalent to rotating the original plane of polarization of the incident light through an angle equal to $\pi(o - e)$. He does not mention the point because that is precisely how selectionists discussed the phenomenon—as a rotation of the original plane of polarization. Fresnel was concerned instead to show how a rather intricate process of wave splitting and recombination can produce by interference in an analyzer the right sorts of expressions. If the analyzer is not there, then the interference does not necessarily occur, *and the light as a whole may not constitute a single polarized wave as it progresses through the quartz or liquid medium.* It may be only a mixture of several differently polarized waves. When these waves are physically forced to the same plane by the analyzer, they interfere in just the right way to mimic a single polarized wave that has been rotated through the angle $\pi(o - e)$. Interference, one might say, imitates polarization.

An undated note that Verdet printed in the *Oeuvres* (which I place sometime during the summer of 1818)[2] shows that Fresnel was thinking about this question—about whether there is a difference between the actual state of the light and its state as revealed by the analyzer. Biot, recall, required the emergent beam from a thin crystal to be polarized at either 0° or at $2i$, or at both, depending on the thickness traversed. Fresnel (in a criticism repeated later, which I shall discuss further below) did not see how Biot could have light simultaneously polarized at both azimuths and insisted that it had, by Biot's own reasoning, to be at one or the other. For my present purposes the important point is that Fresnel's formulas for thin crystals imply that at certain laminar thicknesses the light that emerges from the lamina will behave *as though* it were polarized at either 0° or $2i$.[3] Of this Fresnel remarked, referring to the two orthogonally polarized waves that by his account actually emerge from the lamina: "I will not dispute the appearance [*apparence*] or the reality of this polarization, since in the theory I have adopted the particular system of two waves [here in question] must present all of the properties of light polarized along a single plane, and so probably the one is the equivalent of the other" (Fresnel 1818e, 531). To say, as Fresnel does, that system A (two orthogonally polarized waves) is the "equivalent" of system B (a single polarized wave) because they present the same "properties" implies that they *might not* have been equivalent—had their "properties" been different, they would not have been "equivalent." That they are equivalent therefore has to be demonstrated, which Fresnel had done by means of the principle of interference. Accordingly, Fresnel was still drawing a distinction between the physical conditions of a single polarized wave of light and of a pair of orthogonally polarized beams even when the latter system produces the same effects in an analyzer as the single wave.[4]

Indeed, Fresnel's overall understanding of polarization had not as yet (1818) changed in any fundamental manner, as we can see from the following remarks he made at this time on the difference between polarized and unpolarized light:

> The oscillatory motions of the diverse points of the ether, in ordinary light, are all directed perpendicular to the wave or, if there are oblique motions, take place all round the normal at the same obliquity and with the same degree of energy in all azimuths. What characterizes, on the contrary, the vibrations of polarized

> light is that they do not take place in the same manner in all azimuths, and that the oblique motions I just mentioned do not have the same energy or the same obliquity all round the normal, or perhaps they take place only in a single plane, that of polarization. (Fresnel 1818e, 526–27)

Either natural light contains only a longitudinal motion, or—and this was a new suggestion—it consists of a simultaneous mixture of equally intense oblique oscillations in all directions round the normal. That is, natural light was either an extreme instance or else a mixture of very many independent oscillations, each of which would, if it alone occurred, constitute polarized light. That in certain circumstances two such oscillations, if at right angles to one another, can become "equivalent" to a third oscillation in a different direction has little apparent general significance. Fresnel to this point continued to think in terms of mixtures of waves forming *by interference* a net effect and combinations of conditions forming a general state of polarization. The two—wave mixtures and combinations of states—were not at all the same, since the former involves several independent waves whereas the latter constructs a single wave out of two extreme states. As before, this duality prevented Fresnel from developing a unified understanding of polarization that could open new mathematical paths. Ironically, in view of the vast new ranges opened by the principle of interference, here it blocked progress because of its intimate association with a temporally stable fringe pattern.

8.2 Polarization Set in Motion

At some point between the end of 1819 and the early spring of 1821 Fresnel solved the problem of polarization. Unfortunately we do not, and probably will not, know precisely when or how he managed to do so. There is no documentation of any kind for 1820 or early 1821. Nevertheless, we have been able to narrow the issues sufficiently to understand what Fresnel must at some point have realized, even if we do not know when or precisely how he came to that realization.

By the end of 1819 Fresnel had thoroughly mastered the technique of mixing coherent polarized waves with different phases, and he already understood that a mixture of two coherent waves that are polarized at right angles to one another may be "equivalent" to a single polarized wave. At some point during 1820 or early 1821 Fresnel began to think about the principle of interference in a new way, one that united all these effects by means of a single physical model. And the key to this model was a thoroughly new way of understanding the natures of *unpolarized* and *partially polarized* light. Up to this time Fresnel had thought of a partially polarized wave as one that contains both a longitudinal and a transverse oscillation, whereas an unpolarized wave either is completely longitudinal or else contains equal transverse oscillations all round the ray. The purpose of thinking this way was twofold: first, to explain why unpolarized light can always be made to interfere with itself; second, to explain why partially polarized light exhibits asymmetric behavior on analysis. The existence of interference required oscillations to take place to some extent at least in the same direction, whereas asymmetric behavior required that, also to some extent, they take place in different directions.

At some (unfortunately undocumented) point Fresnel realized that both of these functions can be fulfilled by a single mechanism if one assumes that light of every kind is always *completely asymmetric* and that a distribution of asymmetries over time can always replace the physically distinct categories of unpolarized, partially polarized, and polarized that he had previously deployed. In other words, far from "polarization" in the old sense of asymmetry being an unusual condition, light is *always* "polarized." However, since the eye cannot follow the rapid changes in the asymmetry, waves that behave in this way will seem on analysis to be unpolarized or partially polarized. This kind of light will also be able to produce a stable fringe pattern, since at any given moment the polarization is always fixed in some direction.

Furthermore Fresnel's and Arago's otherwise puzzling third law for the interference of polarized light (that orthogonally polarized waves brought to the same plane of polarization can interfere only if they derive from a polarized wave) can now be understood as an implication of their fifth law (that one must add a half-wavelength path difference when the planes of polarization are brought back together by deviation through an angle of 180°). In Fresnel's words:

> [The new hypothesis] explains why two beams of direct light, which have been polarized at right angles, present no appearance of mutual influence when they are brought to a common plane of polarization by the action of a pile of glass [plates] or a rhomboid of calcspar. It is not that they then exert no action at all on one another, because, independent of mechanical considerations, this supposition would be too contrary to analogy; but it is that the effects produced by the different systems of waves of the direct light compensate and neutralize one another. In effect, one may conceive direct light to be the assemblage, or more exactly the rapid succession of systems of waves polarized in all directions,

> and such that there is always as much polarized light in one plane as in the plane perpendicular to it; now, it results from the rule that we have just expressed [rule 5], that if, for example, one must add a half-undulation to the difference of the paths run through to calculate the extraordinary image produced by the light polarized in the first plane, one must not add it for the extraordinary image that results from the light polarized in the second; so that the two tints that they bring together or successively in the extraordinary image are complementary. The compensation established in this way, and in the same fashion for all directions, forbids perceiving the effects of interference. (Fresnel 1821c, 611–12)

However, with this hypothesis alone one cannot explain any better than before why Fresnel's and Arago's rules work. Moreover, as it stands one would be at a loss to explain why the asymmetry sometimes rotates and sometimes does not—whereas before one could at least understand that a wave might have both a longitudinal and a transverse component. Accordingly Fresnel did not introduce this new hypothesis without also suggesting a unifying physical interpretation for it: namely, that the asymmetry in question occurs *entirely* in the plane of the wave, so that light waves are always completely transverse. Fresnel carefully emphasized this idea in the spring of 1821, and it is worth quoting him at some length because it is the only evidence we have from near the period when he developed his new understanding:

> If the polarization of a luminous ray consists in this, that all its vibrations occur in the same direction, then it results from my hypothesis on the generation of luminous waves that a ray emanating from a single center of disturbance will always be polarized in a given plane, at a given instant. But an instant after, the direction of motion changes, and with it the plane of polarization; and these variations succeed one another as rapidly as the perturbations of the vibrations of the luminous particle; so that, even if we could separate the light that emanates from it from that of the other luminous points, one would never find the slightest trace of polarization. . . . *So that direct light may be considered to be the reunion, or more exactly the rapid succession, of systems of waves polarized in all directions. According to this way of looking at things, the act of polarization consists not in creating transverse motions, but in decomposing them in two fixed, mutually perpendicular directions, and in separating the two components*

the one from the other; because, in each of them, the oscillatory motions will always operate in the same plane. (Fresnel 1821c, 635–36; emphasis added)

Implicit in this physical hypothesis is a new, and difficult, understanding of the function of the principle of interference, one that permits it to be used in a very general manner: it has now become a method for determining simultaneously the direction in the wave front as well as the amplitude of the resultant oscillation rather than solely a method for calculating a fringe pattern. There were two aspects to this new understanding. First, the longitudinal oscillation that Fresnel had for so long compounded with a transverse motion to form partially polarized light, and that predominated in unpolarized light, has been replaced with an oscillation in the front proper. It has, as it were, been rotated 90° from the ray and into the front. Second, the front also contains a second oscillation that is orthogonal to the first one *but that, in general, differs from it in phase.* The two extreme conditions that Fresnel had previously used to constitute by composition partially polarized light have in effect now been replaced by two orthogonal oscillations in the front.

These oscillations have also replaced the independent waves that, before, Fresnel had melded by interference only in the analyzer itself. Two coherent waves that are polarized at right angles to one another and that have a given phase difference, Fresnel now reasoned, always have a resultant in the wave front whose locus in the front as a function of time depends on their phase difference. Conversely, a wave whose polarization varies in a given way over time can always be *decomposed* into two orthogonally polarized waves with a certain phase difference. Where previously the sole function of a phase difference (i.e., of interference) was to alter the amplitude of the resultant in an analyzer, it now determines the direction and magnitude of the resultant in the front at every moment.

And so the intricate procedures of wave mixing Fresnel had previously envisioned could now be understood as methods for decomposing a single wave into components that are affected in different ways in their phase and amplitude. The method of mixing sets of orthogonally polarized waves that Fresnel had developed by the end of 1819 accordingly takes on a new meaning. It no longer requires the intervention of the analyzer. Rather, it reflects the nature of the oscillation as a directed quantity that exists in a given plane and that can be decomposed into components that are independently affected. The mutual independence of the waves in the mixture has been replaced by the differential action of the reflecting surface on the components of the oscillation at the moment it strikes the surface (a point I shall examine further in the next section).

Fresnel's new requirement that the oscillations must always occur entirely in the wave front raised physical questions that were quite troubling (which I will discuss below). Yet that alone cannot explain the tremendous confusion many people apparently experienced in grasping the new function of interference. Part, perhaps most, of the problem lay in the continued influence of the static understanding of polarization that had prevailed for so long. Unlike Fresnel's use (if not his understanding) of interference in diffraction, or even his use of it before 1821 to calculate the effects of mixing orthogonally polarized waves, the new kinetic understanding cannot in any way be reconciled with selectionism. It marks a definitive and irremediable break.

Fresnel did not explain his new conception in any detail in mid-1821. That emerged only a year and a half later when he gave a new explanation of double refraction by quartz.[5] In this work, dated 9 December 1822, Fresnel also begins to introduce a new vocabulary for polarization. Clearly the old terminology had to be modified in rather drastic ways. For Fresnel, just as for selectionists, it had previously required different spatial states to explain the differences between "unpolarized," "partially" polarized, and "polarized" light. Light in one state differed in the nature of its asymmetry from light in another state, and the difference between the two states did not change over time—it was a purely spatial difference. This was no longer true. Light in any of the three states is precisely as asymmetric as light in the other two, but the states are now supposed to differ in respect to what occurs over time. To adapt the old vocabulary to his new physics, Fresnel had to reinterpret it and add new terms. In 1822, after discussing quartz, he accordingly introduced adjectives to distinguish kinds of polarization:

> Considering only the facts, one could always give the name *rectilinear polarization* to what one has for a long time observed in the double refraction of calcspar, and what Malus was the first to remark in the light reflected from transparent bodies, and call *circular polarization* the new modification whose characteristic properties I have just described: it will naturally divide into *circular polarization from left to right* and *circular polarization from right to left*. These denominations, which were suggested to me by the hypothesis I have adopted on luminous vibrations, indicate the very nature of their motions in these two cases; but fearing to abuse the time of the Academy, I thought to limit my-

> self here to justifying the new names that I propose for the simple exposition of facts. . . .
>
> Between rectilinear polarization and circular polarization, there are a throng of intermediate degrees of diverse polarizations, to which one could give the name *elliptical polarizations*, according to the same theoretical views. (Fresnel 1822b, 744–45)

Here, then, "polarization" has been divided into three kinds that, Fresnel remarks, can be distinguished from one another by using the laws of interference. However, these kinds of light are all "polarized"; they are neither "unpolarized" nor "partially polarized," though circularly polarized light behaves in an analyzer precisely like unpolarized light and elliptically polarized light behaves on analysis exactly like partially polarized light. But only the polarized lights can produce colors when analyzed after passing through thin crystals or through quartz. Consequently Fresnel had also to elaborate a new sort of distinction between the *polarized* and the *unpolarized*, one that depended exclusively on the principle of interference. This was without doubt one of the most difficult, if not the most difficult, of the changes Fresnel's work required in traditional ways of thinking, and I shall accordingly pursue it further in the next section. However, to appreciate the full scope of Fresnel's work on polarized light we must also consider his theory of partial reflection, which emerged at this time and which, in its final form, depended strongly upon his new hypothesis. Here Fresnel was able to move from his incomplete first theory, which assumed unpolarized light to be completely *symmetric* at every instant, to a new, complete theory based on the idea that light is completely *asymmetric* at every instant. Appendix 17 presents these developments. They are not directly essential for the remainder of this chapter, but they are important both because they are Fresnel's alternative to the selectionist account and because, as we shall see, the most difficult aspect of Fresnel's theory for many of his contemporaries to accept—the *projective decomposition* of a polarized ray—was the basis for his final account of partial reflection.

8.3 A New Vocabulary for Polarization

The new vocabulary Fresnel had introduced was intimately linked to his understanding of the governing role in polarization played by phase and the principle of interference. The major and most difficult distinction it requires—and the one Fresnel discussed only implicitly at this point—divides the three types of *polarized* light from *unpolarized* light. The two orthogonal components in polarized light of any type have a fixed phase difference and a fixed amplitude ratio;, both remain constant over time. But the phase difference between the two components in unpolarized light, as well as the amplitudes of the components, varies over time.

The distinctions this new vocabulary requires can be very subtle. Consider, for example, how to distinguish partially polarized from elliptically polarized light. Both kinds of light behave in precisely the same way when they are directly analyzed by a crystal. However, only elliptically polarized light can be reduced to linear polarization by metallic reflection at the proper incidence and azimuth, and only it can be used to produce colors by passing it through thin crystals. In Fresnel's new view this means that the components of the elliptically polarized wave must have a constant phase difference and a constant amplitude ratio. The components of the partially polarized wave, on the other hand, must have a phase difference and an amplitude ratio that vary over time. And then the difference between partially polarized and unpolarized light is that, in the latter case, the maximum possible amplitudes of the two components are the same, whereas in partial polarization they are different.

Fresnel's new vocabulary of polarization types and phase relationships accordingly fundamentally alters the way one thinks about polarization. Now the essential distinction between the polarized and the unpolarized is between waves that have components that are fixed in their phase difference and amplitude ratio and waves that do not. In this regard one can say that, for Fresnel, unpolarized is to circularly polarized light as partially polarized is to elliptically polarized light. This was utterly different from any previous understanding, including Fresnel's own. To a selectionist, and to Fresnel before about 1821, light could be unpolarized, partially polarized, or just polarized, but there could not be different kinds of polarization. And this classificatory scheme was closely tied to the limits of human perception.

Consider the selectionist position. Whether a beam is unpolarized or partially polarized depends, in essence, on whether one can form from the rays that compose it at least one set whose asymmetries are oriented in nearly the same way and that are together numerous enough to be visible. It is always a question of grouping rays into sets that can be visibly distinguished from one another. Consider Fresnel's position before about 1821. The polarization of a light wave depended only on the proportion in it of longitudinal to transverse oscillation. If there is very much longitudinal, and very little transverse, vibration, then the wave is unpolarized. If the proportions are more equal, the wave is partially polarized. If the proportions are reversed, the wave is simply polarized. This too is tied directly to the different image intensities that the analyzer reveals, and it is an *approximate* distinction.

For the selectionist, and for Fresnel before about 1821, then, the vocabulary of polarization immediately reflects the approximate character of the phenomenon it is meant to capture—namely, the image intensities formed by light when it is directly analyzed. And so the phenomenon itself escapes an absolute description because there are no absolute physical differences between polarized, partially polarized, and unpolarized beams. The rays in each type of light are differently distributed with respect to their asymmetries, or there is a different proportion between longitudinal and transverse vibrations in a wave, but there are no other differences.

But in Fresnel's new way of thinking the vocabulary of polarization is very nearly *absolute*. Either the amplitude ratio and phase difference for the two waves (components) into which a beam can be divided are fixed over time, or else they are not. The light is either polarized or unpolarized. If the ratio is fixed, then the type of the polarization is fixed also. There is little that is approximate in this. *Polarization,* in the wave theory, is the act of imposing a temporally fixed phase difference and amplitude ratio where none had previously existed.

Although in many situations the fundamental difference in concept between ray theory and wave theory may have no practical effect,[6] it is a barrier to accepting Fresnel's analysis of chromatic polarization, because here a theory existed—namely Biot's—that relied explicitly on ray counts to compute colors and intensities, a method that tacitly presumed that the rays do not influence one another. In chapter 9 I shall discuss the angry controversy that arose over this, but it is also essential to understand that one could remain a selectionist (as Poisson apparently did) even if one accepted Fresnel's account up to his assertion that the difference between the polarized and the unpolarized is a temporal one. All one need do is think of the theory in terms of ray interactions within the analyzer, interactions that are subject to the principle of interference. Even this was not easy for many selectionists, however, and Biot, for example, found it difficult to understand what Fresnel was asserting because he thought exclusively in terms of ray counts, whereas Fresnel's theory, though it can still be thought of in terms of rays, also needs the principle of interference, which fundamentally alters the results of the count. The controversy over chromatic polarization, while certainly not conclusive, was the first point at which selectionism and the wave theory encountered one another in public.

PART 3
Controversy and Unification

9 A Case of Mutual Misunderstanding

9.1 Arago's Attack

On 13 June 1821 Fresnel wrote to his brother Léonor to tell of an angry confrontation between Biot and Arago at the Institute. He wrote:

> My dear friend, there was a great battle at the Institute in the last two sessions, at the occasion of a report by Arago on the memoir I presented to the Academy nearly five years ago, in which I attacked by facts Biot's theory of *mobile polarization* and gave another explanation of the coloration of crystallized laminae. Arago, completely convinced by the experiments I repeated before him, determined to tell the Academy that they completely overturned Biot's theory. Biot responded; the discussion was joined and became very lively, but Arago always had the advantage. It began again last Monday with a written reply by Biot in which he announced, in beginning, that he was going to prove that my experiments, far from overturning his theory, were on the contrary a striking confirmation of it; it's true that he never spoke and that he even found it more convenient to deny one of the facts than to conciliate it with his theory. I'm not angry that he denied it, because it's easy to verify, and that will make the process even clearer. . . .
>
> Something singular was that he reproached Arago for having so long delayed making my report. After refuting his principal arguments, Arago replied to this reproach, saying "that besides the causes of delay that he had explained to the Academy before the reading of his report, there was one that it might perhaps have divined: that being a certain apprehension of the lively discussion that his report could not fail to occasion, and that the Academy had just witnessed. It is altogether natural, he added, that Biot, who had conceived the theory of mobile polarization and had written two large volumes on the subject, could not see his theory attacked without chagrin. I so strongly felt that I couldn't fail to embitter him in giving a faithful account of Fresnel's memoir that I drew back, I acknowledge, for a long time from this somewhat

> distressing task; but Biot must not be angry with me for finally having fulfilled it, because he even finds that I was too long in doing it. For the rest, he added, I do not propose that the Academy adopt the body of the report, that is to say my opinion of Biot's theory, although I am persuaded that it is in error and that Fresnel's explanation is the true one. I know that the Academy cannot make pronouncements on such a subject; but I ask only that it be good enough to adopt the conclusions of the report, that is, to order the memoir [Fresnel's] printed in the *Receuil des Savants Étrangers.*"—Biot demanded that Arago's report be considered simply a memoir and not a report; but this proposal was rejected, and Arago's passed nearly unanimously. Mathieu thought that, had he wished, he could even have had the report adopted; but he did very well not to ask it. (Fresnel, *Oeuvres,* 2:853–54)

The nasty confrontation between Arago and Biot that Fresnel so vividly described marks an epoch in the history of the wave theory.[1] Scarcely two years after Fresnel had won the prize for his diffraction memoir, the Academy ordered printed, over Biot's explicit and public objections, Fresnel's account of chromatic polarization. The members did not accept the report, but apparently only because Arago had not insisted that they do so (and he in any case had the report printed in the *Annales* almost immediately [Arago and Ampère 1821]).

This was a stunning personal victory for Arago and a stinging public defeat for Biot. Years before, Biot had snatched the field away from Arago; now, by proxy at least, Arago took it back. Fresnel's work was somewhat lost in all of this, despite the pride he clearly took in recounting Arago's victory to Léonor. For Arago, despite his pious public disclaimer, was almost certainly much more interested in humiliating Biot than he was in adding another feather to Fresnel's already full cap.

The events that culminated in this bitter confrontation are somewhat tangled, in part because Fresnel's memoir was not printed until the *Oeuvres.* Instead, Arago printed Fresnel's "Calcul des teintes" in three parts in the *Annales.* However, at the end of his report Arago remarks that the first memoir discussed in it was submitted to the Academy on 7 October 1816, and it was followed by "a supplement" on 19 January 1818. There were in addition several notes that Fresnel gave to Arago and Ampère (almost certainly just shortly before the report was written).[2]

In neither of these two papers does Fresnel attack Biot at all strongly—

only noting in the second (where he was first able to provide computations) that Biot had been mistaken in a specific case of crossed axes. Fresnel mentions Biot's theory—having often used his observations—near the end of the 1816 paper, where he remarks only that he, Fresnel, feels that the assumptions Biot must make are "inconvenient." Indeed, in the first paper Fresnel confesses that Biot's formulas have an advantage over his own, albeit one that evaporated once he could combine waves with arbitrarily different phases. In neither paper did Fresnel assert that Biot's formulas had to be rejected: the "supplement" ended with the remark that "the results I have obtained up to now accord as well with Newton's table as those to which Biot was led by a completely different formula."

Arago's "report" was entirely different. It consists of twenty-two sections. The first thirteen attack Biot directly, recounting an experiment that is not mentioned in either of the two papers Fresnel had officially submitted—nor is it even present in any of the extra "notes" that Fresnel wrote about this time and that Verdet included in the *Oeuvres*. The fourteenth section compares Biot's experimental acumen unfavorably with Fresnel's on the basis again of a type of experiment that Fresnel does not discuss in the submitted papers (at least for chromatic investigations with thin laminae). Here Arago goes so far as to assert that "[these are] the experiments with whose aid Fresnel demonstrated the insufficiency of the theory of mobile polarization"—though Fresnel had hardly discussed that theory at all in the submitted papers and had certainly not claimed there to have shown its "insufficiency."

Arago's "report"—for despite the presence of his name on it, Ampère clearly had no direct hand in its composition (though he strongly supported Arago's crusade)—was a polemic directed at Biot. He had probably delayed five years in writing it in part because it took him that long to understand how it might be possible to confute Biot. Fresnel's welfare was at best far from the center of Arago's attention, since to this point Fresnel had not antagonized anyone. Arago even changed the subject of Fresnel's papers, which had originally been "the influence of polarization on the action that luminous rays exert on one another," to what Arago was most interested in, namely "the colors of crystallized laminae endowed with double refraction." And yet Arago certainly knew quite well what Fresnel was doing, since he had collaborated with him in his early experiments on the interference of polarized light.

He went further, claiming in the very first sentence of the report that Fresnel "had proposed . . . first to prove that the ingenious theory of *mobile polarization* . . . is, in many points, insufficient or inexact"—and only second to provide an alternative theory. After quoting at some length from Biot's account of mobile polarization in his *Traité de physique,* Arago argues that Fresnel had shown one of its major assumptions[3] to be true only under very special conditions.

Fresnel made no such claim about Biot's theory in either of the two submitted papers. However, in a "note on mobile polarization" signed 11 June by Delambre, Fresnel does make Arago's point—but he adds the following remark (and recall that this note was written a week *after* Arago delivered the report):

> I must say here that Biot, to whom I long ago remarked that it was necessary to suppose that in most cases the light, in leaving the crystallized lamina, divides between the primitive plane of polarization and the azimuth $2i$, replied to me that this way of envisaging the thing was not contrary to his ideas. I would, however, observe that if we refer to the literal sense of the text of his work, one would believe he had said precisely the contrary, and that even supposing he had intended what I just brought to light, it is quite astonishing that he did not further develop so essential a part of the theory of mobile polarization, on which his silence could induce grave errors, especially when one sees with what detail and clarity he has followed this same theory in all its consequences and experimental verifications. (Fresnel 1821b, 544)

Fresnel's remarks are considerably less damning than Arago's, allowing that Biot might possibly be able to escape the problem. Arago by contrast bluntly remarked that the existence of laminar thicknesses at which the light is definitely not polarized in only one of Biot's two permissible directions "does not apparently accord with the statement of the law that Biot gave."

Arago's major criticism of Biot went beyond this one, to attack the idea that (whether simultaneously present or not) the two directions of polarization that Biot chose are the correct ones—indeed, to question whether *any* two fixed directions can possibly fit the phenomena. To this end Arago discussed an interference experiment that combined Fresnel's mirror with crystals to polarize the beams. In the experiment the polarization of the light that results from the beams' interference is not at all restricted to Biot's two directions. (Details of the criticism are given in appendix 18.)

Arago regarded the experiment as a conclusive argument against Biot's theory. However, the experiment he described does not appear in any of Fresnel's papers he was reporting on, nor did Fresnel himself remark the criticism either in his reply to Biot or in his "Calcul des teintes." He did describe it a year later in "De la lumière" (Fresnel 1822c, 115–17), and he there claimed that it contradicts Biot's hypothesis. Why, if Arago's criticism was so powerful, did Fresnel not mention it at the time, and why did Biot not reply to it? One can only conjecture what the reasons for this contemporary silence may have been, but they may have had to do with the acceptability at that time of relying upon an interference experiment to deduce the polarization of the beams that emerge from the lamina. Biot did not admit that interference had anything at all to do with chromatic polarization—even though, by his own account (see below) in homogeneous light there will in general be two beams emerging from the lamina, one polarized in the original plane and the other polarized at $2i$. Since these two beams are united by the analyzer, they should interfere with one another. Yet Biot never remarked the point—and neither did Arago or Fresnel. It seems more than likely that Biot did not regard the criticism as a fatal one because it required using the principle of interference in what could very well be inappropriate circumstances.[4]

Arago continued with Fresnel's work, describing the experiment with laminae whose axes are crossed at 45°, which had been Fresnel's major emphasis in the submitted memoirs (on which see appendix 4). But again there was a difference from Fresnel's papers. Fresnel had taken the light in this experiment to strike always along the normal to the laminae in order to concentrate on chromatic effects, his purpose being to show that a claim Biot had made (but that Biot had never deduced from his theory) was incorrect. Arago instead began with the beam from a point source that reflects from a mirror to produce a diverging beam whose rays have been reflected very near the polarizing angle. This beam then passes through the crossed laminae, producing on exit two others, each of which also diverges; one beam is polarized in, the other in the normal to, the principal section of the second lamina. With this configuration Arago could concentrate on polarization, again in order to attack Biot's theory at its core (see fig. 9.1).

These two diverging beams will interfere with one another when the analyzer forces their polarizations to lie in the same plane. And the effective paths traversed by each ray in an interfering pair within the analyzer will vary with distance from the laminae, as well as with the laminar thicknesses, because the light diverges. Consequently the type of polarization that the *analyzer* seems to show will also vary with this distance. In Arago's words: "This experiment therefore offers us the singular phenomenon of two beams, polarized at right angles, that first cross one another in space, then reunite at the

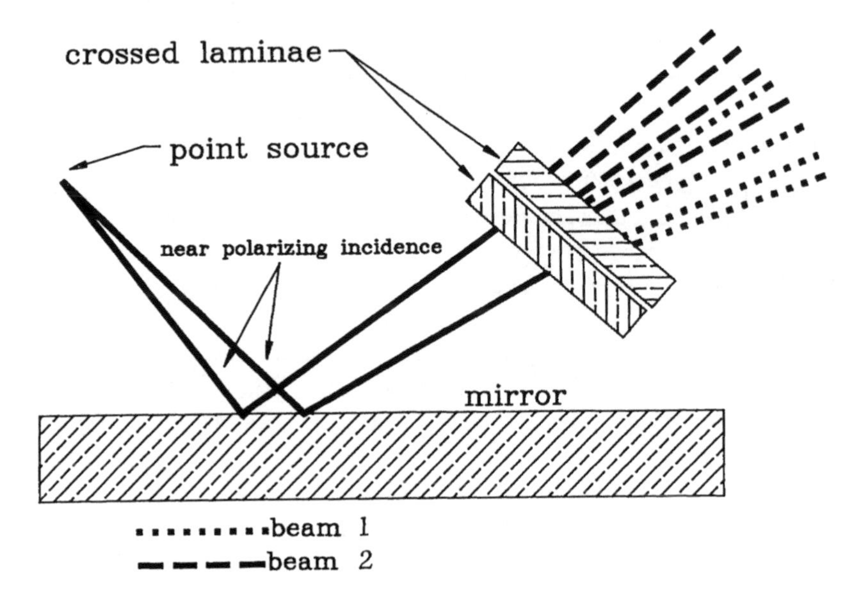

FIG. 9.1 Arago's scheme for attacking Biot's theory; beams 1 and 2 are orthogonally polarized.

base of the eye and form, in sum, a beam now polarized in one sense and now in another, according as the difference of the paths traversed by the two beams that compose it has this or that value" (Arago and Ampère 1821, 564).

Again Arago's emphasis was on something that, he was convinced, cannot be reconciled with Biot's theory of mobile polarization—namely, the apparent variation in the polarization of the emergent light with distance from the laminae. This is essentially the same as the criticism he had begun with. Having devoted most of his report to criticizing Biot's theory on these grounds, Arago turned directly to Fresnel's own explanation of chromatic polarization only in its last three sections. Arago's report was, then, much more a scathing attack on Biot's theory of mobile polarization than it was an exposition of the papers that Fresnel had submitted three and five years earlier. Arago must have understood very well what Biot's response was certain to be like. It came a week later.

9.2 The Polemic with Biot

Biot's elaborate and rather pugnacious attempt to defend his theory strikingly reveals the unbridgeable gulf between selectionism and the wave theory. I shall consider it in a moment. But he would almost certainly not have objected so passionately, nor been so revealing in his responses, had Arago's report been less of an outright attack upon him. Biot was given the papers Arago and Ampère were ostensibly reporting on, and he saw at once what Arago had done:

> In examining these documents, incomplete though they are, I found with pleasure, in many passages, proof that Fresnel had not proposed as the basic purpose of his work to show that what he calls my theory of mobile polarization is, in many points, insufficient and inexact, as the commissioners thought to be able to establish at the beginning of their report; but on the contrary, by a natural progression of ideas, Fresnel first took as his base the laws that I had found, and sought to find the hypothetical conditions that must be introduced in the interferences to satisfy them; precisely as he had also done in another work, where he proposed to represent, by luminous waves, the phenomena of polarization by rotation that I discovered in certain fluids. (Biot 1821, 588–89)

Biot concluded the published version of his reply (printed by Arago in the *Annales*) by complaining that the report "deviates from the rules generally established in scientific societies for assuring the equity of their decisions." There was more, but Biot removed it from the printed version, remarking that the remainder was unnecessary since the Academy had decided to adopt only the "conclusions of the report, and not the report itself." Unfortunately for Biot, he was so stung by Arago's vicious attack that he overreached himself in reply, attempting to take the offensive by quarreling with certain features of the wave theory. That, we shall presently see, brought Fresnel directly into the fray, and he was a much more formidable opponent than Arago.

Arago concentrated most of his reply to Biot on Biot's remark, which was evidently vitriolic in the original discussion, that the Academy regulations had been violated by Arago's and Ampère's not having based their report on the first part of Fresnel's 1816 paper, on their having used various "notes," and on the second paper's being a supplement not to the 1816 memoir, but to a later one. Arago accepted no part of the criticism. Certainly he had ignored the first part of the original memoir, but so what? What regulation forbade an author to withdraw a memoir (not that Fresnel himself had withdrawn anything), and in any case that first part had effectively already been printed in Arago and Fresnel's joint memoir on polarization and interference (Fresnel 1819a). So of course Arago had thought it best not to remark on this part, since that would be the same as acting as his own judge.

As for the two undated "notes" (comprising 1821a and 1821b, both almost certainly written at Arago's direct instigation), they were, Arago could easily prove, "simple developments of the first memoir presented in 1816" (which was no doubt true but beside the point), and in any case he had given them to Biot "only to aid him in his researches, and I was, I attest, far from imagining that he would find in them cause for reproach." As for the supplement's not being to the 1816 paper, well, Arago continued, Fresnel had certainly presented two memoirs, and "the commissioners charged to examine them brought them together in a single report."

Arago's replies were at best disingenuous. Fresnel had written his original papers not to attack Biot directly but rather to present his own theory. Arago turned that around, and Biot knew it. What was worse, Arago replied to Biot's (oral) complaint over the long delay of the report that (as Fresnel reported to his brother) this was only because Arago had long sought to avoid "the discussion in which I now find myself engaged." This must have infuriated Biot, for how could he answer it? How could he show, without reminding the Academy of his own questionable behavior in regard to Malus ten years before, that Arago was subtly, cleverly manipulating the facts to make it seem that Arago was blameless and without guile, while Biot was being petulant and antagonistic to Fresnel? Arago twisted the knife further:

> The memoirs that Biot published on the theory of mobile polarization would form more than two large quarto volumes. This is certainly not too much, if these memoirs establish, as has been claimed, that the molecules of light, in their trajectory across crystals, oscillate about themselves in the manner of a pendulum; while the whole thing could, without difficulty, be reduced to forty pages if Fresnel's objections are founded. One could therefore have presumed that, in speaking favorably of the work of this young physicist, we would not obtain the approbation of our knowledgeable colleague; also I would have delayed even longer, perhaps, in calling the attention of the Academy to this subject, had not Biot himself, very recently, engaged Fresnel to press me to make the report. I thought, I attest, that Biot, to whom the memoir had previously been communicated, would condemn the objections it contains. (Arago 1821, 593–94)

So Biot was upset, according to Arago, primarily because Arago had spoken "favorably" of Fresnel's work—though the bulk of Arago's report was a direct attack on Biot, based on arguments that Fresnel never (and would not have) made or else that he had not regarded as conclusive! Arago's clever polemic had forced Biot into a corner. He could not, and in fact did not, reply any further, since anything he could say would only worsen his predicament.

This incident marks the ascendancy to power in the Academy of Arago, Ampère, and their cohort, a development that had been in the making for several years.

9.3 Salvaging Mobile Polarization

Biot had replied to Arago the next week, on 9 June. He saw only one serious challenge to his theory in Arago's comments: the claim that his approach cannot be reconciled with the fact that the directions of polarization that he allows (0° and $2i$) must frequently both be present after the light emerges from the lamina.[5]

Biot accordingly decided to regenerate his formulas to emphasize that they *necessarily* suppose both polarizations must be present. Because Biot's reply to Arago marks the beginning of the end of the domination of optics by selectionists, I shall discuss it in detail.

He recalls what he takes to be elementary *facts*—that rays can be counted and that they retain their individual identities[6]—and begins his discussion by setting the incident plane of polarization and the principal section of the analyzer parallel to one another. Of the two colored images that are produced under these conditions by a thin crystal, one of them (E) behaves remarkably as the lamina is turned about in its own plane. It is worth quoting Biot at some length on this point, because his comments so remarkably illustrate the tenacious hold of selectionism in the face of an alternative account that violates its elementary principles:

> It [the E beam] is at first null when the principal section of the lamina coincides with that of the primitive polarization; it increases as the lamina turns, attains its maximum when the angle of rotation is 45°, then decreases until it is 90°, where it becomes null again, then to be reborn and to follow the same sequence in the other quadrants that complete the entire revolution the lamina can make. But all these changes affect only the intensity of the image: its color remains invariably constant in all the positions of the lamina in its plane.
>
> Now, I saw that this constancy is a phenomenon entirely characteristic of the mode of action that the thin lamina exercises on the ensemble of polarized rays that have traversed it (i.e., on the beam). Because suppose, to fix our ideas, that at a certain thickness the image E contains, at its greatest intensity, a certain proportion of each simple color, for example 92/100 of all the red light found in the incident light, 82/100 of the orange, 72/100 of

> the yellow, and none of the other colors [note Biot's assumption that "simple" colors are *not* entirely thrown into just one of two polarizations],[7] in which case the tint of this image will be an orange nearly the same as that of the first order in Newton's rings [computable from Newton's color circle]; then when this image attains its *maximum* intensity, the other image, which is given by the ordinary refraction of the prism, and that I call, to abbreviate, O, will therefore contain the rest of the red, the orange, and the yellow, that is 8/100 of the first, 18/100 of the second and 28/100 the third, plus the totality of rays that compose the rest of the spectrum; which will make this image a whitish blue. Now, since we suppose the intensity of the E image to have arrived at its *maximum,* it can no longer, during the motion of the lamina, remove from the O image any portion of the light that composes this whitish blue; it can on the contrary only cede to it a certain number of its own rays, in proportion as its intensity weakens. In this way we may consider the total light transmitted across the crystallized lamina to be composed of two distinct portions, of which the one O, experiencing no change in its primitive polarization, always experiences, in the crystalline prism, ordinary refraction; while the other portion E, always furnishing the elements of the extraordinary image, proves, by this very fact, that it has been deviated from the primitive direction of polarization. (Biot 1821, 572–73)

From the behavior of the analyzer images as the lamina rotates, he deduces the existence of two distinct sets of rays, one doubly refracted by the lamina and the other not. Both sets exist over a range of thicknesses *even using homogeneous light.* Given these sets, Biot's formulas for chromatic polarization (repeated immediately below) for the general case of an arbitrary orientation of the analyzer follow immediately and irrefragably:

$$F_o = \mathrm{O}\cos^2\alpha + \mathrm{E}\cos^2(\alpha - 2i)$$

$$F_e = \mathrm{O}\sin^2 a + \mathrm{E}\sin^2(\alpha - 2i)$$

Since there cannot be anything questionable about these formulas, the ones Fresnel had deduced for homogeneous light (which follow below), though they are apparently quite different, must lead to expressions that "coincide precisely with" Biot's—if Fresnel's are correct:

$$F_o = \cos^2\alpha - \sin(2i)\sin[2(i - \alpha)]\sin^2\left(\pi\frac{e - o}{\lambda}\right)$$

$$F_e = \sin^2\alpha + \sin(2i)\sin[2(i - \alpha)]\sin^2\left(\pi \frac{e - o}{\lambda}\right)$$

And indeed, Fresnel's formulas can be manipulated into a form that is similar to Biot's by expanding the term in $\sin(2i)\sin[2(i - \alpha)]$, yielding in the end:

$$F_o = \cos^2\left(\pi \frac{e - o}{\lambda}\right)\cos^2\alpha - \sin^2\left(\pi \frac{e - o}{\lambda}\right)\cos^2(\alpha - 2i)$$

$$F_e = \sin^2\left(\pi \frac{e - o}{\lambda}\right)\sin^2\alpha + \sin^2\left(\pi \frac{e - o}{\lambda}\right)\sin^2(\alpha - 2i)$$

To obtain the values of F_o and F_e in white light one has to "sum," in Biot's words, the values of $\cos^2[\pi(e - o)/\lambda]$, $\sin^2[\pi(e - o)/\lambda]$ over "all the rays in the spectrum." Denote the cosine sum by O and the sine sum by E. Then Fresnel's formulas for white light become "absolutely identical" to Biot's. The only difference Biot can see between them, one to which I shall return in a moment, is that Fresnel had (in the view of a selectionist) provided expressions for the actual ray counts in O and E, whereas Biot had not.

Biot was utterly convinced that there was only one way to think about these formulas, a way that implicitly assumes ray conservation. He remarked:

> Their [the Fresnel formulas] immediate physical interpretation is also exactly conformable to the principle that I took from experiment, to wit, that the total light O + E, transmitted across the crystallized lamina, behaves, after its emergence, in the prism of calcium carbonate, precisely as though it were composed of two distinct and complementary tints, of which the one O would conserve the polarization primitively imposed on it in the zero azimuth, and the other E would receive a new direction of polarization in the azimuth $2i$. (Biot 1821, 576)

Clearly Biot did not at all see that, according to Fresnel's theory, *it is simply not possible* to divide the light in this way. He did not grasp the deeper significance of the principle of interference. Rather, he seems to have thought of the principle as a way of deducing expressions for O and E—as, that is, a way for calculating how the rays are redistributed between the two allowable sets.

That, he thought, was the single point in which Fresnel had gone beyond him, and he therefore concentrated his attention on it.[8] Fresnel's expressions for O and E, Biot argued, are purely hypothetical, whereas he, Biot, treated O and E simply as *facts*. According to Fresnel's deduction, Biot continued in a note, the sine-squared term must represent the sequence of intensities in the reflected light in Newton's rings. This cannot, however, be true, he argues, because the sinusoidal variation is vastly too slow. The transition from reflection to transmission, though of course continuous, must be much more abrupt than this. For if it were not, Biot argues, Newton's table for the tints corresponding to a given air lamina would be extremely inaccurate—since, Biot was convinced, the table can actually be computed using Newton's color circle together with the assumption that homogeneous light produces rings that alternate nearly rigorously according to a law of arithmetic increase of laminar thickness. If Fresnel's sinusoidal law is correct, then Newton's table cannot be, and Biot confirms the claim by using Fresnel's expressions *together with Newton's color circle* to calculate the tints at various thicknesses. They are not the ones specified in Newton's table. Consequently Fresnel's formulas go no further than Biot's because the intensity law they require must be incorrect.

This at once raises a problem that Biot went to some effort to solve: namely, if the transition between rings is so abrupt, then it seems that the transition between Biot's two azimuths of polarization would have to be just as abrupt to preserve the validity there of Newton's table of tints, in which case how could one ever observe a situation where both azimuths seem to be simultaneously present, which unquestionably does occur? The answer, if we compress Biot's lengthy and obscure discussion to its elements, is simply this: it is entirely a matter of degree.

To preserve the applicability of Newton's table of tints to chromatic polarization, it is not essential, Biot argues, that the alternation between extremes be quite so rapid as it is in the case of Newton's rings. The difference may be so small as to be essentially imperceptible when observing the tints produced by white light. In homogeneous light there will necessarily be certain thickness ranges over which both azimuths of polarization occur. These ranges may be greater than the transition widths in the case of Newton's rings and so can actually be observed—whereas with Newton's rings the alternation between reflection and refraction occurs so abruptly that the transition range cannot actually be measured. If, in addition, the light used is not almost perfectly homogeneous, then the apparent range will be increased.

The entire issue reduced, in Biot's eyes, to the question of the width of the transition range between the two azimuths of polarization. And since Arago reported no data on the point, having only remarked the existence of intermediate thicknesses at which both azimuths must be present if we accept Biot's analysis, the issue remained undecided as far as Biot was concerned.[9] It was also probably undecidable at the time because it hinged critically on the accuracy of both Newton's table of tints and his color circle.[10]

Arago had never penetrated Biot's analogy between Newton's rings and chromatic polarization sufficiently well to see that, though Biot certainly never had previously emphasized the point, he could never have supposed that homogeneous light must at every thickness emerge polarized entirely at only one of the two allowable azimuths. He therefore testily remarked in reply to Biot:

> I have pointed out the inexactitude of the theory of mobile polarization by direct, positive experiments: he opposes a great dissertation on the Newtonian theory of fits about which I never said a word. If I examine the question of the sense of polarization in thin laminae, he replies that empirical formulas of which I spoke neither good nor bad exactly represent the succession of colors. (Arago 1821, 597)

Arago insisted that Biot "in ten different places in his works" says that "a polarized ray of simple [homogeneous] light is polarized entirely at its exit, either in the primitive plane or in the azimuth $2i$." Biot had indeed often written in a way that could easily give this impression, but for him the issue was entirely one of degree—of the breadth of the transition layer between the two allowable azimuths. Since he supposed the breadth to be very small, he had not regarded the issue as worth discussing.

Arago was further incensed at Biot's having ignored the interference experiment (see appendix 18) that, Arago claimed, conclusively demonstrated that even in thin laminae the light is polarized in and normal to the principal section. "If this fact is correct," he insisted, "the theory of mobile polarization is not: because never has there been a more manifest opposition between a system and an experiment." Arago's "fact" was entirely chimerical; the experiment he describes cannot possibly establish it.

Finally, Arago objected to Biot's claim that Biot's formulas for F_o and F_e are identical to Fresnel's—on the grounds that Biot's O and E are merely "substitutes" for Fresnel's "complex coefficients." This remark again can only mean either that Arago had not understood Biot or else that he simply rejected the argument out of hand. For Biot had insisted, in the published reply (to which Arago was here responding), that Fresnel's expressions *cannot* be correct.

Fresnel's own reply to Biot was much calmer than Arago's and right to the point (Fresnel 1821d). He mentions none of Arago's criticisms except the point concerning crossed laminae, about which he effectively supports Biot, but he notes that he himself had not fallen into a similar error.[11] Instead of supporting Arago's rather dubious remarks, he turned to Biot's claim that

Fresnel's intensity formulas cannot be correct, for that was a deep challenge indeed to the theory, and not only of chromatic polarization.

Biot, Fresnel remarked, had attempted to judge these expressions by using Newton's color circle to calculate what tints should appear at given thicknesses in air laminae, had compared the result with Newton's table, and had found marked discrepancies. To this Fresnel replies, first, that Newton's color circle can hardly be regarded as well confirmed by experiment (in particular because there are certain composite colors—"rose" and "purple"—that do not appear in the spectrum, though the color circle allows only for spectral colors). One should not use the color circle to judge the accuracy of an intensity formula for homogeneous light. Furthermore, Fresnel continues, the widths of the dark and light bands in Newton's rings are not the same. Because Biot had argued for a very rapid (Fresnel assumes an abrupt) transition between transmission and reflection—contrary to the requirements of the interference principle—these widths should be equal to one another. But "observing with a lens," they are not: the parts of the dark rings that "present a nearly total absence of light and seem to be of a sensibly uniform black are much narrower than the illuminated parts."

Unlike Arago, Fresnel did not claim that Biot's account fails decisively. Instead he defended his own theory, and he argued that Biot had to make so many physical suppositions concerning the state of a "molecule" in a crystal that it is "hard to conceive" how they could unite in a single entity. As before (in his work on diffraction), Fresnel claimed only that the alternative to his own account was vastly too complicated from a physical point of view and empirically much less fertile.

However Fresnel did not himself accept Biot's position that Biot's theory must be impregnable because it amounts to a statement of the "facts." Fresnel probably took this to be either hyperbole or else a reference to the formulas for F_o and F_e. He did not perceive, or at least he chose to ignore, that it was based on Biot's continuing belief in the individual existence of rays, and not on his advocacy of the emission theory proper. Indeed, when Fresnel read Biot's accounts of mobile polarization he always sought to understand them directly in terms of light particles, which Biot's discussions do make it quite simple to do. However, we have many times seen that Biot himself strongly refused to attach the emission theory in this way to his account of chromatic polarization. As far as Biot was concerned, his reasoning depended exclusively on ray conservation and the kinds of manipulations it makes possible. As a result Fresnel sometimes mistook Biot's arguments, thinking that Biot must have had a reason for doing something that was based in the emission theory when in fact Biot had not made any strong link to it at all. (See appendix 19 for two pertinent examples.)

Biot's arguments in reply to Arago were to little avail. He seems to have been unable to make it clear to Fresnel—to say nothing of Arago—why his way of thinking had to be distinguished from light particles and forces, though he himself had tightly bound the two in his voluminous publications. Here we have a pointed example of how the most elementary terms of a new theory can so thoroughly replace older concepts as to make it very difficult for fruitful communication to occur. Fresnel never admitted that one could build a theory in Biot's fashion, by concentrating on the behavior of rays in groups rather than on what happens to the individual ray. He insisted that the only possible alternative to his own way of thinking was to produce a quantitative account of the behavior of a single light particle. No selectionist had ever insisted on anything like this. They had always drawn connections to particle behavior, but none among them had ever assumed that their quantitative theories *depended* upon the physics of light particles.

Consequently the debate between Fresnel and Biot was not entirely concerned with the comparative merits of the emission and wave theories of light. There was another process at work as well, one in which the elementary terms of optical discourse were themselves being altered in ways that, at the time, were not entirely clear to everyone involved. Biot certainly did not appreciate that Fresnel's principle of interference made it impossible to count rays. Instead, he seems to have understood the trigonometric terms in Fresnel's formulas as specifications of *ray numbers*—numbers he did not believe were correct. For his part, Fresnel shifted his attention in his critique away from Biot's individual light particle to consider how the particles behave in groups. And so he refused even for a moment to consider that one could build a reasonably successful quantitative theory without specifying precisely what happens to the individual ray. This was an impasse that was hardly likely to be resolved by dispassionate argument.

10 Selectionists and Polarization after 1815

10.1 Polarization in the Textbooks during the 1820s and Early 1830s

As late as 1824 very few scientists thoroughly appreciated the difficult issues that were raised by the controversy over chromatic polarization. Two of the texts discussed above provide contrasting examples of the problems that arose. Beudant's qualitative *Essai* was even more firmly selectionist in treating polarization and birefringence than in discussing diffraction (where he at least recognized the claims of the wave theory). Experiments on birefringence, which he discusses at some length, "lead to considering the axis of double refraction as the center of a force that causes a portion of the luminous particles to deviate from their paths" (Beaudant 1824, 515). The two chapters on "mobile polarization" (chaps. 6 and 7 of book 6) present a verbal exposition of Biot's account. It is a good précis of the Biot theory, but since it is devoid of mathematics the theory's empirical power is greatly obscured. There is not even a hint that the theory has difficulties or that there is any alternative to the selection of rays that Beaudant assumes is always involved whenever light beams change their behavior in any way at all.

Péclet's *Cours,* published the year before Beudant's text, provides a remarkable contrast to it. Péclet completely rejects Biot's account of "mobile polarization," arguing that it cannot be correct because Fresnel has shown that thick and thin plates affect light in the same way (Péclet 1823, 497). Biot's theory, he unequivocally remarks, is "without foundation." He recommends reading Fresnel's papers on the subject (which, given the date, must primarily mean the "Calcul des teintes" of 1821 and the "De la lumière" of 1822).

Yet Péclet himself is not altogether a master of Fresnel's explanation. He remarks: "According to Fresnel, polarization consists in the decomposition of the small motions that take place in the waves into two other motions perpendicular to one another and to the direction of the plane of polarization" (Péclet 1823, 497).[1] Taken literally, this remark would betray a failure to grasp Fresnel's meaning. Polarization for Fresnel does not involve a decomposition into two orthogonal components, each one also normal to the "plane of polarization." Rather, a single direction is created out of the temporally varying, random directions in a beam of natural light. One can analyze the process of creation by examining at a given instant what happens to the two components in and normal to the plane of incidence into which any oscillation can be decomposed. Péclet seems confused about this, apparently thinking that in polarization an active process of "decomposition" occurs in which two "other motions" are physically created—much as Fresnel had himself thought for some time, but with the added requirement that both components occur only in the wave front.[2]

This was an extraordinarily difficult thing for many people to grasp. The 1824 and 1831 editions of Baumgartner's *Naturlehre* illustrate the point nicely. In the 1821 edition, recall, Baumgartner had already expressed his sympathy for the "vibration" theory, but we also saw that he had difficulty even discussing diffraction. His remarks on polarization did not at that time deviate at all from selectionist language. They were carried out entirely in terms of rays and had nothing to say about decomposition in Fresnel's sense of the term.

This has changed markedly by 1831, when the text shows a thorough understanding of Fresnel's principles, including such intricacies—so deeply dependent on the idea of decomposition—as circular polarization. He begins the discussion of chromatic polarization with a phenomenological description, followed by a description of the essentials of Biot's theory, which the author regrets cannot be adhered to: "So simple are the inferences to which one is led by the hypothesis of mobile polarization, and so easily explained are the phenomena of chromatic experiments in polarized light; nevertheless [the hypothesis] seems not to be based in nature, but at least Fresnel has given an experiment whose result is not reconcilable with the assumption," (Baumgartner 1831, 507).[3]

Nevertheless, the subject of polarization was not taken up by many people until the 1830s, when Fresnel's ideas were rapidly assimilated. Between Malus's discovery of polarization by reflection and the mid-1820s, for example, only about twenty-odd articles in all the journals discuss the topic at all. Of these, even fewer describe substantive research, and most do not even mention Fresnel. In 1817, for example, the German Georg Muncke wrote a short article describing the colors that are produced in a standard apparatus by incompletely crystallized bodies. His discussion is almost devoid of any kind of theory and contains no calculations (as is usual for German work at the time). However, he at one point remarks: "So many modifications of light polarization take place in these unique experiments, and they are certainly not all. In my opinion one could hardly find a path through this labyrinth without the assumption of light particles . . . and scarcely without the theory of fits" (Muncke 1817, 211). Muncke has evidently been much impressed by Biot's lengthy discussions, but since he attempts no computations, it is difficult to tell whether he understood much of selectionism proper or whether he grasped only the qualitative image of the oscillating particle. Substantive research was done by selectionists other than Biot, however, in particular by David Brewster in the late teens and, a bit later, by John Herschel. Their work is especially interesting because, though intended as a critique of certain aspects of Biot's account, it is deeply selectionist in ways that Brewster for one never fully appreciated.

10.2 A Critique from Within

Biot had not succeeded, he himself felt, in generalizing his formulas to cases in which the light does not strike along the perpendicular to the optic axis of the thin lamina. Or to be more precise, he had succeeded in generalizing the formulas for intensity but not in obtaining a procedure for finding the tints of the two parts, O and E, into which a polarized beam of white light is divided by a given thin crystal. It was no simple matter to determine the general factor to use in multiplying the number that functions as an index into Newton's table of tints. Biot offered $\sin^2\varphi$, with φ the angle between the ray and the optic axis, and this worked reasonably well for small angles but seemed to fail at larger angles, which necessitated further modifications.

The clue to a more accurate generalization had actually been discovered in 1812. At some point late in that year David Brewster had managed to send polarized light through mica or topaz nearly along what Biot had taken to be the axis of double refraction. With his eye placed reasonably close to the analyzing device, or else by first causing the incident light to come to a focus within the crystal, Brewster discerned a series of colored rings that formed a rather intricate, lemniscate like pattern with two poles, the rings seeming to flow around these poles as foci. Then, early in 1813, he found a quite different pattern of rings in "beryl, emerald, ruby &c."; here the rings formed a concentric pattern with the usual optic axis at their common center.[4]

Brewster did not at this time examine Iceland spar. However, Wollaston found in mid-1814 that it too produces rings of the concentric sort, and in December 1815 both Biot and Thomas Seebeck also saw them (Brewster 1818, 213). Since Brewster did not publish his earlier discoveries at the time, Biot—who did not undertake the very demanding task of making similar observations for mica or gypsum—was aware only of the concentric ring pattern, which he assumed to be the general case. That pattern could be easily explained by Biot's theory, and he discussed the phenomenon in 1816 in his *Traité* (4:481 ff.; see appendix 20 for details).

Biot's analysis nicely explained why the rings form, and it could even provide a method for computing the intensity variation throughout a ring (though Biot did not discuss this point). Furthermore, the mutation of the concentric circles into elongated ovals (which occurs when the central axis of the cone of light that reaches the eye no longer lies along the optic axis of the lamina) can also be understood as a result of the breaking of the rays' symmetric relationship to the axis. But can the tints themselves be *calculated?*

Clearly, to calculate the tints one needs to determine the factor by which to modify the index into Newton's table, and that should simply be $\sin^2\varphi$. However, this factor had not worked well in computing the tints at oblique incidences in gypsum, and Biot had in consequence extensively modified it on purely empirical grounds. In 1818 David Brewster read a paper before the Royal Society that corrected Biot on this point. We can easily reduce Brewster's elaborate and empirically detailed paper to this, that the apparent failure of Biot's factor is due entirely to the existence within some crystals of *two* depolarizing axes instead of just one. Very possibly these two axes are also axes of double refraction, since the two kinds of axes are identical in crystals that produce the colorful concentric-ring patterns (on which see appendix 21). The difficulties Biot had had in fitting the $\sin^2\varphi$ law to experiment are all due, Brewster concluded, to the existence in the crystals he had used of two axes rather than only one. Each of the axes produces its own independent set of rings, and the complex pattern that results derives from the superposition of the two sets. In Brewster's words: "No fewer than seven out of the twelve minerals employed by M. Biot, have two or more axes of double refraction. Sulphate of lime [gypsum] itself belongs to this number; and all the irregularities of its action, which M. Biot has represented by empirical formulae, are the legitimate and calculable results of two rectangular axes" (Brewster 1818, 204).

Brewster quite obviously thought of his work as a serious criticism of Biot's account, and indeed it was that—but only from within the confines of selectionist theory. Brewster never questioned Biot's analysis of depolarization proper (viz., its division of rays into O and E sets that have different ray ratios and counts). Quite the contrary; he accepted it implicitly. The rings produced under the proper conditions by depolarization were, for Brewster just as much as for Biot, instances of ray separation and reorientation. Indeed, Brewster took Biot's attempt to compute tints at oblique incidences (and hence for the rings) more seriously than its author had. Biot had erred, Brewster was arguing, only in failing to realize that there are often, perhaps usually, two depolarizing axes.

Though written entirely from a selectionist viewpoint, Brewster's analysis nevertheless sought to go beyond selectionism proper by linking the sine-squared law to the forces that must emanate from the depolarizing axes. Biot had himself done much the same thing. Indeed, the very form of the sine-squared factor was originally suggested to Biot by its appearance in the Laplace-Malus expression for the velocity of the extraordinary ray. Despite his (and Biot's before him) continual references to forces, however, only the factor itself reflected their influence in any substantial way.[5]

During the summer of 1818 John Herschel, then twenty-six years old, turned his attention to the rings produced by depolarization. Within two years he had become Brewster's equal in two respects: first, in experimental acumen, and second, in his mastery of selectionist technique. Herschel was, if anything, even more deeply convinced than Brewster was that depolarization, as well as double refraction, reflects the activity of forces. But once again, the use he actually made of the assumption was minimal. For the most part Herschel, like Biot before him, worked backward from experiment (or from laws based directly on experiment) to the conditions the laws placed on the forces that were by hypothesis their cause.

10.3 Brewster at the Periphery

Herschel had a much suppler, more adaptable mind than Brewster's. He was, we shall see, eventually able to assimilate the essence of the wave theory, though the process was by no means easy for him and resulted in occasional confusion between old (selectionist) and new (wave theoretical) terminology. Brewster himself never succeeded in completely appreciating the core of the new theory. Nevertheless he early realized that the theory had great empirical power for diffraction, and by the late 1820s he had begun to see its strength in dealing with polarization as well.

Brewster's recognition of the wave theory's empirical power combined with his irrefragable selectionism to produce a series of extraordinarily odd, and highly revealing, attempts by him to assimilate some of the new theory's terminology and formulas to selectionist requirements. One question that especially intrigued him was how to understand partially polarized light given Fresnel's formulas for reflection—or rather, given what Brewster was willing to appropriate from Fresnel's formulas.

Perhaps the greatest problem a selectionist had in grasping Fresnel's work on polarization—indeed, the problem Fresnel himself had had in creating the new concept—was the transformation of what had been a static configuration of multiple entities (rays) into the motion of a single thing over time (the change in direction of the oscillation in the wave front). Yet without that change it is impossible to develop a coherent understanding of common and partially polarized light in the wave theory. That is, without Fresnel's kinetic idea, one must necessarily have recourse to something like a selectionist mixture. Furthermore, the reflection formulas that Fresnel deduced make no sense at all when divorced from this kinetic image. They cannot simply be applied empirically, because they do not determine anything except the amplitudes of the components of Fresnel's kinetically polarized wave.

Brewster was always willing to accept formulas that worked well, and this was possible at least in part in diffraction theory. Polarization posed greater problems. Brewster had to take Fresnel's formulas and somehow combine them with a static understanding of polarization to produce a formula he could grasp the meaning of. It seemed to Brewster that he could make sense of Fresnel's formulas by construing in a new, fully selectionist way the structure of common and partially polarized light.

Brewster proposed to alter the usual selectionist understanding of "common light" as a mixture of rays with random asymmetries by reducing the asymmetries to only two: one subset of rays has its asymmetry fixed at 45° to the plane of incidence; the other has it fixed at $-45°$ to the same plane. Given this, he continued, two questions must be answered: first, what happens to the two asymmetries on reflection, and second, what are the ray counts in the reflected subsets (Brewster 1831, 172 ff.)?

Brewster assumed that ordinary reflection can have only two effects on any given set of rays with a common asymmetry: the asymmetry can be changed, and the ray count can be changed by throwing some of the rays into a refracted beam. That is, all rays with a common asymmetry must again possess on reflection (or refraction) a common, albeit possibly different, asymmetry. It can make no possible difference that the rays are accompanied by others in a second subset that have a different asymmetry. To grasp the meaning of Fresnel's reflection formulas, Brewster had to interpret them in this context. He did so (and in this he was very far from being alone) by taking not the component ratios proper, but only the resultant intensity and angle of polarization that they determine.

Consider either of the two subsets in a beam of common light. Each of these, as Brewster understood it, constitutes an independent group of rays with a common asymmetry, a group in no way different from a beam of light in which the other subset is absent—in no way different, that is, from a beam of polarized light. Brewster knew that Fresnel's formulas led to expressions for the intensity and angle of polarization for the reflection of a polarized beam. These expressions, not their component progenitors, now took on meaning for Brewster: they may both, he supposed, he applied independently to the two orthogonally polarized subsets that constitute common light.

The two expressions that Brewster took from Fresnel (or more likely from Herschel's 1827 account of Fresnel) determined I_{refl}, the reflected intensity, and α_{refl}, the reflected azimuth of polarization, as functions of the angle i of incidence, r of refraction, and α_{inc} of incident polarization:

$$I_{refl} = \frac{\sin^2(i - r)}{\sin^2(i + r)} \cos^2\alpha + \frac{\tan^2(i - r)}{\tan^2(i + r)} \sin^2\alpha$$

$$\tan(\alpha_{refl}) = \frac{\cos(i + r)}{\cos(i - r)} \tan(\alpha_{inc})$$

Brewster applied these two expressions to each subset of rays in common light *individually.* This had two results. First, he could at once compute the total amount of light that is reflected at any incidence; second, he could explain (and even quantify) the production by reflection of partial polarization. According to the equation for $\tan(\alpha_{refl})$, reflection below polarizing incidence reduces the original 90° angle between the subsets, which angle then increases for higher incidences. But any change from 90° means that, when analyzed by a crystal, the light will necessarily produce unequally intense ordinary and extraordinary refractions. And since one can calculate the angle between the subsets' polarizations, one could in principle quantify this partial polarization. (But since photometric measurements could not then be made, there was no point in doing so.) In Brewster's words, "partially polarized light is light whose planes of polarization are inclined at angles less than 90°" (Brewster 1831, 176).

This was not the only, or even the most striking, area in which Brewster pressed Fresnel's work into a selectionist mold. In 1830 he read a paper on metallic reflection in which he announced that it involved a new species of polarization that he termed "elliptical" (Brewster 1830b). He was indeed correct; metals do polarize light elliptically. Furthermore he was the first to make the claim, and he even computed the "phase" "of the two inequal portions of oppositely polarized light by the interference of which the elliptic polarization is produced" (Brewster 1830b, 300). All of this might seem to require that Brewster had grasped a considerable portion of Fresnel's concept of polarization.

He had not. More even than his understanding of the composition of common light, Brewster's appropriation of Fresnel's terminology for metallic reflection shows that he failed in basic ways to grasp the central idea underlying the terminology. This is not at first apparent, because Brewster was able to develop a definition of "elliptical" polarization that avoided the pitfalls that are inherent for a selectionist in employing the vocabulary of the wave theory. (Appendix 22 gives the details of Brewster's creative, selectionist appropriation of concepts that are embedded in the wave theory.)

The core of Brewster's theory of "elliptical" polarization preserved the essence of selectionism by conceiving of a beam as composed of other, completely independent beams with their own intensities and polarizations. Unlike Fresnel, who was constrained to decompose his wave vector in and normal to the plane of reflection, Brewster was thereby freed to suppose any polarizations he wished in his beams, subject only to the demands of observation. Accordingly Brewster applied Fresnel's formula for the rotation by reflection of the plane of polarization to the individual subsets that formed his beams (in the ordinary reflection of common or partially polarized light), and superadded another subset, with an appropriate "phase" and polarization where necessary (in the reflection of polarized light by metals).

Brewster thought that this kind of description was entirely unhypothetical, that he had been able to escape all "theoretical reference" in his account. But he had not. His analysis is thoroughly incompatible with Fresnel's "elliptic vibrations" for the same reason that his construction for common light is incompatible with the wave theory: like all selectionist theories, it assumes that beams can be dissected into their constituent rays. Brewster gives no indication at all of having perceived this difference. To him the vocabulary of the wave theory, words like *phase* and *interference,* had to refer to rays conceived as isolatable objects, or else it could have no empirical consequences. If Fresnel had developed formulas that seem to work, then the elements that enter into them had necessarily to be construed in terms of groups of rays—that is, in terms of light beams. There was no other way to think, not even, Brewster and other selectionists implicitly believed, if the wave theory itself were to be accepted. For them the wave theory did not replace rays with waves; it at best provided a new way to calculate with rays. Brewster carried this to such an extreme that it gradually became impossible for the emerging group of wave theorists in Britain to communicate with him in any fruitful way, with the result that he was gradually pushed to the scientific periphery, which he greatly resented. After 1831 he rarely published in the *Philosophical Transactions,* where he had earlier reported many of his discoveries, though his polemical defense of selectionism was just then beginning.[6]

11 Fresnel's Final Unification

By the 1830s startling implications were being drawn from Fresnel's theories for polarization and double refraction. In nearly every respect—physical and mathematical—this work was Fresnel's greatest and ultimately most influential accomplishment. In this chapter we will examine it in some detail in order to grasp precisely what Fresnel had done and, equally important, what he had not done, both in linking the theory to the physics of the ether and in drawing out its mathematical implications. Here, we shall see, Fresnel left a great deal of room for future wave theorists to push his program much further than he himself was able to do before his untimely death in 1827.

11.1 The Link between Polarization and Double Refraction

11.1.1 A New Kind of "Ordinary" Ray

On 19 November 1821 Fresnel presented his first paper on the connection between double refraction and polarization. In it he did more than link the two phenomena; he also argued, and attempted to support empirically, that in crystals with two optic axes there is no such thing as an ordinary ray. This was a startling claim, since no one had ever suspected any such thing and since, of even greater importance, Biot's and Brewster's elaborate discussions of ring formation tacitly assumed that even in biaxial crystals one of the rays always obeys Snel's law. Fresnel was accordingly treading on dangerous ground indeed, and, we shall see, he took great care (as always) to retain as many previous results as possible.

Fresnel relates that "mechanical" reasoning similar to that contained in his "Calcul des teintes" paper (1821c) had convinced him that no ray, in biaxial crystals, can have a velocity that is independent of its direction (Fresnel 1821e, 262). He had apparently come to this realization a year before but had decided not to mention it until he had assured himself that it was "really a necessary consequence of the theoretical views I had indicated." These views concerned the relation between transverse waves and the dynamic structure of the ether.

Fresnel's complete embrace of transverse waves was accompanied by, indeed closely bound to, a change in his understanding of how the ether transmits motion. Before he and Ampère had introduced the idea that waves of polarized light may contain a transverse component, Fresnel had thought of the ether as a rarefied, highly elastic "fluid" akin to a gas. Such a "fluid," Fresnel remarked, had always been thought to transmit only longitudinal oscillations because:

> The geometers who have occupied themselves with the vibrations of elastic fluids had, I think, considered that the only accelerating force is the difference of condensation or dilatation between consecutive layers. At least I don't see anything in their equations to indicate that, for example, an indefinite layer, on sliding between two others, must communicate motion to them, and it is apparent that in this respect their equations don't encompass everything that actually takes place. (Fresnel 1821c, 630)

There are longitudinal oscillations, Fresnel was arguing, but there may also be others, because one must take account of restoring forces that may arise as a result of the lateral displacement between contiguous layers of the fluid as well as the forces that arise from the fluid's compression by volume. This realization perhaps came to Fresnel quite early, about the time he and Ampère thought of adding a transverse component to the longitudinal one. However, now Fresnel needed to argue that *only* the transverse component may exist.

His analysis on this point was in later years considered to be a major weakness of the theory's dynamic foundation. He remarked:

> It remains for me to explain how it can come about that [the ether's] molecules do not undergo sensible oscillations except in the surface proper of the waves, perpendicular to the rays. It suffices for that to suppose a law of repulsion between the molecules such that the force opposing the mutual approach of two slices of fluid is much greater than that opposing the sliding of one of them with respect to the other. (Fresnel 1821c, 633)

Fresnel's argument, it was often pointed out, is at least incomplete, because he does not demonstrate that the force opposing mutual approach can be incomparably greater than that opposing sliding—and in fact it is difficult indeed to satisfy this condition if the ether behaves (as Fresnel supposed) like a point lattice.

Nevertheless, this way of thinking at least made the assumption of pure transversality dynamically reasonable. Furthermore, the underlying mode of connection between dynamics and wave propagation that is involved here forges a direct link between the tendency of the ether to resist an internal shear and the wave speed:

> The greater or smaller rapidity with which the motion propagates depends on the energy of the accelerating force that tends to return the contiguous slices to the same relative positions, and on the masses of these slices, just as the speed of propagation of sound waves in air (as one ordinarily conceives them) depends on the relation between its density and the resistance it opposes to compression. (Fresnel 1821c, 633).

We shall see below that it was this way of thinking that led Fresnel to the conclusion that there cannot be a ray in a biaxial crystal whose speed is independent of its direction of propagation.

Indeed, Fresnel was so convinced by this reasoning that he told Arago that, should his assertion be proved empirically false, he would have had to "abandon all my theoretical ideas on double refraction."[1] Now, Biot had extensively examined the behavior of the biaxial crystal topaz three years before (Biot 1818), and Fresnel proposed to use Biot's measures, together with new ones of his own, to test his startling claim that there is no such thing as an ordinary ray. To do so Fresnel had first to determine the directions of the two optic axes in topaz for his own specimen. The method involved cleaving the crystal (see fig. 11.1).

The single natural cleavage plane in topaz, Fresnel remarked, is normal to a line that bisects the acute angle formed by the optic axes. In the figure, the unmarked z axis lies in the cleavage plane, which is itself normal to the plane containing the optic axes (this being the plane of the figure). The y axis lies in the plane of the optic axes and bisects the angle between them, while the x axis lies in the intersection of the cleavage plane and the plane of the axes. According to Biot's results (and Fresnel's theory), the two planes of polarization for rays that are normally incident on the cleavage plane are in and perpendicular to the plane containing the axes. I shall hereafter refer to this plane as the *axial plane*.

To find the directions of the x and z axes is therefore quite simple. Again suppose normal incidence on the cleavage plane; when one of these two axes lies in the initial plane of polarization, then the extraordinary ray that is produced by passing the light that emerges from the topaz through an analyzing

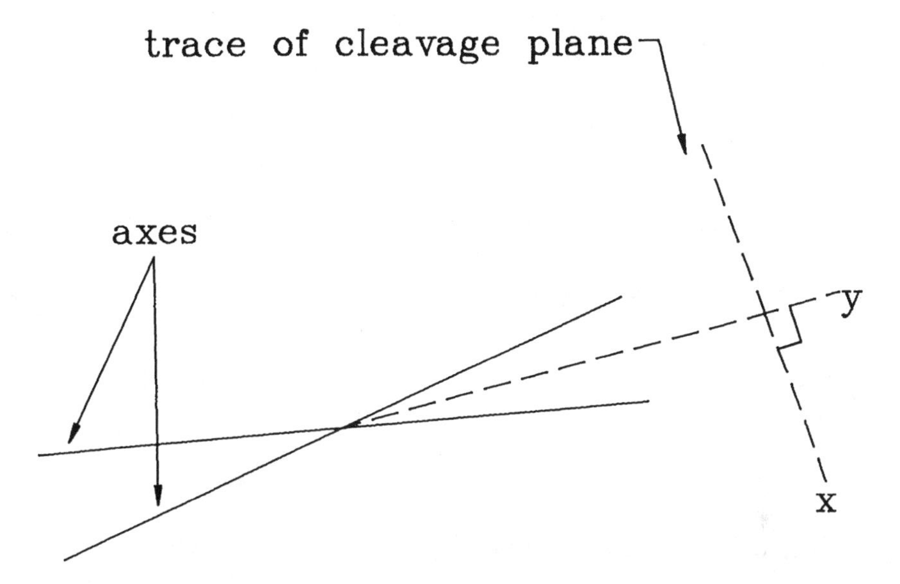

FIG. 11.1 The optic axes in topaz.

crystal will disappear when the analyzer is set with its principal section parallel to the initial plane of polarization. Which of the two axes is which involves a somewhat more complicated procedure, since rings must be observed.[2] From three observations Fresnel determined that the axes in topaz form a mutual angle of twice 30°53′; Biot, using a different method, had measured twice 31°37′. Fresnel placed somewhat greater faith in Biot's results than in his own, since Fresnel's apparatus was not as well adapted as Biot's was for measuring this angle.

Having found the axial plane and the separation of the optic axes in it, Fresnel cut a piece of topaz into two slices of nearly equal thickness and then glued the two pieces together (see fig. 11.2). One piece (A) was cut by planes parallel to the axial plane; the other (B) was cut by planes normal to the x axis. In the figure we are looking down on a pair of cut crystals A and B that are joined together so that the trace of the axial plane in B is normal to the same plane in A and bisects the angle between the latter's optic axes. Rays of light strike from above the plane of the figure. Consequently a ray that is normally incident on A will be parallel to z and so will strike both optic axes along a perpendicular, while a ray normally incident on B must strike in the axial plane itself, parallel to x.

Fresnel set up a double-slit interference experiment to examine the velocities of the light that passes through this device. When both slits are covered

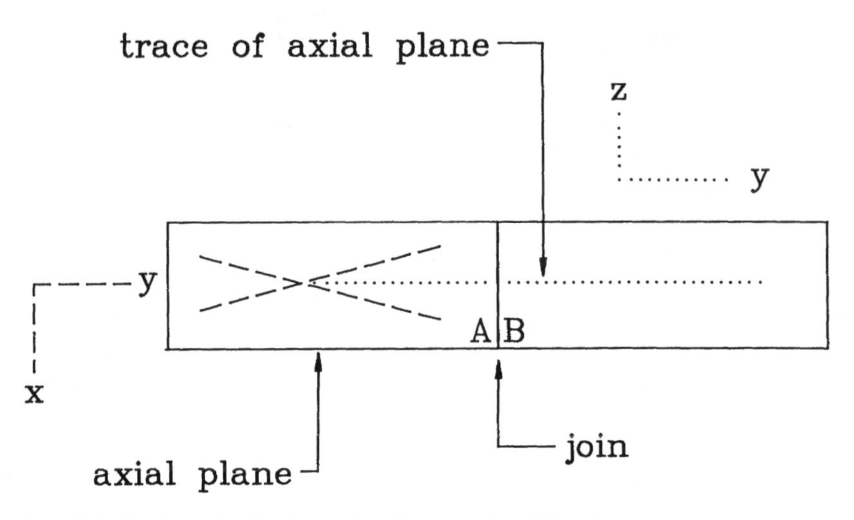

FIG. 11.2 Configuration for the optics of topaz using diffraction.

by the same plate (either A or B) then, as in the plate's absence, a single diffraction pattern emerges. However, when one slit is covered by A and the other by B, two patterns emerge. Fresnel observed that one of the two patterns occupies nearly the same places as the single pattern produced using only one plate; but the other pattern is displaced by about 16.6 fringe widths, which means the rays that form it must have different speeds, since the distances remain the same as before.

Fresnel was at this point primarily interested in demonstrating that the rays that form the "ordinary" image had to vary in speed with direction. To prove this he had to determine which rays form that image. To do so he relied on the polarizations of the fringe patterns, since Biot had previously found that the plane of polarization of "ordinary" rays contains the ray and the y axis—the line that lies in the axial plane and bisects the angle between the optic axes. (This had to be the plane of polarization of "ordinary" rays, since rays polarized in planes normal to this one have speeds that vary markedly with their directions.)

In Fresnel's experiment, then, the "ordinary" rays produced by both plates A and B must be polarized along the normal to the line joining the two plates. Consequently the polarization of the fringe group that is formed by the "ordinary" rays from A and B should itself be normal to the join, and the polarization of the fringe group formed by the extraordinary rays from the two plates should then be parallel to the join.

Fresnel observed that the undisplaced interference pattern is polarized parallel to the join—and so it must be formed by the interference of extraordinary rays. The highly displaced pattern is polarized along the perpendicular to the join, and so it must be formed by "ordinary" rays. Whence the "ordinary" rays in A must have a considerably different speed from the "ordinary" rays in B.

Since the claim that "ordinary" rays have variable speeds was a most "unexpected" result, Fresnel remarked, Arago had insisted that he demonstrate it using a method similar to Biot's, which involved refraction, rather than by using interference—no doubt because Arago well recalled how Biot had just recently brushed aside evidence based on the polarization of interference patterns (Fresnel 1821e, 271). Fresnel accordingly used two topaz prisms glued together to observe the refraction directly (see fig. 11.3). He cut the prism angles to 92.5° and then sandwiched them between two prisms of crown glass so cut that the entry and exit surfaces were parallel to a plane that bisects the angle of the topaz prism. The space between the topaz and the crown glass was filled with turpentine (see fig. 11.4). (The effect of this arrangement was to compensate somewhat for the dispersion the topaz produced by minimizing the refraction that occurs when rays pass through the prism parallel to its base, that is, when rays enter and leave the prism symmetrically.)

The topaz pieces were cut as follows: In piece I, the apex line of the prism is parallel to the x axis; that is, it lies in the plane of the optic axes and bisects the obtuse angle between them. In piece II the z axis parallels the apex; that

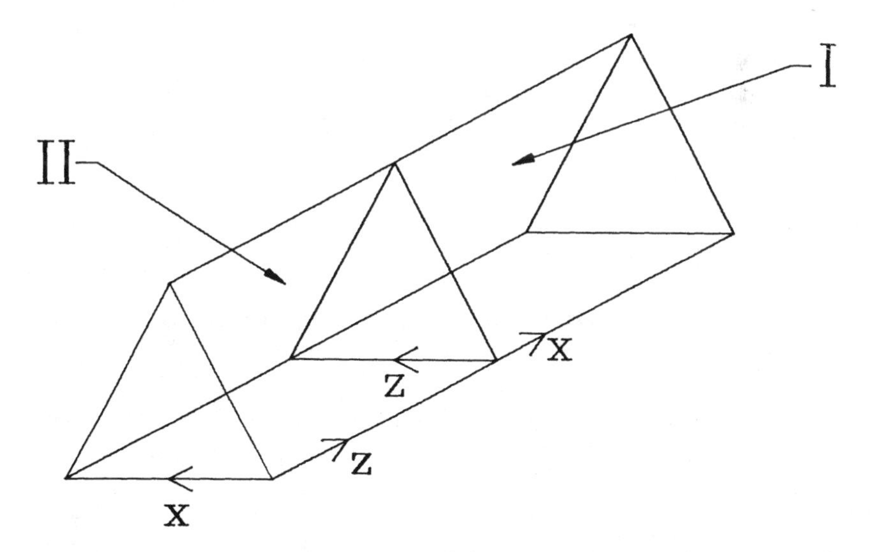

Fig. 11.3 Configuration for the optics of topaz using refraction.

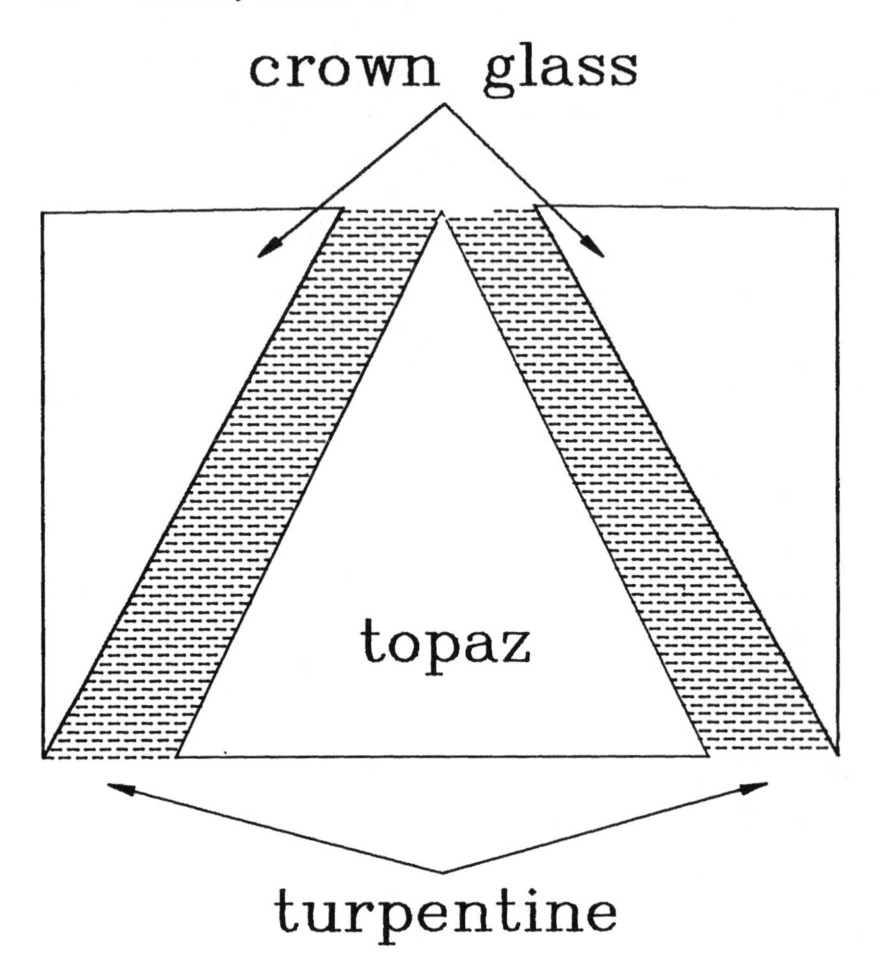

FIG. 11.4 The refraction device in cross section.

is, it lies along the normal to the plane of the optic axes. Consequently rays that enter and exit symmetrically from prism II in this device pass through in the plane of the optic axes, whereas in prism I the rays that enter and exit symmetrically pass through along that plane's normal.

According to Fresnel's theory, we shall see, this should produce the greatest possible difference between the speeds of the "ordinary" rays in the two pieces, whereas the extraordinary rays in the two should here have the same speeds. Consequently, with this device one can observe a line parallel to the base through both prisms and should see, first, a continuous image produced by extraordinary refraction and, second, a broken image produced by ordinary refraction. One can test the polarizations to determine which is which. Biot, Fresnel remarks, had failed to see the difference between the speeds of ordinary rays because he had observed only the divergence between the ordinary and the extraordinary rays in a single prism and not, as here, the comparative refractions from two prisms cut in appropriate ways. These beautiful and delicate experiments strikingly confirmed Fresnel's startling claim that there is no such thing in biaxial crystals as a ray whose speed is independent of its direction. And, Fresnel claimed, he had predicted the fact from "theory."

11.1.2 The Ellipsoid of Elasticity

Fresnel was able to predict that biaxial crystals have no "ordinary" rays in the usual sense of the term because he was strongly convinced that the direction of oscillation (and so that of polarization) must immediately determine the velocity of propagation. His conviction emerged directly from his realization that light propagates because of elastic reactions to shears in the medium. Unlike compressions, shears have directions; so if the reactions to them vary in intensity with their directions, then so must the velocity of propagation of the associated wave. This realization combined with Huygens's construction to suggest a connection between polarization and velocity.

Consider the uniaxial crystal, where Huygens's ellipsoid governs the extraordinary refraction and a sphere governs the ordinary refraction. By convention the ordinary refraction is said to be polarized in a plane containing the optic axis and the ray, and so the extraordinary refraction is polarized along the normal to that plane. Clearly, in order for the ordinary velocity to be independent of direction, the elastic reaction generated by the corresponding displacement must also be independent of direction. Fresnel's reflection theory demands that the oscillation be normal to the plane of polarization (see fig. 11.5). Consequently it follows at once that elastic reactions generated by displacements that are *normal* to the plane containing the ray and the axis must not vary with direction, whereas reactions to displacements in that plane do vary. The next question is how they must vary.

Fresnel operates (incorrectly, as I shall presently discuss) with rays and reasons essentially as follows. Take a plane normal to a given direction. Since in any given direction two different velocities and polarizations are possible, there must, in that plane, be two mutually orthogonal directions along which displacements generate different elastic reactions but remain constant in direction as the disturbance propagates. Furthermore, of the two, one must always be perpendicular to the optic axis. For rays in the principal section

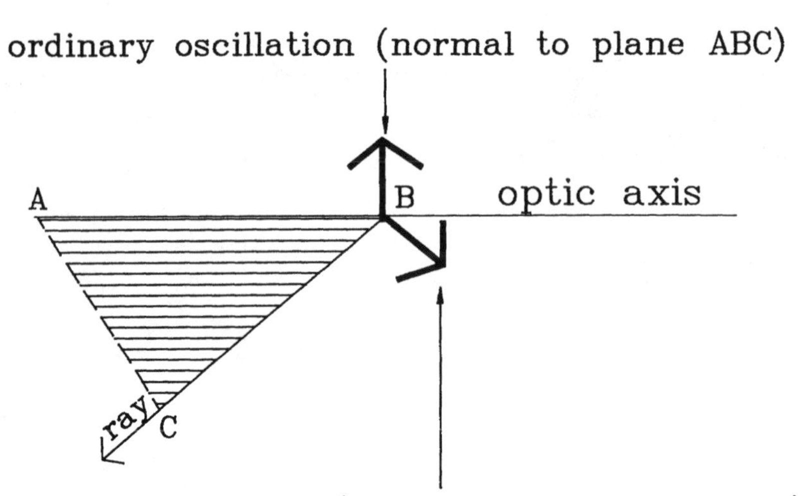

FIG. 11.5 Directions of oscillation in a uniaxial crystal.

proper, it was obvious that the reactions corresponding to the extraordinary rays may be represented by the radii of an ellipse (which I shall call the *conjugate ellipse*) that is in the same plane as, and rotated 90° from, Huygens's ellipse, because the extraordinary oscillation occurs in that same plane. On the other hand, the reaction that corresponds to the ordinary ray must be represented by the radius of the Huygens sphere, and it must be perpendicular to the principal section. Fresnel conjectured that the oscillations corresponding to *any* direction of propagation, whether in the principal section or not, also lie along the semiaxes of an ellipse, with the ellipse being different for different directions of propagation. However, one of these semiaxes—the smaller one—must remain the same in magnitude whatever the direction may be. This, Fresnel further conjectured, can yield the correct dependence of ray speed on direction only if the ellipses in question are produced by the plane sections of an ellipsoid of revolution that is generated by the rotation of the *conjugate ellipse* about the optic axis.[3]

Very little of this reasoning is explicit in Fresnel's first memoir. Rather, after introducing the idea of linking the velocity of propagation with anisotropic elastic reaction, Fresnel simply introduces this *ellipsoid of elasticity* and describes how to use it: to find the velocity of a ray in a given direction, cut the ellipsoid by a plane that is normal to the ray and determine the semiaxes of the resulting section. They are in magnitude directly proportional to the speeds of propagation, and in direction they represent the oscillations.[4]

11.1.3 Retrieving Biot's Results

Fresnel's primary goal at this point was to retrieve from the wave theory results for double refraction that he thought were empirically well established—much as one of his aims in his early analysis of chromatic polarization had been to retrieve Biot's observations. Here the major new result for biaxial crystals was Biot's generalization of the Malus-Laplace velocity law: to wit, that the difference $v_o^2 - v_e^2$ between the squares of the "ordinary" and the extraordinary velocities is proportional to the product of the sines of the angles made by these two rays with the optic axes of the biaxial crystal:

$$v_o^2 - v_e^2 = k \sin(n)\sin(m)$$

However, Fresnel could not use this formula as a goal because it requires that speeds increase on refraction—as they must do in a theory that is based upon the principle of least action. To transform the expression into a form suitable for use in the wave theory, which involves the principle of least time, simply requires taking the reciprocals of Biot's velocities.

The deduction of Biot's formula from the ellipsoid of elasticity was Fresnel's first major result in double refraction. In its style and method it illustrates very well how he attacked problems in this area, and it is therefore worth following in some detail. To obtain it Fresnel had to do two things: first, he had to find an expression for the angles between an arbitrary plane and the two sections of the ellipsoid of elasticity that are normal to the optic axes, since these angles correspond to Biot's n and m; second, he had to obtain the semiaxes of the section cut in the ellipsoid by this arbitrary plane, since these semiaxes determined the corresponding speeds of propagation.

In figure 11.6 the dashed lines AL, AL′ mark the directions of the two optic axes in a biaxial crystal. Fresnel at this point saw no reason to alter the traditional understanding of the optic axis as a direction along which the "ordinary" and the extraordinary rays have the same speeds, an understanding Biot had assumed in his own work. Since the radii of the ellipsoid of elasticity determine the speeds of rays that are normal to them, it follows that the optic axes must be normal to the *circular sections* of the ellipsoid—since then the ray speeds in these two directions will not depend upon the direction of oscillation. The traces of these circular sections in the plane of the optic axes are represented in the figure by NN′ and MM′; i represents the angle between one such trace and the x axis, which we choose to lie along the greatest semiaxis of the ellipsoid, the y direction therefore lying along its smallest semiaxis. Accordingly $2i$ represents the angle between the optic axes LA and L′A.

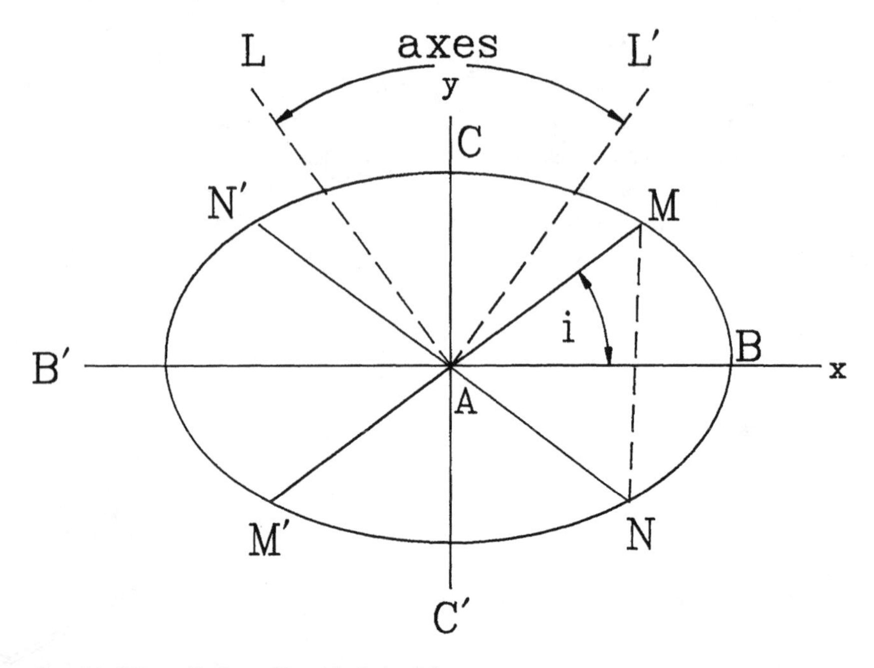

FIG. 11.6 Fresnel's first ellipsoid of elasticity.

Let us assume with Fresnel that the ellipsoid of elasticity has the following equation:

$$fx^2 + gy^2 + hz^2 = 1 \tag{11.11}$$

In order for MM′ to be the trace of a circular section of this surface, the radius AM of the elliptical section MNM′N′ of the surface must be equal to $1/\sqrt{h}$, which is the radius of the ellipsoid along the axis Az that is normal to the plane of the figure. Since AM itself lies at angle i to Ax, this implies that the angle i must satisfy the relation:

$$\tan^2(i) = \frac{f - h}{h - g} \tag{11.12}$$

Accordingly we can at once write down the equations of the circular sections of the ellipsoid, and from these we can calculate the angles between these circular sections and an arbitrary cutting plane whose equation is $z = ax + by$. Let these latter angles be n, m as in Biot's formula; Fresnel finds that they must satisfy the relations:

$$\frac{b}{a} = \frac{[\cos(n) - \cos(m)]}{[\cos(n) + \cos(m)]} \frac{\sqrt{f - h}}{\sqrt{h - g}} \tag{11.13}$$

$$\frac{1}{a^2} = \tag{11.14}$$

$$\frac{-(f - g)(h - g)[\cos(n) + \cos(m)]^2 - (f - h)(f - g)[\cos(n) - \cos(m)]^2 + 4(f - h)(h - g)}{(f - g)(h - g)[\cos(n) + \cos(m)]^2}$$

Fresnel's next step is particularly important, since it typifies the method he always used to find the semiaxes of the section of a surface, which was the central problem to solve in generating the various surfaces required by his new theory. He had to find the semiaxes of the section that is cut by the plane $z = ax + by$ in the ellipsoid, because these semiaxes determine the corresponding directions of oscillation and speeds of propagation. To do so he constructed what he termed the "polar" equation of the section using a method he learned from Alexis-Thérèse Petit (Fresnel 1821e, 296). That method, in essence, replaces the independent coordinates x,y by the expressions αz, βz, so that the square of the radius vector may be written:[5]

$$r^2 = \frac{1 + \alpha^2 + \beta^2}{f^2 + g^2 + h^2}$$

Or since, to obtain the wave translation of Biot's formula, Fresnel needs the *reciprocal* (call it t) of the square of the radius:

$$f\alpha^2 + g\beta^2 + h = t(1 + \alpha^2 + \beta^2) \tag{11.15}$$

Equation 11.15 has the great advantage over the ellipsoid's Cartesian equation that it at once expresses the square of the radius (or here its reciprocal), which is what we must find the extrema of in a given plane.

We are concerned with the radii of the ellipsoid in the plane $z = ax + by$ or, in "polar" terms, with:

$$z = ax + by \quad \text{or} \quad \alpha a + \beta b = 1 \quad \text{whence} \quad \frac{d\beta}{d\alpha} = -\frac{a}{b}$$

Take the derivative of the polar equation 11.15 with respect to α and then use the expression just found for $d\beta/d\alpha$. This yields an expression for $\partial t/\partial\alpha$ that we may set to zero to obtain the extrema (the semiaxes of the section). This gives expressions for α and for β that we place back in equation 11.15, yielding the equation for the semiaxes of the section as:

$$(11.16) \quad \frac{(t-f)(t-g)}{a^2} + (t-g)(t-h) + \frac{b^2}{a^2}(t-f)(t-h) = 0$$

Equation 11.16 determines t—the square of the reciprocal of the velocity of propagation—for a plane wave whose equation is $z = ax + by$. To replace a and b with the angles n and m that appear in Biot's formula we use equations 11.13 and 11.14, which in the end yields:

$$\frac{1}{t_{min}} - \frac{1}{t_{max}} = (f-g)\sin(n)\sin(m)$$

And this is indeed the wave version of Biot's formula, here *deduced* from the properties of the ellipsoid of elasticity. (This was not Biot's only law. He had also provided a law for the polarization of a ray in a biaxial crystal. Fresnel's derivation of it from his first ellipsoid is briefly described in appendix 24.)

Fresnel proceeded to deduce a quantitative formulation of his claim that in biaxial crystals there is no such thing as an "ordinary" ray in the old sense of the term. Recall that (fig. 11.6) Fresnel set the greatest semiaxis of his ellipsoid along Ax, the least along Ay, and so the intermediate one along Az. Consider an "ordinary" ray along Ax and another one along Az. The former must be polarized in the plane of the optic axes, and its speed is governed by the semiaxis of the ellipsoid along Az, that is, by AM. The latter must be polarized in the *yz* plane, and its speed is governed by the semiaxis AB along Ax, which is greater than AM. Consequently in this case the two rays, though both are "ordinary," have different speeds, and Fresnel further showed that this difference is the greatest possible for "ordinary" rays.[6]

Fresnel could similarly show that the speed of an extraordinary ray varies between the limits AM and AC. In consequence of this analysis the difference between an "ordinary" and an "extraordinary" ray changes fundamentally. No longer does one of the two kinds of rays always have the same speed whereas the speed of the other varies. Instead, it becomes a question of degree: both types of rays have speeds that depend on direction, but the range of speeds available for extraordinary rays is greater than that available for ordinary rays.

11.1.4 Testing the Ellipsoid

To test his ellipsoid of elasticity Fresnel performed a new set of experiments, which he used to determine the ellipsoid's semiaxes, and he then compared the resulting values with ones that could be calculated from experiments similar to those Biot had performed. Fresnel accordingly performed two kinds of experiments that were quite different from one another. One kind, like those performed by Biot, used refraction, whereas the other kind used diffraction. The refraction experiment (see fig. 11.3) corresponds to the case of the greatest possible speed difference in the prisms I and II for the ordinary rays, whereas the speeds of the extraordinary rays in the two prisms should be the same. Since the rays travel along the ellipsoid's semiaxes, we have for the speeds:

$$\text{in I: speed} = AB = \frac{1}{\sqrt{f}}$$

$$\text{in II: speed} = AM = \frac{1}{\sqrt{h}}$$

Fresnel proceeded by measuring the divergence between the ordinary and the extraordinary images of a given object at a distance of a meter from the prisms. To minimize refraction (and so dispersion), he turned the apparatus until the image was as high as possible. For prism I he then found that the images diverged by 22.7 mm, whereas for prism II they diverged by 17.0 mm. In the first case the divergence had to be proportional to $\sqrt{g} - \sqrt{f}$, in the second to $\sqrt{g} - \sqrt{h}$ (since the divergence of the images varies as the difference between the reciprocals of the ray speeds). Given that the rays pass nearly symmetrically through the prism, which has an angle of 92.5°, Fresnel could calculate the values for these two differences, obtaining:

Values from Fresnel's Refraction Experiment

$$\sqrt{g} - \sqrt{f} = .00965$$

$$\sqrt{g} - \sqrt{h} = .00723$$

Now from his diffraction experiments Fresnel could calculate the same differences, and here he obtained:

Values from Fresnel's Diffraction Experiment

$$\sqrt{g} - \sqrt{f} = .00922$$

$$\sqrt{g} - \sqrt{h} = .00700$$

Using Biot's own observations (based of course on refraction) the difference $\sqrt{g} - \sqrt{f}$ was 0.0099 (vs. Fresnel's 0.00965 from the same kind of experiment). One could also use the values for the differences to calculate by approximation the angle i from equation 11.12, and the values for i obtained in these ways varied among one another by as much as 50′, a difference Fresnel did not regard as significant because of the difficulty in accurately observing the positions of the optic axes.

But he was somewhat troubled by the discrepancies between the three sets of values for the reciprocals of the speed differences. He remarked:

> One sees that the results of my two experiments differ sufficiently from one another and from those deduced from Biot's observation. Not having taken all the precautions necessary to assure myself that the direction of the luminous rays relative to the crystal axes was very accurately what I supposed, and not having measured the angle of the prisms in the second experiment except rather crudely, I regard these attempts only as a first, approximate verification of the theory. I propose to take them up again in a season more favorable to diffraction experiments [the present being the late fall, which would likely have been cloudy], in bringing to bear all the care necessary and in employing homogeneous light to avoid the mistakes that dispersion of double refraction can occasion in determining the central band. (Fresnel 1821e, 304)

11.1.5 Fresnel's Error

On examining these early experiments in light of his (and also the correct) later theory, one might be tempted to attribute the differences between the results not to experimental error (as Fresnel did), but to a fundamental mistake in theory, a mistake Fresnel could have grasped from the very first and one that he did soon correct, as we shall see. Recall that the semiaxes of a section cut in Fresnel's ellipsoid by a plane normal to the *ray* must be parallel to the directions of the corresponding oscillations. Yet this cannot in general be correct, because the oscillations must occur in the front proper, and this is usually oblique to the ray. Indeed, without this obliquity there could never be any divergence within a crystal between the refractions.

In light of this one might be surprised not at the differences between Fresnel's experimental results, but at their close agreement, for the diffraction experiments in fact measure the wave speeds, whereas the refraction experiments measure the ray speeds. Of course one might argue (and for some weeks Fresnel did, in effect, so argue) that this difference between wave speed and ray speed is itself of the right order to produce the several empirical values. But—as Fresnel himself certainly realized in the end—one would be wrong, because the surprising fact is that, despite Fresnel's conflation at this stage of the ray with the normal to the front, his ellipsoid construction *does* correctly determine the ray speeds, and these speeds are here precisely the same as the wave speeds. Consequently the differences between the experiments were, as Fresnel first asserted, indeed due to errors in measurement and not to errors in theory.

To make clear why this is so, we must anticipate Fresnel's next steps. The proper surface to section by the wave front in order to determine simultaneously the directions of the oscillation and the wave speeds is of the fourth degree:

$$(x^2 + y^2 + z^2)^2 = \frac{x^2}{f} + \frac{y^2}{g} + \frac{z^2}{h} \tag{11.17}$$

The constants f, g, h in equation 11.17 are the same f, g, h that appear in the ellipsoid, equation 11.11. In Fresnel's experiments—diffraction and refraction—the rays lie along the coordinate axes, and in these directions the semiaxes of the sections of the ellipsoid 11.11 and of the fourth-degree surface 11.17 by planes normal to the axes are precisely the same.

Within a month of submitting his first work Fresnel had realized that different surfaces must be used to calculate ray and wave speeds, but he had not as yet also realized that his old ellipsoid remained the correct surface for ray speeds (though not for their corresponding polarizations). He felt at this point only that it was near enough to the correct surface to work well in his early experiments because they had used topaz, which is not a powerful double refractor. The clue to correcting his error did not therefore come from the differences revealed by his early experiments with the biaxial crystal topaz; it came instead directly from Huygens's construction, from the uniaxial crystal.

11.2 Replacing the Ray with the Wave Normal

11.2.1 The New Surface of Elasticity

Through 19 November, at least, Fresnel continued to conflate wave speeds with ray speeds. The ellipsoid construction seemed to him to work reasonably well, both in Fresnel's own experiments and in yielding Biot's sine and dihedral laws. But at some point between the nineteenth and the twenty-sixth of the month, Fresnel realized that he had incorrectly located the oscillation in the plane to which the ray is normal instead of in the front (see Verdet's comments in Fresnel 1821e, 309). This at once raised two questions. First, what must be used instead of the ellipsoid if one wants to use a sectioning plane that contains the oscillations—a plane that must, then, be parallel to the front? Second, why had the ellipsoid worked so well?

On 26 November Fresnel read an "extract" of his double refraction memoir that provides a clue to how he answered the first question, for here he corrects his previous analysis (Fresnel 1821f). In the extract's fifteenth section Fresnel asserts that the ellipsoid must be replaced by a surface of the fourth degree.

> If the elastic constitution of the medium were known, one could immediately find from it the speed of the rays [*sic*] in all directions. But it seems difficult a priori to establish, with any probability, the general law of these elasticities, and it is simpler to recur to experience and to deduce from it the law of the speeds. If this is rigorously represented by the radii vectors of an ellipsoid of revolution in calcspar, as seems to result from Huygens's, Wollaston's and Malus's experiments, it will still be a surface of revolution that gives the law of elasticities; but its generating curve, instead of being an ellipse, will be a curve of the fourth degree. (Fresnel 1821f, 317)

Despite Fresnel's initial reference here to finding *ray* speeds from the surface of elasticity, he in fact meant, without any doubt, wave speeds.[7] That is, for Fresnel to have realized—as this quotation shows he has—that the precise validity of Huygens's construction demands a fourth-degree surface for the elasticities necessarily requires that he has already understood that the latter must yield front, and not ray, speeds if it is to link the speed of propagation with the direction of oscillation.

But how did Fresnel discover this new, fourth-degree surface? He effectively tells us how he did so in this remark. He first refers us to "experience," to Huygens's construction itself. Obviously the next step is to find from it an expression for the wave (front) speeds and then somehow to work backward from this expression to the new surface of elasticity for the degenerate, uniaxial case. Verdet offers a reconstruction of how Fresnel might have done just that. It is, in fact, more than a reconstruction because, in the supplement to the first memoir on double refraction, dated 22 January, Fresnel in effect does very nearly what Verdet suggests he may have done to find the surface in the first place (Fresnel 1822f, 363–64).

One may accordingly follow Fresnel quite directly here. Suppose that we *begin* with Huygens's ellipsoid. Set the optic axis along Ox (center O), and section the surface by the xy plane, yielding the following ellipse, which lies in a principal section (see fig. 11.7).

$$\frac{x^2}{a^2} + \frac{y^2}{b^2} = 1$$

In the supplement Fresnel begins by calculating the tangent $[dy/dx]_{(x',y')}$ at a point (x',y') of this ellipse:

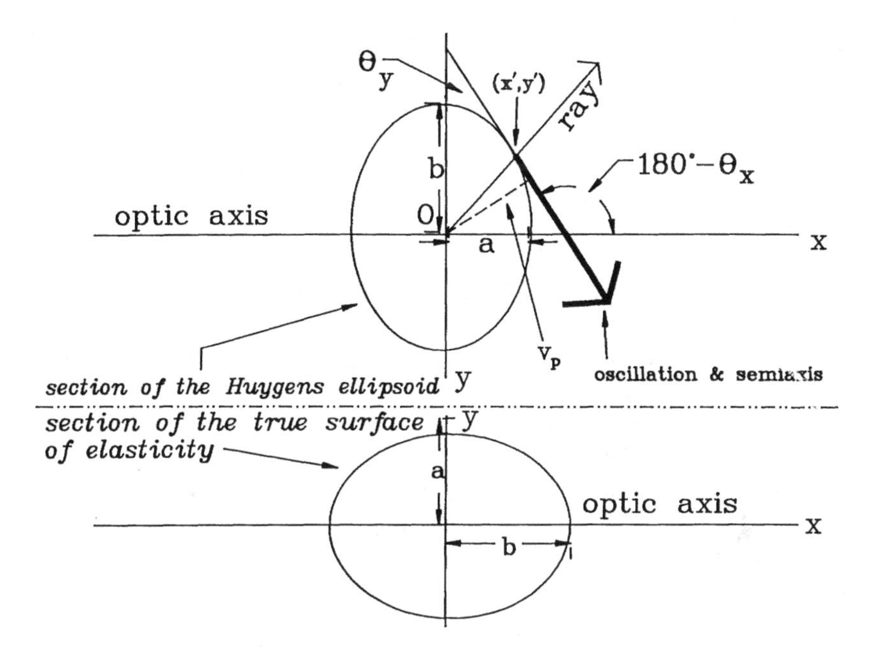

FIG. 11.7 How to find the fourth-degree surface.

$$\left[\frac{dy}{dx}\right]_{(x',y')} = -\frac{a^2x'}{b^2y'}$$

Consequently the tangent to the ellipse at (x',y') has the equation:

$$y - y' = -\frac{a^2x'}{b^2y'}(x - x') \tag{11.2.1}$$

Consider in this principal section a ray that includes the point (x',y'). The trace of the corresponding front in the xy plane will be tangent to the ellipse at (x',y'). To compute the front's velocity Fresnel need only find the distance from the tangent line 11.2.1 to the origin, which is a simple task. That distance is the wave speed v_p:

$$(v_p)^2 = \frac{a^4b^4}{b^4(y')^2 + a^4(x')^2} \tag{11.2.2}$$

This, then, must be used in constructing the *surface of elasticity* instead of the expression for the ray speed that is at once given by Huygens's construction.

Now the angle ϑ_x between the optic axis and line 11.2.1 satisfies the equations:

$$\cos^2\vartheta_x = \frac{1}{1 + \left[\dfrac{dy}{dx}\right]^2} = \frac{1}{1 + \dfrac{a^4(x')^2}{b^4(y')^2}}$$

$$= \frac{b^4(y')^2}{b^4(y')^2 + a^4(x')^2}$$

$$\sin^2\vartheta_x = \frac{a^4(y')^2}{b^4(y')^2 + a^4(x')^2}$$

By Fresnel's definition, the surface of elasticity must have a semiaxis that is parallel to line 11.2.1 and that is equal in length to the velocity v_p of the front—because one of the oscillations in the front must itself be parallel to this line (since *xy* is a principal section, and rays that lie in a principal section are polarized in and perpendicular to it). With this in mind we can quickly see the significance of the following relationship, as Fresnel himself almost certainly did:

$$a^2\cos^2\vartheta_x + b^2\sin^2\vartheta_x = \frac{a^4b^4}{b^4(y')^2 + a^4(x')^2}$$

This is just $(v_p)^2$ according to equation 11.2.2. Replace the angle ϑ_x between the oscillation and the optic axis with its complement, ϑ_y, in the term containing b^2. There results:

$$(v_p)^2 = a^2\cos^2\vartheta_x + b^2\cos^2\vartheta_y \tag{11.2.3}$$

As ϑ_x varies from 0° to 360° equation 11.2.3 will therefore produce the series of radii of the ellipsoid of elasticity that lie in the *xy* plane. That is, equation 11.2.3 must describe the curve that results by sectioning the true surface of elasticity of a uniaxial crystal by any plane that contains the optic axis. Fresnel very likely conjectured that in this degenerate, uniaxial case the curve obtained by a sectioning plane that contains the optic axis must, like the Huygens ellipsoid proper (from which the curve derives and vice versa) be the generating curve for the surface itself. That is, the correct surface of elasticity should result by rotating 11.2.3 about the optic axis (*Ox*), in which case the uniaxial elasticity surface must have the equation:

$$r^2 = a^2\cos^2\vartheta_x + b^2(\cos^2\vartheta_y + \cos^2\vartheta_z)$$

or

$$r^4 = a^2x^2 + b^2(y^2 + z^2)$$

This is the fourth-degree surface mentioned by Fresnel in the "Extrait" that replaces the ellipsoid; it gives wave speeds, rather than ray speeds, and their correlate directions of oscillation.[8] The obvious generalization to biaxial crystals requires three unequal axes:

Fresnel's Fourth-Degree Surface of Elasticity

$$r^4 = a^2x^2 + b^2y^2 + c^2z^2$$

We turn now to the second question posed above, a question of great importance to Fresnel if he was to make a convincing case: Why does even the old ellipsoid seem to work so well? At this point in his work Fresnel could argue only that the ellipsoid had succeeded in "representing the elasticities of the medium" (Fresnel 1821f, 319) because in "most" crystals (but not in calcspar) its axes differ but little from one another, so that the ray and the normal to the front are not sensibly inclined to one another. This then enables him to use the ellipsoid in the remainder of the "Extrait" of 26 November, instead of the new surface, so as to regain the Biot laws, which had therefore to be approximations only (though in fact only the dihedral law is approximate—the sine law for rays is correct).

11.2.2 Fresnel's First Attempt to Deduce the Surface of Elasticity

In the supplement dated 13 January and presented a week later, Fresnel provided his first attempt to deduce the fourth-degree surface of elasticity. It was not a very satisfactory attempt, as Fresnel himself realized (since he later greatly expanded it), but it nevertheless illustrates the efforts he made to provide a physical foundation for his analysis. Reduced to essentials, this first attempt begins with a rather elaborate physical argument designed to prove that, for any triad of axes in the medium, one can construct the reaction to a displacement by compounding the reactions that would be generated by displacements that occur along these axes and that are equal to the components of the actual displacement. This result he in fact regards as "nearly obvious" (Fresnel 1822f, 345–47). Without much further argument he then introduces a special set of these axes—whose existence he simply assumes—such that they are mutually orthogonal and possess the very important property that displacements along them generate reactions that are parallel to the displacements.

Given these axes, then the reaction $\vec{f}$ to a given displacement could be written in the following way (Fresnel of course just gives the components):

$$\vec{f} = l_a a^2 \vec{s} + l_b b^2 \vec{s} + l_c c^2 \vec{s}$$

wherein the $\vec{s}_{a,b,c}$ are unit vectors along Fresnel's special axes, the $l_{a,b,c}$ represent the components of the displacement $\vec{s}$ along these same axes and so, for unit displacement, are the direction cosines of $\vec{s}$, Consequently the magnitude, f, of the reaction will be:

$$f = \sqrt{a^4(\vec{s} \cdot \vec{s}_a)^2 + b^4(\vec{s} \cdot \vec{s}_b)^2 + c^4(\vec{s} \cdot \vec{s}_c)^2}$$

To obtain the fourth-degree surface of elasticity from this expression for the reaction, Fresnel had to adopt a new definition for it. In his previous memoir, wherein he had used an ellipsoid, the surface was said to have its radii equal to the square roots of the *total* reaction generated by a displacement along the radius (though Fresnel had not used the requirement to calculate anything). But though he never mentions the point, it is simple to show that this will lead to a sixth-degree, not a fourth-degree, surface given f.[9] To obtain the proper surface, Fresnel must alter the definition to require instead that its radii be equal to the square roots of the projections along them of the reactions generated by displacements parallel to them; that is,

$$r^2 = \vec{f} \cdot \frac{\vec{s}}{s}$$

This does indeed give the correct, fourth-degree surface.

But why should the plane sections of this surface, constructed by projecting the reaction to a given displacement along the latter's direction, determine by their semiaxes both the possible directions of oscillation and the corresponding wave speeds? Fresnel found the answer to this question when he realized at some point that, for a displacement to propagate unaltered in direction, there must be no component of the reaction in the front that does not lie parallel to the displacement—the only other permissible component must therefore be normal to be front (and he disregarded its effect, which would be a longitudinal disturbance).

Accordingly the surface of elasticity must have the following property, which can be demonstrated analytically to hold at least for the surface described by his fourth-degree equation: to wit, that in any plane section of it there are two directions along which a displacement generates a reaction that lies in a plane passing through the displacement and normal to the section; these two directions must also be the semiaxes of the section. This theorem was, and remained, one of the two major pillars of Fresnel's deductions of the elasticity surface.[10] The other pillar (upon which this one depends, and that he succeeded in constructing later) was the proof that his special axes in fact exist.

11.3 Fresnel's Construction

11.3.1 A Difficult Way to Find the Wave Surface

In the 13 January supplement Fresnel, having justified the new surface of elasticity, turned to a question of direct empirical import that he had not previously addressed: namely, how to compute the locus of a point viewed through a biaxial crystal with the eye set in a given position (Fresnel 1822f, 358–60). That reduces, he remarked, to finding those rays that, proceeding from the point and then emerging from the crystal, first reach the center of the eye's pupil. However, the elasticity surface gives only wave speeds, not ray speeds, and even then only when the front itself is already known. To solve this problem one must find the ray speeds as a function of direction; one must find, in Fresnel's words, "the wave surface," that surface that replaces Huygens's ellipsoid.

This surface, Fresnel continues, must be the locus after unit time of all the plane waves that are emitted initially from a given point, because the rays are drawn from the point to the surface and the fronts are the surface's corresponding plane tangents. Given the direction of a front, one can calculate its two possible speeds from the elasticity surface, and so its locus after unit time. Do this for every possible direction and then form the wave surface as the common locus of these fronts. The problem reduces, Fresnel notes, "to the calculation of an envelope surface" (Fresnel 1822f, 362).

But this explicit method of solving the problem escaped Fresnel. Indeed, he never succeeded with this method of attack. He never, that is, calculated the wave surface from its plane tangents. The task was too difficult (and was only later accomplished for the first time by Ampère). At this point (mid-January) he merely listed the conditions that must be satisfied:

(11.3.1) surface of elasticity

$$r^2 = a^2\cos^2 X + b^2\cos^2 Y + c^2\cos^2 Z$$

(11.3.2) equation of cosines

$$l = \cos^2 X + \cos^2 Y + \cos^2 Z$$

(11.3.3) the wave front, which sections the surface of elasticity

$$\cos X = B\cos Y + C\cos Z$$

(11.3.4) extrema in the front, which determine the front speeds

$$B(a^2 - c^2)\cos X\cos Z - C(a^2 - b^2)\cos X\cos Y$$
$$+ (b^2 - c^2)\cos Z\cos Y = 0$$

(11.3.5) tangent plane at x', y', z' of the surface sought

$$x - x' = \frac{dx'}{dy'}(y - y') + \frac{dx'}{dz'}(z - z')$$

(11.3.6) the square of the distance from origin to tangent plane

$$r^2 = \frac{(x'dy\,dz - y'dx\,dz - z'dx\,dy)^2}{dy^2dz^2 + dx^2dz^2 + dx^2dy^2}$$

Fresnel proposed to use equations 11.3.2, 11.3.3, 11.3.4, and 11.3.5 to eliminate the cosines and then to eliminate the front variables B,C by setting them equal respectively to dx/dy, dx/dz (since the plane tangent 11.3.5 must be parallel to the front 11.3.3 at the point x', y', z'). The result would be a differential equation for the wave surface obtained, in effect, directly from the surface of elasticity by setting the front distances equal to the semiaxes of the sections of the elasticity surface by planes drawn through the origin and parallel to the fronts. It would, to say the least, be tedious to carry out the computation, and Fresnel did not do so, feeling it might "present difficulties in the integration of the differential equation" (Fresnel, 1822f, 363).

11.3.2 The Wave Surface Simply Obtained

By 31 March—scarcely a month later—Fresnel had found a simple way to calculate the form of the wave surface and then to justify it (Fresnel 1822g). From the surface of elasticity he proceeds again to generate a surface whose radii determine the speeds of fronts that are normal to them. I shall refer to this as the *normal surface;* it is the same surface, albeit expressed very differently, as equation 11.3.4. The computation was essentially one Fresnel had many times performed in that its major part involved calculating the extrema of the surface's radii in a given plane using a "polar" equation.[11]

Fresnel's Normal Surface

$$(a^2 - v^2)(c^2 - v^2)n^2 + (b^2 - v^2)(c^2 - v^2)m^2$$
$$+ (a^2 - v^2)(b^2 - v^2) = 0$$

Fresnel next offers two ways to calculate the wave surface, ways that do not differ fundamentally either from one another or from his previous equations. First, he remarks, one can vary the front parameters m,n while holding fixed that front point x,y that touches the wave surface. This corresponds to calculating the intersection of three planes that are tangent to the surface at this point. The basis of the procedure is simple: the distance of the plane $z =$

$mx + ny + C$ (a front that touches the wave surface) from the origin must be set equal to the speed v, yielding the condition:

$$(z - mx - ny)^2 = v^2(l + m^2 + n^2)$$

Combining this result with the equation for the *normal surface,* and varying m,n separately, yields four equations, two containing dv/dm and two containing dv/dn. One could in principle eliminate these derivatives, leaving two equations in v,m,n, and then also use the original pair of equations to eliminate these three variables, leaving the equation of a surface—the wave surface.

But Fresnel did not carry out the task—it would have been no simpler than the previous method. Instead he provided a second route to the wave surface, one that does not directly generate the surface but that provides a condition it must satisfy. Suppose that the front varies in such a fashion that dv vanishes. Then two expressions result for dn/dm; equating these two yields a condition that the wave surface must satisfy:

Fresnel's Condition for the Wave Surface

$$(z - mx - ny)^2(my - nx) + mn(a^2 - b^2)(z - mx - ny)$$
$$+ (na^2x - mb^2y)(l + m^2 + n^2) = 0$$

We have in previous chapters often seen how Fresnel usually sought an easy way to calculate whatever he was interested in; he was not at all fond of mathematical intricacy, and issues of rigorous deduction were for the most part of little importance to him. He usually wanted a quick and direct route to a result that he could be reasonably certain was correct—at least as correct as experiment could test. At some point he realized that he could avoid the daunting intricacies of extensive algebraic manipulation by using the very ellipsoid with which his research into double refraction had begun, a surface he had abandoned only several months before as merely an approximation. He realized that the ray speeds proper are rigorously governed by an ellipsoid whose axes are the same as those of the fourth-degree surface of elasticity:

Fresnel's Ellipsoid for Ray Speeds

$$\frac{x^2}{a^2} + \frac{y^2}{b^2} + \frac{z^2}{c^2} = 1$$

Fresnel did not clearly explain how he realized that an ellipsoid with the same axes as the surface of elasticity determines the ray speeds by its plane sections. It is perhaps not very difficult to understand how he might have come to this understanding (Fresnel 1822g, 386–87), since the reasoning was no doubt very similar to that which had led him to create the ellipsoid in the first place. In the degenerate case of a uniaxial crystal, the ray speeds are governed by the radii of Huygens's ellipsoid of revolution. Quite clearly, a surface that has the correct sectioning properties to yield the uniaxial *ray* speeds in the fashion of Fresnel's original ellipsoid must itself be an ellipsoid of revolution whose generating curve can be obtained by rotating through 90° the axes of a section of the Huygens's ellipsoid that contains the optic axis (i.e., its axis of revolution). If, for example, the Huygens ellipsoid is

$$\frac{x^2}{b^2} + \frac{y^2 + z^2}{a^2} = 1$$

then the ellipsoid to section must be

$$\frac{x^2}{a^2} + \frac{y^2 + z^2}{b^2} = 1$$

in order for the latter to have the correct sectioning properties in the coordinate planes (e.g., the radius along Ox of the Huygens ellipsoid must be the radius of a circle in the yz plane of the ellipsoid to be sectioned, since Ox is the optic axis). But given this, it is almost obvious that the ellipsoid to be sectioned must have the same axes and lengths along them as the fourth-degree surface of elasticity itself, since, in the degenerate case of Huygens's construction, the front and ray speeds are the same for front normals (and so rays) that are parallel to the axes.

Fresnel guessed that a generalization of this relationship would also work for the case of three unequal axes. Accordingly, what Fresnel had to do, in effect, to calculate the wave surface—whose radii, recall, are equal to the ray speeds in their directions—was to follow the same steps from the ellipsoid that he had already followed from his surface of elasticity in generating the normal surface. This was so obvious that he simply wrote down the result without any intermediate calculations:

Fresnel's Biaxial Wave Surface

$$(a^2x^2 + b^2y^2 + c^2z^2)r^2 - a^2(b^2 + c^2)x^2 - b^2(a^2 + c^2)y^2$$
$$- c^2(a^2 + b^2)z^2 + a^2b^2c^2 = 0$$

But how could Fresnel be certain this was the correct surface, since it derives in essence from a guess? That, he felt, was nearly guaranteed by the condition he had previously deduced. In it m,n must be (by its construction using the plane $z = mx + ny$) the partial derivatives $\partial z/\partial x$, $\partial z/\partial y$. Accordingly if we operate with $\partial/\partial x$, $\partial/\partial y$ separately on the wave surface, we obtain two equations that, combined, should yield Fresnel's condition. He remarked that the condition is in fact satisfied, but he did not give any of the calculations, which would themselves hardly be trivial.

11.3.3 Implications of the New Surface

Fresnel realized that his theory could only with great difficulty be made clear to selectionists, so he carefully remarked that it required a certain definition of the "ray" that must be adhered to by selectionists in "translating" Fresnel's theory into their language:

> The word *ray,* in the wave theory, must always be applied to the line that goes from the center of the wave to a point of its surface, whatever might be the inclination of this line to the element it ends at, as Huygens remarked; because this line offers in effect all the optical properties of what one calls ray in the emission system. So, when one wishes to translate the results of the first theory into the language of the second, one must always suppose that the line traversed by the luminous molecules, in the emission hypothesis, has the same direction as the ray taken from the center of the wave to the point of its surface that one considers. (Fresnel 1822g, 393)

Precisely because Fresnel was deeply concerned to make his work as comprehensible as possible to selectionists, he decided to retain the usual definition of an optic axis as a direction in which both ray speeds are the same—even though, when he first discovered the fourth-degree surface of elasticity, he had redefined the axes as the directions in which the *front* speeds are the same (Fresnel 1822f, 354–55). The directions in which there is only one ray speed are not at all the same as the directions in which there is only one front speed, because these directions for the rays are the normals to the circular sections of the *ellipsoid,* whereas the corresponding directions for fronts are the normals to the circular sections of the very different *surface of elasticity.* Quite likely, as late as January 13 Fresnel had not been certain that the ellipsoid works rigorously for rays, so that he did not then have a way of calculating these directions for the rays. He therefore preferred at that time to use the front speeds to define the optic axes, which he could, and did, compute. Nevertheless he was clearly not very concerned with this issue, in major part because for him it is of little empirical or theoretical consequence whether the optic axes are defined as the directions along which there is only one wave speed or only one ray speed. It is an issue only for purposes of "translation" into selectionist language, and in particular for obtaining Biot's sine law, which necessarily presumes the ray definition of the axes.[12] (Fresnel did nevertheless explicitly remark that the single wave and single ray directions do differ from one another, and he drew out the difference in a way that impinges on the sorts of considerations that are necessary to recognize the possibility of conical refraction [Fresnel 1822g, 396; his remarks are discussed in appendix 25].)

11.4 Mechanics in Fresnel's Second Memoir

Fresnel had reached most of his final results—indeed all of the results insofar as the surfaces proper are concerned—by the end of March 1822, having by then produced one memoir, two extracts from it, and two supplements to it since the previous November. It took, then, only four months for Fresnel to move from his original ellipsoid to the fourth-degree surface of elasticity and finally to generate the wave surface itself. Over the next several years he united these various researches to produce an organized, step-by-step presentation of this theory, beginning from mechanical suppositions, which was eventually printed in the *Receuils* of the Académie des Sciences for 1824 (the date of printing being 1827).

The major novelty of this "second" memoir was its firm foundation of the theory in mechanics, a foundation that had a considerable impact after about 1830. Although there were no new results even here, nevertheless Fresnel presented them in a way that was designed to make it appear almost as though the intricate mathematical structure of his theory was first revealed to him from the mechanics of the ether, considered as a system of point masses endowed with forces of repulsion. This was very far from being the case, since Fresnel had from the beginning constructed the mechanics to buttress the analysis, not to derive it. The sole generative effect of mechanics was to convince him in the first place that there had to be something like a surface of elasticity.

We have seen above that an important lacuna in Fresnel's mechanics—one of the two pillars it had to rest on—was that Fresnel had not demonstrated the existence of three mutually orthogonal directions along which the reactions parallel the displacements (see above, sec. 11.2.2). He had achieved a demonstration of their existence by the end of March 1822, one that he again presented in detail in the final, "second" memoir (together with a demonstration of the first pillar—that the semiaxes of a section in the surface of elasticity are the only directions along which the reaction to a displacement lies in a plane that contains the displacement and is normal to the front).

The essence of Fresnel's proof that the axes exist was fairly simple. Assume that the force on a displaced ether particle can be calculated by displacing it alone, holding every other particle fixed. If this assumption is valid, then each component of the reaction will be a linear function with constant coefficients of the three components of the displacement along the coordinate axes, or, in matrix terms, with $\vec{f}$ the force and $\vec{s}$ the displacement:

$$\vec{f} = \begin{bmatrix} a & h & g \\ h & b & f \\ g & f & c \end{bmatrix} \vec{s}$$

where the symmetry of the matrix follows, Fresnel demonstrated, from direct calculation of the several forces—a property that guarantees the conservation of angular momentum.

It is at once apparent to the modern reader that the symmetric matrix can be reduced to diagonal form by an orthogonal transformation, and that this procedure requires solving an eigenvalue problem. This method indeed leads to Fresnel's result that there are precisely three mutually orthogonal directions along which a displacement will generate a reaction that is parallel to the displacement. But Fresnel never used extremely advanced or novel mathematical methods (though an approach through the eigenvalue problem did exist [Hawkins 1975]). Instead he followed the well-known technique of generating third-degree polynomials in the tangents of the angles of rotation that transform the system from one set of orthogonal coordinates to another. Since one of the roots of each of the cubics must be real, the transformation could always be done in such a fashion that the mixed terms in the polynomials vanish, yielding a surface referred to its principal axes, and thence to Fresnel's result. This technique was invented by Euler and had been demonstrated by Jean Hachette and Poisson (Hawkins 1975).

Of course the mere existence of these special "axes of elasticity," as Fresnel termed them, hardly implies by itself the existence of the surface of elasticity in the latter's relationship to the wave speed. It does, however, at once yield the sectioning properties of the surface: that is, given the axes of elasticity, one can build a surface that has the properties of the "first pillar" of Fresnel's mechanical theory. Fresnel constructs the surface, as we saw above, by projecting the reaction onto the direction of the displacement and setting this equal to the squared radius of the surface. (One could also construct the surface directly from the matrix linking force to displacement, since the surface is in fact the characteristic polynomial of the matrix—but then one would have to transform it to principal axes to obtain Fresnel's equation for it.)

In later years, after the wave theory had achieved wide acceptance, two major lacunae were frequently remarked in Fresnel's mechanics. First, and most obvious, his deducing the matrix from the assumption that one can hold fixed all particles but the one displaced was clearly untenable. Second, Fresnel had simply to assume that the reactions along the directions of persistent displacement in a given front determine the speeds of propagation whatever the direction of the front itself might be. He could hardly have done anything else in the absence of a general wave equation (first provided for crystals by Cauchy a decade later).

Despite the certain fact that Fresnel constructed the mechanics—which, in the end, relies primarily on the assumption that the force and the displacement are linked by a symmetric matrix—to support his theory rather than to derive it, one must not underestimate its significance either for him or for subsequent wave theorists. Without this physical underpinning it would have been extremely difficult for contemporary physicists to use the wave theory without feeling that it lacked a firm physical foundation, whereas the alternative, selectionist account (though vastly inferior in generating empirically testable results) did at least have an apparently firm physical basis in the optical particle governed by forces. Indeed, one of the criticisms wave theorists often threw at selectionists was that the ether was a much sounder physical foundation than the emissionist particle—in part because one could, as Fresnel had begun to show (and as Augustin-Louis Cauchy showed with much greater rigor), actually generate from the mechanical structure of the ether theorems that lead to the testable formulas.

11.5 Arago's Report

Fourier, Arago, Ampère, and Poisson were assigned to report on Fresnel's memoir—that is, on the first memoir, together, one presumes, with its two supplements—(though only the first three signed it). The report was printed in the *Annales de Chimie et de Physique* (Arago's journal) in August (Arago 1822). Its style and tone are entirely Arago's. He begins with Fresnel's claim that there is no ordinary ray in biaxial crystals and then briefly describes both kinds of experiments—diffraction and refraction—that demonstrate the point.

Then he turns to Fresnel's theory and discusses the ellipsoid for determining ray speeds. Despite the fact that Arago had almost certainly seen the work Fresnel had done since the first memoir, he apparently failed to understand that the ellipsoid cannot give the planes of polarization as well as the ray speeds—that the fourth-degree surface of elasticity must be used for that purpose. Here again we find that Arago did not abandon the old emphasis upon rays for the wave theory's insistence on fronts. Yet without that rupture one does not have the wave theory but, rather, has a sort of hybrid in which the kinds of interactions that rays may have with one another have been altered but in which the physical role of the ray remains central.

With the exception of Arago's report in 1822, only two short papers were printed concerning Fresnel's theories of double refraction before the second memoir itself finally appeared: Fresnel's short note at the end of 1821 (published in the *Monitor*) that gives the original ellipsoid, and an "Extrait" of the second memoir, describing its major results, which was printed in the *Bulletin de la Société Philomatique* in 1822 and then in Arago's *Annales de Chimie et de Physique* in 1825. Until 1827 there was no other way to learn of Fresnel's theory except by correspondence with him—which was how John Herschel learned the details.

In his own article "Light" (1827, 539) Herschel complained bitterly about the delay in publishing Fresnel's second memoir, though he was not aware that Fresnel was so physically weak by this time that he scarcely had the energy to put the results together until 1826.[13] One can gauge how unknown was Fresnel's work on double refraction (and indeed even on polarization) from the fact, noted by Herschel, that in December 1826 the Imperial Academy at Saint Petersburg posed the following prize question: "To deliver the optical system of waves from all the objections that have (as it appears) with justice, been urged against it, and to apply it to the polarization and double refraction of light" (Herschel 1827, 539).

Fresnel's theory of double refraction—even in the comparatively crude form described by Arago in his report—apparently impressed Laplace far more than his theory of diffraction had. Verdet relates that after the report was read Laplace "proclaimed the exceptional importance of the work . . . he congratulated the author on the constancy and sagacity that had led him to discover a law that had escaped the most able and, anticipating in a way the judgment of posterity, declared that he placed this research above all that had for a long time been communicated to the Academy" (Fresnel, *Oeuvres,* 1:lxxxvi–lxxxvii]. Since Arago's report employed the ellipsoid for rays, Laplace was not faced with the difficulty of accepting the wave theory proper—he needed only to accept the ellipsoid as a method for calculating *ray* speeds, much as he had over a decade before accepted Huygens's construction. Indeed, quite possibly from Laplace's point of view what Fresnel had done was to deduce Biot's law of sines from the properties of an ellipsoid with three unequal axes that, on degeneration into an ellipsoid of revolution, also yields the old Malus-Laplace velocity law (always taking the reciprocal of the radii as the speed).

12 The Emerging Dominance of the Wave Theory

12.1 Herschel's "Light"

Toward the end of the summer of 1826, the year Fresnel's prize memoir on diffraction finally appeared in print, John Herschel posed a series of questions to Fresnel on double refraction. Herschel was at that time engaged in completing a comprehensive article on light for the *Encyclopaedia Metropolitana,* one of whose major goals was to provide as full a treatment as possible of the differences between the wave and emission theories in regard to phenomena of importance. Accordingly he asked Fresnel (evidently via Ampère) the following three questions concerning points that (since the second memoir on double refraction had not yet appeared) Fresnel had not fully covered (Fresnel *Oeuvres,* 2:647–60):

1. The general laws that govern the directions of the so-called ordinary and extraordinary rays in crystallized bodies when they refract an arbitrary ray that strikes their surface.
2. The laws that govern the same rays when the incident ray has an arbitrary polarization, partial or total.
3. The laws that regulate the intensity of the partially reflected rays, at an arbitrary angle, on a crystalline or noncrystalline surface, when the primitive ray has an arbitrary polarization.

In his article Herschel closely followed the answers Fresnel gave to all three questions (though in his account of biaxial crystals Herschel himself generated the normal surface, since Fresnel had not done so in the letters)—and in his answer to the third Fresnel was extremely careful to avoid presenting any results that required the vector decomposition of a transverse oscillation. This, we shall presently see, may have led Herschel into a significantly confused attempt to reconcile selectionist and wave theoretical vocabularies.[1]

Herschel's article, which could be obtained in 1828, was the best account of the wave theory, excepting Fresnel's own (which required at least the prize memoir, printed only in 1826, and the second memoir on double refraction of 1827), available for about a decade. Despite its comprehensive account of the wave theory, on first reading Herschel's "Light" one might conclude that he had not as yet decided to embrace the wave theory fully because he does not do so in so many words. Yet a close reading of the article shows quite clearly that, if Herschel was not as yet prepared to accept the theory publicly, he was at least very close to doing so in private. The following remarks taken from Herschel's discussion of the wave account of double refraction show how deeply the theory captured him:

> The theory of M. Fresnel gives then, as we see, at least a plausible account of the phenomena of double refraction in the case of uniaxial crystals; and when we consider the profound mystery which, on every other hypothesis, was admitted to hang over this part of the subject, we must allow that this is a great and important step. But the same principles are equally applicable to biaxial crystals with proper modifications, and (which is a strong argument for their reality) lead, when so applied, to conclusions which, though totally at variance with all that had been taken for granted before, on the grounds of imperfect analogy and insufficient experiment, have been since verified by accurate and careful experiments, and have thus opened a new and curious field of optical inquiry. Nothing stronger can be said in favour of an hypothesis, than that it enables us to anticipate the results of the experiment, and to predict facts opposed to received notions, and mistaken or imperfect experience. (Herschel 1827, 538)

And:

> If the deduction in succession of phenomena of the greatest variety and complication from a distinctly stated hypothesis, by strict geometrical reasoning, through a series of intermediate steps, in which the powers of analysis alone are relied on, and whose length and complexity is such as to prevent all possibility of foreseeing the conclusions from the premises, be a characteristic of the truth of the hypothesis,—it cannot be denied that it possesses that character in no ordinary degree. (Herschel 1827, 545)

But Herschel's strongest (and best known, since it is quoted in William Whewell's *History*) expression of admiration appeared at the end of his statement of the wave theory's basic principles. He remarks of "Young's" "doctrine of the interference of the rays of light" that it is

> a theory which, if not founded in nature, is certainly one of the happiest fictions that the genius of man has yet invented to group together natural phenomena, as well as the most fortunate in the

> support it has unexpectedly received from whole classes of new phenomena, which at their first discovery seemed in irreconcilable opposition to it. It is, in fact, in all its applications and details one succession of *felicities,* insomuch that we may almost be induced to say, if it be not true, it deserves to be so. (Herschel 1827, 456)

These are strong words of praise from someone whose own scientific reputation was built primarily on a selectionist analysis of anomalies in ring formation by crystals.[2] They were backed by an equally thorough discussion of diffraction, chromatic polarization, and double refraction. It is clear from his account that Herschel, perhaps more than any contemporary besides Fresnel himself, deeply understood the wave theory. He grasped Huygens's principle and the essential role it plays in diffraction; he knew how to compound orthogonal oscillations to find their resultant; and he fully understood that the oscillations must occur in the wave front and, in it, may trace out a path over time that varies quite randomly:

> [In Fresnel's theory] a polarized ray [*sic*] is one in which the vibration is constantly performed in one plane, owing either to a regular motion impressed on the luminous molecule [of the medium], or to some subsequent cause acting on the waves themselves, which disposes the planes of vibration of their molecules all one way. An unpolarized ray may be regarded as one in which the plane of vibration is perpetually varying, or in which the vibrating molecules of the luminary are perpetually shifting their planes of motion, and in which no cause has subsequently acted to bring the vibrations thus excited in the ether to coincident planes. (Herschel 1827, 534)

These remarks show that Herschel had understood what, as I have argued in previous chapters, was one of the most difficult problems that Fresnel himself had had to overcome in creating a theory of polarization—to explain the nature of unpolarized light in a way that avoids the concept of static mixing that underlies selectionism.

And yet Herschel recurs to a novel mixture of ray with wave concepts in discussing partial polarization, and he altogether avoids the crucial decomposition of oscillations that the wave theory must use to analyze reflection. It is worthwhile presenting Herschel's remarks on partial polarization, because they nicely illustrate the difficulties of making the transition from rays to waves:

> We may conceive a partially polarized ray to consist of two unequally intense portions; one completely polarized, the other not at all. It is evident that the former, periodically passing from evanescence to total brightness, during the rotation of the tourmaline or reflector, while the latter remains constant in all positions, will give rise to the phenomenon in question. And all the other characters of *a partially polarized ray* agreeing with this explanation, we may receive it as a principle, that when a surface does not completely polarize a ray, its action is such as to leave a certain portion completely unchanged, and to impress on the remaining portion the character of complete polarization. Thus we must conceive polarization as a property or character not susceptible of *degree,* not capable of existing sometimes in a more, sometimes in a less intense state. A single elementary ray is either wholly polarized or not at all. A *beam* composed of many *coincident rays* may be partially polarized, inasmuch as some of its component rays only may be polarized, and the rest not so. This distinction once understood, however, we shall continue to speak of a ray as wholly or partially polarized, in conformity with common language. (Herschel 1827, 509)

In the first part of the quotation Herschel thinks of a partially polarized "ray" as consisting of two "portions," one "unpolarized" and the other "polarized." Even this—taken literally—violates the sense of the wave theory, but it can be reconciled with wave terminology if (as in Fresnel's reply to Herschel's questions) we treat this as a "representation" solely for purposes of calculation (a representation made possible, Fresnel knew, by the decomposition of the transverse oscillation).[3]

But Herschel continues in a way that diverges markedly from Fresnel's sense: he asserts that the "action" of partial polarization differentially affects portions of the incident light by polarizing one part and leaving the other part entirely unaffected, and he sees this as a "principle" obviously required by Fresnel's representation. But it is most definitely not required, and indeed it makes no sense at all in the wave theory, because no "action" can polarize one "portion" of light and leave the other "portion" unaffected. That we can represent partially polarized light in this way is a mathematical, not a physical, proposition.

Herschel's "principle" would, however, indeed be required on selectionist grounds, since there representation by division necessarily means that an actual physical division has occurred. Consequently the fact that Herschel takes the "principle" to be required by the representation shows that he continues, in this most difficult of situations, to be thinking partly as a selectionist. But not entirely. In the wave theory it is true that, as Herschel says, a "single, elementary" ray is either polarized or not—either the oscillation remains unaltered in direction over time or it does not. However, this makes no sense in the selectionist vocabulary, where polarization must refer to the beam, not to its individual component rays. Herschel's representation of partially polarized light is legitimate in the wave theory (indeed, he obtained it from Fresnel), but his "principle" follows from it only on the selectionist understanding of a beam composed of rays that are selected during the act of polarization; but then it cannot be said, as Herschel does (and as the wave theory requires), that a "single, elementary ray is either wholly polarized or not at all."[4]

Herschel's remarks are therefore *incoherent:* they cannot be consistently interpreted in either theory's vocabulary. And Herschel knows there is a problem, for he remarks in concluding the section that we "shall presently, however, obtain clearer notions on the subject of *unpolarized* light, and see reason for discarding the term altogether"—and indeed he later explains it quite well, as we have seen. The reason for the incoherence of his language at this point in his article must therefore lie not in a lack of understanding of the wave theory, but in the nearly impossible task Herschel set himself of discussing phenomena that seem closely tied to experiment (phenomena like polarization) in ways that avoid committing him to either the wave theory or selectionism (which he identifies with the emission theory). This simply cannot be done in the case of polarization, because the two theories, as we have repeatedly seen, diverge at such elementary levels that it is often impossible for them even to describe what seem to be simple phenomena in mutually acceptable terms. Herschel in late 1826 was in the midst of completing the arduous mental transformation from selectionism to the wave theory. The difficulties this must have caused him reveal themselves strikingly in this passage, where the vocabularies of polarization clash irremediably.

Herschel remained for a number of years unwilling to assert that the wave theory must be accepted and the emission theory rejected. As late as 1830, for example, he constructed properties for optical particles that could deal with interference.[5] Herschel's reluctance to commit himself irretrievably to the wave theory is hardly surprising, since in the "wave theory" he included suitable properties for the "ether," just as in its alternative he included appropriate properties for the optical particle, and he strongly felt these to be "hypothetical" in a way that laws closely related to experiment are not.

Like most of his contemporaries, Herschel would not thoroughly separate the wave "theory" from ether mechanics. Talk about waves meant talk about the ether[6]—or else the propositions one was using had to be interpretable without reference to waves at all. It was possible to ignore waves (and so the ether) in diffraction theory, but polarization unquestionably required that wave fronts, not rays, be the fundamental physical objects. In passing between selectionism and the wave theory one had necessarily to alter the idea of a ray by introducing the front, which seemed tantamount to introducing the ether, or else one risked, in these areas at least, incoherence. Many selectionists never understood how thoroughgoing the change had to be, and many wave men also failed to appreciate it (in that the latter always conflated selectionism with the optical particle, which focused their critical attention on the physics of light rather than on the more primitive issue of whether rays were things or mathematical constructs).

Herschel was more aware than most of the difficulties in translation; however, he chose to avoid the problem where it was most serious by invoking an incoherent vocabulary in discussing partial polarization. Herschel's case shows, I think very clearly, that the transition from *ray as object to ray as mathematical construct* was difficult in two ways: it was difficult to make the transition in the first place, and it was just as difficult to realize that one had made it after the fact.[7]

12.2 A New Mathematical Elite

In 1875 James Challis, Plumian Professor at Cambridge and an eager participant in elaborating the wave theory in the 1830s, remarked: "In the second edition of Airy's *Tracts,* published in 1831, the Undulatory Theory of Light is unreservedly accepted, although, as far as I remember, this view was not so exclusively maintained in the Experimental Lectures by which that publication was preceded" (Challis, 1875, 22). George Biddell Airy had been Lucasian Professor at Cambridge since 1828, when he had begun his course of lectures in "experimental philosophy." According to Challis, then, Airy fully embraced the wave theory sometime between 1828 and 1831. John Herschel had substantially done so by 1827, and his "Light" appeared in print in 1830. By 1832 William Whewell, William Rowan Hamilton, Humphrey Lloyd, and Baden Powell had done so as well, and the 1833 meeting of the newly founded British Association for the Advancement of Science witnessed a direct clash between them and supporters of the emission theory (Brewster, Richard Potter, and John Barton).[8] Morrell and Thackray (1982) argue persuasively that the "Cambridge faction and its Dublin allies" were particularly attracted to the wave theory because it "exemplified the new mathematical physics they sought to promote, and they found in Section A [of the Association] an instrument for propagating their views."

The dominance of the section by the Cambridge faction was sealed in 1834 with Lloyd's "Report" on optics, which advocated the wave theory and compared it in many points unfavorably with the emission theory. The faction's influence had been greatly expanded in 1833 with the prediction by Hamilton from Fresnel's biaxial wave surface of the new phenomenon of conical refraction and its confirmation by Lloyd (on which see appendix 7), to the extent that, as Morrell and Thackray remark (1982, 469), "Brewster was reduced to reporting experiments anomalous to the wave theory, while sniping at Whewell from the columns of the *Edinburgh Review.*"

In part, then, the dominance of the wave theory in Britain by the early 1830s reflects the fact that it, unlike the emission theory, provided the young group of Cambridge-trained mathematicians with a subject that was amenable to analysis. Section A of the British Association provided a convenient vehicle for propagating their views, and the wave men were, by the mid-1830s, able to control almost fully what could be printed in major journals. In Britain there never had been a large, organized group of emission theorists, and so, given its empirical success and ardent advocacy by the young, eager, and ambitious Cambridge mathematicians, there never had been a serious possibility of the wave theory's not ultimately dominating British optics.

But the often bitter exchanges of the 1830s reflect more than the burgeoning power of a new scientific elite, dedicated to the generation of physical laws from highly abstract mathematical analyses. They also reflect the deep conceptual impasse that made communication between the remaining selectionists and the new elite frequently impossible. We have already seen that Brewster never understood that the wave theory makes ray counting impossible, and I know of no wave theorist who, by the 1830s, clearly grasped that this was the central point at issue between the wave theory and its alternative. Apparently the transition from ray counting to the wave front, which Herschel, Airy, Whewell, Hamilton, and Lloyd must all have made in the late 1820s, was not reversible: once accomplished it became extraordinarily difficult to recognize what it was that had been done. For those, like Brewster and Potter, who never did make the transition, there *could not* be any testable statements that the wave theory makes about rays that cannot be directly translated into the language of the emission theory. Consequently, on both sides the points at issue were on occasion obscure and difficult to make precise.[9]

In Whewell's 1837 *History of the Inductive Sciences* we find not even a vestige of selectionism as something to be distinguished from the emission theory, and the latter has been a miserable failure. It is worth quoting Whewell's remarks at some length to gain a sense of how utterly confident the wave men were by this time:

> When we look at the history of the emission-theory of light, we see exactly what we may consider as the natural course of things in the career of a false theory. Such a theory may, to a certain extent, explain the phenomena which it was at first contrived to meet; but every new class of facts requires a new supposition,—an addition to the machinery; and as observation goes on, these incoherent appendages [!] accumulate, till they overwhelm and upset the original frame-work. Such was the history of the hypothesis of solid epicycles; such has been the history of the hypothesis of the material emission of light. In its simple form, it explained reflection and refraction; but the colours of thin plates added to it the hypothesis of fits of easy transmission and reflection; the phenomena of diffraction further invested the particles with complex hypothetical laws of attraction and repulsion; polarization gave them sides; double refraction subjected them to peculiar forces emanating from the axes of crystals; finally di[e]polarization loaded them with the complex and unconnected contrivance of moveable polarization; and even when all this had been assumed, additional mechanism was wanting. There is here no unexpected success, no happy coincidence, no convergence of principles from remote quarters; the philosopher builds the machine, but its parts do not fit; they hold together only while he presses them: this is not the character of truth. (Whewell 1837, 2:464–65)

The wave theory has a very different character:

> In the undulatory theory, on the other hand, all tends to unity and simplicity. We explain reflection and refraction by undulations; when we come to thin plates, the requisite "fits" are already involved in our fundamental hypothesis, for they are the length of an undulation: the phemonena of diffraction also require such intervals; and the intervals thus required agree exactly with the others in magnitude, so that no new property is needed. Polarization for a moment checks us; but not long; for the direction of our vibrations is hitherto arbitrary;—we allow polarization to decide it. Having done this for the sake of polarization, we find that it also answers an entirely different purpose, that of giving the law

> of double refraction. Truth may give rise to such a coincidence; falsehood cannot. But the phenomena become more numerous, more various, more strange:—no matter: the theory is equal to them all. It makes not a single new physical hypothesis; but out of its original stock of principles it educes the counterpart of all that observation shows. It accounts for, explains, simplifies, the most entangled cases; corrects known laws and facts; predicts and discloses unknown ones; becomes the guide of its former teacher, observation; and, enlightened by mechanical conceptions, acquires an insight which pierces through shape and colour to force and cause. (Whewell 1837, 2:465–66)

Note that Whewell does not assert that the "emission theory" must be wrong; he says only that it has "not the character of truth." Unlike the wave theory, he asserts, it requires the continued adding of complicated properties to the original ones in order to encompass phenomena. This is certainly true of the emission theory, but it is not true in quite the same sense of selectionism proper. If we require that rays be understood physically as particles and that any change in their properties requires forces, then Whewell is correct: with the sole exceptions of the laws of reflection and ordinary refraction, the emission theory never generated new formulas. But the same cannot quite be said for selectionism itself: it could, and did, lead to Malus's expressions for partial reflection, and it permitted Biot to deduce formulas for chromatic polarization. But Whewell did not know about Malus's theory, and he felt Biot's theory of "moveable polarization" to be coincident with the oscillations of light particles—though we have repeatedly seen that Biot regarded his basic formulas as direct implications of ray counting and could not understand Fresnel's refusal to see in them the embodiment of pure fact.

To Whewell the power of the wave theory was in substantial part grounded in the suggestive possibilities of the ether—which had no counterpart in the emission theory, since it had proved itself unable to generate new phenomena using particles and forces. Whewell's account of how the wave theorist must be led almost without effort to a theory of biaxial crystals nicely illustrates his views:

> Since the double refraction of uniaxial crystals could be explained by undulations of the form of a spheroid, it was perhaps not difficult to conjecture that the undulations of biaxial crystals would be accounted for by undulations of the form of an ellipsoid, which differs from the spheroid in having its three axes unequal, instead of two only. . . . Or, again, instead of supposing two different

> degrees of elasticity in different directions, we may suppose three such different degrees at right angles to each other. This kind of generalization was tolerably obvious to a practised mathematician. . . .
>
> It was here that the conception of transverse vibrations came in, like a beam of sunlight, to disclose the possibility of a mechanical connexion of all these facts. If transverse vibrations, travelling through a uniform medium, come to a medium not uniform, but constituted so that the elasticity shall be different in different directions, in the manner we have described, what will be the course and condition of the waves in the second medium? Will the effects of such waves agree with the phenomena of doubly-refracted light in biaxal crystals? Here was a problem, striking to the mathematician for its generality and difficulty, and of deep interest to the physical philosopher, because the fate of a great theory depended upon its solution. (Whewell 1837, 2:455–56)

Whewell's point is well taken in this sense, that the wave theory is vastly more amenable than the alternative to quantification once it is posited that polarization controls refraction. We saw in the previous chapter how that assumption combined with transversality almost alone led Fresnel to his first theory for biaxial crystals. Neither the emission theory's forces nor pure selectionism's asymmetries suggested any way to quantify a connection between a ray's asymmetry and its speed.

Nevertheless Whewell's remarks—and those by Lloyd in 1834—could hardly sway a convinced selectionist, since the points they make are persuasive rather than conclusive. One selectionist at least managed to continue publishing until nearly 1860—Richard Potter. Potter, a Manchester merchant tutored by John Dalton (Hankins 1980, 147), began objecting early to the pretensions of the wave theory. He continued to object for the next thirty years. In 1859 the second volume of his *Physical Optics* appeared. In it he attempted to quantify a "corpuscular" theory of light. But though the slim volume contains much algebra, it hardly refers to particles and forces. What it does implicitly assume, however, is that rays can always be counted and that optical intensity means number of rays.

It is hardly worthwhile detailing Potter's "analyses," because they almost all share a single characteristic: where Potter accepts a result from the wave theory he imports it (occasionally surreptitiously, occasionally from "experiment") and molds his algebra around it. For example, Potter claims to deduce Fresnel's law governing the change in the polarization of light by reflection.

He provides a page and a half of algebra and spherical trigonometry that does result in the law. But where does it begin? From the given of "experiment" that (to translate Potter's deliberately obscurantist description) the planes of polarization of the reflected and incident rays intersect in a line that lies at right angles to the refracted ray. This does yield Fresnel's law—because it is algebraically equivalent to it. Obviously Potter had worked backward from the law to something he could give that sounded more closely in tune with ray theory: for he called the line of intersection a locus that the "axis of the molecule" must pass through in moving from its incident to its reflected position (Potter 1859, 27–28).

Potter, like all selectionists, failed to grasp that in the wave theory the front completely replaces the ray as the physical foundation. He recognized, however, that something like a front should be introduced to make the law of interference comprehensible; accordingly he has the luminous point emit sheets of particles at regular intervals. But these sheets are not fronts, because they consist of a countable number of rays (particles): Potter has merely added the requirement that the rays emitted in all directions from a point source are emitted simultaneously at discrete intervals. He has in no way abandoned a selectionist outlook—to the extent that he completely failed to understand the relationship between front and ray that underlies the wave theory.[10]

Potter ended his career a bitter man, convinced that his experiments and criticisms had been unjustly ignored. He wrote in the preface to his "mathematical" brief for ray theory:

> With the reiteration of blunders and the management described above, together with bold assertions of the most complete agreement of the facts of experiment with conclusions from the undulatory theory, it was asserted to be as certainly true as the theory of gravitation.
>
> With such advocacy it was not likely that the author of the present treatise would find companions in investigating critical points where the undulatory theory fails, and he has had the field nearly clear to himself for thirty years. . . . The author is far from considering that he has done more than commence the mathematical discussion of the corpuscular theory of light; but trusts that Physical Optics, recalled from one of its wanderings [i.e., from the wave theory], may by-and-by take a straight course of progress, and that his long perseverance against dogmatic error will not be considered as lost labour by future investigators. (Potter 1859, v–vi)

Potter's brief for a ray theory was marred by his ill-conceived experiments, his ignorance of the wave theory, and his poor mathematics. He had not the ability—by far—of a Brewster or a Biot. Nevertheless, we must not simply dismiss him. He was the last to uphold a way of understanding light that the wave theory had almost eradicated from memory, and he defended the old way as best he knew how. Indeed, he never truly understood that there was an alternative.

12.3 "Optical Activity" between 1800 and 1850

The spread of the wave theory during the 1830s coincides with increasing activity in optics, for both the number of articles produced per year and the number of authors involved in their production take a marked jump between 1830 and 1835, though there are notable, and significant, differences from country to country. Figures 12.1, 12.2, and 12.3 graph the article distribution in absolute value and in percentages by country for every five years between 1800 and 1850.[11]

We can see at once that until about 1845 the vast majority of articles appeared in British, French, or German journals. Further, as late as 1850 the number of German articles was much smaller than the number of British and French articles, though the German numbers increased markedly between 1845 and 1850. A broadly similar pattern emerges as well from figures 12.4 and 12.5, which graph the numbers of publishing authors. Here again we see

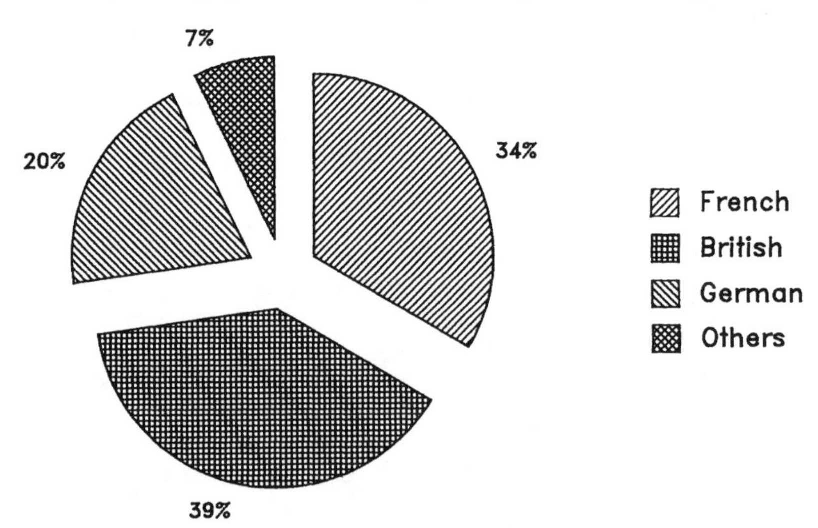

FIG. 12.1 Total number of articles, 1800–1850.

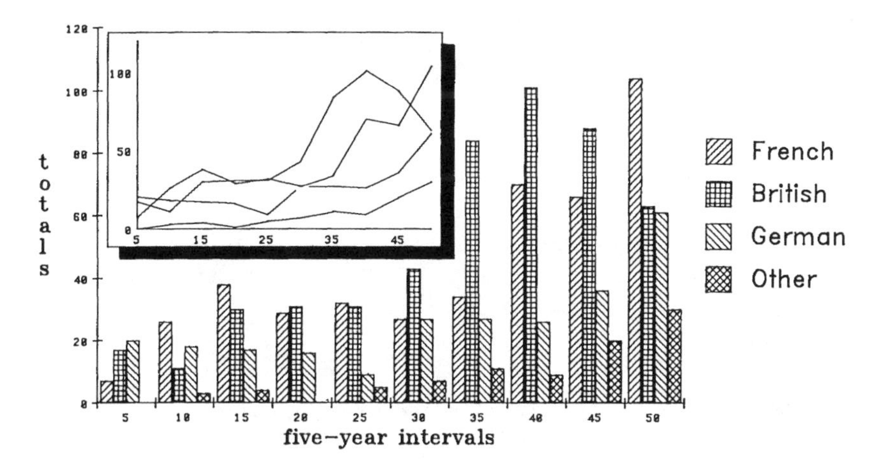

FIG. 12.2 Article distribution.

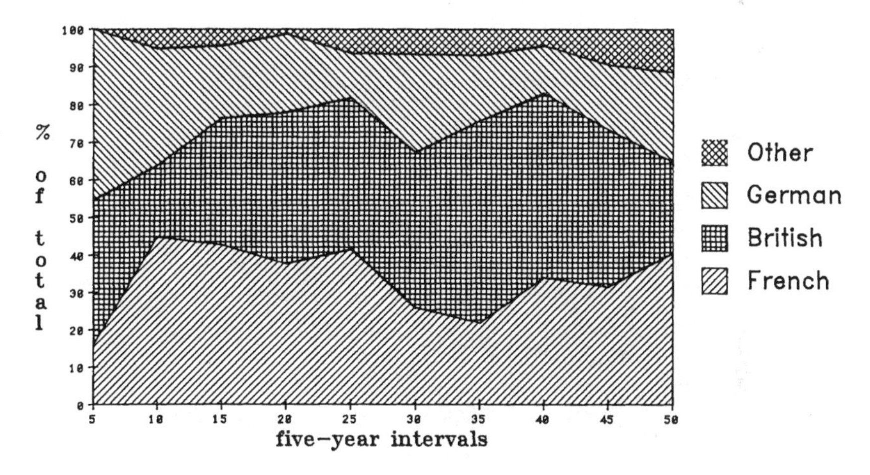

FIG. 12.3 Article distribution.

that British, French, and German authors vastly outnumber all other nationalities, though the disproportion between any one of these three and the "others" is considerably smaller in the case of author numbers than it is for article numbers, reflecting the much greater productivity of the French, British, and Germans.

The graphs enable us to distinguish quantitatively three major periods: 1800–1815, 1815–30, and 1830–50. During the first third of the first period the production of optics articles was dominated by the Germans and the British. During the second five years the French numbers have increased by more

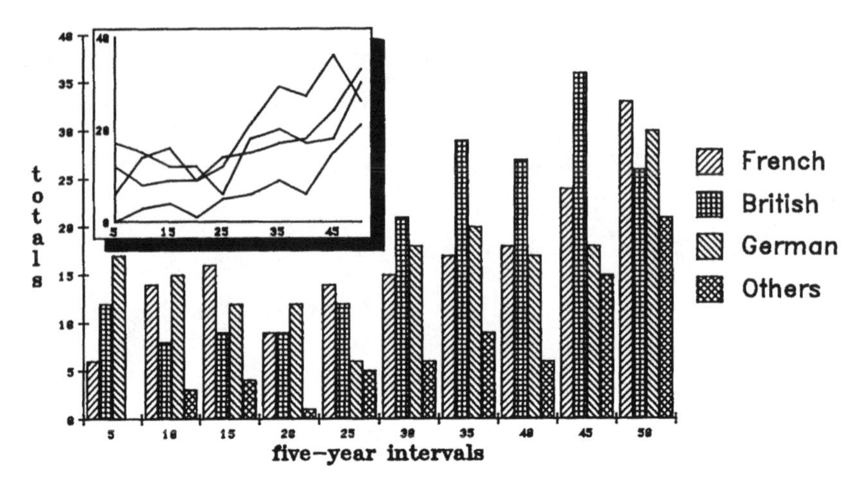

FIG. 12.4 Author distribution.

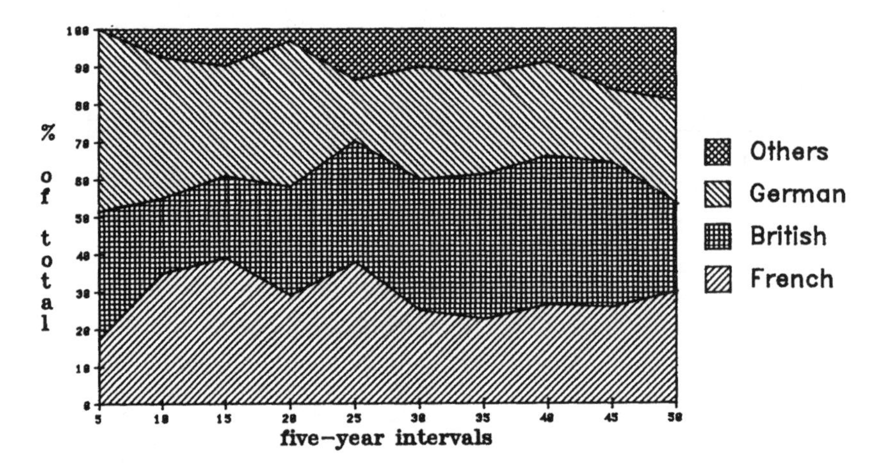

FIG. 12.5 Author distribution.

than 300% in articles and by more than 200% in authors. The French article numbers increase almost linearly during the third five years, whereas the British numbers decrease during the second five years but pick up again during the third. The German numbers drop from a high point during the first quintile.

During the second period, 1815–1830—the years during which the wave theory was developed by Fresnel and began to achieve influence—French article numbers actually *drop* during the first third (1815–20), pick up a bit during the second (1820–25), and then drop again during the third (1825–30). British numbers remain essentially stable until the third five years, when they increase markedly. German numbers continue their drop until the last five years, when they jump to a historical high.

This means that during the fifteen years when the wave theory was being debated in France, activity in optics actually decreased, in France notably so, though much of this decrease was due to Biot. We see little numerical effect of the optical debates in Britain or Germany. By 1830–35 things have begun to change quite radically in Britain, though not in France. British numbers rocket upward, propelled by extensive investigations in the wave theory, while French numbers do not surpass their value between 1810 and 1815 until 1835–40.

We can gain more insight into this data by considering the "productivity" of physicists in optics, that is, the number of articles per author (the article density) (see fig. 12.6). And here we have a surprising parallel: throughout this fifty-year period the British and French productivity curves run close to one another, even during the years between 1820 and 1840 when productivity decreased. I am not certain whether this has a deeper significance, but we can examine the critical years to see how the productivity curves emerge.

From 1820 to 1825 French articles increase very slightly, while the number of authors increases considerably; from 1825 to 1830 French articles decrease somewhat, whereas authors increase much less than they had in the previous

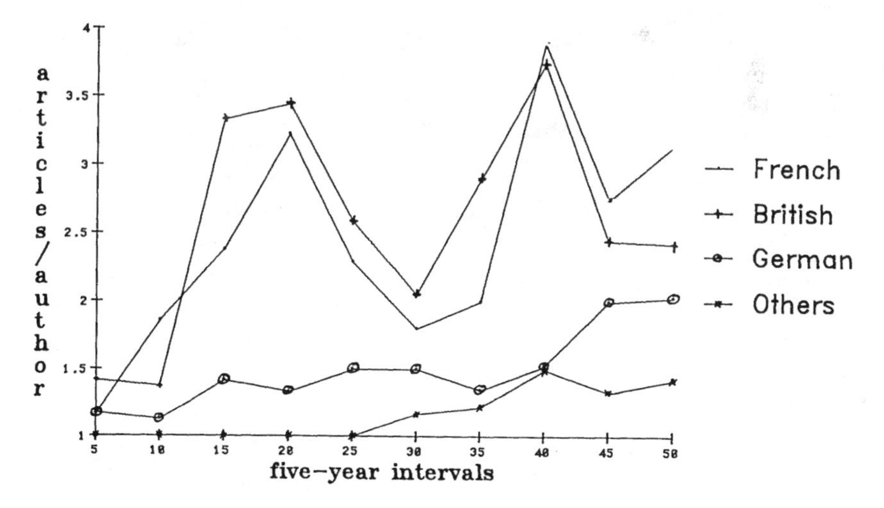

FIG. 12.6 Article density.

five years. Hence the drop in French productivity primarily reflects a slight author increase together with a slight decrease in the number of articles produced.

The British curves show a similar, but more pronounced, pattern during the first five years but diverge from the French during the second. Like the French work, British articles change only slightly between 1820 and 1825; like the French writers, but to a lesser degree, British authors increase during these five years. Between 1825 and 1830, however, British article numbers jump significantly, whereas French numbers decrease; at the same time British authors jump much more than French authors, so that the net result is a similar drop in productivity: but in the British case this primarily reflects a very large increase in authors, whereas in the French case it reflects a slight author increase together with a slight article decrease. Nevertheless, it is striking that the British and French productivities so closely parallel one another.

On the whole, then, the years of transition to the wave theory were also a period during which optical activity, and productivity, actually decreased. The issues certainly did not at once generate an upsurge in investigations or commentary but seem to have depressed rather than stimulated disciplinary activity, particularly in France, but also in Britain and even Germany. Some of this decrease no doubt reflects the diversion of physicists' attention to other areas, in particular to the complex of issues surrounding Hans Christian Ørsted's discovery in 1820 of the electric current's magnetic effect, since both Biot and of course Ampère were strongly concerned with them. However, the decrease in activity begins rather earlier than this, during the years 1815–20.

12.4 The Functions of the Ether

Throughout my account of selectionism and the early history of the wave theory, I have avoided lengthy discussion of the physical models used by selectionists and by wave theorists. I have not done so because these models were not historically significant—they most definitely were. Rather, I have concentrated instead on another aspect of the story, one that enables us to grasp how profound a change in conception was involved in the transition to the wave theory of light. So profound was the change that even if a physicist was prepared to abandon forces and particles (as indeed occurred), he was nevertheless frequently unwilling also to abandon the physical reality of the ray. In this elementary sense the ray, rather than its precise physical structure, constituted an irreducible minimum, a concept that could be abandoned only with the greatest difficulty. The past chapters provide examples of selectionists clinging to the reality of the ray even while distancing themselves from emission principles.

Wave theorists, beginning with Fresnel, were not able to distance themselves quite so much from physical principles as were selectionists. The ray theorist, like Biot or Brewster, could generate formulas whatever rays might be because *formulas* had always emerged almost exclusively from the pure assumption that rays can be counted. The wave theorist, especially in the area of polarization, did not have quite so free a hand. He had no choice but to assume that there exists an ether, or medium for light propagation. This is to assume much more than the selectionist, who could leave the world untouched except for the existence of discrete rays.

Nevertheless I have also argued that Fresnel in fact never directly used properties of the ether to *generate* a quantitative theory—that, indeed, his early understanding of the ether had blocked him from appreciating Huygens's principle and that he created an ether dynamics for polarization only well after he had already built the theory. Fresnel was, however, aided in fully adopting the hypothesis of transverse waves by his physical understanding of the ether as a medium that can transmit waves by shear but not by compression. Moreover, the physical ether was for Fresnel a unifying conception vastly superior to the complicated groupings of particles and forces that were supposed by the emission theorists. For him, then, the ether had at least two major functions: first, to provide an overall physical unity, and second, to underpin the hypothesis of transverse waves. Neither of these functions was *generative* in a direct sense. However, Fresnel was convinced that ether physics could, indeed must, be used to pursue the unanswered questions of the wave theory, and in this he was followed after his death by several generations of physicists.

Fresnel, then, looked to the ether to provide explanations of, and even formulas for, phenomena that escaped the fundamental principles of the wave theory. One could go only so far with these principles, as critics of the wave theory were fond of remarking. One phenomenon in particular seemed to Fresnel to require an intimate knowledge of the ether's behavior: the dispersion of light. In his "De la lumière" of 1822 Fresnel advanced a qualitative explanation of dispersion, based on ether dynamics, which was immensely influential in the 1830s:

> [The usual equations for elastic fluids as developed by Poisson] are founded on the hypothesis that each infinitely thin slice of fluid is repulsed only by the layer with which it is in contact, so that the accelerative force extends only to infinitely small dis-

> tances relative to the length of an undulation. This hypothesis is no doubt perfectly admissible for sound waves, of which the shortest are still several millimeters in length; but it would become inexact for light waves, of which the longest is less than a thousandth of a millimeter. It is very possible that the sphere of activity of the accelerative force that determines the speed of propagation of light in a refracting medium, or the mutual dependence of the molecules that compose it, extends to distances that are not infinitely small relative to a thousandth of a millimeter. (Fresnel 1822c, 89–90)

Dispersion, that is, depended on the spacing and forces between the mutually repelling particles of the ether. The clear implication of Fresnel's remarks was that theory had to address these two factors—spacing and force—in order to deal quantitatively with dispersion. This was precisely what the French mathematician Augustin-Louis Cauchy had begun to analyze, in a different context, in 1827, the very year Fresnel died. Cauchy's mathematics for the ether set a program of research that was pursued in France, Britain, and Germany during the 1830s, and (in Germany and France) through the 1850s. During the 1830s optical theory became very nearly synonymous with ether dynamics.

This turning to the constructive use of ether dynamics marks a point of departure, in both senses of the word, from the major subject of this book. The physical models that first engaged ether analysts, as well as the problems they addressed, have their origins in Fresnel's work, but here we find for the first time physicists attempting to grapple quantitatively with the models themselves rather than relying exclusively on the more fundamental principles that the models are designed to encompass. In this respect the period of the 1830s resembles somewhat the late 1890s and early 1900s, during which quantitative models for matter were built that could yield testable optical formulas. Indeed, I know of no earlier attempt by physicists to develop on this scale the implications of a detailed model of things that cannot be observed. Until the advent of the electron, these kinds of investigations remained unusual and even highly controversial.

By, at the latest, the early 1840s there are scarcely any physicists or mathematicians who dispute the wave theory's fundamental principles, and of equal importance, many were capable of applying it even at the high level of mathematical detail it required. During this period investigations based on the wave theory begin to evolve into two related but distinct areas. Work designed, like Cauchy's, to pursue the implications of ether mechanics continues. However, by the early 1850s many published papers in optics do not directly concern mechanical deductions but, rather, involve the working out of the wave theory's mathematical principles and their applications.

This second area cannot be absolutely distinguished from the first because many papers serve both functions to some degree. One of the first papers of this latter kind, and one of the most influential, was Sir George Stokes's "Dynamical Theory of Diffraction" of 1849. Here Stokes attempted to develop a rigorous theory of the diffraction of a vector wave by starting with the continuum equations for an elastic solid. However, the focus of the paper is not on mechanics but on generating the diffraction integrals. Stokes was followed in this effort during the next thirty years by Kirchhoff, Helmholtz, and several others. In work of this kind few questions were raised concerning the physical foundations of the wave theory, though the issue did arise, at least on the periphery of the investigation (as, e.g., in determining the precise form of the differential equation one had to solve for diffraction theory).

The first area received a great deal of attention between 1830 and about 1880, and aspects of it have been discussed by many historians (most notably by Sir Edmund Whittaker). Here we can very broadly distinguish three main trends. The first derived from Fresnel's speculations on the cause of dispersion and was carried out quantitatively by Cauchy. It involved the deduction of a differential equation of motion for the ether on the basis of lattice dynamics. Many scientists pursued this line of investigation until well into the 1850s (including the Königsberg physicist Franz Neumann in his early career). Indeed, a major German optics text by August Beer of 1853 still goes into some detail concerning the lattice equations. The second trend diverged from the first in Britain during the late 1830s in the work of George Green and James MacCullagh, followed in the 1850s and 1860s by George Stokes and others. Here lattice equations play no role at all; indeed, they are consciously avoided. Instead, the ether's equations are obtained by the manipulation of a macroscopic potential function, which later takes on energetic significance.

These two methods, the molecular and the macroscopic, share the assumption that matter affects the ether by altering the coefficients that determine its equations of motion (though the macroscopic theories diverge from one another over whether to alter density or elasticity coefficients, and so over whether the plane of polarization contains or is normal to the direction of oscillation). However, even during the 1840s some physicists thought this assumption was inadequate and believed ether and matter had to be treated as distinct systems that are dynamically connected to one another. This way of thinking disappeared, or at least became a conviction rather than a program, until the discovery of anomalous dispersion in 1870. Then it gave rise, pre-eminently in the hands of Helmholtz, to a theory I have referred to as the "twin equation" program, which was extensively pursued in Germany through the early 1890s (Buchwald 1985).

These developments, as well as many others that are linked to them, lie beyond the scope of this book, for together they constitute the mature phase of the wave theory of light as it was understood between 1840 and about 1890. Perhaps the major distinction between this mature phase and the earlier period is the absence of deep misunderstanding between the proponents of the several theories. Though the theories differed markedly from one another, nevertheless all of them had to incorporate the same fundamental elements of the wave theory.

APPENDIX 1 Huygens's Construction

1.1 Snel's Law and Wave Theory

To construct a refraction, one supposes with Huygens that each point on an expanding front is the source of secondary "wavelets." The common tangent to the series of wavelets is, at any instant, the front. When the front encounters an interface, secondary wavelets are also generated there; these wavelets expand more slowly in the refracting medium than outside it. One can easily demonstrate, as Huygens did, that the common tangent to these wavelets within the medium obeys Snel's law.

The demonstration leads directly to the following simple method for constructing refractions. In figure A1.1, an incoming ray RC strikes the interface at C. To find its refraction, draw the front that originates at C as it appears after unit time. Then erect a normal CO in the plane of incidence to the ray RC. To CO erect a perpendicular OK, also in the plane of incidence, such that OK touches the interface at K and is equal in length to the distance traveled by light in unit time in the medium of incidence. Through point K, and again in the plane of incidence, draw a line KI that is tangent to the front at I. Then CI is the refraction of RC, and it is simple to prove that, for *spherical* fronts only, the ratio $\sin(i)/\sin(r)$, where r is the angle of refraction and i the angle of incidence, is independent of incidence. To apply the method to Iceland spar, simply replace the sphere with a spheroid.

1.2 Huygens's Measuring Technique

Huygens actually never measured an optical angle directly. It evidently did not occur to him to construct a device akin to a vertical protractor with sights (though Malus did precisely that in the early 1800s), or else he was satisfied with the accuracy of his simple measuring technique. Instead, Huygens perfected a technique that permits angles to be deduced by using only ruler, pen, and paper (and a table of trigonometric functions). Though Huygens did not specify the divisions on his ruler, they could not have been much smaller than a millimeter; nor could they have been much larger, considering the accuracy he attained. The technique is central to the contemporary observational basis of Huygens's theory, and it has also enabled me to replicate Huygens's experiments using crystals 17.5 mm and 39 mm in height (for which I thank Stillman Drake) (see fig. A1.2).

On a "thoroughly flat table" a leaf of paper is fixed, and a line AB is drawn upon it (*Traité,* sec. 12, chap. 5; I have added angles α and β). Two other lines, CD and

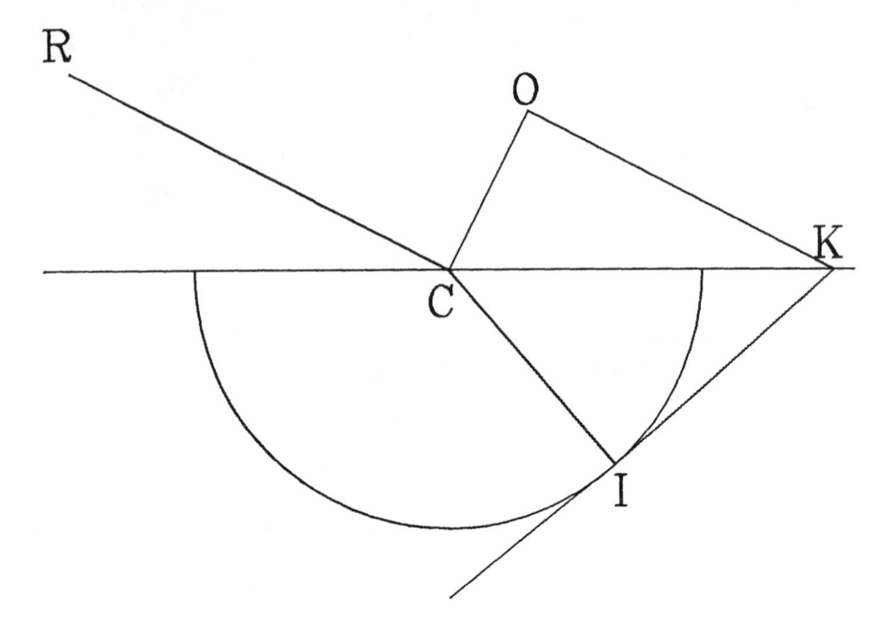

FIG. A1.1 Wave construction for Snel's law.

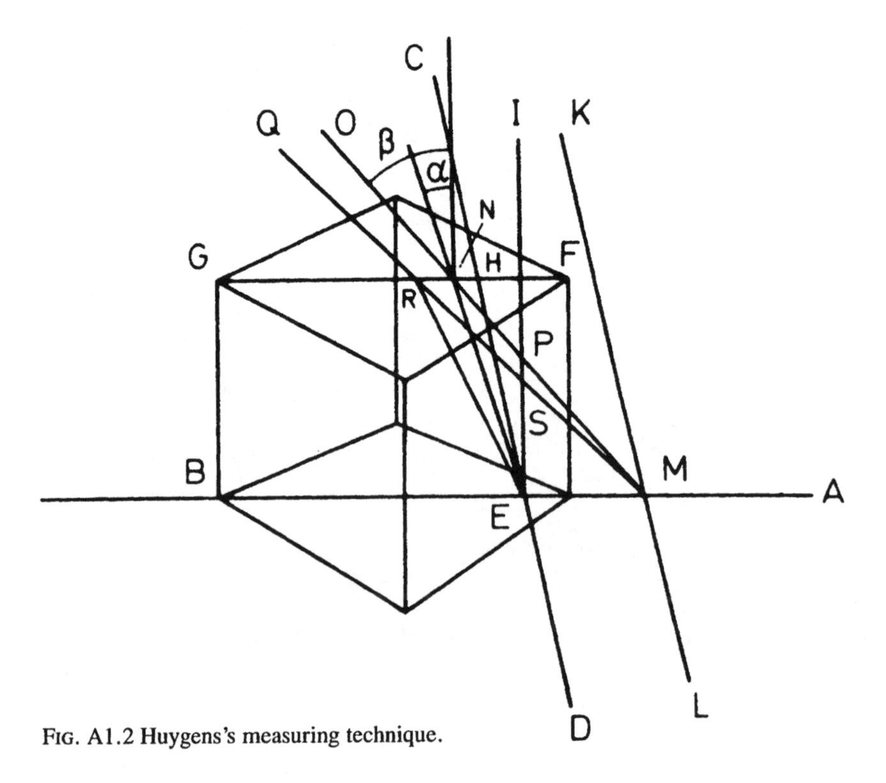

FIG. A1.2 Huygens's measuring technique.

KL, are drawn perpendicular to AB at a small distance from one another. The crystal is so placed that AB either bisects the obtuse angle of the lower surface or is parallel to its bisector; the crystal's right vertical edge lies between CD and KL. To measure the index of refraction of the ray that obeys Snel's law—the "ordinary" ray—Huygens first placed one eye (with the other closed) in the plane of the principal section. He then moved it, always remaining in this plane, until the ordinary refraction of the line CD (which is distinguished by its remaining stationary as the crystal is rotated) became collinear with the segments of CD that lie outside the crystal. This put his eye directly above point E, along the line EI. A point H on the upper facet coincident with the image of E was then marked. Keeping the eye in the principal section, Huygens moved it toward G until the ordinary refraction of CD became collinear with the segments of KL that are visible outside the crystal. He marked the point N on the surface coincident with the image of E. By direct observation Huygens had the distances EM, NH, and he could measure the height EH of the crystal.

On a separate sheet of paper draw the line AB with E and M marked off, and draw EH normal to AB at E. Draw also the line MN, which intersects EH at P, and connect points N and E. The angle α equals the angle of refraction, $\angle$NEP, of ray ON, and β, the angle of incidence, equals (NH/NP)/(NH/EN) = EN/NP. With my small crystal (17.5 mm) I find that EN/NP is about 4.9/3; with my large crystal (39 mm) I find 5/3, as did Huygens, whose own crystal was probably about 40 mm.

Applying the same measuring technique to the extraordinary refraction (here the ray RE), Huygens found that ER/RS, the measure of its "index," is not constant but varies with the angle of incidence so that this ray does not obey Snel's law. However, the simplicity of the technique enabled Huygens empirically to discover a law that applies to the extraordinary refractions of any two rays, incident in the principal section at the same point, and equally but oppositely inclined to the normal (see fig. A1.3). GCFH is the principal section, and C is the vertex of the upper solid obtuse angle. If VK, SK are rays of equal but opposite inclinations to the normal IK, then the respective points X, T at which each strikes the base of the crystal are equidistant from the point M of the base, which is intersected by the extraordinary refraction of a normally incident ray IK. KM is inclined toward the obtuse angle $\angle$GCF of the principal section. This was the first law Huygens gave for the extraordinary refraction, and he later deduced it from his general construction.

1.3 The Proportions of the Spheroid

From the outset Huygens assumed that the spheroid governing the extraordinary refraction was one of revolution. The problem was to determine the orientation and the relative lengths of its axes. Huygens knew that, for a given incidence, the extraordinary ray is always the same if the plane of incidence contains or is parallel to any of the three principal sections of the crystal. His intention was to use the spheroid just as he had previously used a sphere in isotropic bodies; that is, he intended to determine the extraordinary rays by the plane sections of the spheroid. Consequently the only way the three principal sections could produce the same refractions was if each sec-

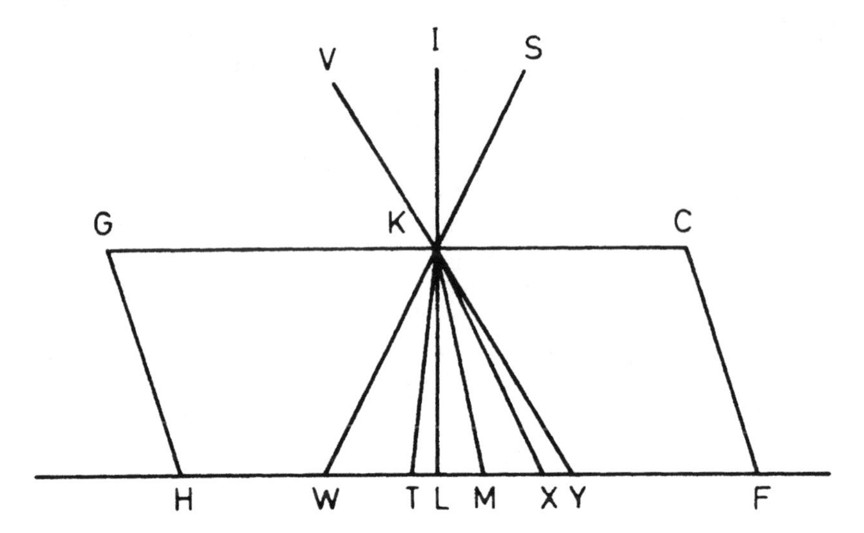

FIG. A1.3 The law of equal deviations.

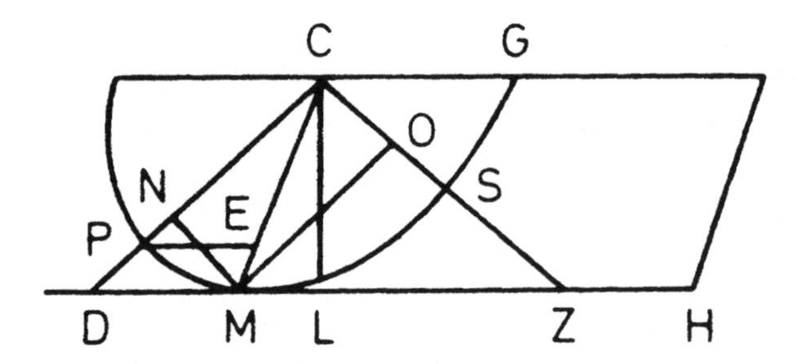

FIG. A1.4 Huygens's semiaxes.

tioned the same curve in the spheroid. For that to be possible, the spheroid's axis of revolution had to lie in all three principal sections. Hence the axis of revolution is equally inclined to each of the edges of the crystal.

In figure A1.4 we have a principal section with CS the axis of revolution—hereafter called the optic axis. The generating ellipse is so constructed that MH, parallel to the crystal facet, is tangent to the ellipse at M, where $\angle$MCL is 6°40′. This is obviously required by the extraordinary deviation of a normally incident ray. Huygens set CM equal to 100000 as a reference for calculating CS, the length of the semiaxis along the optic axis, and CP, its conjugate.

To calculate the proportions of the spheroid, Huygens relied on two simple properties of the conjugate diameters of an ellipse, both of which can easily be deduced from the theorems in Apollonius's *Conic Sections.*[1] Here CM, marking the refraction of a normal ray, is conjugate to the facet diameter CG. Given the angle between CM and the normal CL to CG (6°40′), one can then determine CS, CP, and CG in proportion to CM. On setting CM to 100000 as a unit, there results following his procedure:

$$\begin{aligned} \text{semimajor axis CP} &= 105032 \\ \text{semiminor axis CS} &= 93420 \\ \text{and CG} &= 98779 \end{aligned}$$

Thus, in determining the proportions of the spheroid, Huygens used precisely one crystallographic measurement, one optical measurement, and a property of the refraction for which no number is necessary.

1.4 Refraction in the Principal Section

Consider with Huygens refraction in a principal section. In figure A1.5 line gK is the intersection of the section with the crystal surface; the upper solid obtuse angle is toward g. Take an incident ray RC and produce CO, the intersection of the front with the plane of incidence, until OK, the normal to CO that intersects the surface gK at K, is equal to N, the distance traveled by light in unit time in air. Then, Huygens proved, in this plane of incidence, and only in this plane, the refraction lies in the plane itself. It lies along CI, where I is the point at which a line from K touches the ellipse.

To determine the orientation of the tangent to the ellipse is simple. It is an elementary property that, given a diameter CG, the tangent from a point K on the extension

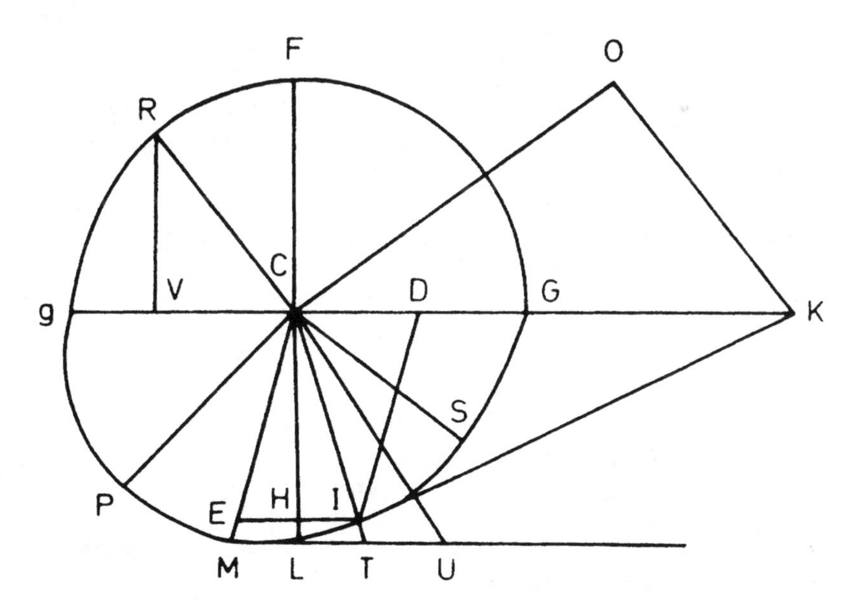

FIG. A1.5 Refraction in a principal section.

of CG will touch the ellipse at a point I where a line DI, drawn parallel to the conjugate CM to CG, intersects the ellipse. The length CD is a third proportional to CK, CG, that is, CK/CG = CG/CD (as found from Apollonius's *Conics,* 1:34). We now have the orientation and parameters of the surface, though only in proportion to CM as a reference of 100000. To develop a set of absolute parameters that can be used to calculate a refraction from a given incidence, Huygens had to determine the value of N in proportion to CM. This determination will incidentally reveal the relation between the radii of the spheroid and the radius of the sphere that governs the ordinary refraction, since we already know the ordinary index, which is equal to the ratio of the radius of the sphere in air to its radius in the medium.

Huygens had, then, experimentally to determine the ratio of OK (that is, N) to some radius of the spheroid, say the facet radius CG. The ratio CK/CG = CG/CD does not immediately give OK. One can nevertheless easily use this latter proportion to find it, which Huygens accordingly did. In figure A1.5, with incident ray RC, and C as center, draw a circle gFG with radius CG, cutting RC in R. Drop RV, the perpendicular on gC, and mark a point D on CG such that OK/CG is equal to CV/CD. Draw DI parallel to CM, cutting the ellipse in I. Then, by use of the relation CK/CG = CG/CD, it is simple to show that CI is the extraordinary refraction. I shall call the important relation OK/CG = CV/CD the *law of proportions.*[2]

The ratio OK/CG can now be determined by experiment. Huygens merely gave its value, but we can easily reconstruct how he might have obtained it. The problem is to measure CD and CV = CG sin ∠RCF. To measure this angle ∠RCF of incidence Huygens would, no doubt, have used the technique described above. Finding CD is more difficult, but it could have been done either by reversing the calculation described below, or by using the equation for the ellipse. Either way CD is determined by the angle of refraction, which can be measured using Huygens's technique. Probably proceeding in one or the other of these two ways, Huygens found that CV/CD is slightly less than 8/5. Since CG is 98779, the law of proportions then gives OK = 156962, and so OK/CS is less than 5/3 by about 1/41. This value is so close to the ordinary index 5/3 that Huygens felt justified in concluding that CS "may be exactly" the radius of the ordinary sphere (in which case along the optic axis the ordinary and extraordinary rays have the same speed). It should be understood that this posited equality was a result not of theory but of experiment. Had the two not been equal, Huygens would not have been worried.

1.5 The First "Confirmation"

As yet Huygens had not provided any measurements to confirm the theory. However, the equality between the radius of the ordinary sphere and the semiminor axis of the spheroid implies that there is one ray in the principal section that will not be divided into two on entry, the single refraction being along the optic axis. Yet Huygens did not test this implication with the natural crystal. The reason he did not is quite simple: no ray can in fact be refracted along the optic axis when the plane of separation is a natural facet. The maximum refraction for an ordinary ray occurs when the angle of incidence is 90°, where the angle of refraction is 36°53′. But the inclination of the optic axis to the vertical for a natural facet is 45°40′, or 8°47′ greater than the maximum of refraction. To test the implication, the crystal must be cut; as we shall see, Huygens did test cut crystals.

To this point, then, the theory is without confirmation, with the exception of the law of equal deviations, for which Huygens gave no sample measurements, and which in any case is independent of the intimate details of the spheroid. The sole measured optical values Huygens has thus far given are the ordinary index, the deviation of a normally incident ray, and the claim that the extraordinary refraction of a ray incident at 16°40′ is not deviated. Huygens used this last as the first measured "confirmation" of the theory. It is of sufficient importance to follow Huygens's calculations here, since this was the only numerical confirmation he ever gave.

In figure A1.5, let a ray be so incident that $\angle$RCg is 73°20′. To find the refraction, CI, Huygens used the law of proportions. First find CV = RC cos $\angle$RCg = 28330. By the law of proportions, CV/CD = N/CG, where Huygens has previously found—from what experiment he did not say—that N is 156962. Hence CD is about 17828. By the properties of the ellipse, $(CG^2 - CD^2)/DI^2 = CG^2/CM^2$. Hence DI is about 98358. In the figure, CE/EI = CM/MT, where CE = DI, and EI = CD. Hence MT is about 18126. ML is CM sin 6°40′ = 11609, and so TL = 29736. Since LC = CM cos 6°40′, we find LT/LC = tan $\angle$LCT = 0.2994, so that $\angle$LCT = 16°40′, precisely as Huygens claimed.

The apparent agreement of Huygens's measurement with his prediction is probably enforced and not experimental. For Huygens's prediction to hold to the minute, the value of N must lie within 200 units of 156962. Huygens had obtained 156962 from some experiment on extraordinary refraction. Though he did not specify the angle he used, it obviously had to be extremely close to 16°40′ to obtain anywhere near the proper value for N, given the accuracy of the measuring technique. Very likely Huygens deduced N from the incidence of this ray and then reversed the calculation to deduce the incidence.[3]

1.6 Refraction outside the Principal Section

In figure A1.6 AEBFH is a piece of crystal whose upper facet AEFH forms an equilateral parallelogram. The section of the spheroid QGqgM by the facet is QgqG. Point E is the vertex of the upper solid obtuse angle. Although the refraction lies in the plane of incidence when the latter is parallel to a principal section, as we shall see this is no longer true in any other plane. To deduce the refraction for other planes of incidence, Huygens found that he needed the following lemma, for which he provided a proof at the end of the chapter: If a spheroid (fig. A1.7) is touched by a line at a given point, and if, at two other points, the spheroid is touched by planes parallel to this line but not to each other, then the three points of contact (b, o, a) all lie on a single section (ToE) made in the spheroid by a plane that passes through its center. The proof cites

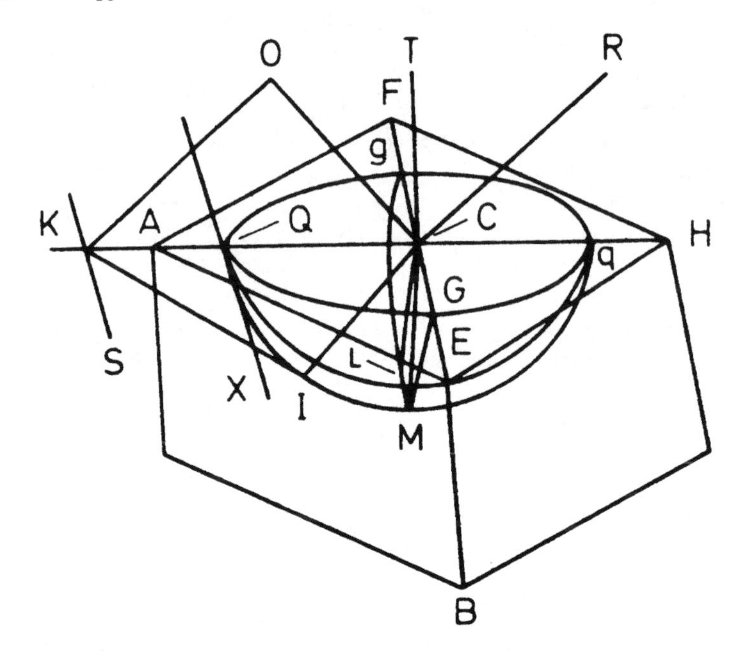

FIG. A1.6 Refraction in a plane normal to the principal section.

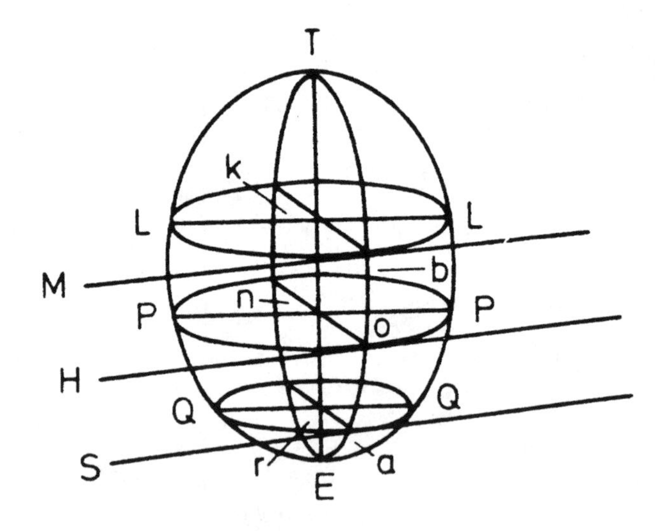

FIG. A1.7 Huygens's lemma.

proposition 15 of Archimedes' *Conoids and Spheroids* (in Rivault's edition of 1615: in the modern Heiberg edition the relevant propositions are 13 and 14, corollary 2).[4]

Using the lemma, Huygens deduced the law of refraction for a plane of incidence perpendicular to the principal section. In figure A1.6 the incident ray is RC, where C is the intersection of AH and FE. As before, we draw OC, the trace of the incident (plane) front in the plane of incidence, and OK, normal to OC and meeting the plane of separation at L, equal to the distance traveled by light in unit time in air.

Let CL be normal to the facet at C, with L lying on the spheroid, and let CM be the radius of the spheroid that lies along the extraordinary refraction for a ray normally incident at C. Through CM and the line KH that bisects the acute facet angles draw a plane; that plane sections the ellipse QMg, and the angle, ∠MCL, between it and the normal CL is 6°40′. Through point K draw a line KS parallel to Cg. We know that the tangent plane that contains KS is parallel to QX, and that QX is parallel to the tangent plane at M. Consequently, the tangent plane containing KS, and the tangent plane at M, though not parallel to each other, are nevertheless parallel to the line tangent at Q to the ellipse sectioned by the facet. By the lemma, the point of contact I of the tangent plane containing KS lies on the ellipse QMq.

Knowing the plane of refracted ray, CI we use the same method as before for finding the position of I in the plane, namely, the requirement that if KI is a tangent to the ellipse at I, and if K lies on the produced diameter qQ, then the point of contact I is the intersection of a line DI parallel to CM with the ellipse. Note that we can now easily prove the law of proportions for this plane of incidence in the same way as before, only here the ratio CV/CD is equal to N/CQ, not N/CG, so that, as Huygens remarked, "the proportion of the refraction for this section of the crystal" is less than the corresponding proportion in the principal section. Huygens claimed to have successfully tested this conclusion, but he again provided no data. We can, however, conjecture how he might have done so.

1.7 Testing the Construction in a Plane Normal to the Principal Section

To begin, draw two mutually perpendicular lines F′E′, A′H′ that intersect at L (see fig. A1.8). Place the crystal on the intersection so that the image of L seen via a perpendicular ordinary ray coincides with point (see fig. A1.6) C of the upper facet. Rotate the crystal about L until, on its upper facet, FE—the bisector of the obtuse angles—is parallel to F′E′ (with F on the side of M toward F′), and so AH is parallel to A′H′. In this alignment a ray normally incident at C will be extraordinarily refracted to the point M′ on the base of the crystal; whence M′ is seen in coincidence with the ordinary image of L, that is, at C.

From the construction for an incident plane AHA′H′ (see fig. A1.6), any point on the line JK, parallel to A′H′ and through M, can be seen in AHA′H′ by looking through C at some given, calculable angle to the plane; for a ray incident at C and in AHA′H′ is refracted into the plane AHJK that is inclined to AHA′H′ at 6°40′. Mark

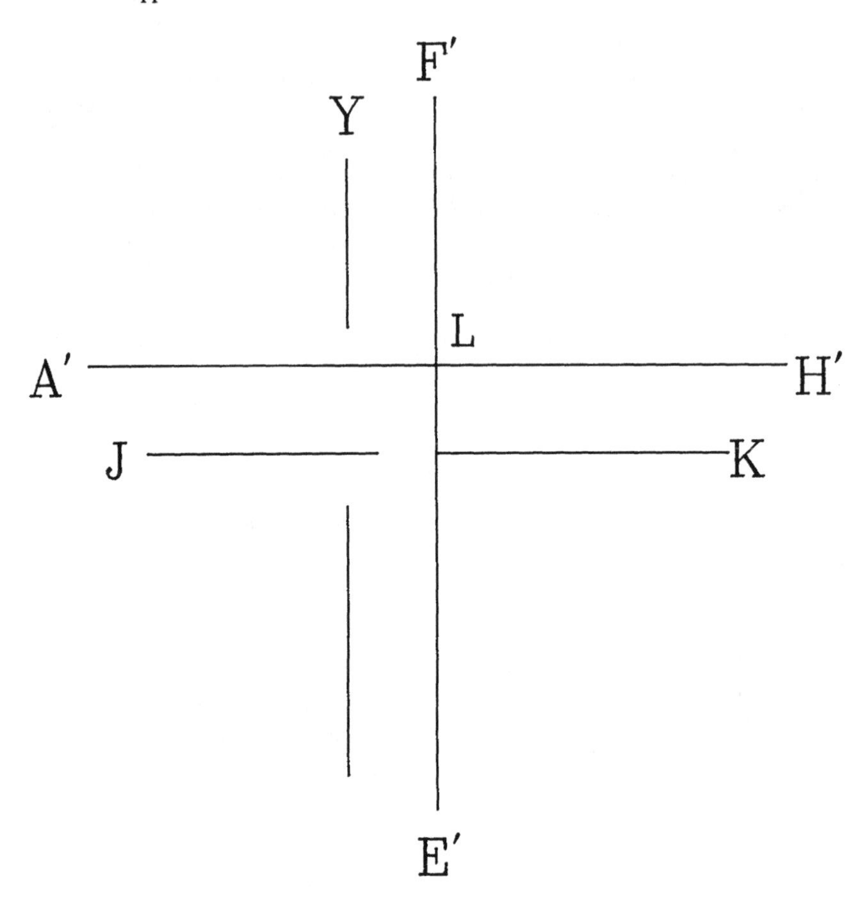

FIG. A1.8 Measuring refractions in a plane normal to the principal section.

a point R on JK and align the eye, keeping it always in AHA′H′, until the extraordinary image of R is coincident with C. Draw the broken line YW, parallel to F′H′, which is seen without refraction and along which the image of R appears to lie. Then the tangent of the angle of incidence is the ratio of the distance from YW to F′E′ to the height of the crystal.

To test the law of proportions, namely, CV/CD = N/CQ, for this section, we need to determine CV and CD. Of these, CV is equal to CQ sin 24°34′, or 43667, while CD can be found in two ways, one of which I mentioned above. Simpler, however, is to find CI, the radius of the spheroid along which the refraction lies, and the angle β between CI and CM, and then to compute CD = CI sin β.

With my own small crystal 17.5 mm high, I marked the point M′ at the distance LM′ = 17.5 tan 6°40′w2 mm from A′H; then along JK I marked R about 5.5 mm from M′. When R is seen in plane AHA′H′ in coincidence with C, the tangent of the angle of incidence is very nearly 8/17.5, since the distance from YW to F′E′ is about 8 mm; whence the angle itself is about 24°34′. Since its tangent is RM′/CM′ = 0.31225, where CM′ = $\sqrt{(LM'^2 + 17.5)}$ the angle β is about 17°20′, while the equation $CI^2\cos^2\beta/CM^2 + CI^2\sin^2\beta/CQ^2 = 1$ yields $CI = 1/\sqrt{[(\cos^2\beta/CM^2) + (\sin^2\beta/CQ^2)]}$ to be about 100418. Hence CD is 1919918. CV/CD is therefore in the ratio of 2.919/2. Since N is 156962, CV/CD should be 2.989/2 according to theory, which is quite close to the measured value considering the size of the crystal. Using my large, 39 mm crystal, I find a ratio of 2.93/2.

1.8 The Elevation of the Images

Huygens described a series of experiments designed to measure the visual heights of the images, which therefore require both eyes for observation. These experiments are significant because, with the exception of his experiments with cut crystals, they are the only ones that provide evidence for the construction without requiring calculation of angles of incidence for set angles of refraction (see fig. A1.9).

Consider the refractions of the point P at the surface Qq; the refracted rays join each eye to the points C, C′ of Qq, where the rays PC, PC′ from P emerge at equal but opposite inclinations to the surface normal. That is, the distances DC, DC′ from a point D directly above P are equal to each other. The image of P will appear above it at S, where the prolongations of the emergent rays intersect. In double refraction the extraordinary ray is, for the natural crystal, always less refracted than the ordinary ray, which implies that the equal distances DC, DC′ are greater for the extraordinary than for the ordinary ray (for a given P and fixed positions of the eyes). Consequently the lines from C, C′ to the eyes are less inclined to the normals in the former than in the latter case, and their intersection lies below the point S of elevation in ordinary refraction; that is, the extraordinary image is always lower than the ordinary image.

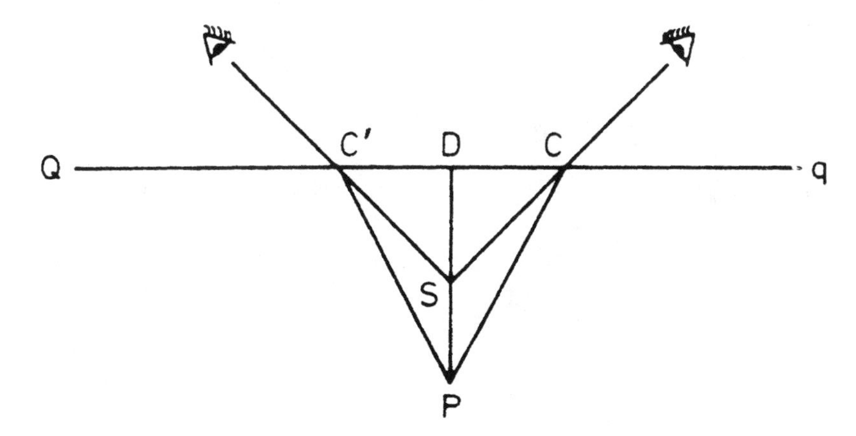

FIG. A1.9 Image elevation.

One can easily use the law of proportions to calculate the elevations if the plane of incidence is either parallel or perpendicular to the principal section, but *only* for those two planes. Mark off a length AB equal to the height of the crystal and divide it at E, D, C such that AB/AE, AB/AD, AB/AC are, respectively, 5/3, 99324/70238, 99324/66163. Since Huygens's crystal was probably about 40 mm in height, the distances AB, AE, AC, AD would be, respectively, 40 mm, 24 mm, 28.3 mm, 25 mm, so that he had to distinguish points whose heights differ by as little as 1 mm. I have unsuccessfully tried to do so with both my crystals. At best the experiment is extremely difficult to perform and highly unreliable.

1.9 Refraction in an Arbitrary Plane of Incidence

The plane of refraction has in effect already been found for an arbitrary plane of incidence, since its position follows, as before, from Huygens's lemma. (Here by plane of refraction I mean that plane in which the refraction always lies for a given plane of incidence; this plane is normal to the crystal surface only when it is parallel to the principal section.)

In figure A1.10, the ellipse HDE is the section of the spheroid by the crystal facet. RC is the incident ray, coming in in that plane, normal to the facet, that intersects the facet in BK. OC is normal to RC, and OK is the distance traveled by light in air in unit time. To determine the plane of refraction, draw a line HF that is parallel to the normal KT to KB in the facet and that touches HDE at H. Join CH and produce it to KT. Consider the plane that contains KT and that is tangent to the spheroid. Both KT, which lies in that plane, and HF, which does not, are parallel to the tangent plane to the spheroid at M. By the lemma, the plane that contains CM, CH is that plane in which refractions from the given incident plane lie.

But to determine the position of the refracted ray in the plane of refraction is not as

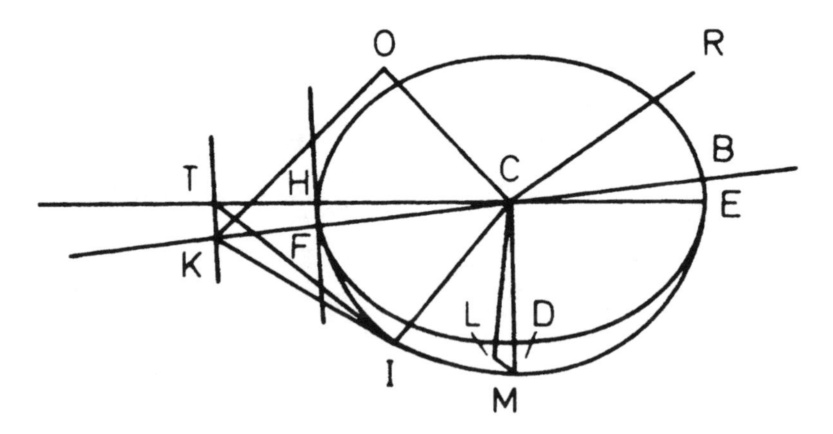

FIG. A1.10 An arbitrary plane of incidence.

easy here as it was before. In the two limiting cases (in which the plane of incidence is the principal section or is normal to it), the point H where the perpendicular KB touches the ellipse sectioned by the facet also lies along the line KB joining K to the center of that ellipse. For any other plane of incidence this is no longer true. Thus HF, parallel to KT, touches the ellipse not at F but at H. Consequently one cannot find the position of I by constructing the tangent line from K to the ellipse sectioned by the plane of refraction. What one does not have is an analogue of the law of proportions. Consequently Huygens could not test his construction in an arbitrary plane of incidence.

He accordingly concluded his test with a discussion of phenomena for which the law of proportions remains valid and that occur when the crystal is cut. These are fifth and final confirmations of the theory. Like all but one of the first four confirmations, they are not supported by experimental data. In the end, Huygens had accomplished the following in his analysis:

A. From two optical measurements and one crystallographic measurement, he deduced the parameters of the spheroid.

B. He deduced the law of proportions for calculating refractions in two limiting planes of incidence.

C. He deduced the law of equal deviations, a law that possibly holds in all planes of incidence but that is limited to two equally but oppositely inclined rays.

D. He had shown how to construct the refraction of a ray for any plane of incidence, but he had not been able to deduce a general law for calculating it.

E. He had provided six confirmations, only one of which (and that one spurious) involves experimental data:
   1. He had tested the law of equal deviations: no data.
   2. He had confirmed the prediction of the angle of incidence for the non-deviated ray; angle given.
   3. He had confirmed that the refraction is not as great when the plane of incidence is normal to the principal section as when it is parallel to the section; no data.
   4. He had tested the law of proportions in both planes for which it holds by calculating image heights; no data.
   5. He claimed that the image heights in other planes of incidence lie between the two limiting cases, and that theory predicts the fact; neither data nor proof of the assertion.
   6. He had cut the crystal at various angles and had confirmed the predicted behavior of normally incident rays; no data.

Of the five confirmations (1), (3), (4), (5), and (6) for which no details were supplied, data are not required in (6), because this implicitly embraces the prediction that a normally incident ray is not deviated for certain crystal shapes, but the remaining four do require data. If Huygens had clearly emphasized the laws of proportion and of equal deviations by removing them from the depths of his theoretical analysis, and if he had provided data and the details of his measuring technique in all cases, then his contemporaries, and the scientists of the eighteenth century, would have been able to replicate his experiments and at least to test the special implications of the general construction. But they would not have been able to test the construction in all cases, because Huygens had not been able to deduce a general analogue of the law of proportions.

APPENDIX 2 Malus's Algebraic Formulas for Huygens's Construction

Let xy be the plane of the facet, xz the plane of the principal section, ϑ the angle of incidence, λ the angle between the plane of the facet and a plane normal to the optic axis, and ω the angle between the plane of incidence and the principal section (Malus 1811b, 342–46). Let the semimajor and semiminor axes of the Huygens spheroid be a, b respectively. If the x axis were along the spheroid's semiminor axis and the center of the spheroid at the origin of coordinates, then its equation would be:

(A2.1)
$$\frac{x^2}{b^2} + \frac{y^2 + z^2}{a^2} = 1$$

Transforming to an x axis in the plane of the facet, we have:

(A2.2)
$$a^2b^2 = Ax^2 + Bz^2 - 2Cxz + b^2y^2$$
$$A = a^2\sin^2\lambda + b^2\cos^2\lambda$$
$$B = b^2\sin^2\lambda + a^2\cos^2\lambda$$
$$C = (a^2 - b^2)\sin^2\lambda$$

The equation of an incident ray is:

(A2.3)
$$x \cos \vartheta = z \sin \vartheta \cos \omega$$
$$y \cos \vartheta = z \sin \vartheta \sin \omega$$

To find the extraordinary refraction we must calculate the point on the spheroid where a certain plane is tangent; this plane is normal to the plane of incidence and contains the point (K) of the facet that lies in the plane of incidence at a distance N from the incident front through the origin, where N is the distance traveled by light in air in unit time. If K has coordinates (x_1, y_1) then we have, using equation A2.3:

(A2.4)
$$x_1 = \frac{-N \cos \omega}{\sin \vartheta}$$
$$y_1 = \frac{-N \sin \omega}{\sin \vartheta}$$

A line through K and normal to the plane of incidence will be:

(A2.5) $$(x_1 - x)\cos\omega = -(y_1 - y)\sin\omega$$

Whence an equation for all planes containing this line is:

(A2.6) $$(x - x_1)\cos\omega + (y - y_1)\sin\omega + \mathrm{I}z = 0$$

We must choose I such that the plane A2.6 is tangent to the spheroid A2.2, which means that the partial derivatives $\partial z/\partial x$, $\partial z/\partial y$ calculated from A2.6 must, at the tangent point, equal the same derivatives calculated from A2.2. Equating the derivatives, we obtain:

(A2.7)

$$\text{a. } \frac{-\cos\omega}{\mathrm{I}} = \frac{-(Ax - Cz)}{Bz - Cx}$$

$$\text{b. } \frac{-\sin\omega}{\mathrm{I}} = \frac{-b^2y}{Bz - Cx}$$

From this we find:

(A2.8) $$b^2y\cos\omega = (Ax - Cz)\sin\omega$$

Since (x, y, z) are the coordinates of the point of tangency, and since A2.8 does not contain the angle of incidence, it follows that all rays incident in the plane ω are refracted into the plane A2.8—a plane that is not perpendicular to the facet in general.

To find the angles of the refracted ray with respect to the z-axis and the plane (xz) of the principal section, we must calculate the coordinates (x, y, z). To do so we use equations A2.2, A2.6, A2.7b, and A2.8. First eliminate I from A2.6 by use of A2.7b:

(A2.6′) $$(x - x_1)\cos\omega + (y - y_1)\sin\omega + \frac{z(Bz - Cx)\sin\omega}{b^2y} = 0$$

Equations A2.2, A2.6′, and A2.8 determine (x, y, z) in function of ϑ, ω and x_1, y_1. However, A2.2 and A2.6′ are of the second degree in z and y (since there are two planes through the line that are tangent to the spheroid—one above the crystal, the other within it). To eliminate the unwanted root, Malus first expanded A2.6′ into:

$$xb^2y\cos\omega + \sin\omega(Bz^2 - Cxz + b^2y^2) = b^2y(y_1\sin\omega + x_1\cos\omega)$$

Using A2.8, we then have:

(A2.6″) $$\sin\omega(Ax + Bz^2 - 2Cxz + b^2y^2) = b^2y(y\sin\omega + x\cos\omega)$$

From A2.2, the coefficient of sin ω is just a^2b^2, so we have an equation linear in y:

(A2.9) $$y = \frac{a^2\sin\omega}{y\sin\omega + x_1\cos\omega}$$

We then use A2.2 to find $Ax^2 + Bz^2$ equal to $a^2b^2 + 2Cxz - b^2y^2$. Using this and A2.9 in A2.6″ then gives an equation linear in x and z and independent of y that we can solve for, say, x. Substituting this result in A2.8 again, we have z. The results are:

$$x = \frac{a^2b^2\cos\omega + C\sqrt{A(y_1\sin\omega + x_1\cos\omega)^2 - a^2(b^2\cos^2\omega + A\sin^2\omega)}}{A(y_1\sin\omega + x_1\cos\omega)}$$

$$y = \frac{\sqrt{A(y_1\sin\omega + x_1\cos\omega)^2 - a^2(b^2\cos^2\omega + A\sin^2\omega)}}{y_1\sin\omega + x_1\cos\omega}$$

The tangent of the angle between the projection of the refracted ray onto the xz plane and the z axis is x/z; the tangent of the angle between its projection into the yz plane and the z axis is y/z. If, then, $\overline{\omega}$ is the angle between the principal section and a plane, normal to the facet, that contains the refracted ray, and $\overline{\vartheta}$ is the angle between the refracted ray and the z axis, we have (setting N to one as a reference and using A2.6′):

$$\frac{x}{z} = \tan\overline{\vartheta}\cos\overline{\omega} = \frac{a^2b^2\sin\vartheta\cos\omega}{A\sqrt{A - a^2\sin^2\vartheta(b^2\cos^2\omega + A\sin^2\omega)}} + \frac{C}{A}$$

$$\frac{y}{z} = \tan\overline{\vartheta}\sin\overline{\omega} = \frac{a^2\sin\vartheta\cos\omega}{\sqrt{A - a^2\sin^2\vartheta(b^2\cos^2\omega + A\sin^2\omega)}}$$

APPENDIX 3 Slit and Obstacle Diffraction in the Fresnel Approximation

Using Kirchhoff's boundary conditions, we may write the general solution to the scalar wave equation ($\nabla^2\psi = -1/c^2 \cdot \partial^2\psi/\partial t^2$) under the approximation r, r_o (fig. A3.1) much larger than the wavelength λ as:

$$\psi(P) = \frac{i}{2\lambda}\int_\sigma \left[\left(\frac{1}{rr_o}\right)e^{i\frac{2\pi(r+r_o)}{\lambda}}\right](\vec{e}_r - \vec{e}_o)\cdot d\vec{\sigma}$$

If we assume that $(\vec{e}_r - \vec{e}_o)\cdot d\vec{\sigma}/d\sigma \approx \vec{e}_z \cdot \vec{e}_{SP} = \cos\delta$ and that we may neglect powers greater than two in the expansions for r, r_o as functions of ξ/r' and η/s, then we obtain:

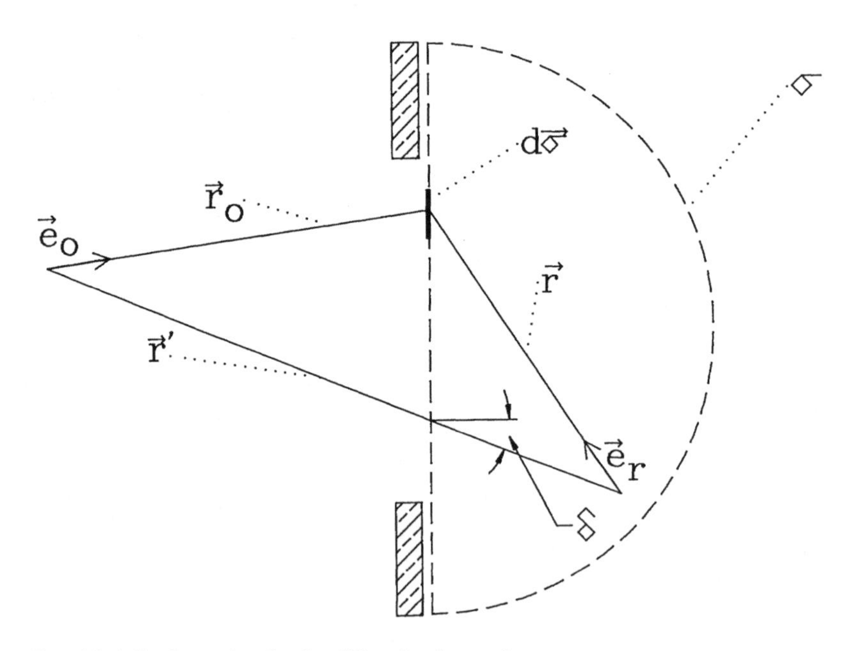

FIG. A3.1 Configuration for the diffraction integral.

$$\psi(P) \approx \frac{-i\cos\delta}{\lambda r's'} e^{i\frac{2\pi(r'+s')}{\lambda}} \iint e^{i\frac{2\pi(r'+s')(\xi^2\cos^2\delta+\eta^2)}{\lambda}} d\xi\, d\eta$$

Here the coordinates in the plane of the diffractor of the front element $d\vec{\sigma}$ are (ξ, η). Next introduce the change of variables:

$$u \equiv \xi \cos\delta \sqrt{\frac{2(r'+s')}{r's'\lambda}}$$

$$v \equiv \eta \sqrt{\frac{2(r'+s')}{r's'\lambda}}$$

Since the limits of v are $\pm\infty$, these yield for the intensity:

Finite Obstacle or Single Slit

$$I(P) = \psi(P)\psi^*(P) = \left[\frac{1}{4(r'+s')^2}\right][(C_1 - C_2 - S_1 + S_2)^2 + (2f + C_1 - C_2 + S_1 - S_2)^2]$$

$$f = 0 \rightarrow \text{single slit}$$

$$f = 1 \rightarrow \text{obstacle}$$

Semi-Infinite Plane

$$I(P) = \left[\frac{1}{4(r'+s')^2}\right]\left[\left(C_2 - \frac{1}{2}\right)^2 + \left(S_2 - \frac{1}{2}\right)^2\right]$$

where:

$$C_{1,2} = \int_o^{u_{1,2}} \cos\left(\frac{1}{2}\pi u^2\right)^2 du$$

$$S_{1,2} = \int_o^{u_{1,2}} \sin\left(\frac{1}{2}\pi u^2\right)^2 du$$

The parameters are determined, with reference to figure A3.2, as follows. The observation point P is at a distance l from the point P′ on the normal to the single slit or screen (HJ) that passes through source S, P′ being the same distance to HJ as P is to HJ. The "variable" origin, or pole, is O. Then we have:

$$u_1 = x_1 \cos\delta \sqrt{\frac{2(r'+s')}{r's'\lambda}}$$

$$u_2 = (x_1 + c)\cos\delta \sqrt{\frac{2(r'+s')}{r's'\lambda}}$$

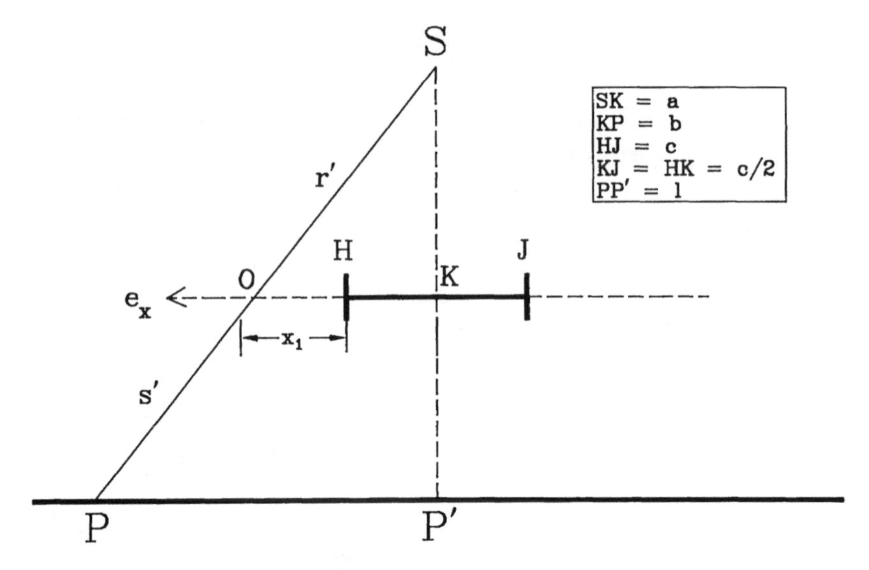

FIG. A3.2 The narrow obstacle.

$$\cos\delta = \frac{a+b}{\sqrt{l^2 + a^2 + (a+b)^2}}$$

$$x_1 = -\frac{1}{2}c - \frac{al}{a+b}$$

$$r' = \frac{a}{a+b}\sqrt{l^2 + a^2 + (a+b)^2}$$

$$s' = \frac{b}{a+b}\sqrt{l^2 + a^2 + (a+b)^2}$$

Figure A3.3 illustrates a typical set of patterns. Notice that all three curves intersect at the locus of the geometric shadow, and that the narrow obstacle curve oscillates about the curve for the semi-infinite plane, the amplitude of the oscillations decreasing with distance from the center of the shadow. Strictly speaking the expression for $\psi(P)$ is limited to front points near the pole, since I have removed the inclination factor, $\cos\delta$, from the integral. This is not a serious restriction unless one computes amplitudes for points that are very far removed from the center of the shadow.

There are two ways to compute the Fresnel integrals. One can proceed, as Fresnel did, by numerical integration, and this is perhaps the simplest method for large, fast machines. I have used a small computer, for which the general series representations of the integrals are most suitable, since the convergence is fairly rapid for the values needed in optical situations. Moreover, there is a closed formula accurate to 10^{-7} when the limit of the integral is greater than or equal to five.

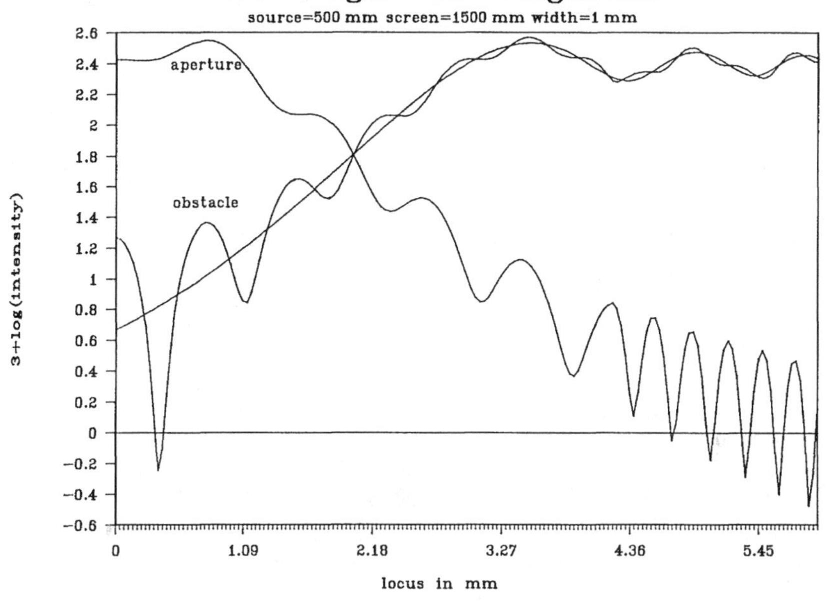

FIG. A3.3 A typical set of diffraction curves.

Cauchy Series for Large u

$$C(u) = \frac{1}{2} - \frac{1}{\pi u}\left[P(u)\cos\left(\frac{1}{2}\pi u^2\right) - Q(u)\sin\left(\frac{1}{2}\pi u^2\right)\right]$$

$$S(u) = \frac{1}{2} - \frac{1}{\pi u}\left[P(u)\sin\left(\frac{1}{2}\pi u^2\right) + Q(u)\cos\left(\frac{1}{2}\pi u^2\right)\right]$$

$$P(u) = 1 - \frac{1\cdot 3}{(\pi u^2)^2} + \frac{1\cdot 3\cdot 5\cdot 7}{(\pi u^2)^4} - \ldots$$

$$Q(u) = \frac{1}{\pi u^2} - \frac{1\cdot 3\cdot 5}{(\pi u^2)^3} + \frac{1\cdot 3\cdot 5\cdot 7\cdot 9}{(\pi u^2)^5} - \ldots$$

Closed Formula for $u \geq 5$

$$C(u) = .5 + \frac{\left[.3183099 - \frac{.0968}{u^4}\right]\sin\left(\frac{1}{2}\pi u^2\right)}{u} - \frac{\left[.3183099 - \frac{.0968}{u^4}\right]\cos\left(\frac{1}{2}\pi u^2\right)}{u^3}$$

$$S(u) = .5 + \frac{\left[.3183099 - \frac{.0968}{u^4}\right]\cos\left(\frac{1}{2}\pi u^2\right)}{u} - \frac{\left[.3183099 - \frac{.0968}{u^4}\right]\sin\left(\frac{1}{2}\pi u^2\right)}{u^3}$$

General Series, Convergent for All u

$$C(u) = u\left[1 - \frac{1}{2!5}\left(\frac{1}{2}\pi u^2\right)^2 + \frac{1}{4!9}\left(\frac{1}{2}\pi u^2\right)^4 - \ldots\right]$$

$$S(u) = u\left[\frac{1}{1!3}\left(\frac{1}{2}\pi u^2\right) + \frac{1}{3!7}\left(\frac{1}{2}\pi u^2\right)^3 + \ldots\right]$$

Because of the factor i in the primitive expression for the amplitude $\psi(P)$ it is often remarked that, considering the elements $\vec{d\sigma}$ as Huygenian emitters, these evidently radiate a quarter-period ahead of the primary wave—though whether they do or not does not affect the pattern. Since Fresnel's solution to diffraction is essentially the same as the one arrived at here via Kirchhoff's boundary conditions, one might conclude that Fresnel thought the secondaries were a quarter-period out of phase with their resultant primary. This conclusion would, however, incorrectly represent Fresnel's understanding of the process. He constructs the diffraction integrals by examining the path length differences between the secondaries and the distance from the pole to the screen point. He then uses these differences to compute the secondary phases relative to the phase at the pole, and finally he employs the purely analytical quarter-wave decomposition to obtain the amplitude factors for the sine and cosine contributions. Since, presumably, the phase of the primary is the same as the phase of the pole, it seems clear that Fresnel assumed the secondaries to be in complete phase accord with the primary. This is an artifact of the way he treated secondaries—essentially as emitters of rays that can be compared for lengths at a given point to determine the effect of their superposition.

APPENDIX 4 Fresnel on Chromatic Polarization

4.1 Crystal Laminae with the Optic Axis in the Facet

Fresnel applied his quarter-wave decomposition to generate formulas for a wave that is first polarized at an arbitrary angle *and* then passes successively through two thin laminae whose axes are crossed at 45°. The resulting formulas can be applied to the usual case of a single lamina simply by removing a term from them. (Indeed, the structure of Fresnel's theory is the same as the modern one, except that it avoids using transverse waves.) Fresnel was interested in this case because he obtained results that conflict with Biot's general claims (claims that, however, Biot had not in fact deduced from his selectionist formulas, as he himself later remarked) (Fresnel 1818c, 495–99; see fig. A4.1).

> PP′ is the plane of incident polarization.
> OO′ is the axis of the first lamina.
> O_1O_1' is the axis of the second lamina.
> i represents ∠PCO′.

The analyzing crystal is set with its principal section in the plane of incident polarization, PP′.

Take the incident intensity to be A and add to A a subscript to denote the sequence of paths the wave has traversed through the two crystals (e.g., A_{o+e} denotes a wave that has traversed an ordinary path in the first crystal and an extraordinary path in the second crystal). Assuming Malus's law (which Fresnel treats here as a hypothesis), then the light emerging from the first lamina will consist of two waves: one, represented by $\cos(i)A_o$, will be polarized along OO′; the other, presented by $\sin(i)A_e$, will be polarized along the plane EE′ which is normal to OO′. Each of these generates two waves in the second lamina, one ordinary, the other extraordinary. However, we must recall Fresnel's general rule that a half-wavelength path difference must be added to the two waves that form a given image if they reach their common plane of polarization by twice increasing the angle between them. This will occur in the second lamina, as one can see from the following sequence:

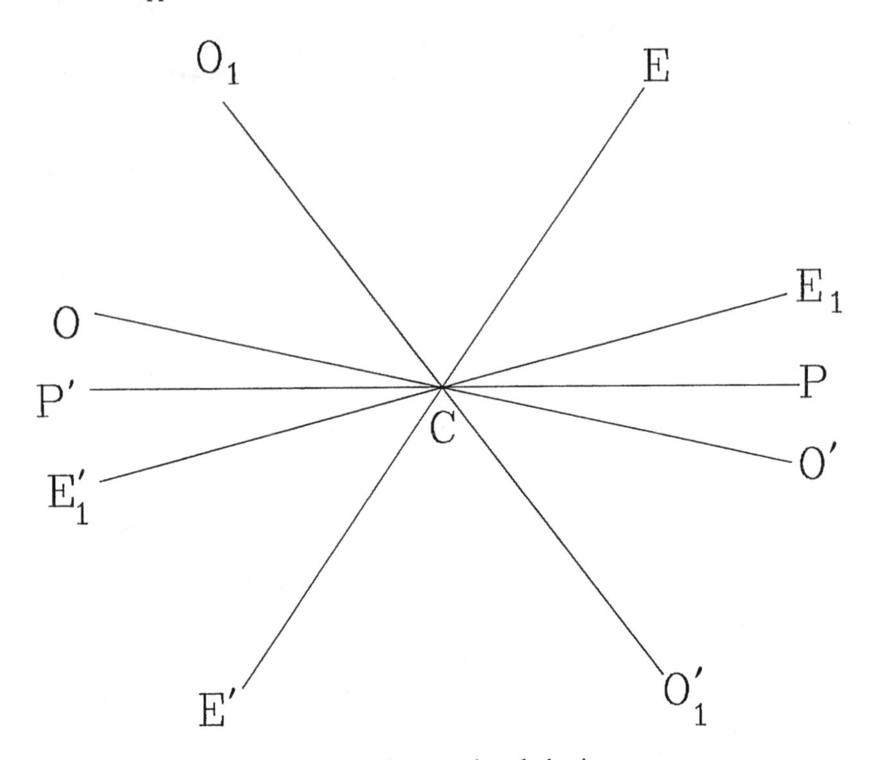

FIG. A4.1 Axes in Fresnel's theory of chromatic polarization.

Emerging from Lamina 1

A_o cos(i) along OO′: P → O

A_e sin(i) along EE′: P → E′

Emerging from Lamina 2

$$\text{from } A_o\cos(i)\begin{bmatrix}1.\ \cos(i)\cos 45°A_{o+o}\text{ along } O_1O_1'\text{: } P\rightarrow O\rightarrow O_1\\ 2.\ \cos(i)\sin 45°A_{o+e}\text{ along } E_1E_1'\text{: } P\rightarrow O\rightarrow E_1'\end{bmatrix}$$

$$\text{from } A_e\sin(i)\begin{bmatrix}3.\ \sin(i)\cos 45°A_{e+o}\text{ along } O_1O_1'\text{: } P\rightarrow E'\rightarrow O_1'\\ 4.\ \sin(i)\sin 45°A_{e+e}\text{ along } E_1E_1'\text{: } P\rightarrow E'\rightarrow E_1'\end{bmatrix}$$

But the two waves (1 and 3) that interfere to form the ordinary wave emerging from the second lamina have reached it by twice increasing their mutual angle and so must have an additional half-wavelength path difference, which can be assigned to either (since it makes no difference to what appears in the analyzer). Consequently we may change the sign of 3 to a minus. (This is the only point where Fresnel's later theory—which is the same as the modern one—adds clarity, in that by using transverse waves one instantly sees the necessity for the sign changes.)

Since the paths traveled by 2 and 3 are the same, we may combine them directly; expanding the resulting expression leaves us in the end with three waves combining to form the ordinary image in the analyzer (which, recall, is set with its principal section in the primitive plane of polarization):

a. $\sqrt{\frac{1}{2}}\cos(i)\cos(45°)A_{o+o}$

b. $\frac{1}{2}A_{o+e}$

c. $\sqrt{\frac{1}{2}}\sin(i)\sin(45° + i)A_{e+e}$

To find their resultant, Fresnel deployed his quarter-wave decomposition. The three waves have different phases because they have been refracted with different velocities, considering the entire journey. Let o now represent the effective number of wavelengths necessary to produce the phase shift undergone by passage through either of the two laminae as an ordinary wave. Similarly, let e represent the number for passage as an extraordinary wave. Then we may represent the amplitudes of the three waves in terms of the incident amplitude A and at any time t as follows (taking the unit of time as a period):

a. $\sqrt{\frac{1}{2}}A\cos(i)\cos(45°)\sin[2\pi(t - 2o)]$

b. $\frac{1}{2}A\sin[2\pi(t - [e + o])]$

c. $\sqrt{\frac{1}{2}}A\sin(i)\sin(45° + i)\sin[2\pi(t - 2e)]$

To apply the quarter-wave decomposition, we have to choose a common reference phase from which each of these deviates by some amount. Fresnel chose $2\pi(o + e)$. Then the phase differences φ will be:

a. $\sqrt{\frac{1}{2}}A\cos(i)\cos(45°)\sin[2\pi(t - [o + e]) - \varphi_a)$ where $\varphi_a = 2\pi(o - e)$

b. $\frac{1}{2}A\sin[2\pi(t - [e + o]) - \varphi_b)$ where $\varphi_b = 0$

c. $\sqrt{\frac{1}{2}}A\sin(i)\sin(45° + i)\sin[2\pi(t - [o + e]) - \varphi_c]$ where $\varphi_c = 2\pi(e - o)$

Each of these three divides into two. One part has the phase $2\pi(o + e)$ and amplitude factor $\cos\varphi$; the other part has the phase $2\pi(o + e + \frac{1}{4})$ and amplitude factor $\sin\varphi$. Combine the three amplitudes for the first phase, and then the three for the second phase. The amplitude of the resultant is the square root of the sum of the squares of the two sums, yielding in the end:

$$A\sqrt{\frac{1}{2} + \frac{1}{2}\cos(2\pi[e - o]) - \frac{1}{4}\sin(4i)\sin^2(2\pi[e - o])}$$

To obtain the case of a single lamina with its principal section aligned at 45° to the primitive plane of polarization requires only setting i to zero (since, recalling the division within the second lamina, this eliminates, in effect, the third and fourth waves and sets $\cos(i)$ to one).

Of course this last does not include the general case of arbitrary analyzer and laminar orientations, but the generalization is simple, and Fresnel gave it (1821c, 615). With s the angle between the plane of primitive polarization and the analyzer's principal section, and i the inclination of the laminar axis, the formula becomes:

$$A\sqrt{\cos^2(s) - \sin(2i)\sin(2[i - s])\sin^2(\pi[e - o])}$$

(which reduces to the previous expression for i equal to 45° and s zero). The formula for the extraordinary image will be:

$$A\sqrt{\sin^2(s) + \sin(2i)\sin(2[i - s])\sin^2(\pi[e - o])}$$

Fresnel immediately pointed out a conflict between his formula and Biot's claims for the two-lamina case. According to Fresnel the tints in the latter case should be the same as those for the case of a single lamina (all else remaining the same) only when the term containing $\sin(4i)$ vanishes, that is, when i is an integral multiple of 45°. However, at any other orientation of the laminar pair to the original plane of polarization this is no longer true: the tints will be different. Biot had claimed that they are always the same however the pair is oriented. The variation, Fresnel remarked, is somewhat hard to see because the colors change slowly, but it unquestionably occurs, making Biot's claim correct only for the octant orientations. This was Fresnel's first (and in fact only) demonstration that Biot had erred in an empirically testable point—but it made little difference because Biot had not in fact deduced the claim from his formulas. (Nor would it be easy to do, since it is hard to see how to select the rays in sets that have already undergone chromatic division when this occurs a second time.)

4.2 Mimicking Optical Rotation

Fresnel (1818c, 505–8) discussed a method by which, he felt, one could "imitate" the effect of quartz on polarized light. Take a crystal lamina (not of quartz) cut with its axis in the facet. Place it between two glass prisms in such a fashion that its axis bisects the angle between the planes of internal reflection within these prisms, these latter being mutually perpendicular.

Total internal reflection occurs twice in each prism. According to Fresnel's understanding at this time, on emerging from the first prism the primitive, polarized wave has been divided in two; the two new waves travel in the same direction but are polarized at right angles (one in, the other normal to, the plane of reflection) and have in addition traversed distances that differ by a quarter-wavelength. From each of these, two other waves will be generated in the lamina, and from each of these four, two more will be generated in the second prism, which adds another quarter-wave path difference. Again care must be taken to alter signs in the laminar images depending upon the deviations in the planes of polarization.

The analysis is intricate but straightforward and leads in the end to the following formula. Let i represent the angle between the primitive plane of polarization and the principal section of the analyzing crystal. Again let o, e respectively represent the numbers of wavelengths traversed by waves ordinarily and extraordinarily refracted by the lamina. Then the ordinary image in the analyzer will have amplitude:

$$A\sqrt{\frac{1}{2} + \frac{1}{2}\cos(2i + 2\pi[o - e])}$$

Clearly, if i remains constant then the tints remain the same. But i is the angle between the primitive plane of polarization and the analyzer section, so this means that the tint will not vary even if the laminar sandwich is rotated as a whole. This imitates the fact that rotation of quartz cut normal to its axis does not change the colors in the analyzer.

APPENDIX 5 Fresnel's Surfaces

I have adapted the following analyses from Schuster (1909) as perhaps the clearest, and simplest, way to grasp how Fresnel's surfaces may be obtained and applied. We distinguish three surfaces:

1. The *optical indicatrix:*	A surface of the fourth degree such that, when sectioned by a plane through its center, the semiaxes of the section give the directions of polarization of, and are proportional to the reciprocals of the phase velocity of, a plane wave that is parallel to the sectioning plane.
2. The *normal surface:*	A surface of the fourth degree whose radii are equal in magnitude to the phase velocities of plane waves drawn normal to the radii.
3. The *wave surface:*	A surface of the fourth degree whose radii are equal to the ray velocities in their directions. It is the generalization of Huygens's spheroid.

5.1. The Normal Surface

We begin with the equation for the optical indicatrix, which we shall simply assume to be given:

The Optical Indicatrix

(A5.1) $$a^2x^2 + b^2y^2 + c^2z^2 = 1 \text{ with } a > b > c$$

(The inequalities merely set the axial conventions for drawing the indicatrix.) In order to find the *normal surface* we must first section the indicatrix by an arbitrarily oriented plane through its center.

Let the plane's unit normal be (s_x, s_y, s_z). Then the plane has the equation:

(A5.2) $$xs_x + ys_y + zs_z = 0$$

Suppose that, in figure A5.1, the semiaxes of the section are OP, OQ. Consider a sphere, centered on O, with radius OP. It will intersect the indicatrix in a "spheroconic" curve (Salmon 1928, chap. 10). The sphere of radius OP has one and only one point (P) in common with the ellipse PQ generated by the section because OP is one

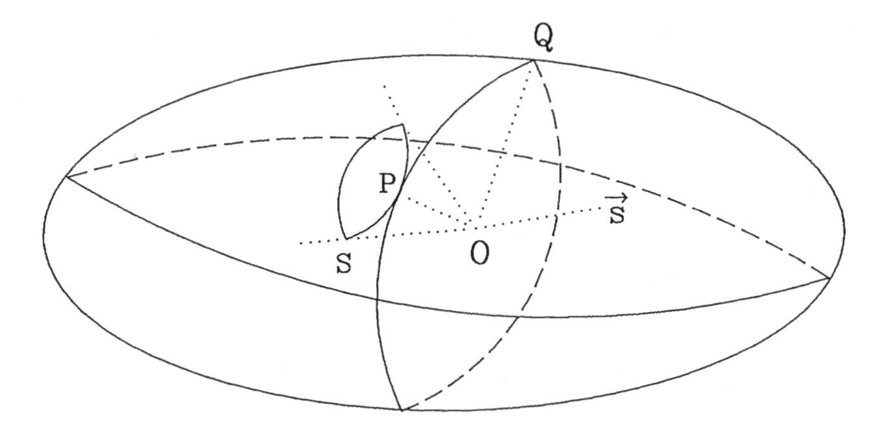

FIG. A5.1 The indicatrix.

of the section's semiaxes. *Point P therefore lies on a sphero-conic that is tangent tc the ellipse PQ at P.*

Draw a cone OPS that includes this curve. Clearly the cone must be tangent to the section at OP, and we can use this fact to establish a relationship between the components of the normal $\vec{s}$ and the semiaxis length OP. First we find the equation of the cone by determining the intersection of the sphere with the indicatrix:

$$(a^2\mathrm{OP}^2 - 1)x^2 + (b^2\mathrm{OP}^2 - 1)y^2 + (c^2\mathrm{OP}^2 - 1)z^2 = 0 \tag{A5.3}$$

We may rewrite this in terms of the phase velocity, v_p, for a plane wave parallel to the section and polarized along OP simply by using the definition of the indicatrix to replace OP with $1/v_p$:

$$(v_p^2 - a^2)x^2 + (v_p^2 - b^2)y^2 + (v_p^2 - c^2)z^2 = 0 \tag{A5.4}$$

This is indeed the equation of a cone.[5]

The normal vector, $\vec{s}$, to the sectioning plane is clearly also normal to the tangent plane to the cone that includes OP. From A5.2, the plane's equation, we may use A5.4 to write, if point P has the coordinates x_p, y_p, z_p:

$$\begin{aligned} s_x &= (v_p^2 - a^2)x_p \\ s_y &= (v_p^2 - b^2)y_p \\ s_z &= (v_p^2 - c^2)z_p \end{aligned} \tag{A5.5}$$

Since $\vec{s}$ is a unit vector that is normal to the line OP, and since OP^2 is just $x_p^2 + y_p^2 + z_p^2$, we find from A5.5, again replacing OP with $1/v_p$:[6]

$$\frac{s_x^2}{v_p^2 - a^2} + \frac{s_y^2}{v_p^2 - b^2} + \frac{s_z^2}{v_p^2 - c^2} = 0 \tag{A5.6}$$

Equation A5.6 determines the normal surface, since it links the phase velocity of a plane wave to its direction of propagation. To find the surface proper we must construct it so that its radii, $\vec{r}$, are equal to $v_p\vec{s}$. We find at once from A5.6:

The Normal Surface

$$-r^6 + r^2\,[x^2(b^2 + c^2) + y^2(a^2 + c^2) + z^2(a^2 + b^2)] = x^2b^2c^2 + y^2a^2c^2 + z^2a^2b^2 \tag{A5.7}$$

Given a wave normal $\vec{s}$ we can now calculate from equation A5.6 the two possible phase velocities v_p and then the two corresponding directions of polarization from A5.5.

The indicatrix is not the same as Fresnel's surface of elasticity, because in the latter the semiaxes are directly, not inversely, proportional to the phase velocities. To obtain Fresnel's surface, first rewrite the indicatrix in terms of the radius, r, and then replace the radius with its reciprocal, which yields:

Fresnel's Surface of Elasticity

$$r^4 = a^2x^2 + b^2y^2 + c^2z^2$$

5.2 The Wave Surface

By the principles of the wave theory, a surface whose radii represent the ray velocities in their directions must be the envelope after unit time of the plane fronts that proceed in all directions from its center. Consequently we must use differential techniques to find the wave surface. We begin with the equation of a plane wave with unit normal s that has traveled the distance v_p in unit time:

$$\vec{r} \cdot \vec{s} = xs_x + ys_y + zs_z = v_p \tag{A5.8}$$

Every point of the wave surface lies on the intersection of two of the surface's plane tangents that are infinitesimally close to one another. That is, for any point on the surface we must satisfy both A5.8 and A5.9:

$$(\vec{s} + d\vec{s}) \cdot \vec{r} = v_p + dv_p \tag{A5.9}$$

Consequently dv_p must be equal to $d\vec{s} \cdot \vec{r}$. Further, since $\vec{s}$ is always a unit vector, $\vec{s} \cdot d\vec{s}$ must vanish. We have, then, three equations to satisfy (obtaining i from the differential of A5.6):

i. $$\frac{s_x\, ds_x}{v_p^2 - a^2} + \frac{s_y\, ds_y}{v_p^2 - b^2} + \frac{s_z\, ds_z}{v_p^2 - c^2} - Kv_p\, dv_p = 0$$

$$\text{where } K = \frac{s_x^2}{v_p^2 - a^2} + \frac{s_y^2}{v_p^2 - b^2} + \frac{s_z^2}{v_p^2 - c^2}$$

ii. $$d\vec{s} \cdot \vec{r} = dv_p$$

iii. $$\vec{s} \cdot d\vec{s} = 0$$

Our first step is to obtain a relationship between the components of $\vec{r}$ and $\vec{s}$; that is, we wish to eliminate $d\vec{s}$ (which we can do because only two of i–iii are independent, since $\vec{s}$ has a fixed magnitude). The procedure is straightforward. To eliminate $d\vec{s}$ first multiply equation ii by a factor A, then multiply i by a factor B and add, the sum being zero. Rearranging terms gives the following equation:

$$\left(A + \frac{B}{v_p^2 - a^2}\right) s_x\, ds_x +$$

ii′. $$(BKv_p)dv_p = \left(A + \frac{B}{v_p^2 - b^2}\right) s_x\, ds_s +$$

$$\left(A + \frac{B}{v_p^2 - c^2}\right) s_x\, ds_x$$

Comparing the factors of dv_p in equations ii′ and ii, we find that B must be $1/Kv_p$, which still contains $\vec{s}$. To obtain an expression for A, again compare ii with ii′, but this time match the factors of the components of $d\vec{s}$. This gives equations for the components of $\vec{r}$ in terms of the components of $\vec{s}$, and from them we form $\vec{r} \cdot \vec{s}$, which, by equation A5.8, is v_p. Using equation i in this result then gives us simply v_p for our factor A. Finally, we calculate r^2 as $A^2 + B^2K$, also using equation i. Combining our expressions for A, B, and r^2 then gives us an expression for B, namely $v_p(r^2 - v_p^2)$, which lacks the components of $\vec{s}$. Substituting this, as well as v_p for A, in our equations for the components of $\vec{r}$ in terms of the components of $\vec{s}$ gives:

$$\frac{x}{r^2 - a^2} = \frac{s_x v_p}{v_p^2 - a^2}$$

(A5.10) $$\frac{y}{r^2 - b^2} = \frac{s_y v_p}{v_p^2 - b^2}$$

$$\frac{z}{r^2 - c^2} = \frac{s_z v_p}{v_p^2 - c^2}$$

Multiply each of A5.10 by x, y, z, respectively, add, and recall that $\vec{r} \cdot \vec{s}$ is just v_p. This result is what we are looking for—an equation for r that contains only the constants a, b, and c:

The Wave Surface

$$\frac{x^2}{r^2 - a^2} + \frac{y^2}{r^2 - b^2} + \frac{z^2}{r^2 - c^2} = 1$$

We may also write this as:

$$r^2(a^2x^2 + y^2b^2 + c^2z^2) - a^2(b^2 + c^2)x^2 - b^2(a^2 + c^2)y^2$$
$$- c^2(a^2 + b^2)z^2 + a^2b^2z^2 = 0$$

Recall how we obtained the wave surface. We used the normalization condition for $\vec{s}$ together with the equation linking $\vec{s}$ with v_p^2 to express the differential changes in v_p for a differential change in $\vec{s}$. This must also be the same difference as that produced by directly considering the intersection of differentially spaced plane waves. The enforced equality yields, after a great deal of manipulation, the wave surface.

APPENDIX 6 **Equations of Motion for the Point Ether**

Briot's 1864 deduction of the Cauchy lattice equations typifies the usual approach to obtaining them. Consider the equation of motion for an arbitrary particle of the ether. Let $\vec{u}$ represent its displacement from equilibrium. The remaining ether particles (with masses m_i) are at distances $\vec{r}^{\,i}$ from this one and exert forces f on it. Their displacements from equilibrium are represented by the $\delta^i\vec{u}$:

$$\frac{\partial^2 \vec{u}}{\partial t^2} = \sum m_i f(\vec{r}^{\,i} + \delta^i\vec{u}) \left[\frac{\vec{r}^{\,i} + \delta^i\vec{u}}{|\vec{r}^{\,i} + \delta^i\vec{u}|}\right]$$

To obtain Cauchy's basic equations for the ether, approximate by assuming that we can neglect all terms in the expansion for f about $\vec{r}^{\,i}$ except for the first two:

$$\frac{f[\vec{r}^{\,i} + \delta^i\vec{u}]}{|\vec{r}^{\,i} + \delta^i\vec{u}|} = \frac{f(\vec{r}^{\,i})}{r^i} + \frac{\partial(f/r^i)}{\partial r^i}\left[\frac{\vec{r}^{\,i}}{|\vec{r}^{\,i}|}\cdot \delta^i\vec{u}\right]$$

Then, since the lattice points are in equilibrium when the displacements vanish, Cauchy's equations at once emerge:

$$\frac{\partial^2 \vec{u}}{\partial t^2} = \sum m_i \frac{f(\vec{r}^{\,i})}{r^i} + \sum\left[m_i \frac{\partial(f/r^i)}{\partial r^i}\left[\frac{\vec{r}^{\,i}}{|\vec{r}^{\,i}|}\cdot \delta^i\vec{u}\right]\right]$$

Next define six new operators as follows:

$$A_{jj} \equiv \sum m_i\left[\frac{f}{r^i} + (\vec{e}_{ri}\cdot\vec{e}_j)\,\frac{\partial(f/r^i)}{\partial r^i}\right]\delta^i$$

$$A_{jk} \equiv \sum m_i\left[\frac{(\vec{e}_j\cdot\vec{r}^{\,i})^2(\vec{e}_k\cdot\vec{r}^{\,i})^2}{r^i}\,\frac{\partial(f/r^i)}{\partial r^i}\right]\delta^i$$

where the $\vec{e}_j$ are unit vectors along the coordinate axes. Then the equations may be written:

$$\frac{\partial^2 \vec{u}}{\partial t^2} = \begin{bmatrix} A_{11} & A_{12} & A_{13} \\ A_{21} & A_{22} & A_{23} \\ A_{31} & A_{32} & A_{33} \end{bmatrix} \vec{u}$$

APPENDIX 7 Conical Refraction

Internal conical refraction: a ray incident at a certain angle will produce a cone of light within the crystal that will emerge as a cylinder.

External conical refraction: a cone of rays with apex on the crystal and with the central axis at a certain angle will refract to a single line within the crystal and emerge to form a cone with apex again on the crystal.

There are several accounts of the deduction of conical refraction by Hamilton and its confirmation by Lloyd.[7] I have little to add to these discussions and refer the reader to them for developments after the discovery (Hankins 1980) and for the details of Hamilton's analysis as presented in the "Third Supplement" (Eichhorn 1986). However, because the discovery was so influential I shall briefly discuss its major features.

Reduced to essentials, Hamilton's discovery amounts, first, to the recognition that of the four tangent lines that touch both sheets of the wave surface in the plane containing the axes of single ray velocity, in fact each lies in a tangent plane that touches both sheets throughout a circle of contact. Second, that the wave surface around the points of intersection of its two sheets looks much like the surface of a plum near its stem, so that there are an infinite number of plane tangents arrayed round each of these points, *approximately* forming by their intersection the surface of a cone. The first property of the wave surface—that of being touched throughout a circle by a tangent plane—is not obvious from the surface's equation. The second property, however, can at least in part be concluded simply from the existence of the four points of intersection between the two sheets of the surface: since the surface is continuous, these four points must each form something like a dimple or cusp and consequently must be surrounded by an infinite array of tangent planes—though it is certainly not at all obvious that these planes must be arrayed approximately along the surface of a cone.

Hamilton related that he first discovered these properties by examining a surface—the "surface of components"—that determines the reciprocal of the *front* velocity as a function of the direction of the front normal. He had first deduced this surface—which is a form of Fresnel's normal surface—because it has the greatest significance for his theory of the characteristic function in that the gradient of that function must be parallel to the front normal and have as magnitude the reciprocal of the front speed (Hamilton 1833).

To deduce this surface of components, as he termed it, Hamilton used Fresnel's elasticity analysis. That surface, Hamilton remarked, can itself be used directly to find the direction and speed of an arbitrarily incident ray, even though this had previously been done only using the wave surface, whose radii determine ray speeds, not the reciprocals of front speeds.[8] Examining the surface of components, Hamilton somehow discovered that it has a tangent cone at a cusp. The tangent cone was sufficient to conclude the existence of *internal* conical refraction from the construction, using this surface, for the ray corresponding to a specific front.[9] It was also clear that the termini of these rays (their lengths being as the ray speeds) must lie on a *curve of plane contact* on Fresnel's wave surface because, by Hamilton's rule, they all have the same front speed (though the form of the curve was not yet determined).

None of this directly suggested to Hamilton the existence of *external* conical refraction, because the latter requires the existence of conical cusps on the wave surface itself, since then a ray traveling in the direction of the cusp corresponds to an array of fronts forming a tangent cone at the cusp, and a conical beam must therefore emerge from the crystal, since each such tangent plane produces its own emergent refraction. Rather, Hamilton remarks, he discovered this phenomenon "from the connexion of the surface of components with the wave surface propagated from a point," in that the existence of conical cusps on the former entailed their existence on the latter (Hamilton 1833).

We find none of this in Hamilton's "Third Supplement" of 1837, wherein he directly proves the existence of the circle of contact and the conical cusps on the wave surface. Instead, he uses the surface of components to generate the wave surface using relationships derived from his theory of the characteristic function (which need not concern us, since this was done merely to show a "simpler" way to find the wave surface "than that which was employed by the illustrious discoverer" [Hamilton 1837, 281]).

The analysis itself, while somewhat intricate, is quite straightforward once the existence of the cusps and the circles of contact are known. In essence, Hamilton rewrites the equation for the wave surface to give the reciprocal of the squared ray velocity as a function of the angles between the ray direction and the directions of single ray velocity (i.e., the cusps). He then transforms this expression by rotating the coordinates in the plane of the single ray speeds until one of the coordinate axes corresponds with a single speed direction. Near the cusp, he then shows, the wave surface approximates to a cone. Transform instead by rotating (in the same plane) until a coordinate axis corresponds with one of the directions of single front velocity, and the circle of contact readily emerges.

APPENDIX 8 Malus's Theory of Partial Reflection Reconstructed

8.1 Partial Reflection at Polarizing Incidence

Malus's division of the refracted bundle into two subsets, one polarized like the incident beam and the other with polarization perpendicular to that of the reflected beam, at once solved the puzzling failure of the cosine law as applied to reflection. The problem was that the constants of proportionality for the reflected and refracted bundles had to be different. But now we see that they do not have to differ if we recognize that it is only ΔI_o, and not the complete refracted bundle $(I_o + \Delta I_o)$, that corresponds to the reflection. In other words, there is a complete analogy between Malus's cosine law for the crystals and for reflection at polarizing incidence. One of the two beams is *always* present and has the same polarization as the incident light, but the other component of the refraction is polarized perpendicular to the reflected light. And so the sum of the intensity of this latter part of the refraction with the intensity of the reflected light must be constant. And this must be the very constant of proportionality that appears in the cosine law for the reflection of polarized light. This recognition permitted Malus to solve fully the problem of partial reflection at polarizing incidence. He did not live long enough to provide the details, but they can easily be deduced from selectionist principles.

For simplicity take the incident beam to have unit intensity, and let I' represent the fraction of the beam that would be reflected at polarizing incidence when the plane of polarization is parallel to the plane of reflection. Then I' must be the constant in the cosine law applied to partial reflection. If ε represents the angle between the plane of reflection and the plane of polarization, then we must have:

$$\text{(A8.1)} \qquad \left.\begin{array}{l} I_{rfl} = I' \cos^2\varepsilon \\ I^v_{rfr} = I' \sin^2\varepsilon \; (= \Delta I_o) \\ I^u_{rfr} = I_o \end{array}\right] \; I_o + I' \equiv 1$$

Here I^v_{rfr} represents that portion of the refracted bundle whose polarization is normal to the plane of reflection, while I^u_{rfr} represents the portion of the refracted light that is always present and whose polarization is the same as that of the incident light.

We now use Malus's ratio law (sec. 2.5.3 in chap. 2). When ε reaches 45° we measure the angle ω' required to rotate the analyzer to minimize the extraordinary refraction. At this angle we therefore have:

(A8.2)
$$\frac{I_o}{I^v_{rfr}} = \frac{I_o}{\Delta I_o} = \frac{2I_o}{I'} = \frac{1}{\tan 2\omega'}$$

(Here we have the probable source of Malus's misprint regarding the erroneous factor of two: at this point he no doubt confused I' with $\Delta I_o = I'/2$.) Since $I_o + I'$ represents the unit intensity of the incident beam of light, we may solve A8.2 for I_o and I':

(A8.3)
$$I_o = \frac{\cos 2\omega'}{\cos 2\omega' + 2 \sin 2\omega'}$$

$$I' = \frac{2 \sin 2\omega'}{\cos 2\omega' + 2 \sin 2\omega'}$$

Equations A8.1 and A8.3 together solve the problem of partial reflection at polarizing incidence. All one needs to know is ω', which can be found by setting the incident polarization at 45° and turning the analyzer until the extraordinary image in it reaches a minimum. One does not need to know the angle of polarizing incidence itself. Then, given ω', we can calculate the intensities of the two refracted bundles (I^v_{rfr}, I^u_{rfr}) and the reflected bundle (I_{rfl}) for any angle of polarization whatever.

Malus's experiments, as we have repeatedly seen, employ crystals of spar to examine the polarization of light. Using A8.1, A8.3, and Malus's cosine law for birefringence, we can compute the way the entire beam of refracted light will divide into O (ordinary) and E (extraordinary) beams when it is examined with a crystal. Suppose that ε represents the angle of polarization of the light before it is refracted, and set the analyzer's principal section at some other, arbitrary angle α (all the angles being taken with respect to the plane of the meridian). Then:

(A8.4)
$$\mathrm{O} = \frac{\cos^2\alpha}{1 + 2 \tan 2\omega'} + \frac{[2 \tan 2\omega'][\sin^2\varepsilon][\sin^2(\alpha + \varepsilon)]}{1 + 2 \tan 2\omega'}$$

$$\mathrm{E} = \frac{\sin^2\alpha}{1 + 2 \tan 2\omega'} + \frac{[2 \tan 2\omega'][\sin^2\varepsilon][\cos^2(\alpha + \varepsilon)]}{1 + 2 \tan 2\omega'}$$

These expressions for O and E directly represent what can be seen with Malus's apparatus, so that they must work reasonably well for him to have had confidence in his theory. And indeed they do work well, just as his ratio law does. But we must be careful to consider what Malus would himself have been looking for. His formulas give the E and O intensities produced by the refracted light for any analyzer orientation and incident polarization. To see whether they worked well, Malus would most likely have set the plane of polarization of the incident light at an arbitrary angle. He would have then rotated the analyzer to see whether the O and E beams within it behave as his theory requires. He could not, however, actually measure intensities, which meant that confirmation could be only qualitative. (In principle he could have found the minimum position for E for the given incident polarization, but that involves an extremely unwieldy formula.)

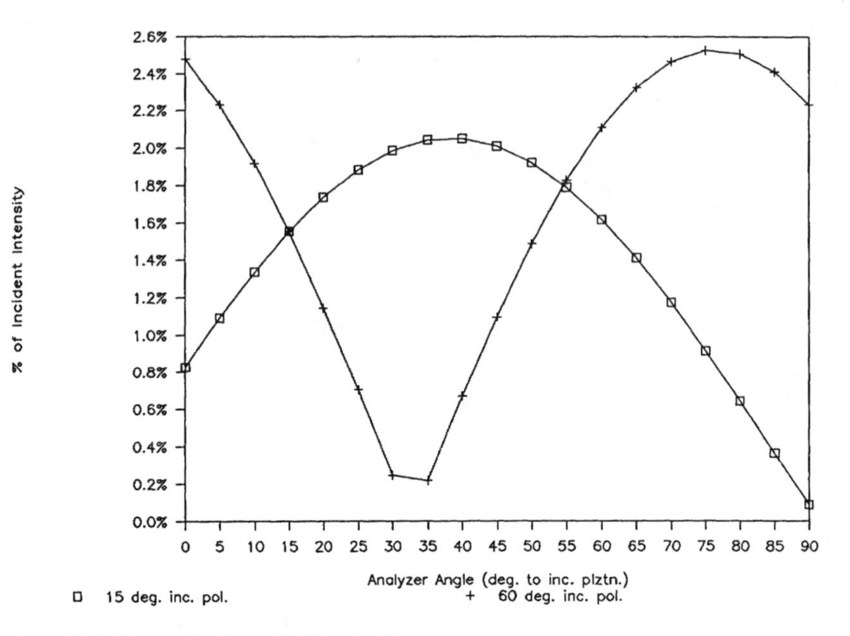

FIG. A8.1 E intensity for polarizing incidence.

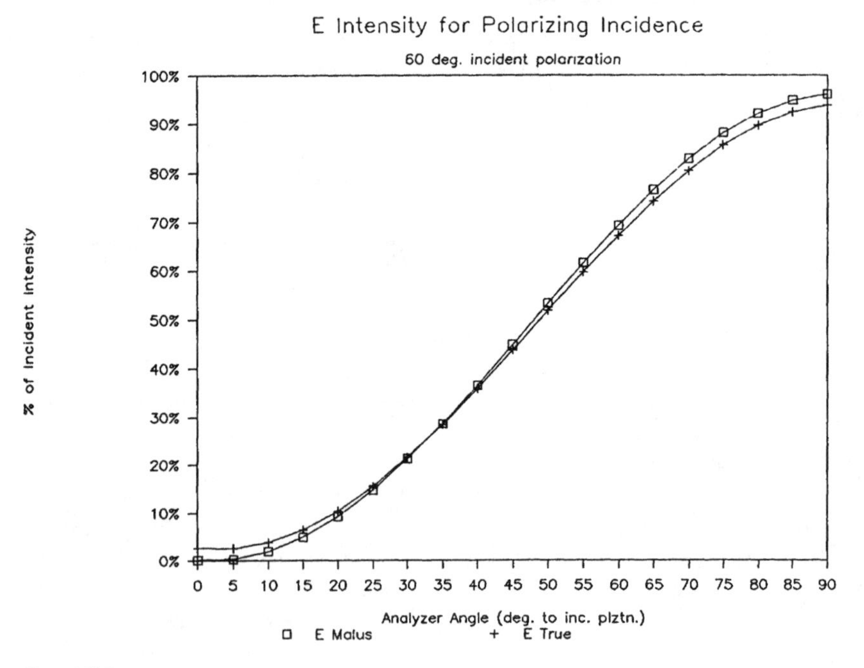

FIG. A8.2

We can chart what Malus could have seen by computing the differences for unit incident intensity between Malus's E values and the E values that would be measured in a highly accurate experiment. (To do so we must assume that Malus's observed value for ω' corresponds nearly to the actual deviation of the refracted polarization when the plane of polarization is set to 45°; see figs. A8.1 and A8.2.) Figures A8.1 and A8.2 show that Malus's formulas work extremely well. Not only does Malus's prediction remain with a few percent distance of the ones that would be measured in a very accurate experiment, but it follows nearly the same curve as a function of the orientation of the analyzer.

And yet Malus's theory, which is based entirely on selectionist principles, conflicts irremediably with what we now believe to be the correct state of affairs. According to his account the refracted beam, except when it becomes 90°, cannot be considered "polarized" in his sense of the term. It cannot, that is, consist solely of rays with the same asymmetry. It must consist of two bundles, one of which (by far the more populated of the two) retains the incident polarization, whereas the other is polarized perpendicular to the plane of reflection. The modern theory, however, requires that the refraction of a polarized beam must itself be polarized, though its plane of polarization does in general deviate several degrees from the plane of incident polarization, as we can see from figure A8.3, which also graphs the ratio of reflected to refracted intensity, all as a function of the plane of polarization of the incident beam of light.

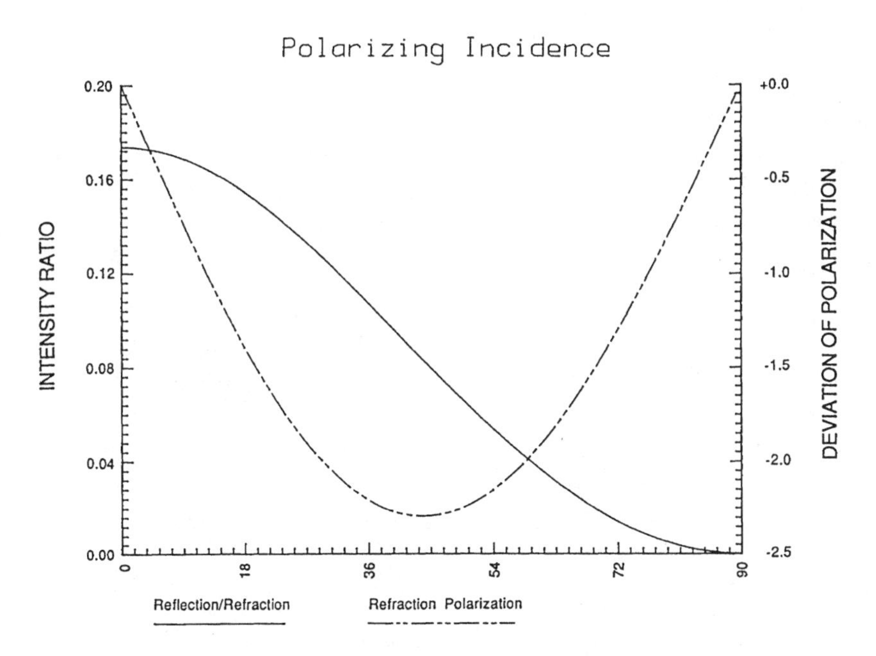

FIG. A8.3 Modern results for reflection.

Precisely because the deviation of the plane of polarization does remain small, it can be interpreted observationally as the addition of a new bundle to the refraction, whose polarization must be normal to the plane of reflection. That, however, means the analyzer can never be rotated to a position where the E beam disappears entirely. One can only minimize E's intensity, because the polarizations of the two bundles in the refraction are not the same. Malus was fully aware of this, and he even saw his observations directly confirm it, since he remarked that the E beam can be made only to "nearly" disappear. Which must no doubt be true: one can hardly have so perfect an arrangement that neither the incident beam nor its refraction is even slightly polarized elliptically.

Of course Malus had not, with this, solved the general problem of partial reflection, because his theory concerned only polarizing incidence, where the reflected beam contains only one, polarized bundle. He claimed, however, that he could solve the general problem, though he again gave few details. But he gave enough to uncover what he must have done and so to reconstruct the first quantitative theory of partial reflection for any angle of incidence whatever.

8.2 The Selectionist Theory of Partial Reflection

Malus knew that reflection of polarized light elsewhere than at the polarizing angle nevertheless still produces "in part" the properties of polarizing reflection (Malus 1811d, 116). He interpreted this to mean that the reflection of a polarized beam will in general consist of two bundles, one of which remains polarized in the original plane, whereas the other is polarized in the plane of reflection. Later selectionists, Biot in particular, would say that the original beam had been "depolarized" in part by the reflection. To construct a general theory of partial reflection, one must be able to compute the number of rays in this additional, "depolarizing" bundle. Malus could do so, and his brief account also reveals that he, and probably many later selectionists, understood the operational meaning of the phrase "polarized with respect to" a given plane in a way considerably different from its definition after widespread acceptance of the wave theory.

Malus begins his experiments with a vertical beam of light that has been reflected at the polarizing angle. He reflects it a second time, at various incidences, from a mirror on which the plane of incidence inclines 45° to the plane of polarization. He described the result (see fig. A8.4):

> When [the mirror] makes an angle of only a few degrees with the horizon, it will reflect part of the incident vertical ray [A in fig. A8.4], and the reflected light [B in fig. A8.4] will be polarized not with respect to the plane of incidence . . . but with respect to the plane of the meridian [the plane of polarization of the incident beam of light]. If we trace in the plane of the mirror a line parallel to the plane of the meridian [D in fig. A8.4], and if we receive the reflected light on a crystal of Iceland

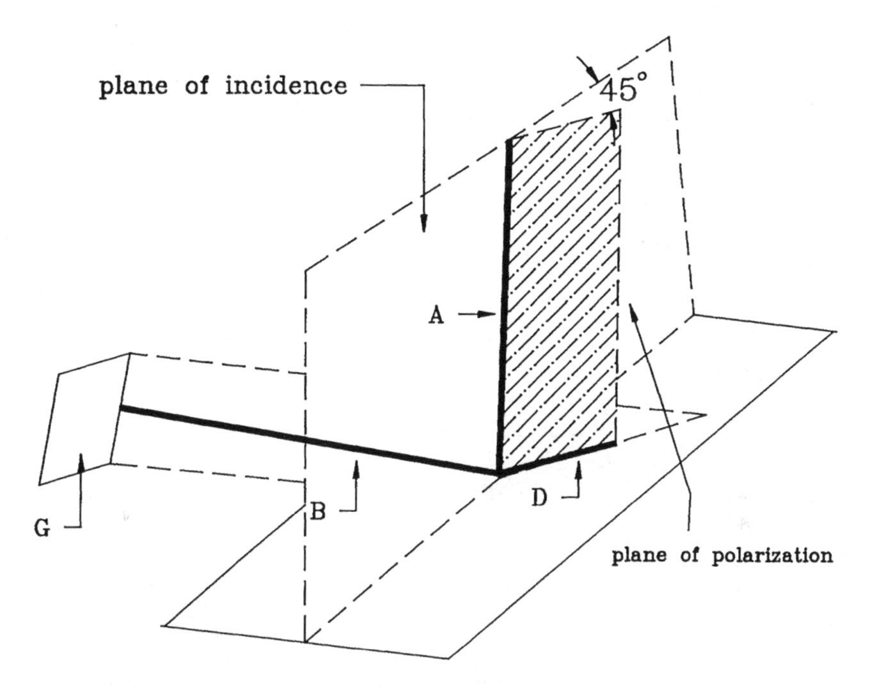

FIG. A8.4 Malus's configuration for partial reflection.

> spar whose principal section is parallel to that line [G in fig. A8.4], the ray will be refracted in a single ordinary ray. (Malus 1811a, 116)

The implication of Malus's remarks is that for small incidences in this configuration of polarized beam and reflecting mirror, the reflection retains in some sense its original polarization, the test of which Malus describes in the second sentence of the quotation (and which I shall discuss in a moment). Yet we now know that the reflection's plane of polarization must, in these circumstances, rotate to the opposite side of the plane of incidence. We would hardly say that, with a deviation of 90° at low incidences, the reflection can be spoken of as still polarized in its original plane. Yet Malus seems to say just that.

We seem, then, to have found here an empirical failure on Malus's part. Or have we? We must pay careful attention to Malus's second sentence and not assume that he means the same thing we do when he writes that a beam is polarized "with respect to" a given plane. One can easily show that Malus's claim, described in the second sentence of the quotation, in fact amounts to this: that the tangent of the angle ε_r of reflected polarization must be equal in magnitude but opposite in sign to the tangent of the angle ε of incident polarization. And that is indeed true for low incidences, according to Fresnel's amplitude ratios (since they yield the formula tan ε_r =

$-\tan[\cos(i - r)/\cos(i + r)]$, with i the angle of incidence and r the angle of refraction).

We would not agree with Malus that this means the reflection remains polarized "with respect to" its original plane. Evidently Malus held a considerably different understanding of polarization as an operational concept. It can be succinctly expressed. A beam remains polarized "with respect to" a given plane if the interface, the given plane, and the principal section of an analyzer set to produce only an ordinary beam from the reflection all intersect in a common line.

We can perhaps understand Malus somewhat better by recurring to his triaxial description of a ray. We have an axis *a* along the ray, and axes *b* and *c* in a plane normal to it. At polarizing incidence the reflection consists of rays with *b* axes (by convention) all normal to the plane of incidence. Consider now reflection other than at the polarizing angle. According to Malus's description the reflection must have its *b* axes normal to the principal section of the analyzer when the latter is set to produce only an ordinary beam. At low incidences that plane in fact does contain both the reflected bundle and the intersection of the original plane of polarization with the interface. Evidently Malus considered such an orientation essentially the same, at least operationally, as the original orientation, because the *b* axes lie at the same angle with respect to the plane of incidence as before reflection—only on its opposite side.

Things change markedly as the incidence increases, however, in particular as the angle of polarizing incidence is approached. For in order to maximize the ordinary beam within it while observing the reflected light, the principal section of the analyzing crystal must be turned gradually toward the plane of reflection until, at polarizing incidence, it must be parallel to the plane. Beyond that point the analyzer must be turned slightly away from the plane of reflection, toward the opposite side of it from its position at low incidences.

To interpret what this means Malus had, on selectionist grounds, to determine what sorts of polarized subsets of rays are in general created by reflection. He came to the following conclusions. Between normal and polarizing incidence the reflection consists of two subsets: one of them is polarized at right angles to the original plane of polarization, whereas the other is polarized in the plane of reflection. As polarizing incidence is approached, the first set of rays decreases in number while the second set increases. At polarizing incidence itself the first set completely vanishes. Above this point an entirely new subset of rays begins to grow, one whose polarization is the same as the polarization of the original beam, while the set of rays that are polarized in the plane of reflection again begins to decrease, vanishing entirely at grazing incidence. Consequently, according to Malus's theory the reflection of a polarized beam of light will *not* in general be polarized except at very low and at very high incidences and at the polarizing angle proper. Elsewhere it will be only *partially* polarized.

In Malus's terms we may say that the reflection consists in general of two bundles, one polarized in the plane of incidence and the other always polarized "with respect to" its original plane of polarization. Below polarizing incidence "with respect to" means in a plane that intersects the line formed by interface and original plane of polarization and, as a little geometry readily shows, on the opposite side of the plane of incidence at an equal angle. At polarizing incidence the second bundle vanishes, but beyond that point it increases, its polarization then being the same as that of the incident bundle. In either case the second bundle remains polarized "with respect to" the original plane of polarization in Malus's sense.

What can be seen in the analyzer depends upon the angle between the planes of polarization of the two reflected bundles. Malus could have easily calculated the angle. Call it β: below polarizing incidence its cosine equals the sine of the angle of original polarization; at and above polarizing incidence it equals the original angle of polarization. We can now generalize the procedure Malus must have used to obtain the law he says he deduced for an arbitrary angle of incidence. To do so introduce $R_1(i)$, $R_2(i)$: these are respectively the numbers of rays in the first and second parts of the reflection for an angle of incidence equal to i. According to Malus's observations $R_1(i)$, which is polarized in the plane of incidence, must vanish at normal and grazing incidence. $R_2(i)$, which is polarized at right angles to the original plane below polarizing incidence or in the original plane above polarizing incidence, must vanish only at the polarizing angle. In an analyzer whose principal section is set at an angle α with respect to the plane of reflection, one will see O and E beams with the following intensities (see fig. A8.5):

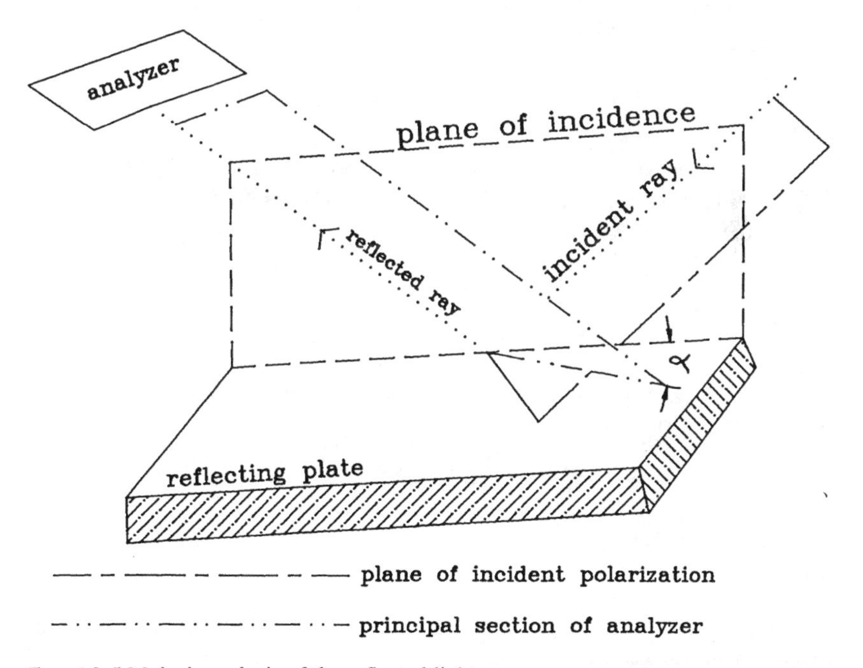

FIG. A8.5 Malus's analysis of the reflected light.

$$\text{(A8.5)}\quad \begin{aligned} \mathrm{O} &= R_1(i)\cos^2\alpha + R_2(i)\cos^2(\beta - \alpha) \\ \mathrm{E} &= R_1(i)\sin^2\alpha + R_2(i)\sin^2(\beta - \alpha) \end{aligned}$$

We must now determine R_1 and R_2 at a given incidence. We can use our previous procedure, but since we are no longer always at polarizing incidence we have an extra bundle to consider, which in the end means we now need two constants instead of one. Our previous equations (equations A8.1) can be applied, modifying them only by including the new bundle, R_2, in the reflection and in the equation for ray conservation. In other words, we do not alter the *composition* of the refracted light: it still consists in general of two sets of rays, one polarized in the original plane and the other polarized in the perpendicular to the plane of reflection. However, since a new subset of rays (R_2) exists in the reflection, the actual numbers of rays in the two parts of the refracted light will necessarily be different from before:

$$\text{(A8.6)}\quad \begin{aligned} &\text{conservation of rays} && \left[\; I_o + I' + R_2(i) \equiv 1 \right. \\ &\text{reflection} && \left[\; \begin{aligned} &R_2(i) \\ &R_1(i) = I'\cos^2\varepsilon \end{aligned} \right. \\ &\text{refraction} && \left[\; \begin{aligned} I^v_{rfr} &= I'(i)\sin^2\varepsilon = \Delta I_o(i) \\ I^u_{rfr} &= I_o(i) \end{aligned} \right. \end{aligned}$$

We were previously able to deduce the ratio $I_o/\Delta I_o$ for polarizing incidence by measuring, at that incidence, the angle ω through which the analyzer must be turned to minimize E when ε is set to 45°. That deduction remains valid for any incidence i, because we have not altered the polarizations of the two subsets of the refracted light. We will obtain from experiment a different value for ω'; call it $\omega'_{rfr}(i)$. We may then use Malus's previous result for polarizing incidence to obtain:

$$\text{(A8.7)}\quad f \equiv \frac{I_o}{I'} = \frac{1}{2\tan[2\omega'_{rfr}(i)]}$$

So the ratio f can be measured directly. Adding ray conservation gives us two relations between three unknowns (I', I_o, and R_2):

$$\text{(A8.8)}\quad \begin{aligned} I' &= \frac{f(1 - R_2)}{1 + f} \\ I_o &= \frac{1 - R_2}{1 + f} \end{aligned}$$

Clearly, we need another empirical constant. It can be obtained by again using Malus's minimum technique, only here we apply it to the reflection instead of to the refraction. Set the incident polarization ε at 45°. Set the analyzer to observe the reflection and rotate it until the E beam reaches minimum intensity at some inclination $\omega'_{rfl}(i)$. At that angle, as we can easily show from A8.5 and A8.6, I' and R must satisfy the following equation:

$$(A8.9) \qquad I' \equiv 2R_2 \frac{\sin(\beta - \omega'_{rfl})\cos(\beta - \omega'_{rfl})}{\sin(\omega'_{rfl})\cos(\omega'_{rfl})} = gR_2$$

With equation A8.9 we have solved the problem of partial reflection, because we can now compute R_2 and I' at any angle of incidence by making two measurements with our analyzer at that incidence: for both measurements we set the incident polarization at 45° and then measure the angle the analyzer must be turned through in order to minimize the E beam produced in it by the refraction (ω'_{rfr}) and by the reflection (ω'_{rfl}).

$$(A8.10) \qquad R_2 = \frac{1}{g\left(1 + \frac{1}{f}\right) - 1}$$

$$I' = gR_2$$

One can again test how well Malus's formulas would work in a very accurate experiment by computing both values of ω' at any given incidence from the Fresnel ratios, for these give us very nearly the angles Malus would himself have measured. We can then substitute them into Malus's formulas to see what the value of the minimum E intensity in the analyzed reflection will be, according to them, at any given incidence for a polarization of, say, 45°. Figure A8.6 shows that the minimum E intensity at any incidence lies below 5% of the incident intensity, so that again Malus could be quite certain of the observational accuracy of his theory. Further, the curve has a cusp at the polarizing incidence, and its shape is roughly the same from 0° to polarizing incidence and from there on up—all of which is entirely consistent with Malus's remarks. Little wonder Malus was convinced he had solved the problem of partial reflection. He remarked: "One may therefore, in this way, determine the quantity of light that is polarized at different angles of incidence, and the measure of this phenomenon is reduced to simple observations of angles, and this considerably simplifies the problem, which until now posed the greatest of difficulties" (Malus 1811a, 117).

The theory had other implications, in particular for the structure of natural light and of light that is partially polarized by reflection above or below the polarizing angle. Taking partially polarized light first, recall that Malus had previously suggested such light might consist of two orthogonally polarized bundles. That can no longer be maintained because of the necessary presence of the second bundle in the reflection, a bundle whose polarization depends upon that in the incident beam.

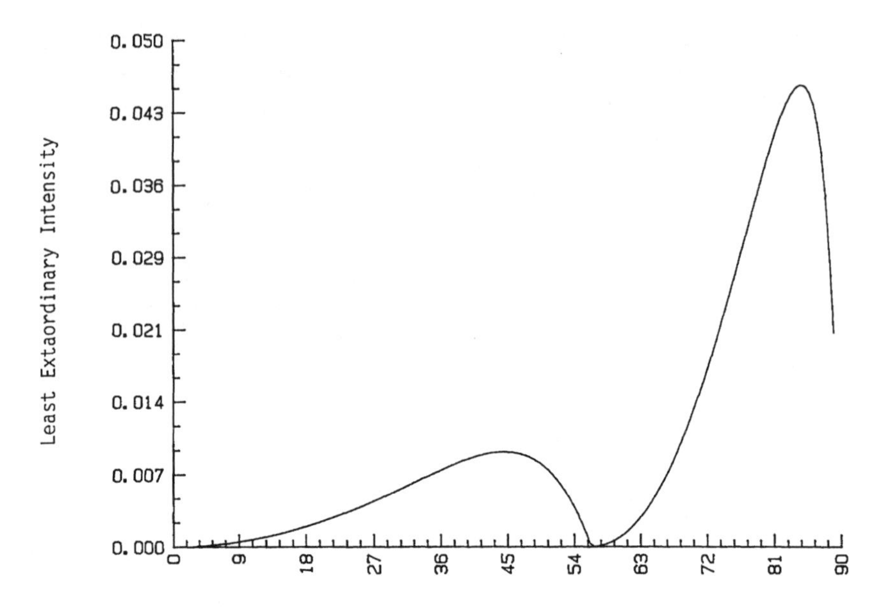

FIG. A8.6 Malus's reflection law.

Natural light consists for Malus (and for almost all later selectionists) of rays with randomly oriented asymmetries. To analyze how it behaves on reflection, gather the rays in the natural beam together in bundles, each bundle's elements having nearly the same asymmetry. Since the rays are completely independent of one another, each bundle will be reflected as a polarized beam, and so it will yield on reflection two other bundles: one will be polarized in the plane of incidence, while the other will either retain the polarization of the incident bundle or else be polarized in the perpendicular to it. But one has many such sets in natural light, each with a different polarization. Consequently the reflected beam remains in much the same state as the incident beam, except that more rays are now polarized in the plane of incidence than before. Malus put it this way:

> The experiments prove . . . that ordinary light, reflected by bodies below and above the determinate [polarizing] angle, lacks the properties of the natural ray, not because it is composed of light polarized in the two senses (as I had also thought) . . . , but because really it has not experienced the modification that produces polarization [since the reflection merely redistributes rays from one set to another when the incident beam contains so many orientations]. (Malus 1811a, 119)

Before Malus's work one could ask without qualification what proportion of light will be reflected at a given incidence from a given transparent body. But his theory makes that way of posing the issue incomplete: one also needs to know the polarizations of the bundles in the incident beam. Once the laws governing partial reflection at a specific polarization have been determined, then the question of what happens to any given beam of light requires specifying the asymmetries of the rays that compose it. This could, and did, lead to controversy between selectionists, who might differ over the composition of a beam of light in respect to the asymmetries of its rays.

APPENDIX 9 The Structure of Natural Light

The structure of natural light was always a central and contentious issue for selectionists. Arago was no exception, and it is worth spending a moment on his remarks in this area because they, like Malus's on the same topic, show selectionist principles powerfully at work. In particular, Arago wondered whether Malus's laws for partial reflection were consistent with the hypothesis that natural light consists of only two orthogonally polarized sets of rays (much as Malus had once suggested—and then rejected):

> In general, this phenomenon of rings shows us rays that are polarized without having been reflected or submitted to the action of forces that impress this modification on them. I was consequently curious to examine if it were not possible to admit that rays of direct [natural] light are composed of many polarized rays. Now, the ensemble of two beams equally vivid and polarized in contrary senses has with respect to double refraction the same properties as rays of direct light. (Arago 1812c, 118)

Arago sought to prove, using Malus's theory of partial reflection, that a refracted beam deriving from such a structure will behave in the same way as a beam that derives from a set of randomly asymmetric rays (which was the obvious alternative hypothesis for natural light).

Let α be the number of rays in either of the two orthogonally polarized bundles in a beam of natural light on Arago's hypothesis. We shall examine the general case, though Arago limited himself to special ones. (In particular Arago considered the case in which the plane of incidence bisects the angle between the two polarizations.) Set the plane of polarization of one of the bundles at an angle ω with respect to the plane of incidence; then the plane of polarization of the other bundle must lie at $90° - \omega$ to that plane (see fig. A9.1).

Suppose that when light is polarized entirely in the plane of incidence a portion α' of the original beam is reflected. Then, according to Malus's laws for the partial reflection of polarized light, when a beam is polarized at the angle ω the quantity reflected will be $\alpha'\cos^2\omega$, and an additional quantity $\alpha'\sin^2\omega$ will be transmitted. On Arago's hypothesis each of the two bundles in the original beam will behave like a polarized beam and so will contribute separately to the reflection according to Malus's law.

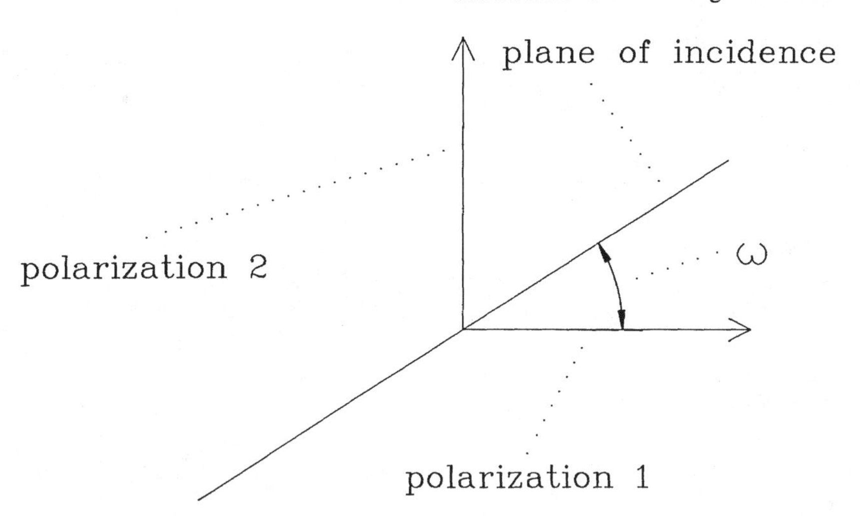

FIG. A9.1 Arago's ray asymmetries in natural light.

Arago considers only polarizing incidence in his discussion, so that the reflected bundle is always polarized in the plane of incidence. Following Malus again, he assumes that the refracted light will consist of two bundles, one polarized normally to the plane of incidence, whereas the other retains the polarization of the incident bundle from which it derives. Assume that the total intensity of the original beam of light is unity, and that this is divided evenly between its two orthogonally polarized parts. Then we find, following Arago's hypothesis and applying Malus's laws to each of the two bundles in the original beam of natural light:

Reflected Light, All Polarized in the Plane of Incidence

$\alpha'\cos^2\omega$ from incident bundle 1
$\alpha'\sin^2\omega$ " " " 2

Transmitted Light

1. polarized along normal to plane of incidence] $\alpha'\sin^2\omega$ from bundle 1; $\alpha'\cos^2\omega$ " " 2
2. polarized in direction of bundle 1] $\frac{1}{2}\alpha'$ from bundle 1
3. polarized in direction of bundle 2] $\frac{1}{2}\alpha'$ from bundle 2

The transmitted light therefore must consist of three parts. Two of these (the second and the third) together form a beam that differs from the structure of the incident light only in being less intense—it has an intensity equal to $1 - 2\alpha'$. The third part (one in our list) is newly polarized along the normal to the plane of incidence. But that is precisely the same result we would obtain if the incident beam consisted of randomly asymmetric rays, which was what Arago had set out to prove. Note also that the third part has the same intensity as the reflected light (α')—whence follows Arago's conclusion that, in the reflection of natural light at polarizing incidence, the refraction consists of natural light mixed together with light that is polarized along the normal to the plane of incidence. This last part has the same intensity (α') as the reflected light, which is polarized in the plane of incidence. Arago was often said to have been the first to make this assertion, which is no doubt correct, but he derived it from Malus's account of partial reflection, and that is never remarked.[10] Arago in fact retained a strong belief in the fundamental character of rays throughout his life, even after he became Fresnel's supporter and sometime collaborator. He never fully accepted that the wave theory destroyed the physical reality of rays, though he did in the end adopt Fresnel's understanding that, in reflection, polarized light remains polarized but its plane of polarization deviates.

APPENDIX 10 Arago's Theory for Quartz

Consider Arago's experimental arrangement. He begins with a polarizing and an analyzing crystal set initially with their principal sections parallel to one another. The O and the E beams that emerge from the analyzer are white until the quartz plate is put between the two crystals. Then both the O and the E beams become colored, and their tints and intensities remain the same however the quartz plate turns about. Turning either the polarizer or the analyzer, however, does alter both tint and intensity.

Arago assumes that each of the two polarized beams that come from the polarizing crystal always splits in two on entering the quartz plate (see fig. A10.1). These two parts into which each of the polarized beams is divided have complementary colors and are polarized at right angles to one another. That is, in the quartz plate we have four beams in all: the two from the original O beam are complementary and polarized

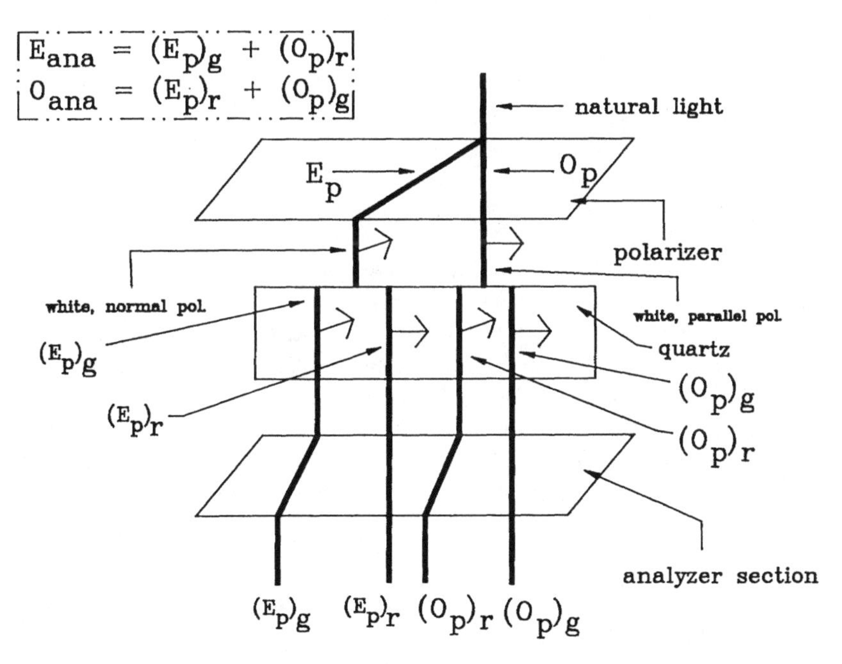

FIG. A10.1 Arago's ray paths and asymmetries in quartz.

opposite to one another, and the two from the E beam are also complementary and polarized opposite to one another. Arago writes:

> On exit from the [polarizer] each image is formed of rays polarized in the same sense, while after having traversed the plate of [quartz], the molecules of the diverse colors of which the two white beams are composed have their poles directed to different points of space. If the green rays [$(E_p)_g$] of the first image [E_p] are, for example, polarized with respect to the plane of the meridian, the complementary rays [$(E_p)_r$], in this same image [E_p], will have received the diametrically opposite modification. In the second beam [O_p], it will on the contrary be the red molecules [$(O_p)_r$] that will have their poles situated in the plane of the meridian, while the green rays [$(O_p)_g$] will be polarized with respect to a plane perpendicular to that one. [Here Arago is very loosely using "red" and "green" to mean colors that are complementary to one another.] (Arago 1811b, 61–62)

The theory Arago was describing here nicely captures the gross differences between mica and quartz. Quartz, he is arguing here, does not disperse the asymmetries of rays in the way mica does. Indeed, Arago thinks that quartz does not alter at all the two orthogonal senses of polarization as they are initially determined by the crystal that originates the light that strikes the quartz plate. Rather, quartz *selectively shifts rays* from one of these two directions of polarization into the other. That is, quartz takes a given beam of polarized light, selects some of the rays in it, and turns their asymmetries into the plane perpendicular to the beam's original polarization.

To see how this works consider figure A10.1, where we have a simple case in which the principal section of the analyzing crystal is parallel to the section of the polarizing crystal. Natural light strikes the polarizing crystal normally and so is split into two beams that travel in the principal section: E_p, which is polarized in the perpendicular to the section, and O_p, which is polarized parallel to the section. Both beams are of course white, since double refraction per se has no effect on color.

When the beams enter the quartz plate both are affected, but in complementary ways. Consider first the E_p beam. From it the quartz selects a certain subset of rays to have their polarization turned around 90°. The rays that are affected in this way do not all have the same color, so that the effect of the action is to create two *complementary colored* beams, $(E_p)_g$ and $(E_p)_r$. The polarization of the subset $(E_p)_r$ has not been changed by the quartz, but the subset $(E_p)_g$ is now polarized parallel, instead of perpendicular, to the plane of the figure (that is, to the principal section).

Consider next what happens to the ordinary beam O_p from the polarizer. Here again the quartz selects a certain subset of rays to have their asymmetries turned through 90°. Only in the beam O_p the colors that *are* turned are the ones that were not turned in the action on E_p. Here, then, we also obtain two complementary-colored beams. One, $(O_p)_g$, has the same overall tint as $(E_p)_g$, but it is polarized perpendicular to it.

The other, $(O_p)_r$, has the same tint as $(E_p)_r$, and it is polarized perpendicular to that beam. In the case we are examining here, these four distinct sets of rays enter an analyzing crystal whose principal section lies in the plane of the figure. Consequently the two beams $(E_p)_r$ and $(O_p)_g$ will be refracted into an extraordinary beam in the analyzer, whereas the two beams $(E_p)_r$ and $(O_p)_r$ will be refracted into an ordinary beam. Since in each case the two combining beams have complementary colors, the result will be that there are no colors in the analyzer images.

Though Arago did not write any equations, we can easily represent what will occur when the principal section of the analyzing crystal forms an angle ω with that of the polarizer. In that case the two beams $(O_p)_g$ and $(E_p)_r$ will be polarized at ω to the analyzer's section, whereas the other two beams will be polarized at $90° - \omega$ to it. Applying Malus's law to each of these beams, we obtain the following expressions for the intensities of the extraordinary (E_{ana} and ordinary (O_{ana}) beams in the analyzer:

$$O_{ana} = [(O_p)_g \cos^2\omega + (O_p)_r \sin^2\omega] + [(E_p)_r \cos^2\omega + (E_p)_g \sin^2\omega]$$

$$E_{ana} = [(O_p)_g \sin^2\omega + (O_p)_r \cos^2\omega] + [(E_p)_r \sin^2\omega + (E_p)_g \cos^2\omega]$$

If ω is any multiple of 45°, then O_{ana} and E_{ana} will clearly be white, but otherwise they must necessarily have complementary colors—and the colors they do have depend only on the angle ω between the principal sections of the crystals. This indeed does capture the gross features of the phenomenon Arago described. It is a simple application of selectionist principles.

However, Arago did not test a quite obvious implication of this theory. If a *single* beam of polarized monochromatic light passes through quartz, then the light that emerges from the quartz should, according to Arago's theory, contain two complementary-colored subsets, one polarized in the original plane and the other polarized in the perpendicular plane. In terms of the equations we just used, suppose that the polarized beam in question is simply O_p—we use only the ordinary beam from the polarizing crystal. Then our equations become:

$$O_{ana} = [(O_p)_g \cos^2\omega + (O_p)_r \sin^2\omega]$$

$$E_{ana} = [(O_p)_g \sin^2\omega + (O_p)_r \cos^2\omega]$$

But in fact the quartz merely rotates the original plane of polarization to some new angle, say ω', so that we may write:

$$O_{ana} = O_p \cos^2\omega'$$

$$E_{ana} = E_p \sin^2\omega'$$

One might think that this discrepancy would at once be fatal to Arago's theory. But it need not be fatal, just as the fact that a polarized beam remains polarized when it is reflected and refracted is not fatal to Malus's theory. If the actual alteration in the plane of polarization of the original beam is small, then the effect can be interpreted as being due to the division of the original beam into two parts. One part contains the vast majority of rays and retains the original polarization. The other, much less intense, part has a different plane of polarization.

In terms of our equations, if $(O_p)_r$ is very small then we can imitate what occurs reasonably well *if* the change in the plane of polarization is in fact small. Arago would, had he done the experiment, have simply begun with the quartz plate absent and the analyzer set to produce only an ordinary ray within itself from the polarized beam. With the quartz plate inserted, there will now be an extraordinary beam in the analyzer owing to $(O_p)_r$, albeit one that must be very weak and not too different in its chromatic composition from its complementary, $(O_p)_g$—since one will in fact observe a very weak extraordinary beam with *no* sensible coloration in these circumstances. Arago would then turn the analyzer slightly to *minimize* the extraordinary beam within it—we would now say to remove it altogether, but Arago would, like Malus in dealing with reflection, have said only to *nearly* remove it, since it cannot be done away with completely on his theory.

APPENDIX 11 Optical Rotation

Biot read the fifth part of his memoir on 31 May. In it he described a new discovery for which he at once developed a selectionist account. He decided to examine quartz crystals, as Arago had already done. Recall that Arago had explained the behavior of quartz (as far as he had looked into it) on the assumption that quartz selectively flips polarization from one of the orthogonal planes to the other—quartz, that is, introduces no new directions of polarization. If a polarized beam entered quartz, the emergent beam would consist of two subsets, one polarized as before, the other polarized in the opposite direction, and the subsets would have different ray ratios. This, we also saw, could not possibly withstand empirical examination, because it leads to a highly inaccurate estimate of the intensity in the E beam of an analyzer when using monochromatic light.

Biot attacked Arago's theory, but he did not himself use a monochromatic beam (Biot 1812, 244–46). Rather, an aspect of the phenomenon in white light demonstrated to Biot that Arago's theory could not be correct. Begin with light striking a mirror at the polarizing angle and then passing into an analyzer set to extinguish the E beam to which the reflection may give rise. Interpose a plate of quartz between the mirror and the analyzer, and a blue-violet E beam appears, while the O beam remains sensibly white. If, Biot noted, this were taking place as in mica or gypsum (or as Arago supposed), then the blue-violet E beam must be formed from the light the quartz had deviated out of the original beam and into the only other (by hypothesis) possible direction of polarization ($2i$ in Biot's theory for gypsum or mica, 90° in Arago's for quartz). The remaining rays would, as usual, all be untouched. Then if we rotate the analyzer to diminish the blue-violet E beam, the E image would necessarily pick up rays from the unaffected beam, since these rays are precisely the ones we do not pick up in the analyzer's original setting. Consequently the E beam's tint would have gradually to approach white as more and more of the unaffected rays are picked up. But, Biot continues, this is "not at all" what happens. Rather, though the E beam does change, it does not grow white but instead diminishes in intensity until it "nearly" vanishes. Further, the angle at which this occurs is very nearly proportional to the thickness of the plate.

In terms of rays this could mean only one thing if we are limited to the usual effects. In Biot's words: "All the axes of the luminous molecules are turned into the azimuth to which this minimum corresponds in each plate" (Biot 1812, 247). That is, all the rays in the original beam must necessarily have been rotated to this azimuth—there is no other way to achieve a minimum near zero. But this cannot be the case because, were it so, then rotation of the analyzer could produce no chromatic difference at all, which it most certainly does. "Consequently," Biot argued:

> one is forced to admit that the action of the rhomboid [analyzer] on the luminous rays so modified is not the same as on luminous rays polarized by reflection; and one cannot explain this dissimilarity by supposing that the luminous molecules of diverse nature that have traversed the plate of [quartz] have, by the action of this plate, turned their axes of polarization in diverse azimuths. (Biot 1812, 242)

Not only do Biot's theory for gypsum and mica and Arago's for quartz fail, but it seems that no distribution whatever of polarizations can alone explain the observation. This forced Biot to a new assumption.

The color of the E beam in the analyzer at its minimum is now called the "tint of passage" (Preston 1901, 429–30) and is always gray-violet. We attribute the minimum to the fact that the luminous intensity in the spectral extremes is small, coupled to the theory—introduced, we shall see in a moment, by Biot himself—that the planes of polarization of the spectral components are dispersed, the rotation of the planes increasing from the red to the blue. Biot, in his analysis, neglected to consider that the ray count (intensity) might vary markedly between the center and the ends of the spectrum (Frankel 1972 also discusses this point).

This phenomenon had to be incorporated into Biot's account. He supposed, in essence, that passage through quartz along the optic axis alters a ray's physical condition in a way that makes it less susceptible, within a certain range, to deviation by the extraordinary refracting "force." Further, the range of this effect depends upon a second action that increases from the red to the blue end of the spectrum. To wit, that the plane of polarization deviates continuously by an amount proportional to the thickness of crystal traversed. (This last is all that remains today of Biot's theory.)

Consider what happens when the analyzer's principal section is set parallel to the plane of polarization of a white beam before it passes through the quartz. In figure A11.1, CZ represents the original polarization, and CV represents the polarization of the most deviated (violet) rays. Between CZ and CV—throughout angle α—the rays of the spectrum are dispersed in their planes of polarization.

According to Biot, no ray can be extraordinarily refracted unless its plane of polarization lies at α or greater to the principal section. Consequently, as the analyzer rotates from CZ to CV the E beam continuously diminishes, until it seems to vanish entirely at CV. But, just past CV, the angle α_r between the deviated red rays and the principal section begins to surpass α, so that they become subject to the usual Malus law. The E beam becomes red with intensity nearly proportional to $\sin^2\alpha$—and α is not small, reaching, for example, 80° in a crystal 3.478 mm thick, so that the red rays appear strikingly as well as suddenly. Then, successively, the remaining colors appear

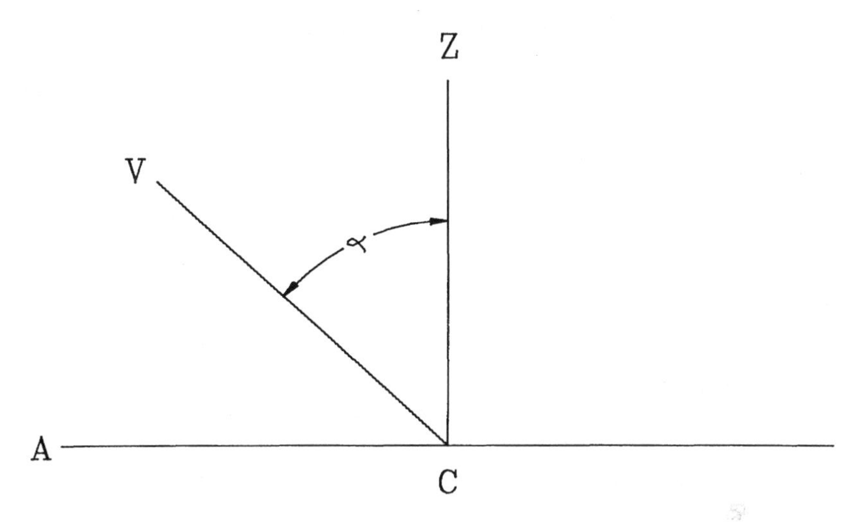

FIG. A11.1 Biot's angle of dispersal of ray asymmetries in quartz.

in the E beam as their rays disappear from the O beam. Finally, when the analyzer reaches CA, the E beam is nearly white, while the O beam must be blue-violet. If, however, the quartz plate is so thick that the dispersion is extremely large, then no minimum point can occur. The light that emerges from the quartz will then be effectively depolarized, and both analyzer images will be white and independent of azimuth.

APPENDIX 12 Biot's Theory for Crossed Laminae

We consider with Biot two laminae with crossed axes. In figure A12.1, CZ is the direction of polarization for the beam incident on the first lamina, whose axis CA lies at angle i to CZ, and the light is normally incident. On emergence from the first lamina the beam contains two subsets, one polarized along CZ, the other along CR. The optic axis CB of the second lamina makes a 90° angle with CA, and so forms $90° - i$ with the direction CR of polarization of one of the two subsets that, having emerged from the first lamina, are incident upon it.

Consider what happens to the bundle that is polarized along CR when it passes through the second lamina. A fraction, say n_2, of the rays in this bundle remain polarized along CR. Another fraction, $1 - n_2$, becomes polarized at $2(90° - i)$ to CR, that is along CZ′—in the same line as the incident polarization CZ, and so indistinguishable in polarization from it. Similarly, in the case of CZ, we find some rays polarized after passage along CZ, and some along CR′, which is the same line as CR.

Now suppose the second lamina is precisely as thick as the first. Consider the rays that, after passage through the first, remain polarized along CZ. Since the laminae are equally thick, these same rays will emerge from the second lamina still polarized along CZ. But consider the rays that have been affected by their passage through the first lamina, the ones that are now polarized along CR. They must also be affected by the second lamina. It will rotate their plane of polarization to CZ′, which is the same line of polarization as CZ. Hence the beam of light that emerges from the second lamina will not produced *colored* iηages in an analyzer, since it is entirely polarized along CZ. It follows at once that any difference in thickness between the laminae will act precisely like an independent, thin lamina. In other words, Biot's ray theory not only is consistent with, it actually *implies* the correct empirical behavior of crossed laminae.

The only issue this raises is its apparent conflict with our first difficulty with Biot's theory: to wit, the first problem seems to require that "partial" polarization cease when the thickness becomes large enough, but the second problem evidently requires that it go on at even very large thicknesses. But there is really no new problem here, because the issue reduces to Biot's questionable answer to the first difficulty. Take two fairly thick laminae that differ by a small amount e. According to Biot, light passing through any lamina, however thick, is always affected by the law of arithmetic alternation, but in some sense as the distance grows a point is reached where the rays begin

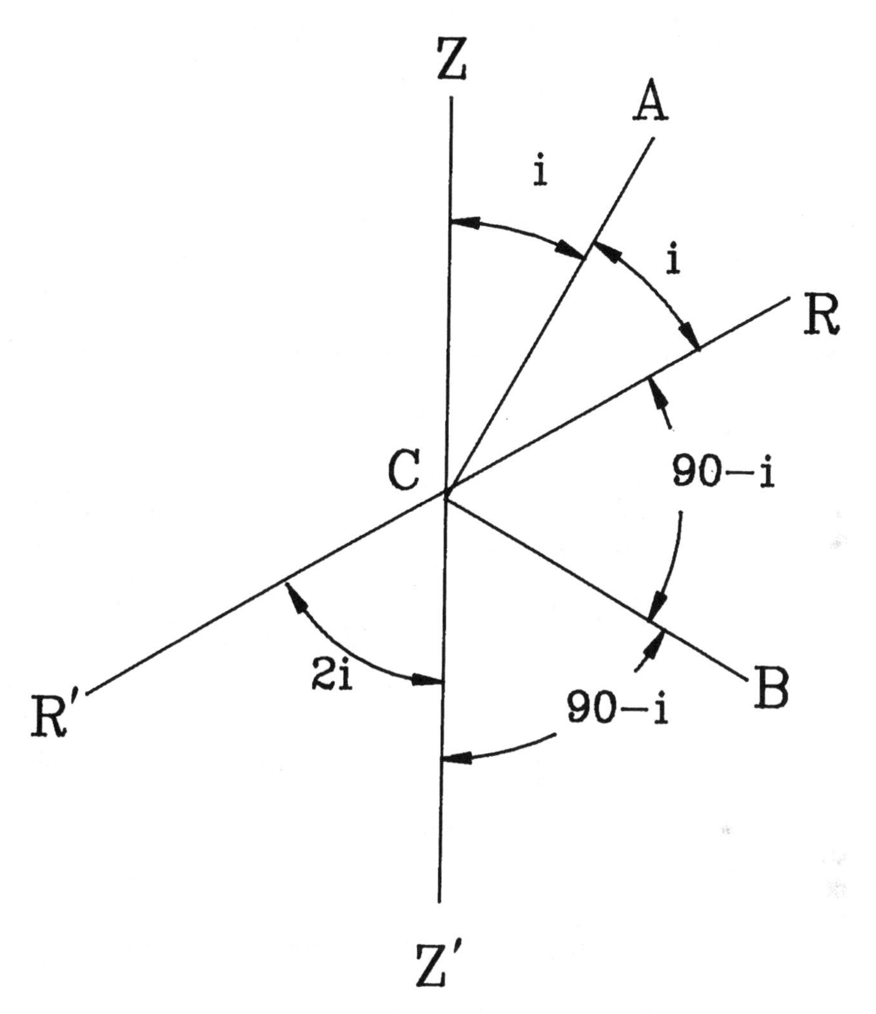

FIG. A12.1 Biot's theory for laminae with crossed axes.

to "mix." Consequently, crossed thick laminae must produce the same effect as *e* alone, since *whatever* happens by a given thickness in the first must be reversed by passage through the same thickness in the second, which will return the light to its state on entry into the first lamina.

APPENDIX 13 The Mirror Experiment

Consider (fig. A13.1) two mirrors, AC and CB, which meet at C, forming there a very obtuse angle $\angle ACB$. Place a source S above them. S will be imaged by I_2 in CB and by I_1 in AC. These images constitute virtual sources, and interference results from the rays emitted by the trio S, I_1, I_2—and with no material edges in the immediate vicinity.

To analyze the situation place a screen GH parallel to the line I_1I_2 joining the virtual sources, and draw the normal ECF to GH and I_1I_2 and passing through C. By construction $\angle I_1SI_2$ is equal to $\pi - \angle ACB$ (since SI_1, SI_2 are respectively normal to CB, AC). Also, by construction CS, CI_1, CI_2 are equal to one another (see fig. A13.2.).[11] That

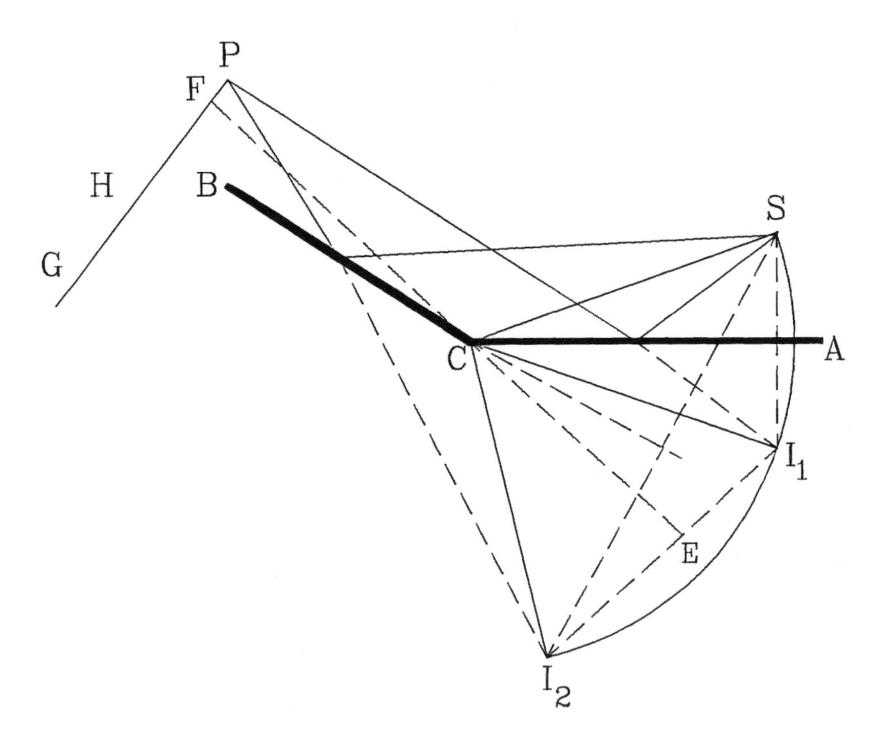

FIG. A13.1 Fresnel's mirror experiment.

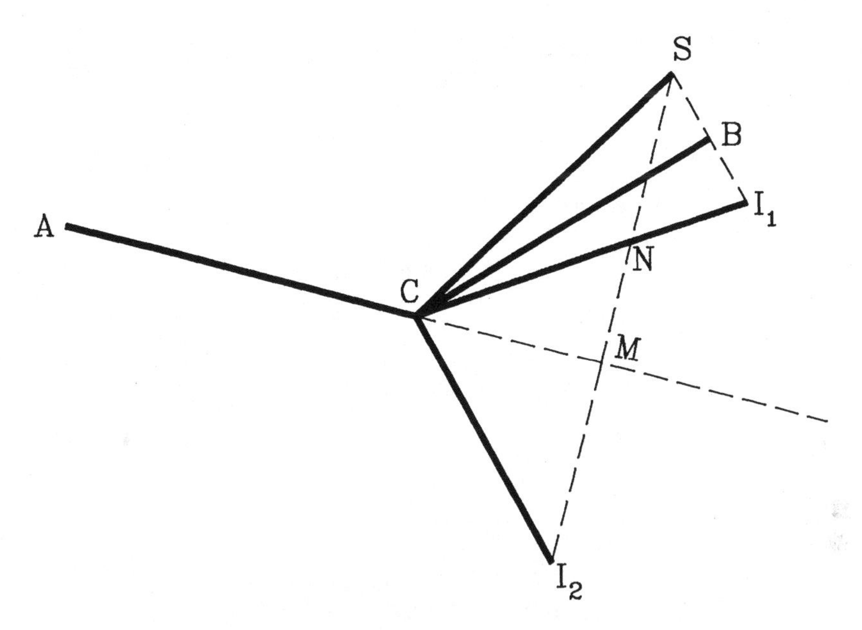

FIG. A13.2 Fresnel's mirror experiment.

is, the three sources lie on a circle centered on the intersection C of the mirrors. This implies that $\angle I_1SI_2$ is equal to $\frac{1}{2}\angle I_1CI_2$.[12] Since ΔI_1I_2 is isosceles and ECF is normal to I_1I_2, then E bisects I_1I_2, whence I_1I_2 is 2CE tan $\angle I_1SI_2$, which is nearly 2CE sin $\angle I_1SI_2$, since $\angle$ACB is very large.

With this (which neither Fresnel nor Arago explicitly gave)[13] one can compute the loci of complete interference. Here—ignoring for the moment the direct ray from the source S—we have two point sources I_1 and I_2. The screen distance is EF. This is precisely the same situation, optically, that we had for the internal fringes within the narrow obstacle, so we may at once apply Fresnel's formula for this case. If y is the distance of a point P on the screen from F, where the path distances I_2F, I_1F are equal, then defining δ as $I_2P - I_1P$, Fresnel's internal formula yields for the dark fringes:

$$y = \frac{EF\delta}{2I_1I_2}; \quad \delta = n\lambda \tag{A13.1}$$

The only restriction is that I_1I_2 must be much smaller than EF, which holds for large mirror inclinations. Substituting for I_1I_2 and $\angle I_1SI_2$:

$$y = \frac{EF}{CE}\frac{1}{4}n\lambda \sin \angle ACB \tag{A13.2}$$

This formula is empirically accurate to several hundredths of a millimeter, as Fresnel found (Fresnel 1816b, 152–53; Fresnel actually used A13.1, not A13.2, because he directly observed $\angle I_1FI_2$).

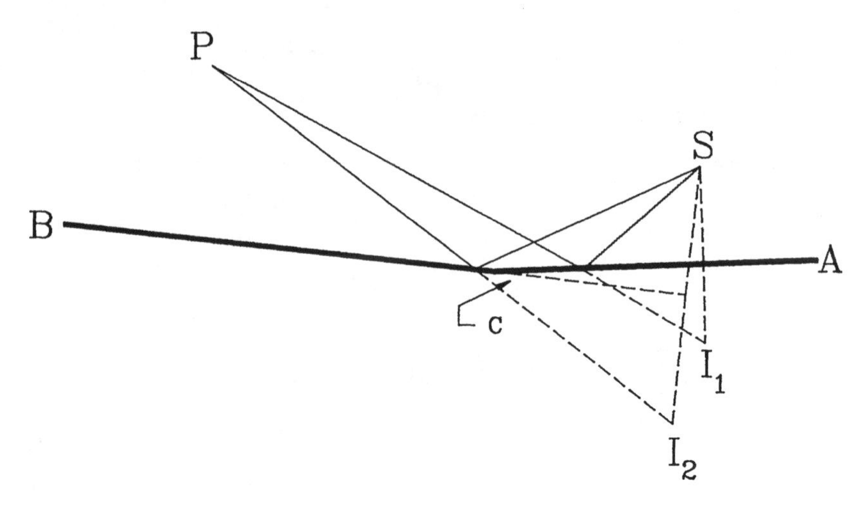

FIG. A13.3 Limiting case of Fresnel's mirror experiment.

If one markedly increases ∠ACB—approaching the case of a plane mirror—then, as Fresnel remarked, the source S itself becomes directly involved in forming the fringe pattern, since the virtual sources are now so closely spaced that the pattern they produce is extremely wide. Fresnel was well aware of this situation; he regarded it as one in which the "internal" fringes due to the virtual sources become mixed with "external" fringes due to interference between the direct ray from S and the rays from its images. Thus consider what occurs when ∠ACB is nearly 180° (fig. A13.3). Here I_1I_2 is so small that the image points alone produce a pattern so large in breadth as to be essentially unobservable. However, considering I_1I_2 to form in effect a single image—the limit case of the plane mirror—at point P we have interference between a direct ray from S and a ray from its image. This situation is closely analogous to the case of points outside the geometric shadow of the narrow diffractor and beyond the first external maximum. Consider S to be the *edge* of the diffractor with the actual source at the image I. Then we have, in essence, a screen normal to the line IS joining "source" with "edge," the "shadow" being IS. If we assume that S and I emit coherently but with a 180° phase difference, then we retrieve Fresnel's "external" formula for locating extrema.

APPENDIX 14 Poisson's Spot

Poisson, Fresnel wrote, had communicated to him the following "singular theorem" drawn from the integral formulas: "That the center of the shadow of a circular screen must be as illuminated as if the screen did not exist, at least when the rays penetrate there at incidences that are not very oblique" (Fresnel 1819b, 365–72). Poisson had notified Fresnel of this result—which Arago verified—shortly before the committee met in judgment, and just after the prize was awarded he also told Fresnel that the integrals can be easily found for circular apertures. Fresnel produced two analyses for the latter case that were added as "Note 1" to the M.C. One, which applies also to the circular screen, depends upon yet another type of zone computation, while the other, which does not apply to the screen, uses an integral formulation. I shall briefly examine both of them, because the integral theory caps Fresnel's work on diffraction, while the zone computation interestingly refers to distance an effect that is later attributed to the inclination factor.

For the zone computation Fresnel first limited the case to points P that lie along the central axis of the circular opening at sufficiently great distances from it that the obliquities of all the rays that reach these axial points can be ignored. To construct zones from P (fig. A14.1), Fresnel drew a series of cones whose apexes are at P and whose bases are on the circular aperture itself—on, that is, the wave front as it passes through. The cones are so drawn that the succeeding lengths PA_1, PA_2, . . . differ from one another by a half-wavelength. Now, Fresnel continued, "it is evident" that all of the zones A_1A_2 that are intercepted on the front by these lines "have the same area." Since the effect of a zone, according to Fresnel, depends solely on its area, we can at once merely count them up to find their net result at the field point. If P is so situated that it sees an even number of zones, then it is at a minimum; if it sees an odd number of zones, then P is at a maximum.

Fresnel's assertion that the half-wavelength zones have the same area is strictly incorrect: their areas vary as $\frac{1}{2}\lambda PA_1$, or very nearly as $\frac{1}{2}\lambda PI \cos \vartheta_1$, where ϑ_1 is the inclination of PA_1 to PI. However, under Fresnel's limitation to small obliquities this last is $\frac{1}{2}PI$, so his assumption is approximately correct—and has the advantage of permitting him to carry out a zone count. Nevertheless, Fresnel was well aware that the zonal areas constructed in this way do in fact depend upon the distance. He made use of the fact to conclude that, when P is at a maximum, the amplitude is twice what it would be if the slit were infinite in extent.[14]

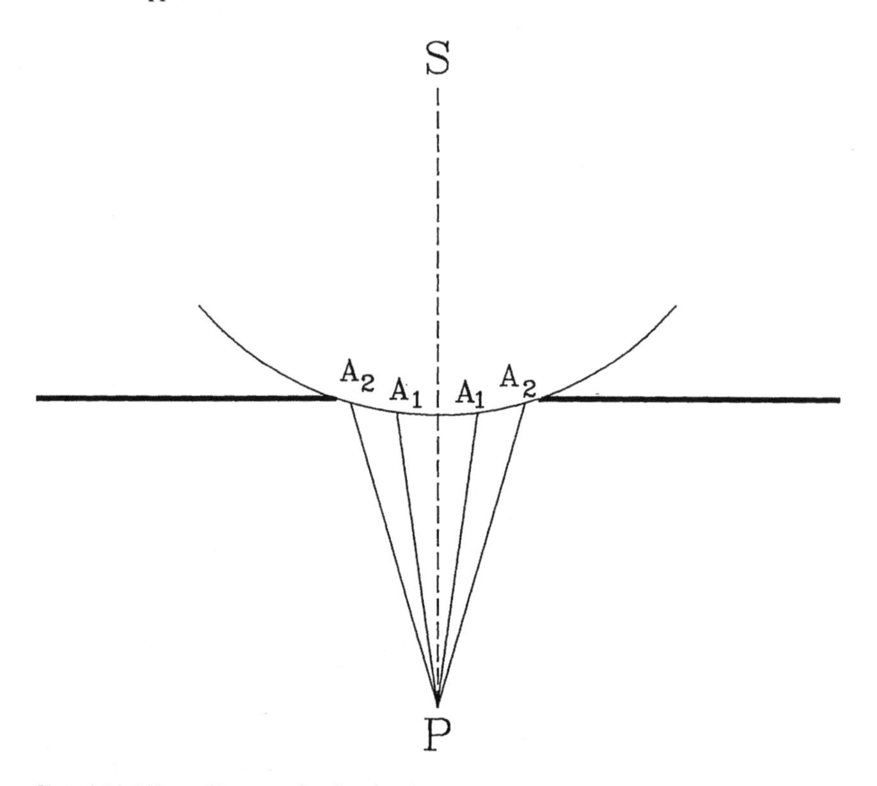

FIG. A14.1 Fresnel's zones for the circular aperture.

Fresnel had in a certain sense reached this same result two years before when he asserted, in his theory of the efficacious ray, that the effect of a given zone is only half destroyed by each of its neighbors acting individually. (For then it is obvious that, in an unobstructed wave, only the halves of the first and last zones are not destroyed, since only they have but one neighbor. Whence, if we ignore with Fresnel the last zone, there remains only half the first.) But in 1816 Fresnel's reasoning on this score was essentially dynamic—the factor of $\frac{1}{2}$ arose because contiguous zones "enfeeble" one another. By contrast, Fresnel now (1896) ignores this dynamic reciprocity between zones and bases his argument on the fact that the zonal areas diminish with obliquity, so that a given zone has slightly less of an effect than its predecessor. Although, Fresnel argues, the areas are approximately equal, there is a first-order difference between the areas of contiguous zones, so that it is more rigorous to say that the rays from a given zone are destroyed by the *mean* of the rays from its contiguous neighbors—this is also an approximation, but it is clearly one of second order.

Fresnel at once points out that this more rigorous method of computing zonal effects (more rigorous, that is, than simply counting them in pairs) nevertheless gives the same result for the circular opening as the pair count, since only the extreme and first zones have but a single contiguous neighbor: if the number of zones is even, then these two extreme zones will interfere destructively with one another; if the number of zones is odd, they will interfere constructively. Consequently in constructive interference the resultant must have twice the amplitude, or four times the intensity, as it does when the wave front is unobstructed. The factor of four in the intensity is therefore a direct result of Fresnel's averaging process, which he based on the decrease in area of the zones, *not,* as we would today, on the inclination factor—which he had to base on the zone areas because he did not assume that the amplitude of a secondary wavelet falls off reciprocally with distance.

Turning now to the circular screen—which had been Poisson's first concern—Fresnel could immediately demonstrate why any point on the axis and behind the screen is illuminated as though the screen were not there. Counting zones as before, only half the effect of the first zone immediately surrounding the screen, and the last zone, whose effect can be ignored, remain. If the screen is not too large, then the area of the effective zone is the same as that of the polar zone for the unobstructed front. Again Fresnel's method works perfectly, because he appropriates as an advantage—the inequality of zonal areas—what at first thought prevents estimating the effects of zonal

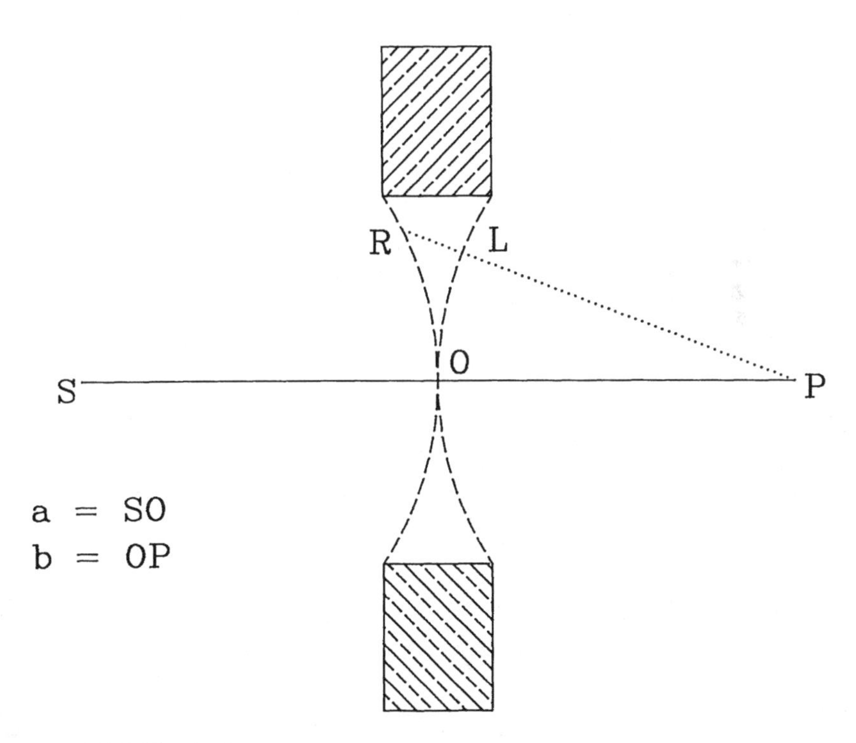

FIG. A14.2 The circular aperture.

superposition. The method works precisely because Fresnel ignores any distance dependence of the secondary amplitudes.

I come finally to Fresnel's integral analysis for the circular opening. For an axial source the theory is simple. Fresnel approximates by using a plane front and considers the path difference RL between OP and RP (see fig. A14.2). If z represents OR, then RL is nearly $z^2[(a + b)/2ab]$ as usual, and by symmetry this same difference obtains throughout the annulus of area $2\pi z\, dz$ centered on O. The integrals therefore become, for an opening with a radius r:

$$\left[\int_0^r \cos\left(\pi z^2 \frac{a + b}{ab\lambda}\right) 2\pi z\, dz\right]^2 + \left[\int_0^r \sin\left(\pi z^2 \frac{a + b}{ab\lambda}\right) 2\pi z\, dz\right]^2$$

These are easily integrable, yielding:

$$2\left[\frac{ab\lambda}{(a + b)^2}\right]^2 \left[1 - \cos\left(\frac{\pi(a + b)^2 r^2}{ab\lambda}\right)\right]$$

Whence extrema occur at:

maxima $\quad b = \dfrac{ar^2}{(2n + 1)a\lambda - r^2} \quad n = 0,1,2, \ldots$

minima $\quad b = \dfrac{ar^2}{2na\lambda - r^2} \quad n = 1,2, \ldots$

As b varies a succession of tints will appear in white light, as Fresnel discussed in some detail.

APPENDIX 15 The Fresnel Rhomb

Fresnel took polarized light and set the initial angle of polarization at 45°—where the "perturbing action [of reflection] is greatest"—and turned his attention to the effect of reflection at the *lower* (internal) surface of a glass plate on the polarization of the light. He found that when the light strikes the upper surface of the plate near grazing incidence, then the light that reflects within the plate from its lower surface will be polarized opposite to the light that reflected from the upper surface (see fig. A15.1). Thus, using a plate of glass as in figure A15.1, the incident light, which is polarized at 45°, reflects from the upper surface of the plate at point C. The reflected light, CR, retains nearly the original plane of polarization. Inside the plate the refracted light CP strikes the lower surface at point P, where from it is generated an internally reflected beam PQ and a beam PT that leaves the plate. PQ in its turn reflects and refracts at point Q, the beam QS emerging from the plate's upper surface. Fresnel discovered that QS is *perpendicularly polarized* to IC and CR. And so, since the polarization of IC was 45°, the polarization of QS arises from the flipping of IC's polarization across the plane of incidence.

This was simple to understand. If the glass has a fairly high index of refraction, then the internal beam CP will strike its lower surface at a low angle of incidence. If we assume that CP retains nearly the original polarization, then the polarization of PQ

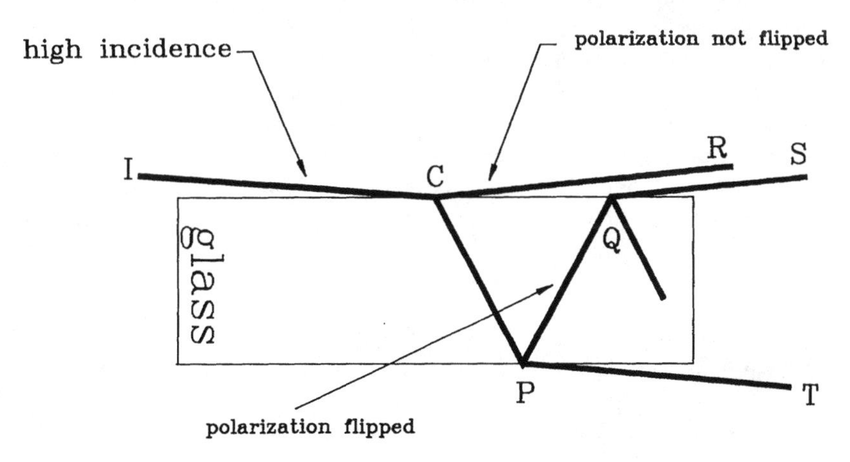

FIG. A15.1 Polarization and internal reflection.

will lie on the other side of the plane of incidence from the polarization of IC—so that, since QS presumably retains the polarization of PQ, then it must be perpendicularly polarized to IC and CR. If, however, the plane of polarization of IC is not 45°, then QS will not be perpendicularly polarized to it.

Given these results, Fresnel realized that he could use the internal reflection to produce interference colors that alter in unusual ways. On a metal mirror, place a transparent film thin enough to produce the usual interference fringes. Illuminate it near grazing incidence with polarized light. If the plane of polarization of IC is *not* 45°, then the polarizations of beams CR and QS will not be mutually perpendicular. Consequently they can be made to interfere with one another. But if the polarization of IC is 45°, then the interference should very nearly disappear, because CR and QS will be polarized at right angles to one another. Using turpentine to produce the film, Fresnel found that this does indeed take place.

Fresnel tried several variations of this experiment, now generating, now obliterating interference by manipulating the initial plane of polarization. He then decided to use a glass prism, or rhomb, to see what would happen if the light within the rhomb underwent *total,* and not just partial, reflection. And here he found, to say the least, unusual effects. Always using an incident polarization of 45°, he at once discovered that a single internal reflection in these circumstances apparently "partially depolarizes" the light. When it is analyzed by a crystal, one cannot turn the crystal to any orientation such that there is within it only a single ordinary or extraordinary beam. With the crystal set to produce a maximum ordinary beam, it still produces an extraordinary beam, albeit one whose intensity is at a minimum. Under these conditions the crystal's principal section is aligned in just the same way that it must be to produce within it just an ordinary beam from the internally reflected light immediately before the internal reflection becomes total (Fresnel 1817b, 452–53).[15] (See fig. A15.2.)

Fresnel soon found other peculiar facts. A second reflection within the rhomb at the same angle as the first but in a perpendicular plane to it restores the original polarization. By contrast, a second internal reflection within the rhomb but in the same plane as the first one makes the depolarization complete: the images in the analyzer are always equally intense no matter how the analyzer is turned. A third reflection in the same plane restores the partial depolarization; a fourth restores complete polarization, but in a plane at right angles to the initial polarization. Fresnel was particularly surprised that a beam that had been completely depolarized by two successive internal reflections in the same plane could nevertheless still be used to develop colors by passing it through a crystal lamina and then analyzing it—because unless it is first passed through a lamina it behaves in an analyzing crystal precisely like natural, unpolarized light. It shows no signs at all of polarization. The colors it produces, Fresnel observed, differ from what would be produced using the original beam, which had been polarized at 45° to the plane of incidence. In these experiments the optic axis of the lamina lay in the plane of reflection. Fresnel described his observations in the following words:

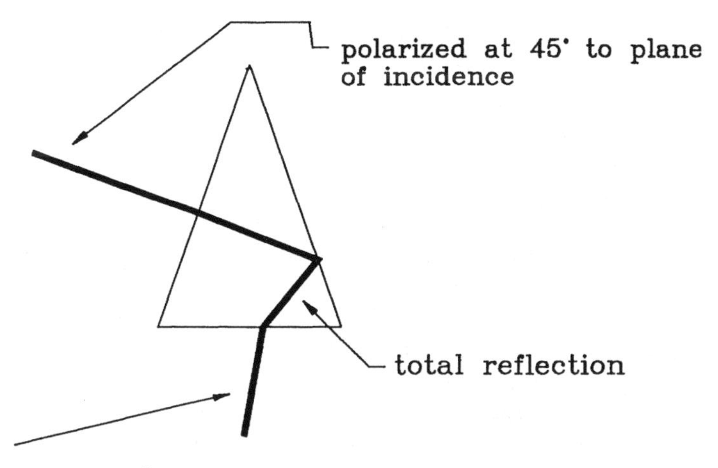

FIG. A15.2 Fresnel's rhomb.

> It's noteworthy that the tints produced by light modified in this way are intermediary and exactly the mean between the complementary colors developed by ordinary polarized light. By a mean tint between two others I intend that which corresponds to the middle of the arc that separates them on the circular figure Newton used to represent the return upon themselves of the colors of the spectrum. (Fresnel 1817b, 456)

He investigated how the tints change when the optic axis of the lamina lies in or perpendicular to the plane of reflection and what happens when, fixing the lamina, the analyzer rotates or, fixing the analyzer, the lamina rotates.

The nature of the tints in these circumstances could be explained using the principle of interference if, Fresnel reasoned, the ordinary and extraordinary beams that are produced within the lamina from this unusual kind of light have lost a path difference of a quarter-wavelength compared with what they would have had if the original polarized light had been used instead:

> The tints of the reflected images are means between those of the corresponding direct images and their complementaries, and such that, by analogy, would result from a change of a quarter-undulation in the interval that separates the two systems of luminous waves that concur in their production. Furthermore, the colors of the reflected images are higher in the order of [Newton's] rings than those of the direct images, when the axis of the lamina is parallel to the plane of incidence, as I have supposed, which indicates that the interval between the two systems of waves has diminished by a quarter-wavelength. (Fresnel 1817b, 466)

Fresnel's reasoning here is entirely straightforward. If the colors produced on analysis of the light after it has passed through the lamina were, for example, complementary to those produced by ordinary polarized light, then there would have to be a half-wavelength change in the path difference between the ordinary and extraordinary beams within the lamina. To change the O tint to its complementary (see fig. A15.3), F_{eo} and F_{oo} must have an additional half-wavelength path difference, and similarly for F_{ee} and F_{oe} in E. This requires that, in the lamina, the beams L_e and L_o must have had this difference as well, since it exists in both O and E beams within the analyzer (so that the analyzer proper could not have produced it). However, in fact the tints observed are not complementary to the usual ones; they are "means" between the usual ones and their complementaries.

Here Fresnel was thinking of the colors that correspond to the different gaps in Newton's scale. A half-wavelength addition to the gap shifts a tint on the scale to its complementary color. Fresnel had observed a tint that seemed to him to lie on the scale halfway between a tint and its complementary. That is, it occurs in Newton's rings where the gap width lies halfway between the value it has for a given tint and the value it has when that tint has been changed to its complementary. Furthermore,

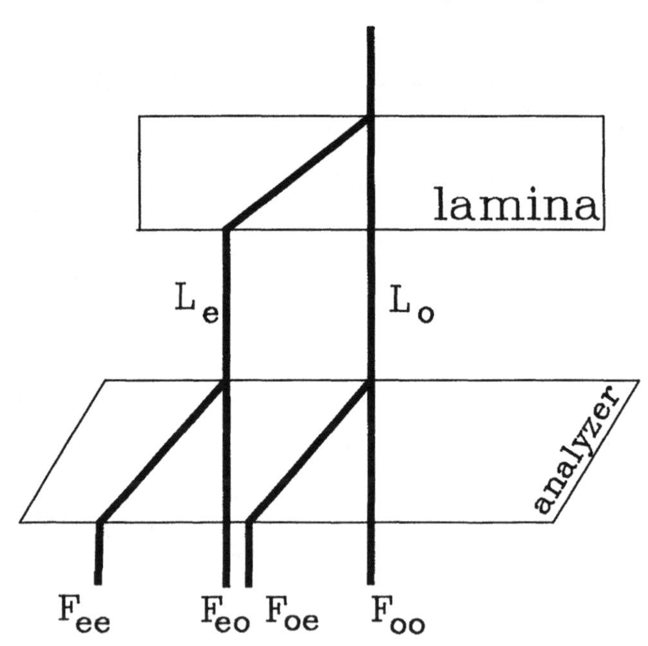

FIG. A15.3

the tint he found seemed to lie *above* the usual one on the scale, so that a quarter-wavelength *decrease* in the path difference was called for. And so, since two successive reflections within the rhomb are necessary to produce this quarter-wave difference, one reflection should, presumably, produce an eighth-wave difference. That is, for some reason light that has been totally reflected within the rhomb divides when doubly refracted into two beams that *start out with* an eighth of a wavelength path difference between them. Obviously this property must have been added to the light by the total reflection itself.

But what then is the path difference produced between when the light is reflected? After all, there seems to be only a *single* internally reflected beam of light. Fresnel's answer to this question illustrates both his inventiveness in the face of a puzzle and how very far he still was from a thoroughly new understanding of polarization. In effect, Fresnel thought there had to be two *distinct* waves in the reflected light, just as there apparently are in doubly refracting crystals. These waves are polarized at right angles to one another, and they must differ in phase by an amount that would be produced had they traveled paths that differ by an eighth of a wavelength. In the noncrystalline glass these two waves must travel in the same direction. But on entering the crystalline lamina, one of the two waves will be refracted ordinarily, while the other must be refracted extraordinarily. Consequently, when the light has been totally reflected two times, thereby producing a quarter-wave difference between the perpendicularly polarized waves that compose it, one will obtain the colors Fresnel had observed, since the ordinary and the extraordinary beams within the lamina will also have this difference:

> One may therefore admit, as a general principle, that whenever a polarized ray is reflected in the interior of a prism, at an incidence giving total reflection, and that is sufficiently far both from partial reflection and from parallelism to the surface, it divides into two others, one of which is polarized parallel to, and the other perpendicular to, the plane of incidence, the *first having been reflected somewhat closer to the surface,* so that on leaving the prism it is behind by an eighth of an undulation. (Fresnel 1817b, 466–67; emphasis added)

We must take Fresnel's words here quite literally. He does not assert that the totally reflected light may be *decomposed,* in the projective sense, into two others that are polarized at right angles to one another. Not at all. Rather, from a given polarized wave the total reflection creates *two new and physically distinct* (though coherent) waves. One of these is polarized in the plane of incidence, whereas the other is polarized in the perpendicular plane. These two waves have been produced by reflections that take place *at different distances* from the interface, which explains their phase difference: it is due literally to a path difference. In Fresnel's words, the "depolarization" produced by the internal reflection "causes to be born two waves polarized in opposite senses" (Fresnel 1817b, 470n).

The best way to see how thoroughly Fresnel distinguished these two waves from one another is to follow his diagrammatic description of what happens when one analyzes the waves that are produced by a single total reflection within the rhomb. In figure A15.4, RR′ is the plane of incidence, and PP′ is the plane of incident polarization, which must always be at 45° to the plane of incidence. After a single internal reflection the wave splits (*se partagera*) into two others: one is polarized along RR′, while the other is polarized along SS′. One is an eighth-wavelength in advance of the other, but they have the same intensities. Take an analyzing crystal and set it with its principal section either in the plane of incidence (RR′) or else in its normal (SS′). In either position it will produce two equally intense images, since all of the RR′ and all of the SS′ polarized waves go into either the ordinary or the extraordinary refractions.

Suppose instead that we set the principal section along PP′, the original plane of

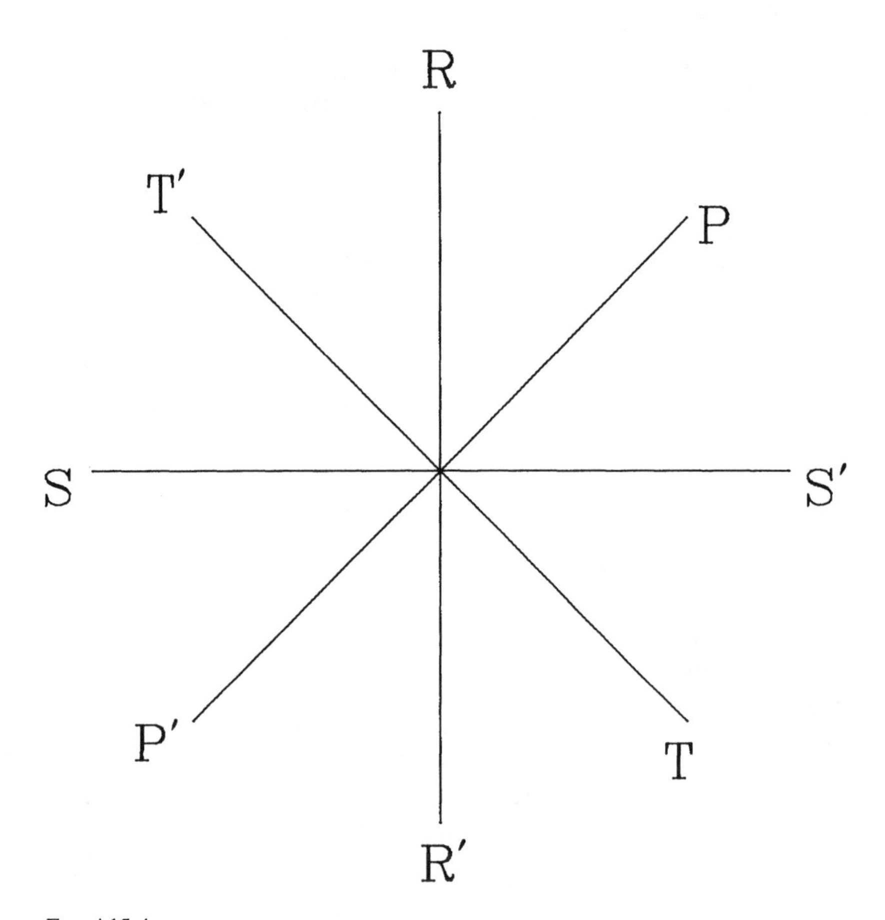

FIG. A15.4

polarization. Then the ordinary and the extraordinary refractions in the analyzer will each contain equal amounts of light from the two waves. However, the two waves that conspire to form the ordinary image are brought to the same plane of polarization after having first been deviated from, and then returned to, PP′. The two waves that form the extraordinary image have been twice deviated in the same direction from PP′. Consequently, in addition to the eighth-wavelength path difference between the two waves in each image, there must, in the extraordinary image, be an additional half-wave difference according to Fresnel's general requirement. And so, Fresnel concluded, the extraordinary image must be less intense than the ordinary image because there will be more interference between the waves that form it than there will between the waves that form the extraordinary image—though neither image will ever vanish since the path difference is either $\frac{1}{8}\lambda$ or $\frac{5}{8}\lambda$.

APPENDIX 16 Fresnel's First Analysis of Chromatic Polarization

Fresnel applied his theory to the following questions:

1. Under what conditions on α and i are both analyzer images white?
2. For α equal to 0° (meaning that the principal section of the analyzer lies in the original plane of polarization), for what values of i will the image be "maximally" colored?
3. For α equal to 45°, when are the images white, and when are they "maximally" colored?

Biot had considered these questions as well, but he had in addition been able to solve problems concerning image intensity. It is instructive to compare Fresnel's discussion with Biot's, since he himself no doubt did so. That way we can see the senses in which the theories, or rather the expressions to which the theories lead, are at all comparable.

First recall Biot's formulas for the images in the analyzer:

ordinary	$U \cos^2\alpha + A \cos^2(2i - \alpha)$
extraordinary	$U \sin^2\alpha + A \sin^2(2i - \alpha)$

From these, for Fresnel's case 1 Biot had found that i may be equal to an integral multiple of 90° added to α. Fresnel obtained the same result by a simple argument. For the two images to be white, one of the two beams that form each of them must vanish to preclude interference. The conditions on α and i then follow at once from Fresnel's expressions without any consideration of the actual intensities. Biot instead had had to require that the factors that multiply U and A in his expressions must be equal to one another in order to produce two beams that have the correct number and ratio of rays to compound to white. That is, where Fresnel in effect required the *absence* of a beam to produce white, Biot required the presence of a *precisely* calculated one.

Fresnel's case 2 had been a stimulus to Biot's discovery of his equations because his first expressions could not, he felt, accommodate his observations. Biot had found that the images are "maximally" colored, in this case, when i is equal to 45° plus any integral number of right angles. His final expressions accommodate this fact by means of Newton's color circle, for under these conditions the positions of the colors of the two images in the circle are as far from one another as they can be.

Fresnel's argument was quite different from Biot's. According to Fresnel's formulas, each of the two images (the ordinary and the extraordinary) is produced by the interference of two beams: F_{oo} with F_{eo} produces the ordinary image, and F_{oe} with F_{ee} produces the extraordinary image. Each image, Fresnel assumed, will be "maximally" colored if the two beams that form it are equally intense—since then the effect of their interference will be most pronounced. Consequently, Fresnel's condition for "maximal" coloration when α is equal to 0° (case 2) requires that $\sin^4 i$ must be equal to $\cos^4 i$, and so i must be equal to 45° plus any integral number of right angles—which is precisely what Biot had observed. Following a similar line of argument, Fresnel also reached the correct observational results for his case 3.

Nevertheless Fresnel's argument seems rather weak in comparison with Biot's. Unlike Biot, Fresnel could not claim at this point actually to calculate what colors will be produced under given conditions, because he could not compound waves. One might therefore put the operational difference between Fresnel's and Biot's theories in the following way. For Fresnel the equality of the factors in his expressions produces the maximum possible coloration, though he could not say what color is produced. For Biot the equality of his factors produces no coloration at all, but when the factors are not equal to one another then Biot could calculate the resulting color.

Fresnel's theory was also weak in another respect. I remarked in chapter 7 that the differences between the phases of the two beams that interfere to form each image should, it seems, be the same—that F_{oo} should differ in phase from F_{eo} by the same amount that F_{oe} differs from F_{ee}—but that this cannot be so, since the images are complementary in color. Fresnel felt that this fact could not be "completely" explained until we have "discovered the causes of double refraction and polarization." Nevertheless, he remarked that the images *must* be complementary in color as a result of the conservation of vis viva. (Fresnel in fact used the word "energy" here.) He wrote: "If one species of rays is weakened in one of the images by the discordance of their vibrations, [then], in order for the sum total of their undulatory movements to remain constant, the intensity of these same rays must receive an equal increase in the second image, which will in consequence be complementary to the first" (Fresnel 1816c, 401. Such an increase obviously requires constructive interference where, in the other image, destructive interference took place. And so it follows at once from the conservation of energy as Fresnel understood it (an understanding we shall examine in more detail when we consider his theories of reflection) that there must be an additional half-wavelength path difference between the constituent beams in one of the analyzer images.

But in which one? That Fresnel could not answer, because he had no theory of polarization. However, he could provide an empirical reply to the question by interpreting Biot's observations. The only case for which Biot had examined colors was for α equal to 0°—where, that is, the principal section of the analyzer contains the primitive plane of polarization. However, Biot actually had not used an analyzer at all to examine chromatic effects—he had used a second polarizing reflection at 90° to the first one, which immediately gave the tint of the *A* constituent in his formulas. And as it varied with the thickness, *A* ran through the reflection tints in Newton's scale. This experiment, in which one uses a second polarizing reflection, corresponds to setting α equal to 0° when, instead, one uses an analyzing crystal: for in the latter case Biot's formulas imply that the extraordinary image in the crystal will consist solely of the *A* beam. With α equal to 0°, then, one may say that the extraordinary image should correspond to Newton's reflection rings as the thickness of the lamina changes.

If, therefore, the extraordinary image corresponds to Newton's scale of *reflection* tints, then by energy conservation the ordinary image must correspond to Newton's scale of transmission tints. Fresnel had earlier shown that to explain Newton's reflection rings one must add a half-wavelength path difference. By analogy, for α zero one must add a half-wavelength to the actual path difference between the constituents of the extraordinary beam. If one changes α by 90°, then by the same reasoning this addition must be made to the path difference between the constituents of the ordinary beam.

Fresnel went a bit further. What, he apparently asked himself, distinguishes the case of α equal to 0° from that of α equal to 90°, where one must switch the additional half-wavelength from the extraordinary image to the ordinary one? In the first case, he noted, α is smaller than i, whereas in the other it is larger. That is, in the first case the analyzer's principal section lies between the original plane of polarization and the optic axis of the lamina; in the second case the lamina's axis lies between the analyzer's principal section and the plane of polarization. Fresnel then assumed—quite without justification in Biot's observations, since Biot never altered α at all—that *in general* one must add a half-wavelength to the path difference between the constituents of the extraordinary beam when α is less than i, and to the difference for the ordinary beam when α is greater than i.

Fresnel insisted on generalizing from Biot's observations, because he was reaching for some sort of understanding of how polarization operates in conjunction with the principle of interference. Accordingly he did not stop with this claim, but noted that it is equivalent to another one, which he expressed in the following words: "The image whose tint corresponds exactly to the thickness of the crystal lamina is that whose two constituent beams have each undergone two opposite motions in their plane of polarization; while in the two beams that produce the complementary tint, the plane of polarization, on the contrary, always deviates in the same direction from its original position" (Fresnel 1816c, 402). The extra phase difference, then, arises whenever the planes of polarization of the constituent beams twice increase the angle between them—and so come to lie again along the same line, after having deviated from one another through 180°. Fresnel later remarked, we have already seen, that he and Ampère had early realized that this fact could be understood if the waves were entirely transverse, since then an angle of 180° between the planes of polarization equates to direct opposition between the directions of oscillation.

APPENDIX 17 Decomposing Reflected and Refracted Light

17.1 Energy and Momentum Conservation for Waves

We have already seen how Fresnel used energy principles to obtain results that could not be found directly. His first demonstration that the images produced by chromatic polarization must have complementary colors was based on energy conservation, and he often recurred to the principle when he could not draw a result directly from calculation.[16] And his first attempt to derive reflection formulas, in mid-1819, employed energy conservation as a check against results obtained from another conservation principle—that of momentum.

In the *Oeuvres* (1:649–53) Verdet reproduces a note, itself undated, contained in a folder marked 12 July 1819. Given its contents, the note was almost certainly written about this time. In it Fresnel attempts to derive reflection formulas by deploying momentum conservation and then checking the results against conservation of energy. Throughout the analysis he assumes that the polarized wave consists entirely of a *longitudinal* oscillation—which therefore seems to show that he had not by this date developed the idea that even unpolarized light can be completely asymmetric (see fig. A17.1).

In Fresnel's figure GA and HC are two rays from the incident wave front, which we assume to be plane. AM and CN are their respective reflections and AK, CL the refractions. Fresnel's goal was to obtain expressions for the reflected (v) and refracted (u) amplitudes in terms of the incident amplitude (V), the angle of incidence (i), and the index of refraction (n). Obviously, to do so two equations are necessary, and Fresnel generated them by adapting the requirement of momentum conservation to waves.

The problem was how to choose the "masses" to use in computing the momenta. Fresnel assumed that the momentum a given mass in the incident beam has at a given time will be transmitted in a time δt, during which the incident front traverses this mass, to the masses in the reflected and refracted beams, which are themselves traversed by the reflected and refracted fronts during δt. In the figure, during the time that point E moves to C on the interface, the volume AEC is traversed by the front AE. During this same time the volumes ADC (equal to AEC) in the reflected, and ABC, in the refracted, beams are traversed by the fronts DC and BC.

If the masses of AEC and ADC are m and the mass of ABC is μ, then momentum conservation (or what Fresnel termed the "conservation of the motion of the center of

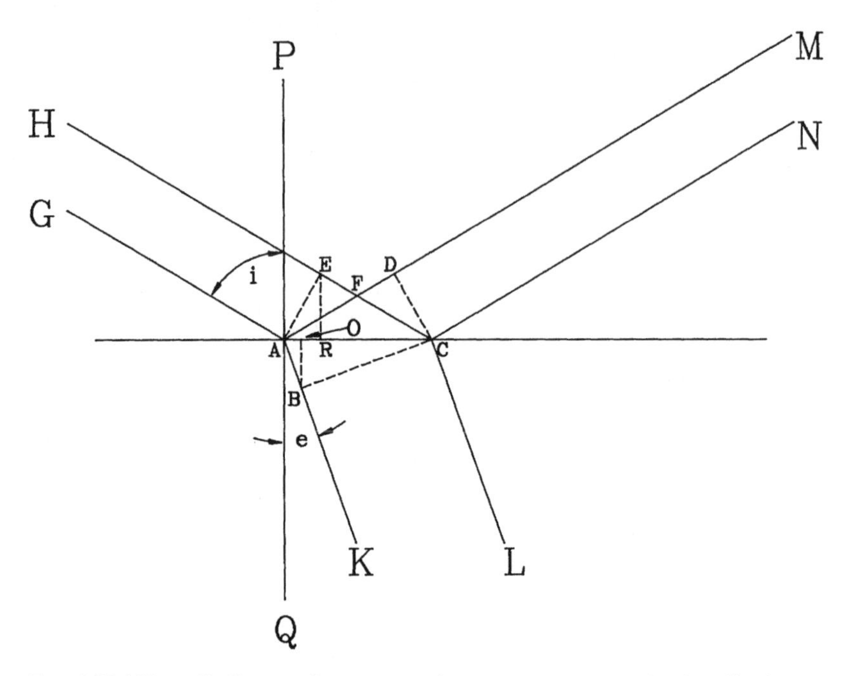

FIG. A17.1 Fresnel's diagram for energy and momentum conservation in reflection.

gravity") yields two equations, one for the vertical, the other for the horizontal component of the momentum:

$$Vm \sin(i) = u\mu \sin(e) + vm \sin(i)$$

$$Vm \cos(i) = u\mu \cos(e) - vm \cos(i)$$

The angle of incidence, i, is given, and the angle e of refraction follows from Snel's law. Consequently only the masses m and μ must be eliminated. Fresnel assumed that the density of the refracting medium is proportional to the square of its index of refraction.[17] Then geometry yields $n \cos(e)/\cos(i)$ for the mass ratio μ/m. Substituting in the momentum equations and solving yields in the end:

$$v = V\frac{\sin(i - e)}{\sin(i + e)}$$

$$u = V\frac{\cos(i)}{r \cos(e)}\frac{\sin(2i)}{\sin(i + e)}$$

Note that these expressions represent the amplitudes of oscillations that are normal to the front.

Fresnel next checked his results against energy conservation. As he understood it the latter required the energy (vis viva) in AEC to equal the sum of the energies in ADC and ABC, or:

$$mV^2 = mv^2 + \mu u^2$$

Combining our three equations with Snel's law yields in the end:

$$\cos^2(e) = \cos^2(i)$$

which, unless the index of refraction is unity, can only be satisfied when i is equal to e, that is, when the angle of incidence vanishes. Consequently Fresnel's equations are consistent with energy conservation only at zero incidence.

Either, Fresnel concluded, Snel's law itself does not hold generally or else the waves may "also" contain transverse oscillations, in which case the entire analysis must be redone:

> With the help of transverse motions one could satisfy both the general principle of the conservation of the center of gravity and that of the conservation of living force, which must be satisfied in all the vibrations of elastic fluids, and one might be able, in thus determining the transverse motions of waves, to define that singular modification of light to which we have given the name *polarization*. (Fresnel, *Oeuvres*, 1:652)

That is, by working backward from momentum and energy conservation one might, Fresnel seems to reason, be able closely to specify the oscillatory structure of the wave. That had become a moot point by 1821, but Fresnel preserved the essential features of this early derivation of the sine law ratio in his first analysis for polarized light.

17.2 Decomposition

In his new analysis, which was based directly on the requirement that the oscillation must occur entirely in the wave front, Fresnel continued (in effect) to rely on momentum conservation, but he had also to use energy principles because the oscillation is no longer parallel to the ray. If the oscillation is along the ray, then momentum conservation, as we have seen, provides the two necessary equations. But now the oscillation must be normal to the ray. Suppose it to occur entirely in a plane parallel to the reflecting interface. Then momentum conservation will not apparently involve the angles i and e at all, because here we have no component in the plane of incidence itself.

Fresnel had accordingly to modify his previous analysis. In 1821 he provided the first of two ways to do so, this one being (in his eyes and in those of later wave theorists) the less satisfying of the two.[18] The essence of both procedures requires decomposing the incident oscillation into two components. One lies in the plane of incidence; the other is perpendicular to it. At first Fresnel was able to solve only the case in which the oscillation occurs in the normal to the plane of incidence, whereupon it is parallel to the interface. He again used momentum conservation, but he had to alter his previous approach, which had applied to longitudinal oscillations.

In this case of an oscillation parallel to the interface we have, Fresnel assumed, to deal with the elastic impact between two bodies along the line joining them. The effect of this assumption is to combine energy with momentum conservation (though Fresnel would not have thought about it in quite that way). In the previous case of an oscillation along the ray we had to consider independently the two components of the impact, since the direction of the oscillation alters as a result of the interaction. Here the direction of the oscillation remains the same; only the direction of propagation changes. Suppose that the masses and velocities are represented as above: that is, we treat the portions of the incident, reflected, and refracted beams that Fresnel had previously delineated as colliding "masses." Then the usual formulas for an elastic collision require that, after the impact is complete, the velocity (amplitude) ratios must be:

$$\frac{v}{V} = \frac{m - \mu}{m + \mu} \quad \text{and} \quad \frac{u}{V} - \frac{2m}{m + \mu}$$

Since the mass ratio remains $n(\cos(e)/\cos(i))$ we have for the optical intensity ratio (the square of the amplitude ratio):

Fresnel's Sine Law
oscillation normal to the plane of incidence

$$\frac{\text{reflected intensity}}{\text{incident intensity}} = \left[\frac{\tan(i) - \tan(e)}{\tan(i) + \tan(e)}\right]^2 = \left[\frac{\sin(i - e)}{\sin(i + e)}\right]^2$$

This is precisely the same as the ratio Fresnel had previously found for longitudinal oscillations.

To this point Fresnel could not compare his formula with experiment because he was not as yet able to deduce the corresponding formula for an oscillation that occurs in the plane of incidence. He remarked that, from two measurements Arago had made of the deviation of the plane of polarization by reflection, one could use the new formula to deduce for both instances the amplitude of the plane-parallel component of the reflection, and thence the total intensity of the reflected light. However, the only photometric measures available were old ones generated by Pierre Bouguer, and they were too inaccurate for quantitative comparisons.

Within a few days at most after finishing the "Calcul des teintes" paper, in which he detailed his new hypothesis that light is always asymmetric, Fresnel remarked in a postscript that he had "by a mechanical solution" obtained the formula for an oscillation that takes place in the plane of incidence, finding it to be:

Fresnel's Tangent Law
oscillation in the plane of incidence

$$\frac{\text{reflected intensity}}{\text{incident intensity}} = \left[\frac{\sin(2i) - \sin(2e)}{\sin(2i) + \sin(2e)}\right]^2$$

(Fresnel later reduced the ratio to the form $[\tan(i - e)/\tan(i + e)]^2$. Note that in this analysis he did not specify the sign of the amplitude ratio itself, since he gave only its square.)

The first formula, for light oscillating parallel to the interface, must hold for polarization in the plane of incidence, because the reflected intensity cannot vanish at any angle (excepting the trivial case of normal incidence). Consequently, on the basis of this analysis the oscillation takes place along the normal to the plane of polarization. The second ratio, for light oscillating in the plane of incidence, can vanish. From the equivalent tangent formula we at once see that this will occur when the angles of incidence and refraction are complementary (thereby satisfying Brewster's law that the light will be completely polarized when the tangent of the incidence is equal to the index of refraction—a point Fresnel noted in his table comparing theory with experiment).

Fresnel did not describe his "mechanical solution" here, but it was no doubt the same as the one he provided a year and half later in his last, and most detailed, analysis of reflection (Fresnel 1823a). The new method applies as well to the sine law as to the tangent law and makes no use at all of elastic collisions. Instead it combines energy conservation *with a constraint on the continuity of the medium at the interface* and continues to assume, as before, that the density ratio varies with the square of the refractive index. In other words, Fresnel for the first time used a true boundary condition. Previously he had not actually considered what happens to the oscillation at the interface; he had assumed only that, whatever happens, energy and momentum are conserved. To do this he had used Snel's law to determine the cross sections of the incident, reflected, and refracted beams. He had then, in effect, pressed the energy and momentum ahead of a front in the incident beam into the reflected and refracted beams. He continues to deploy this method here, but he now combines it with a constraint on the actual oscillation at the boundary.

Thus Fresnel first used energy conservation together with his expression for the density ratio to obtain an equation that connects the amplitudes with the angles of incidence and refraction. The analysis is the same as his earlier one for the case of a longitudinal oscillation, since the only things to be determined are the corresponding volumes in the reflected and the refracted fronts that, in a given time δt, receive the energy from a given volume traversed during δt by the incident front. Consequently the energy equation is, for unit incident amplitude:

Fresnel's Energy Equation

$$\sin(e)\sin(i)(1 - v^2) = \sin(i)\cos(e)u^2$$

For a boundary condition Fresnel assumed that the components of the oscillation that are *parallel* to the interface must be continuous across it. That is, the medium does not support slip. Since we take the incident amplitude as one, this gives us, for the case of an oscillation that is itself parallel to the interface:

$$1 + v = u$$

Combined with the energy equation, this condition then yields the sine law by easy manipulation, including now the sign of the amplitude ratio proper:

$$v = -\frac{\sin(i - e)}{\sin(i + e)}$$

The boundary condition also suffices to solve the case for an oscillation that is parallel to the plane of incidence, for here we need only take the oscillation's component parallel to the interface, which gives as boundary condition:

$$\cos(i) + v\cos(i) = u\cos(e)$$

With the energy equation we obtain the tangent law in the form:

$$v = -\frac{\sin(i)\cos(i) - \sin(e)\cos(e)}{\sin(i)\cos(i) + \sin(e)\cos(e)}$$

Many readers will at once recognize these two laws as the very ones that are today obtained from electromagnetic principles (a deduction first explicitly carried out by H. A. Lorentz in 1875). One might be surprised by this, since the principles involved seem so vastly different. However, the deductions are rather closely related. Fresnel's energy equation expresses what, in electromagnetic theory, one deduces from the Poynting equation for the flux of energy in the field. And Fresnel's boundary condition corresponds to the electromagnetic requirement that the tangential component of the electric field must be continuous at the boundary. These two equations—Poynting's and the continuity of the tangential electric field—suffice to deduce the Fresnel amplitude ratios.

These two laws were extremely important historically because, as we shall see, anyone who accepted them both had also accepted, and understood, the deepest recesses of the wave theory. Indeed, for many years they remained *individually* quite difficult for many of Fresnel's contemporaries to grasp. They could not in any case be tested in themselves, since accurate photometric techniques were not available, so that until the late 1820s even those (like Arago himself) who were otherwise sympathetic to the wave theory tended to ignore them. What could be examined, and what could be understood by most everyone, however, was the angle of deviation for the incident direction of polarization that the two expressions together entail.

Suppose the incident wave is polarized at an angle α to the plane of incidence—which means, on Fresnel's analysis, that the oscillation occurs at an angle $90° - \alpha$ to it. Assume that the amplitude of the incident wave is unity, which therefore has components $\sin(\alpha)$, $\cos(\alpha)$, respectively, in and perpendicular to the plane of incidence. The tangent of the angle of polarization, α_{refl}, of the reflected wave will therefore be:

$$\tan(\alpha_{refl}) = \cot(\alpha)\frac{\cos(i - e)}{\cos(i + e)}$$

This formula could be tested directly (indeed Fresnel had, in effect, already done so for α equal to 45°), and by 1830 it had come into common use even among selectionists like Brewster.[19] But it was often taken—like Fresnel's diffraction formulas by Poisson—as a purely empirical rule without a proper theoretical basis.

17.3 Phase Shifts Where the Equations Fail

Fresnel's mastery over the process of composition and decomposition appears to its best advantage in his new interpretation of total reflection, an intepretation that builds on his previous understanding. If the index n is less than one (the incidence then being internal), then beyond an incidence of arcsine(n), the value of cos(e), and with it the Fresnel ratios, must become complex.[20] This is also the incidence beyond which total internal reflection occurs—all the light is reflected. This seems to have provided Fresnel with a clue for interpreting the meaning of complex amplitude ratios, and his work here was of great importance during the 1830s, when the wave theory was developed in great analytical detail.

The amplitude ratios must have become complex, Fresnel thought, because a condition that had gone into deriving them had been violated. That condition, he assumed, was the assumption that the phase is continuous at the interface (Fresnel 1823h, 783). Suppose instead that when the light strikes its coherence is momentarily interrupted. This will yield a reflection that is shifted in phase from the incident wave. Can we, Fresnel in effect queried, interpret the complex ratios in a way that would be consistent with such a phase shift? And if so, what must the new phase be?

To understand Fresnel's reasoning, first rewrite the ratio for light polarized perpendicular to the plane of incidence in a way that separates the real from the imaginary part:

$$v = -\frac{\sin(i - e)}{\sin(i + e)} = \frac{1 + n'^2 - 2n'^2\sin^2(i)}{n'^2 - 1} - \frac{2n'\cos(i)\sqrt{n'^2\sin^2(i) - 1}}{n'^2 - 1}\sqrt{-1}$$

(defining n' as the reciprocal of the index n from the more to the less refracting medium). Fresnel noticed that the sum of the squares of the real part (R^2) with the real factor (I^2) of the imaginary part is equal to one. That very likely suggested to him the following interpretation.

The light must be totally reflected, and it must undergo a phase shift φ_1. Since the reflected wave has some new phase, it can, by Fresnel's usual quarter-wave decomposition, be separated into two parts that differ in phase from one another by 90°. One of the parts has the same phase as the incident wave, and the amplitudes of the parts must be the cosine and sine of the phase of their resultant. Now here we have two terms, R and I, whose squares sum to one. Suppose that each of these represents the amplitude of one of the parts of the usual quarter-wave decomposition. Then the light will be completely reflected, and it will be shifted in phase by an angle φ whose tangent is R/I if we assume that R is the part with the same phase as the incident wave.[21]

One can do precisely the same thing for the component of the incident light that is polarized in the plane of incidence, computing for it a phase shift φ_2. Then, Fresnel concluded, the reflection will consist of two orthogonal components that differ in phase from one another by the quantity $\varphi_1 - \varphi_2$:

$$\cos(\varphi_1 - \varphi_2) = \frac{2n'^2\sin^4(i) - (n'^2 + 1)\sin^2(i) + 1}{(n'^2 + 1)\sin^2(i) - 1}$$

Furthermore, the amplitudes of the orthogonal components are each equal to one.[22] This result, Fresnel demonstrated in some detail, fully explains the depolarizing behavior that he had observed much earlier in totally reflecting light two times within his rhomb. In terms of his new vocabulary, this light has been *circularly polarized.*

APPENDIX 18 Arago's Critique of Biot on Chromatic Polarization

Set up a mirror experiment, but look only at the images formed by rays that reflect from each mirror near polarizing incidence. Then the fringe pattern is created by light that is polarized in the plane of incidence, as attested by an analyzer. Now, Arago continues, take two gypsum laminae and place each in the path of one of the two beams that interfere to produce the fringe pattern. Orient one of the laminae so that its axis lies at 45° to the plane of incidence; orient the other lamina's axis at right angles to this (see figs. A18.1 and A18.2).

If Fresnel's theory is correct, each thin lamina will behave precisely like a thick crystal and so will generate from the polarized beam that strikes it two others: one polarized for each lamina in its principal section, the other polarized in the section's normal. By construction the ordinary beam from the first lamina will have the same polarization as the extraordinary beam from the second lamina, and vice versa. Thus, in figure A18.2 the ordinary beam KMF from lamina A will interfere with the extraordinary beam NOF from lamina B, while the extraordinary beam KLF from A will interfere with the ordinary beam NPF from B. But the path difference between KLF and NPF is different from the path difference between KMF and NOF, so that the fringe patterns will not be coincident. There must accordingly be two patterns, one of which should be polarized at 45° to the plane of incidence, whereas the other should be polarized in the plane normal to the first. This, Arago remarks, is "perfectly confirmed" by experiment.

But if so, then Biot's theory must be wrong. For according to Biot, the light that emerges from both laminae is polarized in part in, and in part along the normal to, the plane of incidence (since here $2i$ is 90°). Consequently the fringes should be polarized in these two planes. Since they are not, Biot is wrong.

The gist of Arago's criticism is that one cannot possibly assume, as Biot did, that each of the two beams consists of light polarized in and perpendicular to the plane of incidence—because the fringe pattern shows completely different polarizations than this would require. This is indeed a powerful criticism, and Biot never answered it (but see my remarks in sec. 9.1).

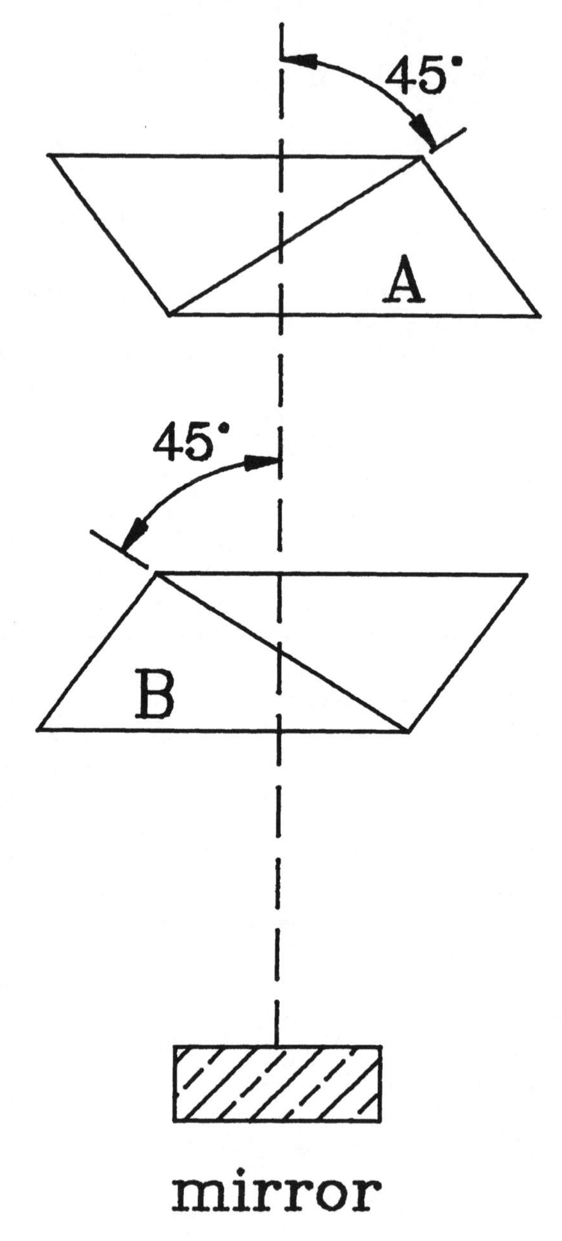

FIG. A18.1 Arago's experiment to critique Biot's theory; view in the plane of incidence.

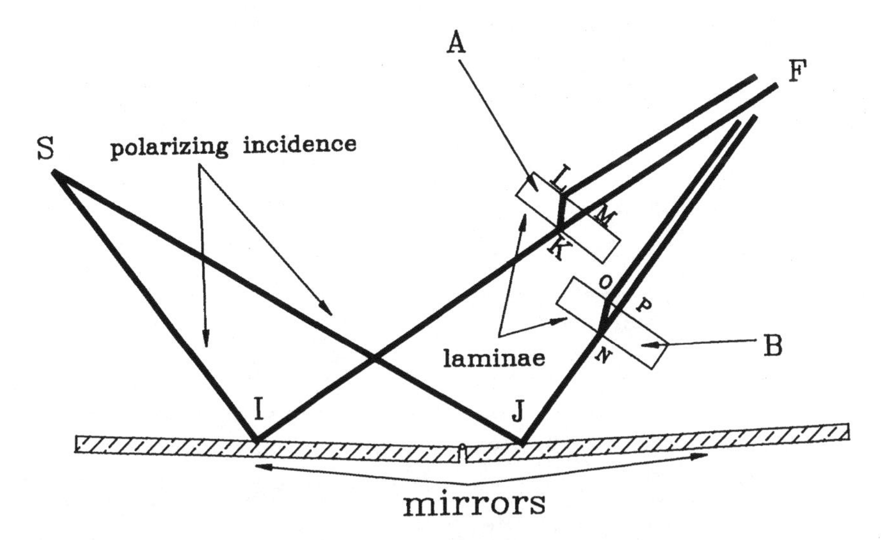

FIG. A18.2 Arago's experiment to critique Biot's theory; view at right angles to the plane of incidence.

APPENDIX 19 Fresnel Mistakes Biot's Meaning

We find two examples of how Fresnel did on occasion mistake Biot's meaning in the unpublished Fresnel (1818e). In the first, Fresnel's concentration upon the "luminous molecule" leads him to assume that Biot *could not possibly* have allowed for situations in which the light in chromatic polarization is not polarized entirely in one of the allowable azimuths. Fresnel remarked:

> The theory of interferences shows that there cannot be unique and absolute polarization in the $2i$ azimuth or the zero azimuth, for a given type of rays, except when the difference of the paths traversed is equal to zero, $n\lambda$, or $[n + \frac{1}{2}\lambda]$. In the other cases there is only partial polarization, and even the total absence of apparent polarization, when the difference of paths traversed is equal to $[n + \frac{1}{4}\lambda]$. Biot has on the contrary supposed, in his mobile polarization, that each type of luminous molecule is entirely polarized in the plane toward which it was carried by its last oscillation, whether or not it is completed at the moment it leaves the crystal. This is not a simple difference of theoretical opinion, but a question of fact on which experiment can pronounce, in making it with homogeneous light. (Fresnel 1818e, 531)

Biot, as we have seen (sec. 9.3), had never made any such absolute requirement, though Fresnel insists he must have done so. This was because Fresnel, unlike Biot himself, insisted that the *physics* of Biot's theory had to be identical to its *quantitative structure*. Fresnel would not allow Biot to make any suppositions that could not be directly supported by arguments based on the behavior of light particles. Yet Biot had never himself insisted on so tight a connection, in major part because he, like all selectionists, concentrated mainly on the behavior of the beams that are formed by groups of rays, asking what happens to the dispositions of the rays in the beam. He was not directly concerned in his *quantitative* work with the behavior of an individual ray, which he felt usually could not be calculated: only the behavior of the groups could be. Certainly, Biot would admit, the individual ray must have its asymmetry along one of only two azimuths, but whether the beam contains rays with only one kind of polarization or contains both kinds is an open question (rather, it is open to the requirements of other facts, such as Biot's analogy to Newton's rings).

Fresnel, on the other hand, did not think at all of groups of rays, even when he was attempting to criticize the emission theory. He concentrated instead on what mobile polarization apparently requires of the *individual light particle*. Here, for example, Fresnel tacitly assumed that what holds for a single particle must also hold for the beam as a whole. Biot insisted that one could not a priori make any such assumption because the essence of selectionism is to *discover* how the rays are distributed over the allowable sets in given circumstances. Fresnel was so far removed from thinking of rays as basic physical objects that he refused to grant the essential terms of selectionist analysis—despite Biot's repeated insistence on using them.

A second example follows immediately in this same note. Recall that Biot had quite easily shown that, when the axes in two superposed laminae are crossed at right angles, the effect in the second lamina reverses the effect in the first: it is entirely a matter of the preservation of the individual identity of the ray. Fresnel, however, remarked of this:

> Biot's theory seems little satisfying, especially in the case of crossed laminae. He supposes, without showing how this results from his fundamental hypothesis, that the oscillations of luminous molecules in the second lamina must be subtracted from the oscillations they executed in the first, when the two laminae, having the same kind of double refraction, have their axes at a right angle. (Fresnel 1818e, 531)

Biot made no "supposition" that he had not already made. Fresnel missed the point because he was concentrating on what happens to the individual light particle in its "oscillations" and not—as Biot himself had—on the requirements of ray preservation.

APPENDIX 20 **Biot's Explanation of Brewster's Rings**

In this adaptation of Biot's figure (see fig. A20.1), M_2 is a mirror set to reflect light at polarizing incidence to a plate S of Iceland spar *that is cut with its optic axis normal to its triangular faces*. The light from the center of the mirror strikes S normally. Most of the rays that pass through the plate therefore form only small angles with the axis. M_1 is a second mirror that is set in such a way as to extinguish (or, realistically, nearly extinguish) the rays that are reflected from the first mirror in the absence of the plate S. But with the plate inserted between the two mirrors, light reflects at the second. With the eye at O, fairly close to M_1, a series of colored concentric rings will be visible. The ring pattern is crossed by two mutually perpendicular dark lines, one of which lies in the plane of polarization of the light reflected from M_1.

Biot quite easily explained the overall features of this phenomenon. According to him a ray will be polarized either in its original plane, after passage through S, or else in a plane that lies at an angle $2i$ to the original plane, i being the angle between that plane and a plane formed by the ray within S and the optic axis.[23] The existence of the two dark lines that Brewster had observed follows from this requirement in the following way. Rays that pass in a plane that is parallel to the original plane of polarization

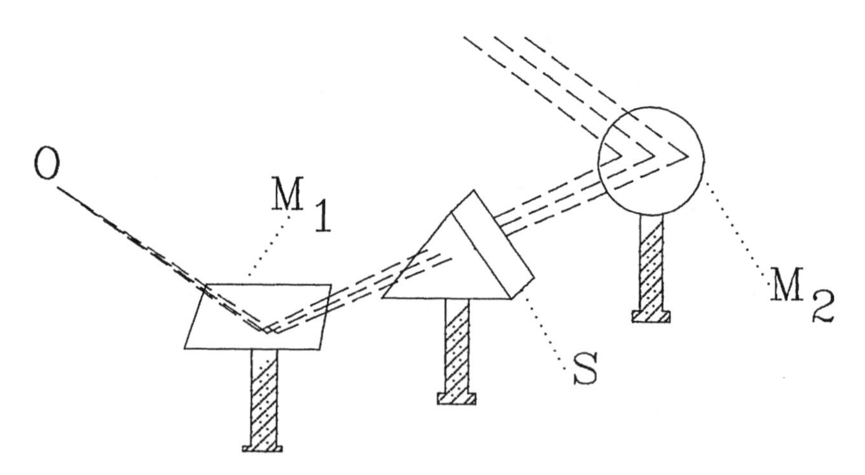

FIG. A20.1 Biot's experiment for Brewster's rings.

will not have their polarization changed, since for these rays the two possible azimuths are coincident with one another. Similarly, rays that pass through in a plane normal to this one will also not be altered. Hence both sets of rays will be extinguished by mirror M_1, producing Brewster's dark lines. A beam that does not pass through in either of these two planes will undergo depolarization because, for it, the plane at $2i$ differs from its original plane of polarization. The rays in such a beam will therefore divide into sets O and E, and a portion of the E rays will be reflected by M_2, since they no longer lie in the original plane of polarization.

To explain the form of the pattern, consider that every ray that lies on a given circle drawn on the incident face of S and with the optic axis of S as its center forms precisely the same angle with respect to the axis—and so travels the same distance through the lamina—as every other such ray. Or better put, every *beam* originating from some point on the given circle has the same relation to the optic axis of S as does every other such beam. Consequently, whatever chromatic division affects the rays from one beam on the circle must similarly affect the rays in the other beams, which means that, along this circle, the same division into O and E subsets occurs, and hence the circle has the same tint throughout when observed in the mirror M_1. Of course, the angle i differs from beam to beam, but that angle determines only the total number of rays in the E beam that are reflected by M_1, not the ray ratio in E. Consequently the intensity must vary from point to point throughout a given ring, but the ring's tint must remain constant—which is essentially what Brewster had observed.

APPENDIX 21 **Brewster Links Birefringence with Polarization**

Brewster sought to discover whether two depolarizing axes entail two axes of double refraction. This presented acute empirical difficulties:

> The want of a transparent mineral with two powerful axes, which, like calcareous spar, could be obtained in large pieces and cut with facility in every direction; and the necessity that it should have its resultant axes considerably inclined to each other, in order to obtain a measurable separation of the images at several points, between these resultant axes, rendered all my experiments for a long time completely unsuccessful. The discovery, however, of crystals which possessed, in some degree, the most important of these requisites, has enabled me to resume and to complete the investigation. (Brewster 1818, 267–68)

Brewster succeeded in observing a double refraction that clearly involved both depolarizing axes in that the colored images were single for rays transmitted along either axis but double elsewhere.

He even essayed a law for the phenomenon, one similar to a law he provided for calculating the tints produced by two depolarizing axes. The latter abstracted from the effects of the path-length differences on the tints, preferring to concentrate instead on what he thought was most important—the relation of the tints to the depolarizing axes. Imagine the crystal cut into a sphere and the polarized beam to pass through the sphere's center. The depolarizing axes intersect in the sphere's center, and the rings formed by the axes' action can be drawn on the sphere's surface. With a single axis the rings are bands bounded by concentric circles with the single axis as pole. They are much more complicated with two axes, but Brewster offered the following law, which he obtained by combining the separate effects of the two axes according to a "parallelogram" law: "The tint produced at any point of the sphere by the joint action of two axes, is equal to the diagonal of a parallelogram, whose sides represent the tints produced by each axis separately, and whose angle is double of the angle formed by the two planes passing through that point of the sphere and the respective axes" (Brewster 1818, 239–40). Brewster tested this construction on gypsum and found it accurate indeed.

Similarly, to find the *refraction law* when there are two axes, he used a "parallelogram" whose sides are the squares of the velocities each axis would produce if it acted alone in accordance with the Malus-Laplace velocity formula required by Huygens's construction in conjunction with the principle of least action: "The increment of the square of the velocity of the extraordinary ray produced by the action of two axes of double refraction, is equal to the diagonal of a parallelogram whose sides are the increments of the square of the velocity produced by each axis separately, and calculated by the law of Huygens, and whose angle is double of the angle formed by the two planes passing through the ray and the respective axes" (Brewster 1818, 270). Brewster provided no data at all to test the refraction law.

APPENDIX 22 Brewster's Theory of Elliptical Polarization

Recall that very little was known about metallic reflection save that it seems to partially depolarize light.[24] Brewster had examined metallic polarization early in February 1815 and had written to Biot of his discovery of "the curious property possessed by silver and gold of dividing a polarized ray into complementary colours by successive reflections." Brewster then obtained a set of excellent metal plates, and experiments with them seemed to indicate that "a single reflexion from a metallic surface produces the same effect upon polarized light as a certain thickness of a crystallized body" (Brewster 1830b, 287–88). He did not, however, pursue the topic further at that time.[25]

Brewster was stimulated to renew his inquiries fifteen years later by Fresnel's discovery of circular polarization, which he almost certainly studied primarily in Herschel's 1827 "Light." Brewster began his new account of the phenomenon with the following two properties that characterize light polarized at 45° to the plane of incidence when it is reflected from a metal (at this angle of polarization the reflection seems most strongly affected by the peculiar properties of the metal):

1. Light reflected once at this angle is not polarized because it always produces two images on analysis, albeit with different intensities that vary with the azimuth of the analyzer.
2. But it is also not common (or even partially polarized) light, because a second reflection at the same angle restores it to the state of "light polarized in one plane."

Brewster soon decided that the metallic reflection cannot be constituted like light which has passed through thin crystals because the analogy of metallic reflection to chromatic polarization is incomplete: two reflections should, he reasoned, correspond to doubling the thickness of the crystal lamina, but in fact the property disappears entirely after two reflections.[26] "Having thus ascertained," he continued,

> that light polarized +45°, and reflected at the maximum polarizing angle of metals, is neither common light nor polarized light, nor light constituted like that which passes through thin crystallized plates, I conceived the idea of its resembling circularly polarized light—that remarkable species of light which comports itself as if it revolved with a circular motion during its transmission through particular media. (Brewster 1830b, 292)

Brewster seems to have understood circular polarization as a common rotation about the beam as axis of the asymmetries of the rays that compose the beam. Fresnel had discovered that this "remarkable" property can be produced by twice totally reflecting light initially polarized in an azimuth of +45° within a glass rhomb, as we have seen in some detail. A second pair of reflections at the same angle, Fresnel had discovered, restores plane polarization, but at an azimuth of −45°—whatever the angle between the first and second sets of reflection planes may be. Suppose, Brewster reasoned, that a single metallic reflection at the proper angle (chosen to produce the maximum effect) corresponds to a *pair* of total reflections within Fresnel's rhomb, with the beam in both cases having a primitive polarization of +45°. Then a second metallic reflection at the same angle should restore plane polarization, as it does in the rhomb—and indeed the plane polarization is restored in this fashion.

However, there is a difference from the circumstances in which the rhomb produces circular polarization, in that the angle of incidence on the metal that is necessary to restore the plane polarization varies with both the incident and the restored azimuths. Or put another way, there are pairs of original and restored azimuths of polarization, as well as a corresponding angle of restoring incidence, such that the two azimuths are functionally related in some way.[27] In the circular polarization produced by Fresnel's rhomb, the angle of incidence and the azimuths of polarization are fixed. How then can this apparently new kind of polarization be characterized? That posed a delicate problem that Brewster resolved in the following way:

> In circular polarization, as we have seen, the ray has the same properties in all its sides; and the angles of reflexion at which it is restored to polarized light in different azimuths are all equal, like the radii of a circle described round the ray. Hence, without any theoretical reference, the term circular polarization is from this and other facts experimentally appropriate. In like manner, without referring to the theoretical existence of elliptic vibrations produced by the interference of two rectilineal vibrations of unequal amplitudes, we may give to the new phenomena the name of elliptic polarization, because the angles of reflexion at which this kind of light is restored to polarized light may be represented by the variable radius of an ellipse. (Brewster 1830b, 293)

We see that Brewster—from having read Herschel's "Light"—knew that Fresnel had discussed the possibility of elliptical oscillations, so he took care to distinguish his own conception of "elliptic" polarization as entirely unhypothetical, as depending solely on the experimental fact that the azimuths and incidences for restoring plane polarization vary in metallic reflection but not in circular polarization. The polarization is said to be "circular" or "elliptical" in the following sense. Describe a curve in polar coordinates whose radius varies as the restored azimuth and whose angular coordinate varies as the incident azimuth. If the curve is an ellipse, then the polarization is "elliptical." If, as may happen with some metals, the ellipse is nearly a circle then the polarization is "circular."[28]

There was more. Brewster felt he could quantify the effect. He first built a table (using silver) of the azimuths of restored polarization for various angles of incident polarization. He wished to be able to predict these azimuths, and he gave the following formula for doing so:

$$\tan \varphi = \tan \varphi_{45} \tan x$$

where φ is the restored azimuth, x is the incident azimuth, and φ_{45} is the restored azimuth when x is 45°. The formula worked to within 20′ in φ for the eleven entries in Brewster's table. He applied it at some length to series of reflections using silver and steel plates.

Thus far Brewster had not appropriated any of the wave theory's vocabulary. But now he did so. Fresnel, and Herschel in his discussion (which was based directly on Fresnel's), had used the word "phase" in describing the resultant of two orthogonal oscillations. That alone would certainly not have made an impression on Brewster. What did apparently impress him was that Fresnel had, for circular polarization, provided a formula for computing the "phase" on total reflection, and Herschel (1827, 553) had remarked that the formula works very well because one can use different incidences, and corresponding numbers of reflections, to calculate from it when the phase becomes 90°, and hence when the polarization becomes circular.

Brewster felt he could do something similar for metallic reflection. Take a steel plate, fix the incident azimuth of polarization at 45°, and set the planes of reflection all parallel to one another. Instead of fixing the first incidence at the one appropriate to producing the maximum effect—which requires only a single reflection to produce "elliptic" polarization—let it vary, but let all subsequent incidences be the same as the first one. Then several reflections will be necessary to produce the effect, which, as Brewster understood it, is only "partial" until a sufficient number of reflections have occurred. The restored azimuth will vary with the angle of incidence.

Brewster conceived that if the total number of reflections required to go from rectilinear polarization back again is n, then the elliptical polarization becomes complete at $\frac{1}{2}n$ reflections. When this "completely" elliptic light is analyzed with a crystal, one can turn the crystal to produce within it a maximum ordinary beam, and so a minimum extraordinary beam. The angle to which the analyzer's principal section must be turned to do this, Brewster argued, is governed by the usual Fresnel law for the rotation of the plane of polarization, namely,

$$\tan(\varphi_{refl}) = \frac{\cos(i - r)}{\cos(i + r)} \tan(\varphi_{inc})$$

where φ_{inc} is the incident azimuth—here fixed at 45°—and φ_{refl} is the orientation of the analyzer to produce a maximum ordinary ray. Of course, since metals are opaque the angle r of refraction cannot be observed. Here, however, we have φ_{inc} set at 45°, so that $\tan[\varphi_{inc}]$ is one, and we can then easily use several observations of φ_{refl} for given i to determine the corresponding r, and from that we can calculate an "index" as $\sin(i)/\sin(r)$.[29]

Here, as he had in his analysis of "common" and "partially" polarized light, Brewster had appropriated a formula from the wave theory in a way that completely strips it of its original significance. But he did not stop with this. He next introduced "phase" into his account in the following words:

> This [the rotation formula, with φ_{refl} representing Brewster's φ below] is a very important relation, and enables us to determine the phase P of the two inequal portions of oppositely polarized light, by the interference of which the elliptic polarization is produced. It may be expressed by
>
> $$P = 2R$$
>
> But $\quad R = 45° - \varphi,$
>
> Hence $\quad P = 90° - 2\varphi$
>
> (Brewster 1830b, 300)

This definition of "phase" was demanded by Brewster's wish to be able to use the word operationally in the way that, Herschel had indicated, one can use Fresnel's "phase." A single reflection at a given angle of incidence with a given index of refraction produces, according to Fresnel's formula for total internal reflection, a certain "phase." For circular polarization the total "phase" change must in the end be 90°. In Brewster's understanding, I remarked above, "elliptical" polarization is produced either by one reflection at the incidence for maximum effect for a given metal or by several reflections at other angles. At these latter angles the polarization is only "partially" elliptical until the full effect is achieved. When it is finally achieved in full, Brewster decided, the "phase" must be the same as it is for circular polarization, namely 90°.

Make a table listing in one column the number of reflections that are empirically necessary to produce elliptical polarization and, in the other column, 90° divided by these numbers. These latter numbers are, for the data Brewster gives, just what his formula for the "phase" produced by a single reflection predicts they should be. Like Fresnel for circular polarization, he too could calculate for a given incidence a "phase" from which the number of reflections that are necessary to produce "elliptical" polarization can be found by dividing the "phase" into 90°.[30]

Brewster actually referred to "the two inequal portions of oppositely polarized light, by the interference of which the elliptic polarization is produced." He thought that two "inequal" and "oppositely polarized" beams have a phase, and that this must be 90° for elliptical polarization to be observed as a result of their "interference." In the wave theory, if they do have a 90° phase *difference* they must be "inequal" to show the asymmetry Brewster observed. But if they are unequal, then they certainly need not—and generally will not—have a 90° phase difference. It may be anything except 0° or 180°. Or rather, it may be such if the "inequal" beams Brewster refers to are, as Fresnel had them, in and normal to the plane of incidence.[31] If he did agree with Fresnel on this point, then he obviously had no clear conception of what he meant by the word "phase."

Brewster was hardly incompetent, and Herschel had rather clearly explained how phases work in calculating the resultants of interfering beams.[32] Recall that in discussing the composition of common and, especially, partially polarized light Brewster had constructed both kinds out of two beams. In common light the beams are orthogonally polarized, but in partially polarized light the polarizations form an acute angle with one another. If we carry this over to elliptical polarization, then Brewster begins, in *his* terms, to make sense. Suppose, for example, that when a metal reflects a polarized beam of light it creates a second, orthogonally polarized beam in the following way. The original beam is affected much as it would be by a nonmetallic reflection; that is, it is merely rotated. To it one can apply, as Brewster did, the usual formula for calculating the reflected azimuth. The second, newly created beam is much smaller than ("inequal" to) it and has an orthogonal polarization.

Suppose first that the phase difference between the beams is *not* 90°. Then the maximum intensity visible in the ordinary beam of an analyzer that receives the two beams will not occur when the analyzer's principal section lies along the direction of the more intense beam—the one that is rotated per the usual law. If, on the other hand, the phase difference is 90°, then the maximum intensity will occur in that direction. And very nearly the first law Brewster gave for elliptically polarized light was this very one, that the analyzer's principal section lies along the *rotated* direction. If the two beams are polarized in this way, and if this law is correct, then the phase difference *must* be 90°. Brewster can require that the phase be 90° in all circumstances because he uses a selectionist division of the original beam that makes the assumption actually necessary. All of this is far from clear in Brewster's account, but that is almost certainly because Brewster took it very nearly for granted. Like his analysis of common and partially polarized light, it attempts to subvert the vocabulary of the wave theory by appropriating it to selectionist ends, and it does so rather well.

APPENDIX 23 **Biot's Discussion of Tints**

We have already seen that Biot had so thoroughly adopted Newton's color mixing theory that in consequence he had rejected his own first hypothesis for chromatic polarization. He was equally convinced by the discussion in the *Opticks* of Newton's rings, and he based his theory upon a close analogy with it:

> Let *e* be the lamina's thickness expressed in thousandths of a millimeter. Multiply it by the number 0.10917, which gives the product .10917*e:* with this product consult the table found on page 266 of the French translation of Newton's *Opticks,* and in the column that expresses the thicknesses of the thin laminae of glass that reflect this or that color, you will find next to your number the color on which the thin lamina of [gypsum] exerts its action at perpendicular incidence. Suppose, for example, that the thickness of the lamina expressed in thousandths of a millimeter is 82.44036; on multiplying by 0.10917, the product is 9, which in Newton's table corresponds to blue of the second order: it will therefore be on blue of the second order that the lamina will exercise its action, and consequently it will be this blue that will be represented by [*A*] in our formula. [*A*] being known, the complementary tint [*U*] is also; it is the one that is complementary to blue of the second order, or, what comes to the same, it is the one opposed to [second-order blue] in the order of colored rings, and is consequently yellow. (Biot 1811, 158; partially translated by Frankel 1972, 246) (See fig. A23.1.)

Biot's claim stands in direct conflict with Arago's assertion in his unpublished notes, which he read six months after Biot read his own paper, that "the rule Newton gave for ordinary rings must not be applied to those I discovered in crystals, because these latter do not depend solely on the material thicknesses of the bodies" (Arago 1812b, 92). Since Biot's memoir had not yet been printed, we cannot be certain whether Arago was disagreeing with him or whether Biot did not have the chromatic law at this time. In any case, Biot mercilessly attacked Arago's opinions in the printed article without mentioning him directly. Recall that Arago had argued that internal partial reflection is the cause of chromatic depolarization. Biot in several ways showed Arago's hypothesis to be "inadmissible" (Biot 1811, 272–74), which no doubt exacerbated Arago's already considerable anger with him.

	COULEURS RÉFLÉCHIES.	ÉPAISSEURS DES LAMES		
		d'air.	d'eau.	de verre.
1er ORDRE.	Très-noir	$\frac{1}{2}$	$\frac{3}{8}$	$\frac{10}{31}$
	Noir	1	$\frac{3}{4}$	$\frac{20}{31}$
	Commencement du noir	2	$1\frac{1}{2}$	$1\frac{2}{7}$
	Bleu	$2\frac{2}{5}$	$1\frac{4}{5}$	$1\frac{11}{20}$
	Blanc	$5\frac{1}{4}$	$3\frac{7}{8}$	$3\frac{2}{5}$
	Jaune	$7\frac{1}{9}$	$5\frac{1}{3}$	$4\frac{3}{5}$
	Orangé	8	6	$5\frac{1}{6}$
	Rouge	9	$6\frac{3}{4}$	$5\frac{4}{5}$
2e ORDRE.	Violet	$11\frac{1}{6}$	$8\frac{3}{8}$	$7\frac{1}{5}$
	Indigo	$12\frac{5}{6}$	$9\frac{5}{8}$	$8\frac{2}{11}$
	Bleu	14	$10\frac{1}{2}$	9
	Vert	$15\frac{1}{8}$	$11\frac{2}{3}$	$9\frac{5}{7}$
	Jaune	$16\frac{2}{7}$	$12\frac{1}{5}$	$10\frac{2}{5}$
	Orangé	$17\frac{2}{9}$	13	$11\frac{1}{9}$
	Rouge éclatant	$18\frac{1}{3}$	$13\frac{3}{4}$	$11\frac{5}{6}$
	Ecarlate	$19\frac{2}{3}$	$14\frac{3}{4}$	$12\frac{2}{3}$
3e ORDRE.	Pourpre	21	$15\frac{3}{4}$	$13\frac{11}{20}$
	Indigo	$22\frac{1}{10}$	$16\frac{4}{7}$	$14\frac{1}{4}$
	Bleu	$23\frac{2}{5}$	$17\frac{11}{20}$	$15\frac{1}{10}$
	Vert	$25\frac{1}{5}$	$18\frac{9}{10}$	$16\frac{1}{4}$
	Jaune	$27\frac{1}{7}$	$20\frac{1}{3}$	$17\frac{1}{2}$
	Rouge	29	$21\frac{3}{4}$	$18\frac{5}{7}$
	Rouge bleuâtre	32	24	$20\frac{2}{3}$
4^e ORDRE.	Vert bleuâtre	34	$25\frac{1}{2}$	22
	Vert	$35\frac{2}{7}$	$26\frac{1}{2}$	$22\frac{3}{4}$
	Vert jaunâtre	36	27	$23\frac{2}{9}$
	Rouge	$40\frac{1}{3}$	$30\frac{1}{4}$	26
5^e ORDRE.	Bleu verdâtre	46	$34\frac{1}{2}$	$29\frac{2}{3}$
	Rouge	$52\frac{1}{2}$	$39\frac{3}{8}$	34
6^e ORDRE.	Bleu verdâtre	$58\frac{1}{4}$	44	38
	Rouge	65	$48\frac{3}{4}$	42
7^e ORDRE.	Bleu verdâtre	71	$53\frac{1}{4}$	$45\frac{4}{5}$
	Blanc rougeâtre	77	$57\frac{3}{4}$	$49\frac{2}{3}$

FIG. A23.1 Newton's table of tints as presented by Biot.

Nearly a fourth of this, Biot's first memoir on the subject, concerns a highly specific configuration: polarizing incidence with the plane of incident polarization set at 90°, and the optic axis set at 45°, to the plane of reflection. Yet Biot asserted that his law applied to normal incidence, and he did not say that it depended upon the orientation of the lamina's optic axis. The reason for this apparent contradiction between the statement of the law and Biot's actual experiment derives from the fact, which Biot had discovered very early in his work, that it is extremely difficult accurately to observe light transmitted through the lamina, whereas it is very simple to observe light reflected by it. Hence Biot invented a clever alternative to transmission observations that permitted him to use the reflection of skylight on a uniformly cloudy day (which produces an excellent white).

Take a thin lamina and place it flat atop a blackened surface. Reflect the cloud light at polarizing incidence. Receive the reflection on glass blackened on one side, again at polarizing incidence, but in a plane normal to that of the first reflection. Then look at the lamina's image in the blackened glass, and (see fig. A23.2)

> You will see it entirely and uniformly tinted with a vivid color, which will change intensity and nuance [perhaps saturation] on turning the lamina in its plane. The reflected color will become null when the axis of the lamina coincides with the light's plane of incidence on its surface or is perpendicular to it; it will attain its absolute *maximum* of intensity when

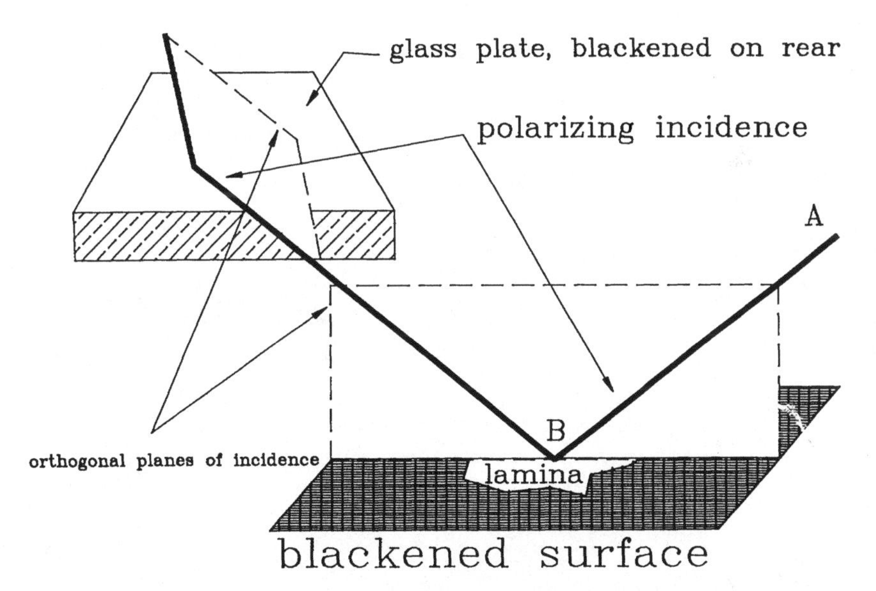

FIG. A23.2 Biot's reflection method for observing chromatic polarization.

> the axis of the lamina forms an angle of 45° with this plane; and *the tint reflected in that position will be precisely the one on which the lamina will exert polarization by transmission at normal incidence.* (Biot 1811, 171; emphasis added)

Two partial reflections occur in the lamina, one at its upper and the other at its lower surface. Biot reasoned essentially as follows. The first reflection (BC) is the same as if the lamina were simply a plate of glass, and so it will be white and polarized in the plane of incidence. Consequently, when it strikes the reflecting mirror, on which the plane of incidence is *normal* to that on the lamina proper (fig. A23.3), it will be entirely transmitted through it.

But the second reflection, at the lamina's lower surface, behaves in a different way. The beam within the lamina, BD, derives from what was an *unpolarized* beam AB. According to Biot's theory a *polarized* beam will divide, on entry into the lamina, into two subsets: one, *U*, is unaffected by it, whereas the other, *A*, has its plane of polarization altered. The beam BD may be considered to have a very large number of subsets, each with its own polarization—since the rays in it have their asymmetries pointing in essentially random directions. Consequently, when BD passes through the lamina each of these many subsets will be acted on in the fashion required by Biot's theory, but the result will be no net change in the composition of the beam.

However, when BD strikes the inner surface of the lamina it will be polarized by

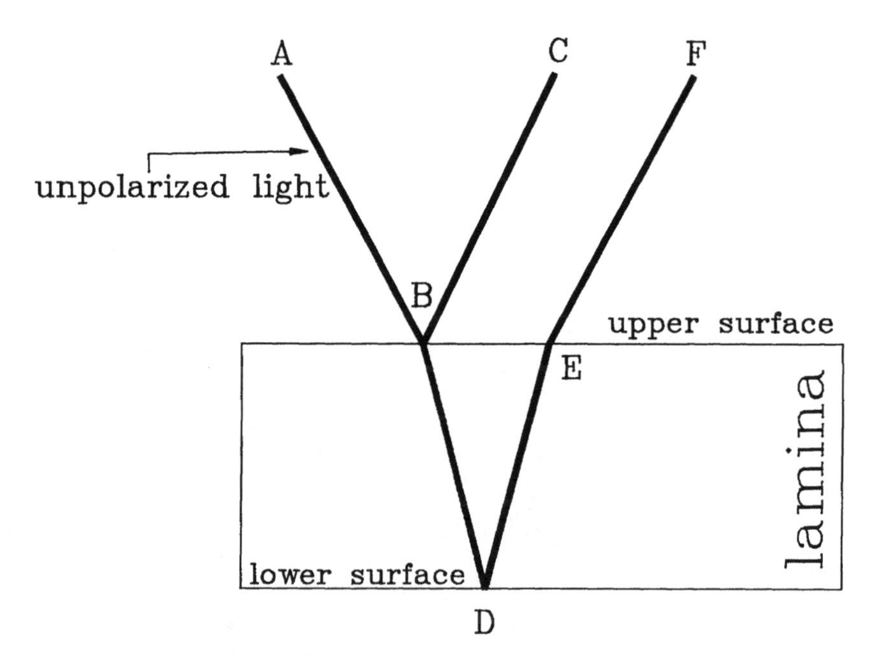

FIG. A23.3 Ray patterns in Biot's experiment.

reflection, since it derives from a beam that was incident at the polarizing angle. Consequently its reflection, DE, will be acted upon by the lamina in precisely the same way that the lamina would act upon a beam of light polarized in the plane of incidence. That is, beam DE will divide into two subsets. One, *U*, will not be affected by the lamina and so will, on striking the reflecting mirror, be completely transmitted through it. But the other subset, *A*, will have its polarization flipped across the lamina's optic axis. And since in Biot's experiment the angle between the axis and the incident plane of polarization is 45°, then after the lamina acts on subset *A* of beam DE its polarization will be parallel to the plane of incidence on the mirror—and so it will be very strongly reflected by it.

With this configuration, then, Biot was actually observing the tint of the *A* subset in a beam that has passed through the lamina. However, the incidence was most certainly not in the perpendicular to the lamina's lower surface—it was polarizing (i.e., about 55°). He apparently discovered purely empirically that the *A* tint in this arrangement will appear to be the same as the *A* tint at normal incidence (Biot 1811, 223–24), a possibility that was suggested to him by the following observation. When the optic axis lies in the plane of incidence, then the colors move through Newton's table with increasing incidence from the normal as though the lamina were becoming thinner; when the axis is normal to the plane of incidence, on the other hand, the colors change in precisely the reverse fashion. Whence, Biot reasoned, setting the axis between the extreme positions—at 45°—should cancel the chromatic changes produced by any deviation from normal incidence. In other words, with this orientation of the lamina's optic axis, one should obtain the same tint as at normal incidence, and in fact Biot found that he did.

Biot performed an extensive series of experiments with this configuration using multiple layers of laminae to investigate quantitatively the effect of thickness on color. And he found that the colors follow the reflection tints in Newton's table with this difference only: that instead of using Newton's gap width as an index into his table, one must multiply it by a factor that depends on the substance of the lamina.

Biot's experimental technique and presentation strikingly parallel those Malus had used in confirming Huygens's construction (see above, chap. 1). First comes the development of a precise method of observation based on reflection. Then follows the construction of a device with a calculable accuracy for measuring the variables of interest—laminar thickness in Biot's case. Biot's device was originally built by the instrument maker Jean-Nicholas Fortin for the optician Cauchoix (Biot 1811, 177–78). (See fig. A23.4.) It was designed to measure lenticular curvature, but it could be adapted to Biot's needs—and it was accurate to 0.03 mm. The device was simple but precisely constructed and achieved its accuracy from the central screw's fine step height. In fact, some of Fortin's instruments achieved accuracies of better than 0.01 mm, much higher than required. Of it Biot remarked:

> The extreme perfection that M. Fortin knew how to achieve in the construction of the metallic screw, is a certain guarantee of the exactitude of

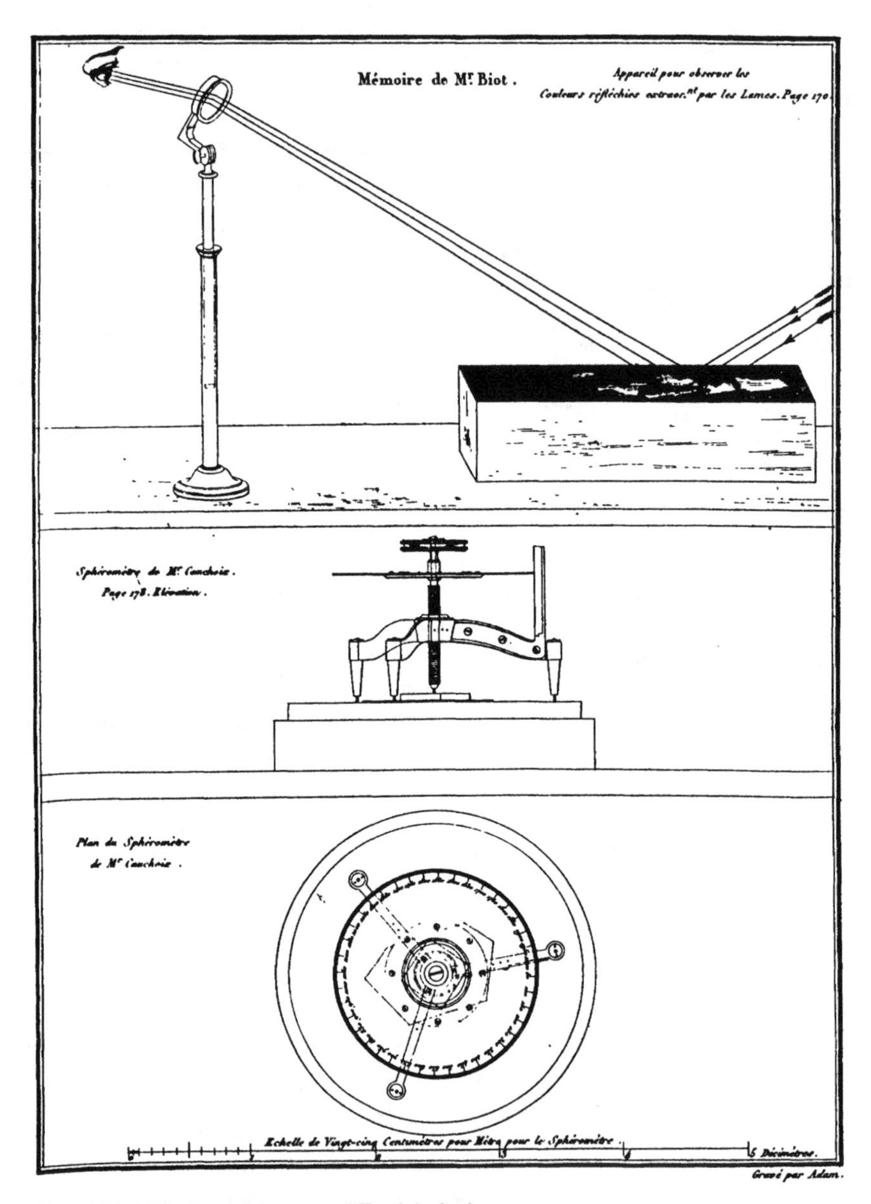

FIG. A23.4 Biot's experiment and Fortin's device.

> that which he adapted to this instrument; but for those who might not be aware of this exactitude, I will add that the consequences to which the measurements lead me are independent of the absolute values of the screw step; it suffices that [the step] be regular and, at that, only in a very small interval, since it has only to measure or rather compare among themselves very small thicknesses. (Biot 1811, 180–81)

Using this device, Biot determined the factor to be used in Newton's table by averaging the ratios obtained from a series of experiments, much as Malus had worked in confirming Huygens's construction.

Biot's method was quite simple. Observe the color at a given laminar thickness. Then multiply this by the factor that must be used to transform the thickness into an index for Newton's table. Then look up what the tint should be for this number in the table. This procedure certainly quantifies the phenomenon, but it is badly limited in accuracy by the small number of entries in the table. Nevertheless, for the colors he did examine Biot found extremely good agreement, amounting to about a 1.6% difference between the thicknesses that are predicted and those that are observed for a given color.

Biot also attempted to apply his procedure to oblique incidences, which, though the precise formulas were of only passing importance for him (since he soon found them inadequate), is nevertheless worth briefly examining to see how difficult it was for him to generalize. Oblique incidence posed considerable difficulties for both calculation and experiment, because if the angle of incidence is an important factor, then Biot could no longer use the reflection method—he had instead to observe the transmitted colors directly, and this is inherently less accurate.

In figure A23.5 we have a hollow tube to which a mirror (M) is fastened at the lower end, while the lamina (L) is set into a plate at the upper end. The mirror and the plate that holds the lamina can each pivot about an axis perpendicular to the plane of the figure, and so the angle of incidence on each of them can be adjusted. Furthermore, the plate that holds the lamina can be rotated about the axis of the tube itself, and this will change the plane of reflection of the light that strikes the lamina. In Biot's experiment a beam of natural light (NO) strikes the mirror at polarizing incidence—so that it becomes polarized in the plane of the figure. It then passes up the tube and strikes the lamina. The geometry of the device is not entirely simple, and Biot had therefore to introduce a rather intricate set of angles. However, for our purposes we need only Biot's variable i, which represents the angle between the optic axis of the lamina and the intersection of the plane of polarization with the lamina.

By examining extreme cases and retaining the overall structure of his formulas, Biot could generalize them in the following way. First, since the tints and the intensities change with the incidence, Biot assumed that U and A must also change, becoming U^1,A^1. He further hypothesized that the affected bundle—A^1—rotates in a different fashion when the incidence is not normal. That, in general, A^1 rotates through an angle

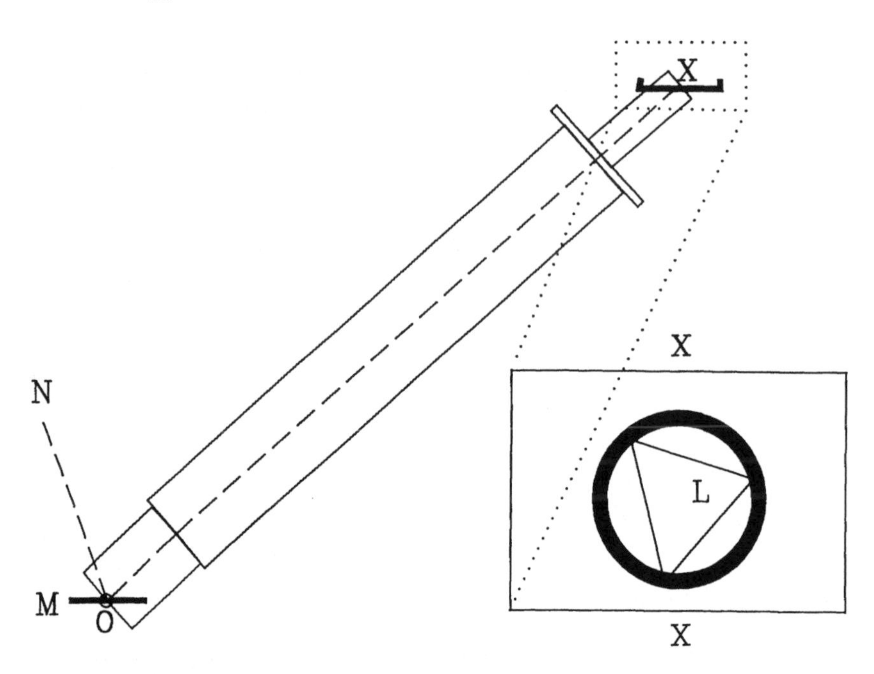

FIG. A23.5 The structure of Biot's experiment.

$2i + 2k$ instead of just through $2i$, as occurs at normal incidence. Since Biot always set the analyzing crystal with which he examined the light transmitted through the lamina in the plane of initial polarization, his generalized formulas may be written:

$$F_o = U^1 + A^1 \cos^2[2(k + i)]$$

$$F_e = A^1 \sin^2[2(k + i)]$$

He had by experiment to do two things: first, he had to determine the values of k, which determines the angle through which the polarization of the rays in A^1 are turned; second, he had to determine the tints (ray ratios) in U^1, A^1.

It would be tedious to detail Biot's intricate, but essentially empirical, considerations of the many possible factors that can affect tints and intensity. We need only remark that he became convinced that the tints of U^1 and A^1 are entirely independent of the initial plane of polarization. Only the angle of incidence, the position of the optic axis with respect to the plane of incidence, and the thickness of the lamina determine them. Biot succinctly summarized his view of the matter in the following words:

> As the axis of crystallization [optic axis] inclines to the refracted ray, the force that makes the luminous particles oscillate diminishes, and the number of oscillations made in the same space decreases as the square of

> the sine of the angle this axis forms with the refracted ray; but at the same time the trajectory of light in the plate augments with the obliquity; and the oscillations become more numerous in the same space [laminar thickness]. These two elements, modified by a factor nearly constant, which probably depends on the speed, determine in all cases the tints the plates must present. (Biot 1812, 247)

Biot was in essence assuming that the number of fits per unit length that are involved in chromatic depolarization must be modified by the same factor ($\sin^2\varphi$, with φ the angle to the optic axis) that appears in the Malus-Laplace expression for the velocity of the extraordinary ray (viz., $v_e^2 = v_o^2 + d^2\sin^2\varphi$ [see Frankel 1972, 257–58 for Biot's several formulas]). The result of his assumption is this: in applying Newton's table to chromatic polarization at oblique incidences, multiply the usual thickness factor by $a(\sin^2\varphi/\cos\varphi)$, where a is an empirical adjustment: the $\sin^2\varphi$ term derives from Biot's assumption that the fits are themselves altered, whereas the denominator represents the effect of an oblique ray's having to travel a greater distance through the lamina than a normal ray.

APPENDIX 24 Fresnel on Biot's Dihedral Law

Fresnel remarks, paraphrasing Biot (drawn from Biot's *Précis élémentaire*):

> Conceive a plane drawn through each of the axes of the crystal and through the ray that undergoes ordinary refraction. Conceive, through this same ray, a third plane that divides in two parts the dihedral angle the first two form. The luminous molecules that have undergone ordinary refraction are polarized in this intermediary plane; and the molecules that have undergone extraordinary refraction are polarized perpendicular to the intermediary plane through the extraordinary ray under the same conditions. (Fresnel 1821e, 582)

In figure A24.1 AL and AL′ are the two optic axes, and AO is some ordinary ray. Construct with Biot three planes intersecting in OA: one contains the axis AL, the

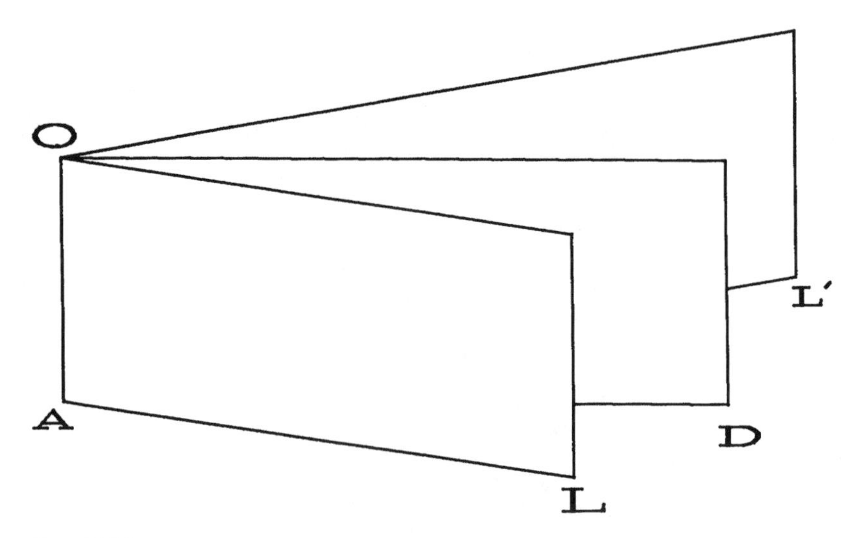

FIG. A24.1

other contains the axis AL′, and the third, OAD, bisects the angle between the two planes OAL and OAL′. According to Biot, the bisecting plane OAD is the ordinary ray's plane of polarization. Construct three planes in the same way for the extraordinary ray by replacing the point O with a point E on the extraordinary ray. Then it will be polarized *perpendicular* to the bisecting plane EAL. I shall call this Biot's *dihedral law.* It is a reasonably simple generalization of the corresponding uniaxial relationship, since when the dihedral angle—the angle between the planes OAL and OAL′ or between the planes EAL and EAL′—vanishes, then bisecting planes reduce to the planes that contain the optic axis and the rays.

Fresnel was able to derive this law directly from his ellipsoid, which, recall, is sectioned by planes whose normals lie along the rays. He provides a compressed but straightforward argument, based on spherical trigonometry and the definition of the planes of polarization as the planes that are formed by the ray and the semiaxes of the section cut by a plane normal to the ray in the ellipsoid.[33]

APPENDIX 25 Fresnel on Biaxial Refraction

Fresnel recognized that the coordinate planes section a circle and an ellipse in the wave surface (indeed, he seems to have realized this even before he deduced the wave surface) and that the two are, in the xz section (recall that by convention $a > b > c$), touched simultaneously by a plane TS (see fig. A25.1). If this plane represents a front, then to it correspond a pair of rays, OA and AE, with different directions (call this case 1). The plane TS itself is parallel to a circular section of the surface of elasticity. And the vector LA, which points in a direction for which there is but a single ray speed, must therefore be normal to a circular section of the ellipsoid for rays (call this case 2). Neither case occurs in the degenerate, uniaxial situation, because there the sphere shrinks to touch the ellipse at only two diametrically opposite points, whereupon EA and OA become coincident with AL.

In case 1, which is produced by a beam that strikes the crystal at a certain angle, we have two internal rays but only one front, so that on emergence the two form a pair of rays that are separate from but parallel to one another. Fresnel (hardly surprisingly) failed to realize that the front TS is in fact tangent to the wave surface throughout a circle of diameter EO, so he did not suspect the possibility of producing a *cylinder* of emergent light from an appropriately incident ray. Conversely (case 2), to the single internal ray LA there correspond two different fronts, namely the tangents to the sphere and the ellipse at L. Hence LA, which may be formed by a certain *pair* of rays that are incident at a given point, emerges as two rays that diverge from a point.

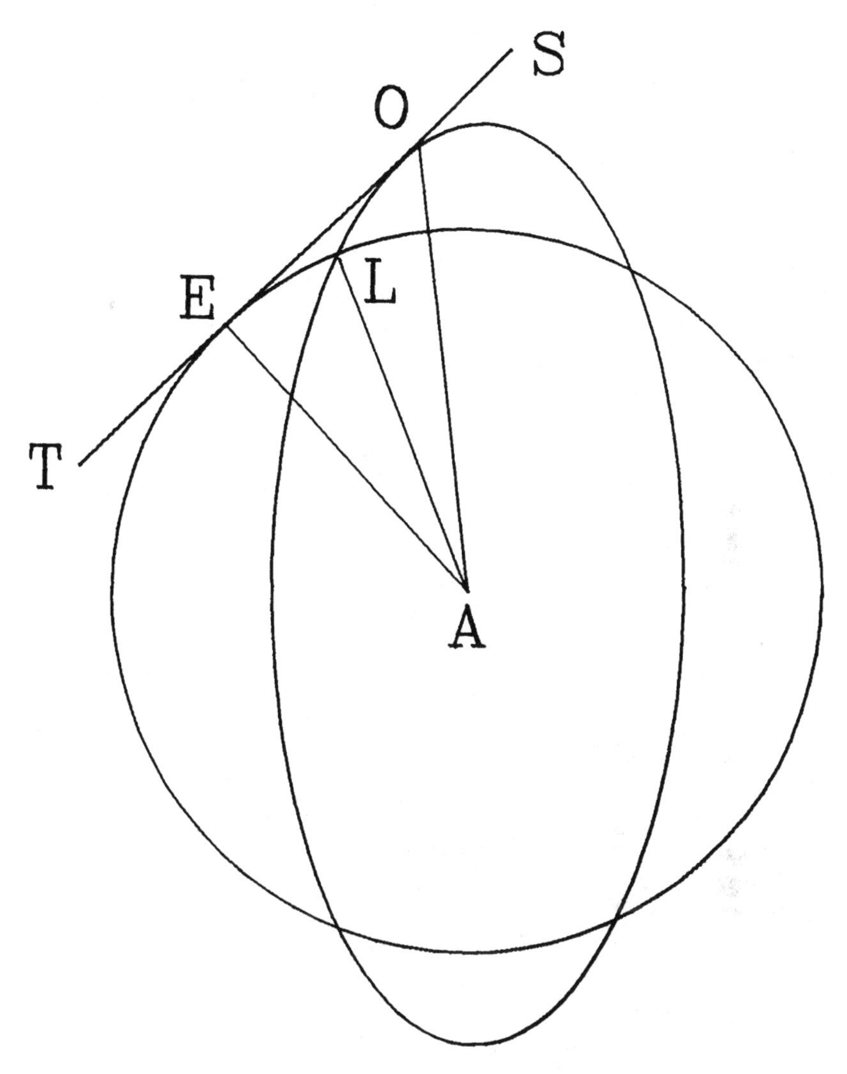

FIG. A25.1 A section of Fresnel's wave surface.

Notes

Introduction

1. The differences between wave and ray theory can be very subtle, and it is possible for isolated effects nearly to confound the two (e.g., one might imitate the wave theory's unpolarized ray in the ray theory by considering a sequence of particles whose asymmetries vary randomly from one particle to the next). However, there is an irremediable difference that inevitably reveals itself. Take a plane-polarized ray and freeze it at a moment in time. Project the ray's asymmetry at any point along it into two orthogonal planes. In the wave theory these projections simultaneously determine both the polarization and the intensity of the light. But in ray theory the projections determine only the polarization, since the intensity is a function of collections of rays. Similarly, one could imitate circular polarization in ray theory by considering a single ray to involve a succession of asymmetries pointing randomly in many directions. Again, however, there would never be any connection between intensity and the orthogonal components of the asymmetry for a single ray, nor could there be any difference at all between circularly polarized and unpolarized light—precisely because in ray theory orthogonal components are projections of a rigid thing, whereas in wave theory they are projections of a vector that represents an oscillation.

2. For example, Huygens's principle could (and in fact was by Arago) be put in the following way: take a point source of light and draw about it a sphere that may be partially interrupted by an obstacle. To find the optical effect at any point beyond the sphere, draw rays to that point from every point on the sphere and apply the principle of interference to the collection of rays. Of course in reality there will not be an indefinite number of rays to add up at the observation point, as is evidently assumed by Fresnel's integral formulas, but there are so many that what are in reality summations (according to selectionist theory) can be replaced by Fresnel's integrals. There is no need to assume that the sphere has any intrinsic physical significance other than as a locus from which rays may be drawn for purposes of computing interference patterns.

3. Indeed, we shall see that during a controversy with Fresnel Poisson invented a way to have physical rays even if light were a propagated motion.

Chapter 1

1. Euclid assumes that visual radiation moves from the eye outward, but there was considerable diversity of opinion on this point. Lindberg (1976) discusses the "intromissionist" (from without in) and "extramissionist" (from within out) accounts of ray behavior.

2. Cohen and Drabkin (1975, 259), from the English translation by H. E. Burton of the Heiberg edition.

3. Lindberg (1976) discusses many aspects of these developments through the work of Kepler. See also Lindberg (1971) for a discussion of Alkindi on physical rays.

4. What I shall say about the seventeenth century derives in major part from Shapiro (1974). This is a careful and detailed study of the intricate development of the concept of a front and its relationship to the ray, culminating in Huygens's complete solution to the problem.

5. Shapiro (1974) discusses in detail the work (besides Hobbes's) of Maignan, Barrow, Hooke, Pardies (through the work of Ango), and Huygens. In every account except for Pardies's and Huygens's the ray retains a distinct physical identity. Further, even in Pardies's work the fronts themselves are not carefully analyzed (Shapiro 1974, 209–18), and the rays are not treated as independent of one another (Shapiro 1975, 197).

6. It has been remarked that Huygens's principle should be considered in two ways: first, as a means for defining the pulse front, and second, as a way to explain rectilinear propagation (Shapiro 1974, 224–25). The pulse front itself is found at any instant by constructing the common tangent to these secondary pulses at that moment. Rectilinear propagation requires limiting the spreading of the pulse when it passes through an aperture or past a barrier. To do so, Huygens assumed that the secondary radiation is visible only at the front itself, that is, at the common tangent to the secondaries at some moment.

7. According to Shapiro (1974, 246–47) until the time of Fresnel only La Hire, Papin, and Leibniz accepted the principle, and Huygens's theory was, in at least one case, confused with Pardies's and Hooke's (neither of which is identical even to the other, much less to Huygens's, since Hooke uses physical rays whereas Pardies defines the rays as the normals to the fronts).

8. Because any part's contiguous neighbors in the front are "as high as it is itself" and so prevent it from having an effect except along the normal (Shapiro 1974, 211).

9. The crystal facets are parallelograms, and the "principal section" is a plane normal to a facet that bisects an obtuse angle. See appendix 1.

10. A line that in geometric optics is called a "ray" therefore marks the path of molecules of light in the emission theory. Since the molecules are discrete things, the paths they follow can also be counted up.

11. Young in fact distinguishes carefully between the very little light that can diverge into the shadow and the much greater light that is responsible for interference patterns. The latter actually reflects from the edge of the slit and then interferes either with the direct light that passes with little diversion through the slit or with the light reflected from the opposite edge of the slit. He remarks: "Whatever may be the cause of the inflection of light passing through a small aperture, the light nearest its centre must be the least diverted, and the [*sic*] nearest to its sides the most: *another portion of light* falling very obliquely on the margin of the aperture, will be copiously reflected in various directions; some of which will either perfectly or very nearly coincide in direction with the unreflected light, and, having taken a circuitous route, will so interfere with it, as to cause an appearance of colours" (Young 1802, 42–44; emphasis added). Consequently, for Young in 1802 (and for Fresnel in his early work) diffraction has little to do with the divergence of the original pulse into the shadow.

12. One must not, however, push this claim too far. Since for both Young and Fresnel the front was the physical basis of light, the ray had to be thought of in terms of the front—in particular as a delimited portion of it. Consequently the ray can no longer be treated as an isolatable, individual thing; rays should not be counted because

the front is a continuous surface (though one could, and Poisson apparently later did, count rays by conceiving that one never has a continuous front but has only a set of laterally discontinuous pulses, each of which forms a ray). But this had little immediate effect, since both Young and Fresnel (at first) built their diffraction theories using ray interactions, and we shall also see that whether rays can or cannot be counted has little importance in computing interference patterns.

13. If ω represents the angle between the plane of incidence and the principal section, i is the incidence, h the crystal height, n^1 La Hire's index, and r the refraction, then the coordinates (x,y) of the point on the crystal base struck by the refraction will be:

$$x = \frac{ha}{c};\ y = \frac{hb}{c}$$

$$a = -\cos(r)\sin(\mu) - \frac{\lambda}{n^1}$$

$$b = \frac{-\sin(\omega)\sin(i)}{n^1}$$

$$c = -\cos(r)\cos(\mu) + \frac{\lambda \sin(\mu)}{n^1}$$

$$\lambda = \cos(\mu)\sin(i)\cos(\omega) - \sin(\omega)\cos(i)$$

$$\cos(r) = \frac{1}{n^1}\sqrt{(n^1)^2 - 1 + [\sin(i)\cos(\mu)\sin(\mu) + \cos(i)\cos(\mu)]^2}$$

In the principal section b vanishes. This formula is algebraically intricate, and its geometric equivalent would be more complicated for computational purposes. However, the geometric equivalent would not require a trigonometric table.

14. It is particularly interesting that Haüy's variable amplitude of aberration arises naturally from Newton's rule if the distance between the refractions is simply inclined to the base. Consequently any theorem deduced from Haüy's law that does not depend upon this inclination must also be an implication of Newton's. Haüy himself demonstrated that Huygens's law of equal deviations—see appendix 1—follows from Haüy's law. Yet he expressed "surprise" that Newton had advanced his own rule when he (Newton) "ought to have been struck by a remarkable property of the amplitude of aberration discovered by Huygens," to wit the law of equal deviations. Haüy obviously thought that law was incompatible with Newton's rule. But if it were, then it would also have been incompatible with his own law, which he, of course, did not think it was. This illustrates a point I shall return to below—that formulating laws for double refraction entirely in geometric terms often masked significant relationships that would at once be obvious or simple to calculate analytically.

15. We can translate Haüy's geometric construction for La Hire's law into the following algebraic expression for the aberration:

$$e^0l^0 = el = \frac{ms}{mh}eh$$

$$= \frac{ms}{mh}(nh - em)$$

$$= b\left[\tan(i) - \frac{\sin(i)}{\sqrt{n^2 - \sin^2(i)}}\right]\frac{\tan(\mu)}{\tan(i)}$$

Here the crystal height is b, the incidence is i, n is the ordinary index of refraction, and μ is the inclination of La Hire's axis to the vertical. Note that, according to this expression, the aberration would have to be independent of the value of the "index" for the extraordinary ray in La Hire's law, which it most certainly cannot be, since the point struck by the extraordinary ray, for a fixed incidence, varies as a function of the "index." This independence is obvious from the algebraic expression but not from the geometric construction.

16. This is not too surprising since, using modern parameters in Huygens's construction and Haüy's parameters in his law, they are remarkably similar:

$$\text{Haüy:} \quad \tan(r_e) = \tan(6°20') + \frac{1.06415 \sin(i)}{\sqrt{2.8 - \sin^2(i)}}$$

$$\text{Huygens:} \quad \tan(r_e) = \tan(6°12') + \frac{.9955 \sin(i)}{\sqrt{2.464 - \sin^2(i)}}$$

17. In later years Haüy recognized all three of his errors, for they are not present in his *Traité* of 1803. He recognized that Newton's rule is a limit of his own law (and so it too implies the law of equal deviations) and that Huygens's construction is as accurate as his own law, and he did not reiterate his incorrect construction for La Hire's law. In the *Traité* he disposed of La Hire's law by linking it to George-Louis Buffon's physical hypothesis and then drawing an unacceptable conclusion from the latter. If, as Buffon reasoned, there are two sets of laminae inclined at an angle to one another in the crystal, then an ordinary ray, proceeding normally upward from a point marked below the crystal, would necessarily undergo a refraction each time it encountered one of the oblique layers that produce extraordinary refraction. One could not possibly see a point below a crystal by looking from directly above it.

18. Construct with Wollaston the lines ef, ed, meeting in e, where d is the point of emergence from the prism of the ray bd. Set ef/ed equal to n^1. Proceed next by proportions. From the figure, bd/dh is in the same ratio as 1 to sin(∠hbd), where the latter sine is to unity as n is to n^1. Again from the figure sin(∠efg)/sin(∠edg) is as ed to ef. Since ed emerges from the prism, then sin(∠edg)/sin(∠bdh) is as 1 to n^1. From these proportions we find at once that ∠efg must equal ∠bdh. Hence, again from the figure, ef is to gf as bd to dh, whence ef/gf = n^1/n.

19. Wollaston's device measured the angle of emergence from the crystal and into the glass prism of a ray that had passed nearly parallel to the crystal facet. One can envision the situation at the prism/crystal interface as one where the ray observed, which is actually reflected from the surface of the crystal, in effect produces a refraction of 90° within the crystal. Consequently the tangent of the angle of refraction is infinite, and the denominator in Malus' formulas (see appendix 2) must vanish:

$$1 - [1.583]^2\sin^2(i)[CG^2\cos^2(\omega) + CP^2\sin^2(\omega)] = 0$$

The numerator in the formulas may be anything at all—Wollaston's device cannot tell the difference. One can use this formula, together with Wollaston's measurements, to gauge his observational accuracy.

20. Chappert (1977, 50–52) discusses other letters to Lancret, also from 1800, that continue these themes.

21. Malus's deduction of this proposition relied solely upon the laws of reflection and refraction coupled to an intricate infinitesimal analysis (Chappert 1977, 102–9). He did not, that is, use the principle of either least time or least action. In the late 1820s William Rowan Hamilton developed his "theory of systems of rays," which in effect began by demonstrating the general validity of Malus's theorem in a very simple way (see Hankins 1980, 70–71) by using the principle of least time. That demonstration makes Hamilton's analysis less than purely geometrical but has the advantage of great simplicity, showing almost at once that Malus's theorem can be generalized to any number of reflections.

22. With the index replaced by the velocity ratio, "Snel's law" in the emission theory means only that the velocity components parallel to the interface remain the same. That remains true whether the bounding substance is opaque or transparent. But one must also consider the actual variation of the velocity with distance from the interface. In the emission theory the velocity does not change abruptly. When we use Snel's law to calculate the angle of total internal reflection, we set the angle of external refraction to 90° and interpret the calculated internal incidence as the angle at which total reflection abruptly begins. But in the emission theory this angle represents the point at which a particle will just curve through and back out of the transition layer, and this will not be possible if the bounding substance captures every particle that enters it. Hence the angle implied directly by Snel's law represents the smallest incidence at which total internal reflection can in principle occur, not the angle at which it will actually first be observed.

23. Frankel (1974, 231–32) argues that Malus was thinking about double refraction before the announcement of the prize memoir, that Malus's techniques in the *Traité* "could clearly be extended to that problem," and that "the prize contest would appear to have been an attempt on the part of the Laplace and Lacroix to reward Malus for work he had already begun." As primary evidence Frankel cites Delambre's eulogy and a comment by Lacroix in the commission report on the *Traité*. The eulogy did claim that the *Traité* "brought attention" to double refraction; however, as published it neither discusses the phenomenon nor contains anything of direct use for it. Lacroix remarked that the *Traité* would be followed by two other memoirs "comprising the details of the application of the general formula and the very complete exposition of a purely physical result. That which [Malus] has communicated to one of us [probably Laplace] of the second [memoir] will give incontestably new worth to the first" (cited and translated, with insertions, by Frankel 1974, 231). Lacroix's remark is the strongest evidence cited by Frankel for Malus's early interest in double refraction, but it is both inferential and ambiguous. If Frankel is correct in arguing that Lacroix was referring to Laplace—and I think he is—then Lacroix had probably not personally seen Malus's work. Moreover, Lacroix did not mention double refraction. Finally, his reference to a "purely physical [as opposed to analytical?] result" is remarkably obscure.

24. The device, originally used in surveying, consisted of three major components. A rigid column, J, was inserted in a metal plate; the plate was attached by two short legs to a larger metal plate, QR. Adjustable feet GG were used to set the angular position of QR on a metal stand ST, which sat on the ground (to compensate for irregularities in the terrain). A rectangular metal cage rr was fastened to column J (which was tapered for the last fourth of its length) by screws gg. Two lenses—an objective and an eyepiece—at C and D, respectively, were attached to a flat metal plate (CD) that was rigidly fastened to rr through a framework held in place by screws. (One screw, h, is shown; the other is hidden behind AB.) The second component of the device was a circle, commonly graduated in tenths of a degree, and attached by spokes l to a pivot (below F) on the plate CD; the circle could be rotated. Finally, a metal plate (F), attached by a pivot to the same center (below F) as the graduated circle, bore a glass AB with objective at A and eyepiece at B. Although the glass was not centered through F, it carried, below A and B, two fine wires (k, k) that overlay the circle and whose prolongations passed through the center.

25. Frankel (1974) provides a detailed analysis of Laplace's and Malus's deductions of the translated Huygens's formulas from least action. I shall have very little to add to his account, and I refer the reader to it for details of the mathematics involved. We differ for the most part only in matters of emphasis and timing.

26. Frankel (1974, 238) quotes Arago's remarks in his biography (Arago 1857, 2:168–69) that Malus always insisted that he "had himself first conceived the possibility of this investigation, and that he had spoken of it publicly before the publication of that note [by Laplace on the same subject]." There is no reason to doubt Malus, except that I think it likely Laplace would have conceived the same idea independently since he, like Malus, would have been looking for a way to interpret the equations without relying directly on Huygens's spheroid.

Chapter 2

1. Arago (1857, 2:154) gives the following account: "One day, in his house in the Rue d'Enfer, Malus happened to examine through a doubly refracting crystal, the rays of the sun reflected by the windows of the Luxembourg palace. Instead of the two bright images which he expected to see, he perceived only one,—the ordinary, or the extraordinary, according to the position which the crystal occupied before his eye. [Clearly an exaggeration: the reflection was certainly not fully polarized, so Malus would have seen marked intensity variation but not complete extinction.] This singular phenomenon struck him much; he tried to explain it by supposing some particular modifications which the solar light might undergo in traversing the atmosphere. But when night came, he caused the light of a taper to fall on the surface of water, at an angle of 36°, and found, by the test of a double refracting crystal, that the light reflected from the water was also polarized, just as if it had emerged from a crystal of calc spar. The same experiment made with a glass reflector at the incidence of about 35°, gave the same result."

2. Malus's discovery was sufficiently startling that Young at first refused to believe it. He remarked that "we are inclined to think, that the portion usually reflected is in this case wholly absorbed, if not destroyed" (Young 1810, 250). Malus soon demonstrated experimentally that this cannot work (Malus 1811a).

Young greatly admired Malus's discovery of polarization by reflection but he was also depressed by it. He remarked: "The discovery related in these papers, appears to us to be by far the most important and interesting that has been made in France, concerning the properties of light, at least since the time of Huygens; and it is so much the more deserving of notice, as it greatly influences the general balance of evidence, in the comparison of the undulatory and the projectile theories of the nature of light" (Young 1810, 247–48).

The "balance" Young refers to was not favorable to the wave theory, though the point is a bit subtle. Polarization with crystals had, of course, been known since the time of Huygens and had been urged by Newton as an objection to the wave theory. Young (1809) had previously suggested that Newton's objection was not impregnable because it requires that the "curvature" of the undulation "which has passed through a crystal" be "alike on all sides" (Young 1810, 229). Evidently Young felt that this need not be so after passage through a crystal, presumably because the extraordinary wave in the crystal is not "alike on all sides," that is, is not spherical. Somehow the ellipticity of the wave in the crystal might, Young seems to be asserting, affect the wave that emerges from it. (Perhaps by altering its amplitude asymmetrically?) In any case the fact that polarization can be produced by reflection destroys this possibility, since passage through a crystal is not necessary to produce the asymmetry.

3. The factors m_{refl} and m_{refr} might be different from one another even though we assume both to be independent of ε, because I_0 might itself depend on ε. I_0 must be equal to the difference

$$I - [m_{refl} \cos^2\varepsilon + m_{refr} \sin^2\varepsilon]$$

This will obviously depend upon ε unless m_{refl} is equal to m_{refr}. All we can say to this point is that when ε vanishes I_{refr} must not vanish, which it will not because I_{refr} then reduces to I_0 equal to $I - m_{refl}$. If m_{refl} were equal to m_{refr}, then the implication would be that I_0 has always the same value, equal to $I - m$, where m is the common constant.

4. In his article "Chromatics" (1817) Young did not distinguish, as most people did not at the time, between selectionist and emissionist positions, but he used the word "selection" in essentially the sense I have adopted: "In the theory of emission, the resemblance of the phenomena of polarization to the selection of a certain number of particles, having their axes turned in a particular direction, supposing these axes, like those of the celestial bodies, to remain always parallel, will carry us to a certain extent, in estimating the quantity of light contained in each of the two pencils, into which a beam is divided and subdivided" (Young 1817, 334). Apparently Young was well aware that selectionism could be quite powerful, for he seems to have recognized that one could build a theory of partial reflection upon it.

5. Allowing physical rays to have successive parts merely means that the number of elements in the group that is considered must be the product of the number of physical rays and the number of parts in some standard segment of each physical ray (since the greater the latter number the more particles strike per unit time). That in the emission theory the parts of a physical ray succeed one another simply adds a new dimension to the count, but the possibility of counting does not depend on the existence of *successive* parts, since the physical rays themselves suffice. I shall hereafter always use the word "ray" to mean the least element in a beam of light, whether this is a part of a physical ray or the physical ray itself. I thank C. Hakfoort for helping me clarify this point.

6. As before, k_x and k_y may differ from one another, since we have not required that the same amount of light be always present in the refracted beam that is there when no light reflects.

7. Before 1811 Malus himself thought otherwise. He remarked: "It is probable that all the light produced by partial reflection is polarized like that which has been acted on by a crystal; but since the reflected ray contains at once molecules that are polarized in one sense and those that are polarized in the other, in its decomposition by calcspar it presents the same properties as a natural ray; and it is only at the angle of reflection, where all the molecules are polarized in the same sense, that this modification becomes sensible and incontestable" (Malus 1811b, 447).

One might be tempted to read Malus here as asserting that reflected light always consists of perpendicularly polarized components of different intensities. But one would be badly mistaken, because Malus's bundle does not have "components" in the sense that a vector does. Rather, it has parts in the sense that a divisible whole has parts. For Malus, but not for Fresnel, the statement just quoted could in principle be false; it depends on the action of matter upon light. For Fresnel the statement cannot be wrong because it is a tautology, merely restating that polarization is a directed quantity akin to a displacement.

8. Specifically, Malus notes that it is "an unmodified portion that conserves the characteristics of direct light" (Malus, 1811c, 108).

9. This led at once to the idea of using a pile of plates to polarize light by refraction, since successive plates will necessarily reduce the proportion of natural light in the refraction.

10. Note that my comparison here is essentially a photometric one: I am com-

paring ratios of intensities whatever may be the polarization of the light that contributes to the intensity. Consequently this gives only a very weak check on the overall empirical accuracy of Malus's claim, since neither he nor anyone else at the time could perform anything but the crudest photometric comparison.

11. Malus's observational claims square well with what one would see even in a very accurate experiment. At the "polarizing" angle for a metal, the difference in amplitudes between the two components (in the modern sense) of the reflected optical vector reaches a maximum, but neither vanishes. (Indeed they differ by only about 30% at most.) The phase difference becomes 90°. So the reflection shows circular polarization: in an analyzer no difference arises between the O and E refractions.

Chapter 3

1. Note that Arago was in effect looking at a very small region in one of the rings that are seen in the usual experiment—namely, the ring that derives from light reflected at this specific incidence.

2. Arago mentions Malus's reaction in a note (Arago 1811a, 17, note), even though the paper containing it was read only the following March. Arago therefore probably discussed his discovery directly with Malus.

3. Arago's use of a metal mirror (while necessary to produce two sets of reflection rings) makes a precise analysis of this experiment difficult. However, we can broadly understand why he saw what he did. The upper laminar surface generates a reflection polarized in the plane of incidence. The lower surface, bounded by the metallic mirror, reflects at polarizing incidence both components in and normal to the plane of incidence, but with a 90° mutual phase difference. The component polarized in the plane of reflection has a very small phase angle at this incidence, and so interference occurs between it and the light reflected at the upper laminar surface.

Roughening the metal surface simply produces additional reflections, much as Arago thought. They will of course exhibit the same polarization as the main reflection. In all cases the polarizing incidence on the lens determines at what incidence the reflection from the upper laminar surface becomes completely polarized in the plane of reflection.

4. Kipnis (1984, 93–96) discusses Young's work on thin films. He also remarks (p. 173 and pp. 224–28) Arago's comment on Young and Hooke and argues (correctly I believe) that Arago had not in 1811 adopted the principle of interference. However, in view of Arago's remarks on the effects of roughening a metal mirror, even if he did not adopt the interference principle he nevertheless seems to have had some idea that both surfaces of the lamina are involved in ring formation, and he considered this a difficulty for the emission theory. We must be careful to distinguish selectionist from emissionist positions, and here we have our first example of the difficulties this can present.

Arago believed that the *polarization* of the light that may exhibit rings occurs at the upper surface of the lamina, at the first locus of partial reflection. But this does not necessarily mean that rings are actually *formed* there, only that the light that may *give rise* to rings has been abstracted and thrown into the transmitted beam. In other words, quite possibly the ring-forming property is latently imposed on the light at the upper surface along with the polarization, but it is the lower surface that activates the latency. This would be of no great further significance were it not that Arago shortly built a theory, for which he had great hopes, based directly upon the distinction between the selection and polarization of the light that can form rings and the actual formation of rings by that light.

Nevertheless, we must not conclude that Arago in any firm sense "abandoned" the emission theory early in 1811, having turned instead to the wave theory. He was not very concerned with this sort of issue (or even fully aware of what Young had done), because it had nothing apparent to do with constructing a general theory based directly upon the distinction between the selection of light that can form rings and the formation of rings by that light, and we shall shortly see that this was his major concern.

5. On, that is, the angle of its principal section if it were birefringent, which Arago thinks it "natural" to believe, though it "does not have" the property in the usual sense in thin laminae (Arago 1811b, 50).

6. Though perhaps Arago regarded this last issue as too trivial to discuss. As the mica plate becomes thicker, the process of ring formation will repeat many, many times. Eventually the effect will be the same as a random redistribution of the asymmetries of the rays in the original beam of light, so that the chromatic effect will no longer be detectable.

Chapter 4

1. The University of Toronto's copy has the note. Grattan-Guinness (1981, 676) has also found a copy with the note and gives the original French. Grattan-Guinness interprets the date at the end of the note—which is separated by a period from the body of the text—as referring to the time when the paper was read. However, the copy at Toronto gives the date of reading as 1 June 1812 at the very beginning of the memoir, so I interpret the date at the end of the note as the time when the note itself was added. (The portion of the note between asterisks was printed in Arago's *Oeuvres* and translated by Crosland 1967, 334.)

2. Biot also does not tell us when he began to work on the subject, but he does say that, when he decided to begin, he publicly (in front of the Bureau des Longitudes) asked Arago to deposit Arago's unpublished material, which he did. This probably occurred shortly after Biot's return to Paris. In the 11 August paper Arago remarks that he had submitted it early for publication, before even completing the work, because "several people" were also looking into the subject—presumably Biot (Crosland 1967, 225).

3. Thus, in the case of Arago's setup, had he found that the tint seen in the lower mirror did not change as the lamina was rotated, he could have come to either of two conclusions: that the same rays as before are still deviated, and no others, or else that some previously undeviated rays are now deviated and vice versa, but in such a way that the ray ratio of the set reflecting from the lower mirror remains the same.

4. By "separation" Biot did not mean spatial distance. Spatial separation depends entirely on incidence, which is fixed in these experiments along the perpendicular, so that all the refractions in the analyzer lie in its principal section and are always separated by the same distance independent of either α or i.

5. When, that is, the principal section of the analyzing crystal is parallel to the original plane of polarization and both that plane and the section lie at 45° to the lamina's optic axis.

6. Newton's circle represents the results of color mixing in the following way. Arrayed along the circumference of the circle are the various colors of the spectrum. Each color occupies an arc whose length represents the "weight" Newton attributed to that color. To find the result of mixing a set of these colors together, first compose a set of "weights" in the ratio the intensities of the compounding colors have to one another. Then hang each of these weights at the center of the corresponding arc in the

color circle. The position of the resulting color lies at the center of gravity of this system of weights. Draw a line from the center of the circle to this resultant and extend it to the circumference. The color of the compound will be the same as the color of the intercepted arc on the circle; and the distance of the resultant from the circle's center represents how much white must be mixed with the pure tint to produce the compound.

7. Suppose we arbitrarily specify that tint U is situated at a distance U_T from the center of Newton's color circle. Then A must lie at some distance A_T on the same diameter but on the other side of the center, since the center of gravity of the compound formed by mixing U and A together must be white and so must lie at the center of Newton's circle. Now F_e consists entirely of the tint U, whereas F_o is formed by a mixture of the tints U and A. Consequently the distance between the tints of F_o,F_e can be found simply by adding to the distance U_T of U from the center, the fraction of the distance of A_T from the center that represents the amount of A contained in F_o:

$$\text{distance} = \frac{1}{U_T + A_T}\left[U_T + A_T(\cos^4(i) + \sin^4(i))\right] \qquad (4.2.4)$$

The distance accordingly reaches a minimum when i is 45°, precisely as Biot claimed. Here, then, the problem raised by experiment depended upon accepting Newton's color-mixing formalism, which Biot always did.

8. Biot goes even further. In general, he notes, the intensities will be the same if:

$$\tan(2\alpha) = -\frac{\{(U + A\cos(4i)\}}{A\sin(4i)}$$

From this we can explore several exemplary situations. First, we see that, whatever A, U, and i may be, for any triad (U,A,i) there exist four values of α ($\alpha + n{*}90°$, where $n = 0,1,2,3$) which produce equal ray counts F_o,F_e. Second, we may derive information about the ratio U/A. When α vanishes we must have $\cos(4i)$ equal to $-U/A$. Then A must be greater than U for i to have (in Biot's words) a "real" value—which means that, when the optic axis is in the plane of incident polarization and F_o,F_e have equal intensities, then the ray count of A must be greater than the ray count of U.

Finally, we can determine the conditions for producing white images whatever α may be. For this the factors in equations 4.25 that multiply U,A must be equal, which requires that $\sin(i)\cos(i)\sin(i - \alpha)\cos(i - \alpha)$ vanishes. This gives four values of i that are independent of α and four that are functions of α ($i = n{*}90°$ and $i = \alpha + n{*}90°$ for $n = 0,1,2,3$). These are the conditions under which, as Biot put it, "the lamina imparts no rotational movement to the luminous molecules" (Biot 1812, 155).

9. Biot explicitly remarks this necessary partial reflection at intermediate thicknesses: "One must consider the variable intensity of light reflected in the width of one and the same ring formed by the simple [homogeneous] light; because the reflection is not invariably limited to thicknesses 1,3,5,7 . . . , nor the transmission to the intermediate thicknesses 0,2,4,6 . . . ; but reflection and transmission also occur before and after these limits with a continuously decreasing intensity; so that near the thickness 3, for example, a portion of the light is already transmitted, while the greatest portion is reflected, and so on, through continuous degradation" (Biot 1812, 64).

10. Since the emergent light is, in general, elliptically polarized. But it would be a difficult task indeed to adjust the ray counts to accomodate the ellipticity at any given thickness.

11. Biot was aware of this limitation.

Chapter 5

1. In detail, (1) algebraic analysis included: properties of polynomials, curve representation, progressions, series summations, generation of sine and log tables by constant difference methods, etc.; (2) differential and integral calculus included: Taylor's series, maxima and minima, curve normals and tangents, integrability, separable first- and second-order differential equations, simple logarithmic and trigonometric integrations, etc.; (3) mechanics included: primarily the motion of a mass point, but also statics, the concept of pressure, machines and motors, and some sort of hydrodynamics.

2. 1. extension, impenetrability, porosity, elasticity
   2. capillarity
   3. dilatation by heat; thermal conductivity
   4. astronomy and gravitation
   5. barometry and hygrometry
   6. meteorology
   7. acoustics, including:
      i. propagation
      ii. vibrating cords and wind instruments
      iii. uniformity of speed of propagation
      iv. variation of intensity with distance and density
      v. echos, acoustic horns, etc.
      vi. frequency of vibration of cords
      vii. terminated cords: harmonics
      viii. "principle of coexistence of small vibrations"
      ix. physiology of the voice
      x. diatonic scale
      xi. harmonics of wind instruments
   8. light, including
      i. laws of propagation, methods of speed determination, variation of intensity in transparent and uniformly dense media.
      ii. catoptrics including caustics
      iii. dioptrics including caustics
      iv. prismatic dispersion
      v. the rainbow
      vi. double refraction and "inflection"
      vii. retinal images
      viii. optical instruments
   9. electricity including galvanism
   10. magnetism

3. Fresnel, *Oeuvres,* 2:830–31. I do not know precisely what Ampère had in mind here. He was perhaps referring in part to Arago's early objections (see chap. 4 above) to the emission account of Newton's rings, but it seems likely that Fresnel's remarks went beyond this point, so that by 1814 Arago must have developed other objections as well. Since they were, according to Ampère, similar to Fresnel's (which were probably akin to the concerns expressed in the letter to his brother Léonor quoted above), they probably concerned such things as the speeds with which particles are emitted and so forth. They would not have involved considerations of empirical adequacy at this point.

4. A report after the hundred days (dated 28 October 1816) from a royalist

functionary reads: "Attestons qu'aussitôt que la formation d'une armée sous les ordres de S.A.R. Mgr le Duc d'Angoulême fut connue dans le département de la Drôme M. Fresnel ingénieur ordinaire des Ponts et chausées à la résidence de Nyons se rendit au quartier général de S.A.R. alors à Nimes pour y prendre de service en qualité d'ingénieur . . . que cet acte de dévouement fut justement apprécié par cet officier général et par nous, mais que le mauvais état de santé de M. Fresnel empêcha qu'il fut utilisé" (Brunot and Coquand 1982, 124).

5. Kipnis (1984, 190–91) calculates that Fresnel's fringe measurements imply an accuracy for the wavelength of 6.8% to 7.5%. Young's observations (Kipnis 1984, 293) yield an accuracy of 4.7% to 6.0% if a phase change of 180° is introduced for the light that strikes the edge of the diffractor (on which see below), or of 13.9% if, as in Young's original paper, the phase change is not used. Young, that is, thought even so great a discrepancy as this was reasonable.

6. Young remarks: "In making some experiments on the fringes of colours accompanying shadows, I have found so simple and so demonstrative a proof of the general law of the interference of two portions of light, which I have already endeavored to establish, that I think it right to lay before the Royal Society, a short statement of the facts which appear to me so decisive" (Young 1804, 1). Young, however, did not glue a fairly sizable piece of paper to the diffractor's edge as Fresnel did. He instead took a small "screen" and placed it "a few inches from the [diffractor] so as to receive either edge of the shadow on its margin." Kipnis (1984, 113–14) remarks that this method of observation, unlike Fresnel's, introduces a number of ambiguities. Furthermore Young, again unlike Fresnel, never generated a formula for these internal fringes that he tested against measurements of them. We shall see below that Fresnel's first modification of the binary theory resulted from his observation that the formula governing these fringes should also hold in a different situation but does not.

7. Kipnis (1984, 107–8) discusses this point. Also see Cantor (1970). Young at first attributed the rays' deflections to changes in ether density rather than to reflections from the diffractor's edge.

8. This was actually unnecessary because a 1 mm wire requires that the source be less than 10 mm away—much closer than Fresnel ever placed it—to produce a 0.001 mm change in its effective diameter, and it requires about 0.01 mm change in diameter to produce a fringe shift of significant dimensions.

9. I do not know precisely how Fresnel discovered the necessity for the phase shift. However, I can assert with some confidence that he did not discover it by observing very close to the diffractor edge. The original formula implies a maximum at the edge, whereas the phase-shifted formula implies a minimum there. Which occurs? Strictly speaking there is a major maximum nearest the edge, followed by the first major minimum. Suppose we take a, b as, respectively, 342 mm, 0.001 mm, with c as 1 mm. Then I find that the maximum is only 0.005 mm from the edge, and the minimum is 0.005 mm from the maximum. Since the intensity increases rapidly with relatively slow oscillations as one approaches the edge of the shadow from within, Fresnel would naturally have located this maximum within the shadow, while the minimum would seem to be indistinguishably close to the edge. However, it is obvious that one cannot by measurement distinguish the loci of this maximum and minimum, so that here it makes no difference whether one uses the phase-shifted formulas or not—the difference between them lies well outside the range of his observational accuracy. Clearly Fresnel must have discovered the phase shift by looking elsewhere, and it seems likely that he did so by measuring the first major minimum past the major

external maximum—since, as Naum Kipnis remarked to me, the difference between the original and the phase-shifted formulas is greatest for the first minimum.

10. Which Young had already recognized.

11. Consider, for example, the first experiment: a = 1,490 mm, b = 385 mm. Here I find a minimum of logged intensity 2.303 at 1.29 mm and a second of logged intensity 2.465 at 1.37 mm. Fresnel places the minimum at 1.32 mm; his formula implies 1.34 mm. At 1.32 mm the logged intensity is about 2.33 (again we have about a 1% intensity difference). Note that here, as in every other case I have examined, the observation is shifted away from the shadow and toward the predicted locus.

12. This is an extremely accurate approximation: since r^2 is actually equal to $b^2(\frac{1}{2}c - y)^2$, and c, y are both about 1 mm, while b is 500 mm or more, the error involved is about 0.000333 mm in r, which produces a computational error of approximately .01% of a wavelength in the locus of the minimum.

13. Fresnel omitted the geometric proof, but it is simple. Draw GO, DM parallel and respectively equal to EF, KL. We assume a plane front and construct KL such that K, L have the same phase, that is, ED + DL = FG + GK. We want to show that KL is normal to GK, DL—that it is the reflected front—only if ∠FGA = ∠KGD = ∠LDB. We have ∠KGB = ∠FGA = ∠EDA, since the front is plane, and GM + MK = OD + DL, so that KL is the normal only if GM = OD. But GM, OD are respectively GD cos(∠KGB), GD cos(∠EDA), so the result follows.

14. If the incident front (fig. 5.7) is spherical, then the rays FG, ED proceed from the center J. FE is the incident front. We construct K, L such that ∠FGA = ∠KGB and ∠EDA = ∠LDB, with FG + GK = ED + DL. We want to show that KL is the reflected front under these conditions, that is, that KG, DL meet at the reflected front's center, C, which requires CK = CL. By construction we have FG + (CK − CG) = ED + (CL − CD). We need to show that FG − CG ≐ ED − CD, or JG − CG = JD − CD, since FE is the incident front. We have ∠CGA, ∠ADC respective equal to ∠JGA, ∠JDA, whence, since GD is common to both ΔJGD, ΔCGD, we have JG − JD = CG − CD and so, transposing, we have our result. Note that every step is an equality, so we have shown that the law of reflection holds if and only if points of equal phase on the incident front are transformed by reflection to the same front.

15. For example, take the wavelength at 5176Å. Then for a source distance of 750 mm, the differences between the true locus and that required by Fresnel's formula are, for screens at 3 m and at 6 m, 0.05 mm and 0.75 mm. Examination of a range of source and screen distances indicates that, unless the observational accuracy is better than about 0.05 mm, it would be difficult to find Fresnel's formula in error unless the source distance surpasses 1.5 m or the screen distance surpasses 3 m.

16. Fresnel (1815b, 44–45). In particular, Fresnel deduced the following formula for the maximum path difference, δ; here r is the radius, i the angle of incidence, and n the index of refraction:

$$\delta = r\left[\frac{2 - n}{\sqrt{n^2 - \sin^2(i)} - \cos(i)} - \frac{1 - \cos(i)}{n}\right]$$

If, for example, we take an incidence of 10°, a wavelength of 5176Å, n as 1.5, and r as 10 mm, then δ/λ is about 0.3. This is at the edge of significance. Note that increasing the index decreases the path difference.

17. He did test the formula for a rather different situation: two thin, neighboring threads, which he treated as a transmission grating. He arranged the silk threads

to meet at an angle of about a third of a degree, thereby forming a device similar to his micrometer. Using the diffraction formula for a plane wave incident at 90°, Fresnel had sin(r) equal to ($n\lambda$/AB) − 1, where r is the angle to the image and AB is the distance between the wires. He measured r, computed AB from it and compared with observation, obtaining an accuracy of 6% to 11%. The fairly large error is due to Fresnel's assimilation of what is actually a dual obstacle diffractor to a line grating: see Verdet on this point (Fresnel, *Oeuvres*, 1:49).

18. It is trivially simple to understand what occurs if we allow each point of the front to send rays in all directions—which is precisely what Fresnel *did not* allow in his early work. Draw a line from any given point in the illuminated field to the center of the slit and to the slit's edges. Suppose that the line to one of the edges is a half-integral number of wavelengths longer than the line to the center. Assume further that *every point on the front* sends rays to the field point. Then construct a series of rays to that point and starting from the edge that differ from one another by a half-wavelength. Obviously the approximate result will be to produce a dark fringe at the field point if we pair rays to the right of the central ray with rays to its left. However, the rays to the edges differ from one another by an integral number of wavelengths, so that according to the binary formula the field point should be bright. That this simple count leads to an immediate conflict with the binary formula is of no historical consequence at all, because Fresnel discovered the anomaly empirically. He did not, that is, discover that two methods of computing fringe positions—one involving only a pair of rays, the other involving rays from all points on the front—lead to different results. He had no idea what the latter method would lead to, because he did not believe that the front sends rays in all directions. Nor, I shall argue in a moment, did he recognize this conflict immediately after discovering the anomaly, because he continued to use only two rays to form fringes.

19. Of course the transition must actually occur continuously. Analysis indicates that, at the shadow proper, the difference betwen FC and FA is an eighth, not a quarter of a wavelength, so that the transition from a difference FC − FA of $+\frac{1}{4}\lambda$ to one of $-\frac{1}{4}\lambda$ takes place somewhat past the shadow.

20. This kind of computation, though simple, in fact requires an additional assumption that I have not mentioned, and that Fresnel never explicitly made: namely, that the amplitude due to a region on the front at a distance r from the region varies reciprocally with r. Without this assumption it is not generally possible to construct half-wave zones that have equal effects at a given point, and this would prohibit conclusions based on a simple zone count. I shall return to this point below, for it is related to the difficulties Fresnel had in understanding how to use Huygens's principle.

Chapter 6

1. It is difficult to see how eighteenth-century work could have been of much help to Fresnel, because the physical problems it treated do not have precise analogues in his theory of light. Fresnel had to envision a set of coherent point sources of luminous waves that differ only in the times at which the disturbances reach given values for the same distance from each source (i.e., they differ only in their phases). Nothing similar occurs for the vibrating string or membrane except at the boundaries, where the waves are reflected. But here one does not have to consider the points that constitute the boundary to be emitters, because the configuration of the string or membrane at any time follows from the initial configuration, which includes the boundary conditions, if we have a general solution. By contrast, until the development of the three-dimensional wave equation and then the introduction of the complicated boundary

integrals its solution requires, wave optics had to make physical assumptions where, it was later realized, the results can be obtained purely analytically. Indeed, Huygens's principle is itself a physical representation of the general diffraction integral.

2. For two coherent waves with an arbitrary phase difference, Fresnel simply generalized the procedure. If a, a' are the amplitudes of two waves whose centers are a distance c from one another (which means they have a phase difference $2\pi(c/\lambda)$, then the amplitude of their resultant is the same as that of two forces with magnitudes a, a' that are inclined to one another at an angle $2\pi(c/\lambda)$. The phase of the resultant wave (relative to the phase of the original pair for which c is zero, that is, relative to the source from which x is measured) is:

$$\tan^{-1}\left[\frac{\sin\left[2\pi\dfrac{x}{\lambda}\right]}{1+\cos\left[2\pi\dfrac{x}{\lambda}\right]}\right]$$

Combining a series of waves, each with its own phase, cannot conveniently be done using this decomposition, because one would have to combine the waves in sequences of pairs: one would have to calculate the resultant of the first pair and then compute its resultant with the next wave, and so on. Precisely this situation arises in Fresnel's integral diffraction theory, and so here he had direct recourse to the quarter-wave decomposition. In essence, to find the resultant of a set of waves with given phase differences, choose one among them as a reference and decompose each of the others into sine and cosine components. Then group together all the sine terms and all the cosine terms. The members of each set differ from one another in amplitude but not in phase, so their resultant will be the square root of the sum of the squares of the component amplitudes.

3. To see how this follows, produce in figure 5.1 the line AB until it intersects SP_e at some point m′. Then very nearly we have:

$$\frac{y}{\mathrm{m'A}} \approx \frac{a+b}{a}$$

To effect the transformation replace m′A with z and use equation 5.2.1, in which d now becomes m′s′.

4. Fresnel did not reveal his source for this result, but he most likely obtained it simply by using the error integral and substituting in it an imaginary for a real exponent, as follows:

$$\int_{-\infty}^{+\infty} e^{-\alpha z^2} dz = \sqrt{\frac{\pi}{\alpha}}$$

$$\alpha = \frac{1}{2} i\pi$$

$$\int_{-\infty}^{+\infty} e^{-(1/2)i\pi z^2} dz = 1 + i$$

Whence $\int_0^\infty \cos(\frac{1}{2}\pi z^2)dz = \int_0^\infty \sin(\frac{1}{2}\pi z^2)dz = \frac{1}{2}$.

5. Kipnis (1984, 249–50) argues that the unknown author was Honoré Flaugergues, who had written on diffraction in 1812 and again in September 1819, immediately after the competition. "Since," Kipnis remarks, "Flaugergues is the only physicist to publish on diffraction both before and after the contest, with improvements corresponding to its program . . . it is most likely that he is the one who competed with Fresnel."

6. If ϑ is the angle between an oblique ray and the normal to the front, then the correct (albeit still approximate) inclination factor is not Fresnel's cos ϑ but $-\frac{1}{2}(1 + \cos \vartheta)$, as Stokes demonstrated in 1849.

7. Young did not indicate with great clarity what his symbols meant, so he may perhaps not have erred in estimating the path difference. He considered two rays to combine somewhere along the axis of the circular aperture: one direct from the source, the other from some point in the circumference. He writes that, with x representing the area of the "orifice," then the path difference, if the indirect ray comes from the circumference, will be ax/d, where d is "the distance of the object" (i.e., the field point on the axis), and he gives no explicit meaning to a.

From the usual Fresnel approximation the path difference will be nearly $[r^2 \times (a + b)]/2ab$, where r is the radius of the aperture, a is the distance from the aperture center to the source (assumed to be axial), and b is the distance from the aperture center to the axial field point. This is hard to reconcile with Young's estimate. In his reply to Young (see below) Fresnel simply set Young's d equal to a product $1d$ (which would correspond to ab) and included a factor of $\frac{1}{2}$—but Fresnel also did not specify the meaning of his symbols. In answering Fresnel, Young remarked he had been too hasty.

8. Kipnis (1984, 264) remarks that Jöns Berzelius in 1820 completely misunderstood interference, thinking it equivalent to a loss of vis viva (with a possible transformation of energy into heat).

9. Poisson's remarks on this point are discussed in Arnold (1978, 189–92). Arnold also discusses Poisson's lengthier critique of the forms Fresnel tacitly presumed for the distance and wavelength dependence of the secondary waves (193–98), though Fresnel did not reject Poisson's positive conclusions, but rather sought to justify them.

10. I cannot here discuss in any detail Poisson's innovative work in deducing and solving the three-dimensional wave equation. But it is important to grasp the outlines of what he did and did not do. He did not determine the effect at a point P by integrating some function over a surface that contains a given distribution of velocity and condensation. Nor did he even consider a wave train. Poisson instead proceeded more or less in the traditional manner by obtaining special solutions to the equation, which he rewrote in spherical coordinates. In effect, Poisson examined the case of an isolated spherically symmetric pulse, paying no attention to how it had originally been generated.

He found first that the solution for what was later called the velocity potential φ of the fluid (from which the velocity v and condensation s of the fluid at a fixed point in space are obtained) consists of two parts. One part vanishes if the fluid is initially in equilibrium and constitutes a wave of condensation. The second part vanishes with the initial speed and constitutes a velocity wave:

$$\varphi = \varphi_{con} + \varphi_{vel}$$

If r is the distance from the center of the pulse and c is its speed, then, he demonstrated, the pulse will expand outward *only* if the initial value of φ^0 of φ satisfies

$$\frac{1}{c^2}\frac{\partial \varphi^0}{\partial t} + \frac{\partial r \varphi^0}{\partial r} = 0.$$

If this is not the case then the pulse generates *both* divergent and converging waves. See Baker and Copson (1939, 8–20) for details.

For Poisson, then, the unit of analysis was the entire spherically symmetric solution, and the absence of retrograde radiation meant that the initial conditions of velocity and condensation over a sphere were such that only one of the two solutions exists. Fresnel was proposing to remove the retrograde wave, it seemed to Poisson, simply by fiat in assuming that his secondary emitters could have the right sorts of velocity and condensation, given the demands placed upon them by the diffraction integrals, to obviate retrograde radiation.

11. In the M.C. he remarks: "I will make another hypothesis concerning the nature of these disturbances: I will suppose that the speeds given to the molecules are all directed in the same sense, perpendicular to the spherical surface, and are also proportional to the condensations; so that the molecules cannot have retrograde motion" (Fresnel 1819b, 293–94).

12. In a sense the distinction between a velocity wave and a wave of condensation is an artificial one, since all waves are displacements of the medium's particles. What Fresnel and Poisson meant, and what is meant today, concerns the initial conditions over a given front. If the particles on the front are initially at rest but displaced from their positions of equilibrium, then the resulting wave may be called a wave of condensation. If, on the other hand, the points are initially undisplaced but have velocities, then a velocity wave results. These two kinds of waves, Poisson had shown, appear as distinct solutions to the wave equation. In general both occur (since the initial front's particles may have both velocity and condensation); both must occur if there is to be no retrograde radiation.

13. Baker and Copson remark that "to justify Huygens' principle for isolated waves [i.e., for pulses], we must have recourse to analysis and take into account the dynamics of the medium in which the wave-motion occurs" (1939, 19).

14. He remarked: "The reason is easy to grasp; in effect, the particles of luminous bodies must experience frequent perturbations in their oscillations as a result of rapid changes that occur about them; now we cannot suppose that these perturbations act simultaneously and in the same way in separate and independent particles; consequently, two systems of waves having left two different sources will experience, in their interference, anomalies that will cause the regularity [of the pattern] to disappear" (Péclet 1823, 492).

15. As a fourth example one may take Cesar Despretz's 1825 *Traité élémentaire de physique,* designed as an elementary book. It was dedicated to Gay-Lussac and to Arago, probably more because of their importance (Arago being editor of the *Annales de Chimie*) than because Despretz was entirely aware of Arago's role in advancing the cause of the wave theory.

Despretz was quite favorable to the wave theory, which he apparently learned from Fresnel's *Annales* article. He even went so far as briefly to remark that, in it, rays are lines to the wave surface. Nevertheless Despretz continued to employ selectionist language in his discussion of interference, which involves only Fresnel's mirror expériment. He discusses double refraction almost entirely observationally. Polarization seems to puzzle him somewhat, and he occasionally deploys the language of beam division in discussing it. With the exception of his brief remark on the definition of a ray in the wave theory, Despretz's work would not have disabused many readers of their belief that rays remain fundamental elements in optics.

Chapter 7

1. Verdet included enough of the original, unpublished documents in his edition of the *Oeuvres* to permit reconstructing Fresnel's early thoughts on polarization. There are essentially two variants of Fresnel (1816c). Verdet constructed the first variant from three sources: (1) an undated but signed manuscript in Fresnel's papers; (2) a signed manuscript dated 30 August 1816 in Biot's papers; (3) a second copy of (2), also in Biot's papers and with many annotations. The second variant derives from two

sources: (1) a manuscript that completes many passages lacking in variant (1); (2) a copy dated 6 October 1816, submitted to the Academy on 7 October, which lacks the first twenty-two paragraphs of (1).

2. Fresnel later developed a sophisticated experiment to prove the converse point (Fresnel 1819c, 518–21)—that interference does take place if the interfering beams were originally polarized in the same plane. He took a gypsum lamina cut with the optic axis in the facet and shined *polarized* light normally on it. The angle between the plane of initial polarization and the optic axis was 45°. The bottom of the lamina was covered with a copper plate pierced with two openings. From each opening two beams with equal intensities emerge, with one in each pair polarized in, and the other perpendicular to, the optic axis. Observe the two holes through a doubly refracting prism. The resulting fringe patterns require that all the ordinarily refracted beams in the prism interfere with one another, and that all the extraordinarily refracted beams also interfere with one another.

3. To a selectionist, recall, an "unpolarized" beam consists of a set of rays with completely random asymmetries. "Polarized" light consists of rays that all have precisely the same asymmetry. "Partially polarized" light contains many randomly oriented rays as well as at least one set of rays that all have the same asymmetry. In fact, one could in principle even create a partially polarized beam by physically mixing polarized with unpolarized light.

4. As we proceed through Fresnel's early work, we will be able to see how he constructively deployed assumptions based on the image of a temporally stable polarization, assumptions he was later able actually to deduce once he had set polarization in motion. One might, very roughly, describe the alteration this way: before 1820 or 1821 Fresnel thought polarization itself had to be explained; after that he thought that only the effects of reflection and birefringence *on* polarization needed explanation. Here we shall be examining a particularly subtle sort of conceptual change, one in which the empirical and theoretical connotations of terms related to "polarization" had to be disentangled and reconstructed in new ways. I do not think Fresnel himself thoroughly realized how fundamentally he had altered past ways of thinking. It was, I shall argue, more this change than Fresnel's quantitative theory of diffraction that marks the emergence of a theory that was thoroughly incompatible with selectionist ways of thinking.

5. By "temporally fixed" I do not mean that the light vector ceases to oscillate. Rather, I mean that it points to the same direction in space at every instant. This involves a point I will be discussing in some detail below, but which is worth making here. To wit, that for some time Fresnel did not connect phase differences with anything but amplitude variations in a given direction. Or better put, for some time he insisted on interpreting phase differences always as path differences.

6. Frankel (1972, 313–16) discusses Fresnel's and Young's early work on chromatic polarization and remarks the deficiency of Young's theory in not explaining why polarized light must be used and what the role of the analyzer is. Both questions are obviously answered even in Fresnel's first account. Frankel also discusses the difficulty, just noted, concerning the complementary colors of the analyzer images.

7. Frankel (1972, 317–18) remarks the difficulty with thick crystals, but he views the difficulty as "a major contradiction," whereas Biot unquestionably saw it merely as an unsolved problem.

8. He even mentioned his astonishment in a letter to his brother dated 22 October: "Many observations reported in my memoir have already been made, but those to which I attach the most importance are entirely new. I confess that I could scarcely believe how the first phenomenon of this kind I noticed, the one that put me on the

path to the others, could have escaped Malus and Biot, who both speak of the case where it occurs and set up a principle that is not at all in agreement with the facts I observed" (Fresnel, *Oeuvres*, 2:843).

9. In fact Fresnel first applied his newly found method for calculating resultants to a special case where Biot's claims differed from those he obtained. I will return to the point below, since Fresnel emphasized it heavily in his critique of Biot. See appendix 4 for details of the calculations.

Chapter 8

1. Fresnel had to use interference to detect the birefringence. In optically rotating liquids the "birefringence" cannot separate the paths of the two waves; it only produces a phase difference between them. Consequently the effect cannot be directly observed, as it can in crystals like Iceland spar.

Instead, Fresnel first reflected light at polarizing incidence from a pair of prisms set to produce Newton's rings. The light then passed through a tube filled with turpentine. Observing the light coming from the tube with just a lens, he saw no more rings than before it had passed through. But using a gypsum crystal as well produced many more, albeit fainter, rings than before. These, he argued, can only mean that the phase differences between portions of the light reflected from the superior and inferior surfaces of the air lamina have been reduced (yielding more rings). This very limited claim says nothing about the polarizations actually produced, only that there must be some effect that produces waves differing in both polarization and speed.

2. The major purpose of the note was to contrast Fresnel's explanation of chromatic polarization with Biot's. It contains formulas that result from Fresnel's January calculations, and it nicely supplements the January paper by directly addressing the issues raised by chromatic polarization. Fresnel would naturally have written such a note shortly after he discovered how to produce a theory that could compete adequately with Biot's at all points.

3. Fresnel had therefore developed the formulas for the general case of chromatic polarization by this time. Suppose first that, in the formulas of appendix 4.2, $o - e$ is an integer—that the path difference is an integral number of wavelengths. Then the extraordinary image vanishes for s equal to zero, which means the light behaves precisely as though it were polarized in the primitive plane. If, on the other hand, $o - e$ is an odd multiple of $\frac{1}{2}$, then the extraordinary image vanishes for s equal to $2i$, which means the light acts as though it were polarized in the azimuth $2i$.

So much is clear given only orthogonal polarizations and path differences. If, in addition, we adopt transversality, then this result is at once obvious. The entering wave is split into components $\cos(i)$, $\sin(i)$ along and normal to the lamina's principal section. If the components emerge with the same phase, then obviously the original wave is simply reconstructed. If they have opposite phases then the normal component has flipped across the axis, so that the reconstructed oscillation inclines at i to the axis, but on the opposite side from its progenitor. These are the only two possible directions for linear polarization; any other phase difference yields an elliptically polarized resultant.

4. My point is not that Fresnel failed to realize that a single polarized wave can always be mimicked by a pair of orthogonally polarized waves with no phase difference and the correct ratio of amplitudes. Fresnel could easily have known it, indeed probably did, because it follows at once from applying the principle of interference. However, as long as one thinks of the orthogonally polarized waves not as components but as independent waves, then it is entirely possible that they are only *artificially* equivalent to a polarized wave: the analyzer may be necessary to force the two to

interfere, in which case what their resultant is in the analyzer's absence remains an open question.

5. See Fresnel (1822b). There is no need to discuss this work in detail because the points it illustrates are best shown through Fresnel's new analysis of partial reflection. In essence Fresnel began by using his new method of orthogonal decomposition to demonstrate that any linearly polarized wave can be decomposed into two circularly polarized waves that rotate in opposite directions ("right" and "left"). The angle of polarization determines the phase difference between the similarly directed components of each circularly polarized wave. Optically active media, he supposed, propagate the circularly polarized waves at different rates, resulting in new phase differ ences and hence, on emergence, in an altered direction for the resultant plane of polarization.

6. Consider the principle of interference. Many selectionists did accept the principle but thought of it exclusively in terms of rays. Or rather, they accepted a very limited form of the principle, to wit, that two rays may, under the proper conditions, apparently destroy one another's visible actions. The existence of intermediate intensities was paid little heed. Beudant (1824, 556) is a good example. He remarks only that "if we cause two rays of light to meet in any way, their encounter gives rise to a series of dark and light fringes"—not "fringes of varying intensity," say. This way of thinking was furthered by Fresnel's method of zone computations, which are adapted solely to calculating loci of total interference. But even if intermediate intensities are considered, still the ray-based emphasis of selectionism is not a substantial barrier to accepting Fresnel's analysis of diffraction.

Chapter 9

1. Frankel (1976, 171) notes that rivalry between Arago and Biot had already flared up again toward the end of 1820 over a position as inspector of studies, which Biot succeeded in acquiring. That was perhaps the precipitating factor in Arago's decision to write his report several months afterward, but Arago's bitterness and longstanding anger hardly needed much prompting, despite an "accord" between the two men worked out by their wives.

2. The 1816 memoir is Fresnel 1816c, whose original title said nothing about chromatic polarization and which dealt with that subject only toward its end. The "*Supplement*" of 1818 was Fresnel 1818c, in which Fresnel showed how to combine waves with arbitrary phase differences. This was in fact a *supplement* not to 1816c, but to Fresnel 1817b on the effect of polarization on reflection. Much of what it had to say about chromatic polarization concerned the special case of laminae with axes crossed at 45° and Fresnel's model for optical rotation (see appendix 4).

3. Namely, that a beam of *homogeneous* light emerges from the lamina entirely polarized either in the original plane or at the angle $2i$ to it (i being the angle between the laminar axis and the original plane), but not simultaneously in both.

4. There is another problem with this experiment, though Fresnel himself never remarked it. It is a dubious enterprise to deduce the polarizations of the compounding waves from the polarization of the fringe pattern they form when the waves that emerge from the thin laminae are in fact polarized elliptically, not linearly. Furthermore, in general the ellipse axes are *not* in and perpendicular to the principal sections of the laminae.

Fresnel perhaps realized this difficulty, because when he did finally discuss the experiment in his "De la lumière" he began with crystals thick enough that they do not produce chromatic polarization—and so they separate the beams sufficiently that elliptical polarization does not occur. These, he showed, form well-separated fringe patterns with polarizations in and normal to the principal sections. Having established this fact, he then discussed crystals thin enough to produce colors, and these, he explained, have essentially the same effect—the only difference being that the two fringe patterns overlap somewhat. Here he relied implicitly on a principle of continuity—that if thin laminae show the same kind of phenomena as thicker laminae, then there can be no essential difference between them.

But it may not be so. In this experiment we have two elliptically polarized waves interfering with one another. The axes of the ellipses lie in and perpendicular to the respective principal sections of the laminae. These are also (according to Fresnel) the directions along which the beams within the laminae are polarized. But suppose for a moment that the directions of polarization on emergence are as Biot supposes them to be. It is nevertheless possible to produce precisely the same elliptical resultant that Fresnel's polarizations produce by adjusting the phases and amplitudes of the constituent beams. But if this is so, then the fringe patterns and polarizations produced must also be the same. Consequently Arago's and Fresnel's critique holds only if we also assume that Biot's beams in thin laminae have the same intensities and phases as Fresnel's. Or to put it in a way that avoids transverse waves entirely, Fresnel's directions of polarization, intensities, and phases can be replaced by an alternative set with polarizations as required by Biot.

5. Biot paid no attention at all to Arago's lengthy discussion of the fringe-production experiments, in which Arago claimed decisively to have proved that Biot's chosen azimuths (0° and $2i$) are incompatible with the results of many experiments. Biot was clever to have ignored Arago on this point, because, as I remarked above, Arago's experiments prove nothing of the kind. However, Biot's silence on the point certainly does not reflect his own mastery of the issue; rather, it probably reflects his view of interference as a special, and highly limited, process of ray interaction that can be used to account for fringe formation but for nothing else. Biot, that is, almost certainly thought of interference primarily as a constructive method rather than as a fundamental physical principle.

6. Though, in so doing, he often uses the word "ray" to refer to a "beam," writing, for example, of the division of a "ray" into parts.

7. In these circumstances Biot's allowable directions of polarization are 0° and 90°—that is, in the principal section of the analyzer and perpendicular to it. Consequently, if only one of these two polarizations existed in homogeneous light at every laminar thickness, then the 0 beam and the E beam could not possibly share rays of a given color, as Biot supposes they may.

8. Biot added a note to the printed article that considerably clarifies the argument as it was originally presented to the Academy. I shall consider the note an integral part of his remarks.

9. There was more in this vein, again revealing Biot's deep-seated selectionism. Arago had called the light at intermediate thicknesses of the lamina "partially polarized." Biot objects. It is not partially polarized, because it consists of two sets, each completely polarized, one at 0°, the other at $2i$. *Partially polarized* light must have many randomly asymmetric rays as well as rays with common symmetries.

Arago had also made two other objections that Biot felt he had to answer. The first concerned Biot's mistaken claim regarding the behavior of two laminae with axes crossed at 45°. That claim, Biot repeated (having so told Fresnel long before) resulted from an inaccurate application of his own laws: he had simply not applied his own theory correctly. The other issue was more difficult: it involved the circumstances in which chromatic polarization is transformed into fixed polarization as the lamina becomes thicker.

Biot adds a new element here. He still does not attempt any explanation of how the change occurs, but he tries to mitigate the difficulty by postulating an intermediate stage. Suppose that, as the thickness of the lamina increases, the rays begin after a point to distribute themselves more uniformly between the two azimuths of polarization, eventually reaching an equal distribution. The analyzing crystal will not then be able to distinguish between this state of the beam and a state in which the rays are distributed equally between any other two orthogonal azimuths. Consequently the transition from the 0°, $2i$ azimuths of chromatic polarization to the azimuths in and normal to the principal section that obtain in thick crystals could occur without being detectable "by this method of observation"—the beams must separate sufficiently to be separately examinable, and this might not occur until well after the transition between the two modes of polarization has already taken place.

10. I have perhaps inordinately compressed Biot's argument in a way that makes it seem somewhat clearer than it actually was. Biot did not attack the issue head on but decided first to explain why it is at all possible for there to be light at both azimuths. In essence, he argued that each ray in a set of homogeneous rays need not be in the same "phase" of its tendency to be at a particular azimuth. The spectrum of phases cannot, however, be too wide (or the number of rays at the extremes too large), or else at *every* thickness both azimuths would occur. The narrower the range of phases, the narrower the transition range over which both azimuths are present in comparable amounts, and so the more accurate Newton's table of tints will be. Biot gave the argument in detail for Newton's rings, but the principle holds as well for chromatic polarization.

11. Fresnel writes: "Biot recalls a conversation in which he explained to me how the forumlas that had led him into error about the tints produced by two laminae of equal thickness crossed at 45° were not a necessary consequence of the theory of mobile polarization. I avow that I did not very well understand what he did me the honor to tell me on this subject, and that I still do not divine how this knowledgeable physicist can *deduce from his* theory the general formulas for the case where the axes form an arbitrary angle. But I never cited the error into which he had been led by his first formulas, and *from which I had been warned by mine,* as a decisive proof of the inexactitude of his theory; I only wanted to show by this example that I had chosen a better guide than his" (Fresnel 1821d, 606–7).

Chapter 10

1. In the original: "D'après M. Fresnel, la polarisation consiste dans la décomposition des petits mouvemens qui ont lieu dans les ondes, en deux autres mouvemens perpendiculaires entre eux et à la direction du plan de polarisation."

2. Péclet's wording here is in fact nearly incoherent on its face, because the motions cannot be mutually perpendicular and also be perpendicular to the *plane* of polarisation. He might perhaps have meant that each of the two motions is orthogonal to its corresponding plane of polarization, which would mean he considered the motions so independent as to, in effect, acquire each its own polarization plane. This obscures in a very deep way the sense in which decomposition was used by Fresnel. In Fresnel's understanding there is always, at every instant, a single direction, and the usefulness of any particular decomposition is determined entirely by the physical actions of reflection and refraction on the amplitude and phase of the oscillation rather than by the creation in any meaningful sense of two new and mutually independent systems of waves.

3. The experiment Baumgartner refers to, which he regards as conclusive, is a typical case of the usual criticism—that in homogeneous light only one of the two possible azimuths should be observed.

Even Baumgartner has not thoroughly emancipated himself from selectionist *terminology*—although he has indeed abandoned selectionist concepts. In presenting the laws for partial reflection he does not give the Fresnel amplitude ratios but, instead, provides an expression only for the reflected intensity (Baumgartner 1831, 490–91) and for the angle to which the plane of polarization is rotated by the reflection. He believes one can consider a partially polarized beam to consist of two parts, one fully polarized and the other in the "natural state," with the latter itself consisting of two polarized "parts" at right angles to one another. These two characteristics—the ignoring of Fresnel's amplitude ratios and the dissection of partially polarized light into "parts"—persist for a considerable period of time even among avowed wave men, and they contribute to a great deal of confusion among those selectionists who tried to accommodate various elements in the wave theory's vocabulary.

4. Good (1982, 284) discusses the chronology of Brewster's discovery.

5. The other effect of the assumption that forces were at work was Brewster's predilection for "parallelogram"-like laws, though this was at the very best a loose analogy to the composition of forces, since the legs of the figure were either tints or squared velocities and not forces. Good (1982, 134–41) discusses especially Brewster's vocabulary of forces, including his (Brewster's) critique of Biot's similar vocabulary.

6. On which see Morrell and Thackray (1982, 466–72) and Hankins (1980, 138–54). I shall return to the polemics of the 1830s in chapter 12.

Chapter 11

1. In a remark that should please those philosophers of science who follow Sir Karl Popper, Fresnel insisted "that the more perfect a theory, the less it is indifferent to the replies of experience," so that his new theory of double refraction would be particularly "perfect" if it could withstand this dangerous test.

2. Fresnel inclines the facet in the two planes $x_e y$ and yz_e, that is, he rotates it about the x_e or z_e axes (an *e* subscript indicating that the axes are thus far empirically undifferentiated). When the axis of rotation is z—that is, when the plane of incidence contains the optic axes—then, Fresnel writes, as the inclination increases concentric rings appear. (Since then we have, very nearly, a situation similar to the case of a uniaxial crystal cut at right angles to its single axis, because the axial separation in topaz is rather large, which leaves one of the two axes out of play.)

3. Here, in this first attempt to produce a theory of double refraction, Fresnel did not explicitly prove that the ellipsoid will, by sectioning, yield Huygens's construction for any plane of refraction. However, in his later work Fresnel did indeed prove the point when he generated the wave surface for biaxial crystals (of which Huygens's is a degenerate case).

4. Fresnel does offer a reason, very weak at this point, why these two must be the only possible *permanent* directions of oscillation in the front. An oscillation normal to the optic axis must generate the minimum possible reaction, whereas one parallel to the axis must generate the maximum reaction, as one can see by considering a beam normally incident on a crystal cut with the axis in the facet. Consider now an oscillation oblique both to the optic axis and to its normal. Precisely because such an oscillation has components along and normal to the axis, it cannot over time remain parallel to itself because the component reactions differ. In this case of a ray incident along the normal to the axis, then, we see that the oscillations must lie along the

semiaxes of the ellipsoid of elasticity, or else they cannot remain unaltered in direction over time. Fresnel could not at this point demonstrate that this remains true for an arbitrary direction; he merely conjectured that it did. Later, we shall see, he generalized the relation between reaction and elasticity in a way that leads directly to this result.

5. Fresnel's method here nicely avoids the difficulty of transforming a quadratic form to principal axes—though, we shall see, he did precisely that several months later for a different problem. That is, it leads almost instantly to an explicit expression for the radii of the surface, which can then be subject to an extremum condition in a given plane.

6. Which is a simple matter to show. The intersection of the plane of polarization of any ordinary ray with the plane of the optic axes must, by Biot's dihedral law, lie within ∠LAL′. Whence the diameter of the ellipsoid's section that is normal to the plane of polarization has a projection in the plane of the axes that lies within ∠MAN and ∠M′AN′. It follows that the corresponding semiaxis cannot be less than AM, that is, it cannot be less than the semiaxis that governs the speed of an ordinary ray along Ax. Hence the difference between the speeds of ordinary rays along Ax and Ay—both in the axial plane—is the greatest possible for ordinary rays.

7. Because he already knew very well that, though the ellipsoid seems to work well in yielding Huygens's ray speeds, nevertheless it cannot also give the directions of oscillation since the ray is not generally parallel to the normal to the front.

8. My reconstruction here of Fresnel's route to the new surface differs only slightly from Verdet's in that I have based mine directly on Fresnel's recognition that $(v_p)^2$ can be written as $a^2\cos^2\vartheta_x + b^2\cos^2\vartheta_y$.

9. If we were to use the total reactions, then the radius of the surface would have to satisfy:

$$\vec{r} = \sqrt{f}\,\frac{\vec{s}}{s}$$

$$r^2 = \sqrt{a^4\cos^2 X + b^4\cos^2 Y + c^4\cos^2 Z}$$

where X,Y,Z are the angles between s and Fresnel's special axes, so that (x,y,z) is $r(\cos X, \cos Y, \cos Z)$, in which case the equation may be written:

$$r^6 = a^4x^2 + b^4y^2 + c^4z^2$$

10. The proof of the theorem was not difficult (Fresnel 1822f, 351–53). Begin with a sectioning plane $x = By + Cz$. If it contains the radius vector that makes the angles X,Y,Z with the coordinate axes, then we have the following equations:

1. $\cos X = B\cos Y + C\cos Z$
2. $r^2 = a^2\cos^2 X + b^2\cos^2 Y + c^2\cos^2 Z$
3. $\cos^2 X + \cos^2 Y + \cos^2 Z = 1$

Fresnel calculates the extrema of r given these three equations by setting dr to zero in equation 2 and varying with respect to X. Equation 11.3.4 results. He then demonstrates that 11.3.4 also gives the radii along which displacements generate reactions in planes that contain the radii and that are normal to the section.

11. Fresnel first rewrote the elasticity surface in "polar" form, with (as usual) x,y being replaced by αz, βz:

$$r^2(1 + \alpha^2 + \beta^2) = a^2\alpha^2 + b^2\beta^2 + c^2$$

Taking a plane $z = mx + ny$ and requiring dr to vanish in it gives the extrema as:

$$r^2\left(\alpha + \beta\,\frac{d\beta}{d\alpha}\right) = a^2\alpha + b^2\beta\,\frac{d\beta}{d\alpha}$$

Simply replacing $d\beta/d\alpha$ with $-m/n$ then gives the normal surface in a form that contains only the parameters m,n, which specify the direction of the front, and r^2, the square of the wave speed. To interpret this as a surface, simply require r to be the radius that points in the direction of the normal to the plane $z = mx + ny$—that is, just replace m^2,n^2 respectively with $(x/z)^2$, $(y/z)^2$.

12. However, Biot's dihedral angle law, unlike his sine law, had clearly to be an approximation that was worse the greater the differences between the constants a,b,c that appear in the surface of elasticity—a point Fresnel remarks in some detail in his final complete paper (Fresnel 1827, 581–84). The correct law, Fresnel demonstrates, must replace the axes of single ray velocity, which appear in Biot's dihedral law, with the axes of single wave velocity, and the rays must be replaced by the wave normals. For most crystals the effect is not large, and so Biot's statement remains reasonably accurate. (And to minimize his discontent Fresnel entitled the section in which he proved that the law cannot be accurate "The rule given by M. Biot for determining the planes of polarization of ordinary and extraordinary rays agrees with the theory developed in this memoir" [Fresnel 1827, 581]—but the "agreement" is only approximate.)

13. In a letter to Young dated 4 September 1825 Fresnel writes: "When you asked me, about a year ago, to make you privy to my theoretical views on polarization and double refraction, I had the honor to point out to you the Extract of my memoir on double refraction which was published in the two Bulletins of the Société Philomatique in the months of April and May 1822: not having had reprints made, I couldn't send you any. . . . I regret not having as yet found the time or the occasion to have the entire memoir printed" (Fresnel, *Oeuvres*, 2:776). In a footnote to the letter Fresnel's brother Léonor adds that Fresnel was too ill to finish the memoir until 1826.

Chapter 12

1. Fresnel's theory of partial refraction, like his theory of double refraction, was unavailable at the time (excepting the incomplete accounts in 1819c and 1821c). In replying to Herschel, Fresnel gave only the resultant intensity for the reflection of an initially polarized ray. Further, he permitted partial polarization to be "represented" by two beams, one of completely polarized light, the other of natural light. One can "also," he continued, always "decompose" light into two "beams," one polarized in, the other in the normal to, the plane of incidence (Fresnel, *Oeuvres*, 2:651–52). Herschel followed Fresnel very carefully here.

2. Good (1982, chap. 4) discusses Herschel's work on ring formation in detail. Herschel (1820) in essence discovered (in selectionist language) that the depolarizing axes in some crystals are different for different colors. Brewster had tried to explain the anomalous colors that Herschel thereby accounted for by postulating that each depolarizing axis in reality consisted of two others—that is, Brewster attempted (without calculation) to multiply the axes held in common by all colors, whereas Herschel assigned different axes to each color and calculated the effect (see Good 1982, 276 ff. for the controversy between Herschel and Brewster).

3. Fresnel gives no details, but the analysis is quite simple. Compound the amplitude of a partially polarized wave from two amplitudes (a_{pol} and a_{nat}): one directed at some angle α, the other corresponding to the amplitude of an unpolarized wave. The latter has the same components in and normal to the plane of incidence as a wave polarized at 45° (though its components, unlike those of the polarized wave, have no fixed phase relationships). Accordingly, the partially polarized wave has a component $a_{pol}\cos\alpha + \frac{1}{2}a_{nat}$ parallel to the plane of incidence, and $a_{pol}\sin\alpha + \frac{1}{2}a_{nat}$

normal to that plane. Then the amplitude in an analyzer inclined at angle β to the plane of incidence will be:

$$\cos \beta(a_{pol} \cos \alpha + \tfrac{1}{2}a_{nat}) + \sin \beta(a_{pol} \sin \alpha + \tfrac{1}{2}a_{nat})$$

which reaches a maximum when

$$\tan \beta = \frac{a_{pol} \sin \alpha + \tfrac{1}{2}a_{nat}}{a_{pol} \cos \alpha + \tfrac{1}{2}a_{nat}}$$

One can always imitate any given partial polarization by adjusting the angle α and the factors a_{pol}, a_{nat}.

4. One might, however, object that Herschel should not be interpreted literally here—that he intended the act of polarization to be interpreted only "as if" it operated in this differential fashion. This possibility is, I think, excluded by the purpose of his remarks, which was to reach the conclusion that polarization either is or is not, that it is a "property or character" that has no "degree": the surface, in partially polarizing a beam, must then make part of the beam "polarized" and leave the other part unaffected. One has to think of the polarization at a point in the front as not susceptible of degree (either the oscillation remains fixed in direction or it does not), as Herschel well understood, but he could not, or did not as yet, fully see that this made it impossible any longer to use the old vocabulary in the same way it had been used since the time of Malus—though he knew it made that vocabulary unclear.

5. Good (1982, 405–7); see also Good's chapter 6 for further discussion of Herschel's "Light." Good provides enough evidence to show that Herschel remained ambivalent for some time about committing himself to the wave theory.

6. For example: "The undulatory system especially is necessarily liable to considerable obscurities; as the doctrine of the propagation of motion through elastic media is one of the most abstruse and difficult branches of mathematical inquiry, and we are therefore perpetually driven to indirect and analogical reasoning, from the utter hopelessness of overcoming the mere mathematical difficulties inherent in the subject when attacked directly" (Herschel 1827, 450).

7. Ray terminology was extremely tenacious; one can see vestiges of it as late as the 1890s, for example, in Thomas Preston's *Theory of Light.* In it [sec. 212] he remarks that, at the Brewster angle, "the reflected light alone is completely plane-polarized, and the two beams [the reflected and the refracted] contain *equal quantities [sic] of polarized light.*"

This assertion follows a detailed presentation of the Fresnel amplitude ratios and (though Preston writes nothing more than this) gains its meaning from them. With i the angle of incidence and r the angle of refraction, at the Brewster angle ($i + r = 90°$) we will have for the intensities of the components in (||) and normal ($\perp$) to the plane of reflection:

	inc	**reflected**	**refracted**
$\parallel$	1	$\sin^2(i - r)$	$\sin(2i)\sin(2r)$
$\perp$	1	0	1

Preston's assertion, in the language of the wave theory, means that, if we take from the normal component of the refraction a quantity equal to the reflected light, then what is left in the refraction should satisfy the same condition as natural or common light. And indeed it does: for the difference between the normal refracted intensity (1) and the reflected intensity ($\sin^2(i - r)$) is equal to the parallel refracted intensity ($\sin(2i)\sin(2r)$) when $i + r$ is 90°.

8. The best recent accounts are in Morrell and Thackray (1982, 3466–72), Hankins (1980, 138–54), and Cantor (1975, 1978). Also see Wilson (1968, passim). Morrell and Thackray briefly describe several of the criticisms made by Brewster,

Potter, and Barton, while Hankins goes into more detail. I shall not repeat their work here because I do not think that these debates had a major influence on whether the wave theory was to become the new orthodoxy. That issue was settled by 1832 at the latest, and the only effect of subsequent criticism (theoretical and experimental) was to force wave analysts further to develop their work, or to explain the possibility of developing it.

9. As a particularly compelling example of a wave theorist's failure to grasp the core of a selectionist analysis, consider Lloyd's (1834, 371–72) account of Brewster on metallic reflection (Brewster 1830c). Brewster, recall, had combined a selectionist account with a creative use of formulas deriving from the wave theory, a combination that enabled him to calculate with a "phase" without in any way abandoning selectionist principles.

Lloyd read Brewster carefully, and he saw clearly that Brewster had used the phrase "elliptic polarization" in a sense that does not require the wave theory. But Lloyd did not understand how Brewster could use the word "phase" without necessarily "acknowledging" the wave theory. Lloyd remarks: "In other parts of his memoir, however, Sir David Brewster seems to acknowledge that theory [Fresnel's]; for he speaks of elliptic polarisation as produced by the interference of two unequal portions of oppositely polarized light, and even calculates their difference of phase for any incidence" (Lloyd 1834, 372).

Here Brewster used a word that derives from the wave theory, so that at least Lloyd was able to recognize what seemed to him to be an oddity. But often a word common to both theories might be used though it has entirely different meanings in the two. For example, Lloyd gave "Arago's" law that reflection generates equal "quantities" of polarized light in the reflected and refracted "pencils." For Lloyd, however, quantity meant the square of an amplitude, and Arago's statement had to be interpreted in terms of the vector decomposition of the incident light. Quantity, that is, had nothing to do with rays. But to a selectionist like Brewster, saying that there are equal "quantities" is to say that the beams contain subsets that have the same number of rays.

10. Potter's incomprehension nicely shows itself in his discussion of an experiment that, he believes, makes Fresnel's wave surface untenable (Potter 1856, 23–25). Looking at a candle through a piece of aragonite cut with its facets normal to the bisector of the optic axes, Potter claims that, when the crystal is turned so that the image of the candle passes nearly along an optic axis, then the image has a large "hole" in it. His description of the conditions (as usual with him) is too vague for us to grasp precisely what was occurring, but very likely what he saw was a form of external conical refraction, with the width of the "hole" being a section of the emergent cone.

What is interesting, however, is not Potter's observation or his vagueness in describing it, but his conviction that it must be inconsistent with Fresnel's wave surface. For, he argues, that surface consists of two continuous sheets, whereas his observation of a "hole" in the candle's image means there must be an "eyelet-hole" in the "luminiferous surface" about an optic axis. Obviously he had no understanding whatever of the relationship between fronts and rays. For him Fresnel's surface merely marked the loci of rays, and if rays were missing from a region about a given line then the surface had to have a hole there. Because he was so vague in describing his experiments, and because he clearly never understood much of the wave theory, he did not pose a serious challenge to the wave men.

11. The time intervals are taken in five-year increments, so that the label "5," for example, refers to the total number of articles or authors during the period 1800–

1805. I tabulated the data for these graphs from the index to the *Royal Society Catalogue*. Random checks convinced me that the index is generally quite reliable (though dates are occasionally off a year or two and page references can be widely incorrect) as a guide for five-year intervals. I have counted all entries pertinent to optics except those involving microscopy or telescopy.

For the most part the French, British, German, and Others numbers reflect the nationalities of the authors, but not invariably. The labels refer to the language of the journals and not to the authors of the articles, so that in a fair number of cases an author of one nationality, or an article by the author, will be counted under another nationality. (In, e.g., Fresnel's case the German translations of his articles appear under the German counts.) I have chosen this method of counting because it reflects the state of the discipline as it was intended by the editors to be seen by readers of a given nationality.

Appendixes

1. Namely, Book 1, propositions 21 and 36. Following Huygens (fig. A1.4):
   1. To find CP, from the center C draw a line that intersects the tangent at M in the point D and the ellipse in the point P, and from M draw a line parallel to an axis (here CS) of the ellipse to intersect CD in N: then by conic geometry CP = $\sqrt{(CD \times CN)}$. Similarly we can find CS = $\sqrt{(CZ \times CO)}$.
   2. To find CG, from P (the endpoint of CP) draw PE, parallel to DM, the tangent at M, meeting CM in E; then by conic geometry CG = PE [MC/$\sqrt{(MC^2 - CE^2)}$], where MC, CE, PE are readily found.

2. The proof is simple. By similar triangles, CK/OK = RC/CV, which is CG/CV by construction. If CI is the refraction, then CK/CG = CV/CD. Consequently OK/CG = CV/CD. This is equivalent to the following law of refraction, which, of course, Huygens never wrote analytically:

$$\tan(r_e) = \tan(\delta) + \frac{CG^2\sin(i)}{N \times CM \times \cos(\delta)\sqrt{1 - \left[\frac{CG}{N}\right]^2 \sin^2 i}}$$

Here δ is the deviation of the normal ray, i is the angle of incidence, and r_e is the angle of refraction. In Huygens's parameters CM is set to be a unit of 100000 in terms of which N has the experimentally determined value of 156962.

3. Huygens did not, I think, intend to conceal his procedure: in section 30 he remarked that N had to be set at 156962 "having regard to some other observations and phenomena of which I shall speak afterwards"—the only ''observation'' later mentioned is that for the nondeviated ray.

4. I thank Alan Shapiro for pointing this out to me.

5. The semiaxes of the indicatrix have lengths whose reciprocals are a, b, c. OP must be less than or equal to the greatest of these three lengths and greater than or equal to the least of them. Consequently only two of the factors in the equation have the same sign, which means it must be a cone.

6. To obtain equation A5.5 most simply, first define a unit vector $\vec{s}_p$ along OP and replace the coordinates of P by the products of the components of $\vec{s}_p$ by OP. Then, in the last step, recall that $\vec{s}_p$ is normal to $\vec{s}$.

7. The two most recent are Hankins (1980, 88–95) and Eichhorn (1986, 143–90). Eichhorn presents the details of Hamilton's final analysis. See also Sarton (1932). Graves (1882, 1:623–38) reprints correspondence between Airy, Lloyd, Hamilton, and Herschel on the subject.

8. In Hamilton's words: "Construct first the two successive surfaces of components, which would be here a sphere for the air, and a spheroid [Hamilton considers a uniaxial case] (not the Huygenian) for the crystal, the common centre of both being at the point of incidence; and then, after determining the point of the hemispheroid within the crystal, which is on the same ordinate to the refracting face as the point where the incident ray prolonged meets its own interior hemisphere, we should only have to draw a tangent plane to the spheroid at the point thus determined, and to let fall a perpendicular on this plane from the point of incidence; for this perpendicular is, in length and direction, the radius vector of the Huygenian spheroid, and therefore represents the undulatory velocity and the direction of the extraordinary ray" (Hamilton 1833, 302).

9. We can easily see this from the rule described by Hamilton in the previous note. Take a ray incident at such an angle that, when prolonged to intersect the air sphere in the crystal, the point of intersection lies the same distance below the plane of refraction as does the cusp on the surface of components. Then, by Hamilton's rule, the corresponding refraction will be the perpendicular to the tangent plane at the cusp. But since there is an infinite number of such planes, forming a cone, there will also be "an infinite number of refracted rays," which must also form a cone, since the planes to which they are the normals do. These must emerge to form a cylinder of light.

10. See, for example, Lloyd (1834, 359). Lloyd, however, did not understand that Arago had originally derived this result from Malus's account, since he did not know there was such a thing as selectionist theory, separable from emission theory. Lloyd in any case knew of Arago's claim from the *Encyclopaedia Britannica* article of 1824 on polarization written by Arago and translated into English by Young. That article was also a major source for dating the early discoveries concerning polarization. In it Arago claimed that "this proposition" had escaped Malus. Arago subsequently extended the equality to all angles of incidence, which is again compatible with Malus's account, since the claim concerns natural light. It is, indeed, merely an esay application of Malus's selectionist theory. Arago did not publish the result at the time, but he did communicate it to Biot. See Arago (1824, 378–80).

11. Construct the normal SM to AC, and the normal SB to CB (see fig. A13.2). Then $I_1B = BS$ and $I_2M = MS$, whence we have:

$$CI_1^2 - CS^2 = BI_1^2 - SB^2 = 0$$
$$CI_2^2 - CS^2 = MI_2^2 - SM^2 = 0$$

12. We have (fig. A13.1):

$$\angle I_2CI_1 = \pi - \angle SI_1C + \angle I_2SI_1 - \angle CNI_2 = -\angle CSI_1 + 2\angle I_2SI_1 + \angle CI_1S$$

But $\angle CSI_1 = \angle CI_1S$. N is the intersection of SI_2 with CI_1.

13. Fresnel (1816b, 151) remarked only that the "internal" formula is to be used with I_1I_2/EF (which is approximately $\angle I_2FI_1$) representing the angle under which the wire—here I_1I_2—is seen from screen point F.

14. To reach this conclusion with a proper estimation of the dependence of amplitude on distance, we need to assume that the inclination factor continuously reduces the effect as the obliquities of the lines from P to the zones increase. Suppose Σ is the resultant of the amplitudes a, produced by the zones. Since the zones alternate in phase:

$$\Sigma = \sum_{\substack{\lim \\ n\to\infty}} (-1)^{n+1} a_n$$

Since the areas of the zones vary directly with the distance, but the amplitude a zone generates varies reciprocally with distance, the a_n all have the same magnitude as far as distance is concerned. But the inclination factor requires that a_{n+1}/a_n be less than one. Now either a_n is greater than $\frac{1}{2}[a_{n-1} + a_{n+1}]$ or it is less—both are compatible with this requirement of the inclination factor. On either assumption one can show that, whether n is odd or even, Σ must be very nearly equal to $\frac{1}{2}[a_1 - a_n]$ if the factor varies sufficiently slowly that $a_n \approx a_{n-1}$. If the front is unobstructed, then n is infinite, and so a_n can be neglected. Whence the illumination at P from an unobstructed wave is due almost entirely to half the effect of the first zone.

15. Almost from his first look at the reflection of polarized light Fresnel discovered the most complicated phenomenon that can occur except for reflection from metals—the phase effect of total internal reflection. Though I would prefer that readers not think of the phenomena in modern terms, that is, in the way Fresnel himself invented three years or so later, nevertheless they are sufficiently confusing that it may be difficult for many to form a coherent picture without some clear idea of what is occurring.

Reduced to essentials, at and above the critical angle the components of the oscillation in and perpendicular to the plane of incidence acquire a phase difference, φ, such that:

$$\cos\varphi = \frac{1 - (n^2 + 1)\sin^2 i + 2n^2\sin^4 i}{(n^2 + 1)\sin^2 i - 1}$$

If the index, n, is greater than about 1.4, a single reflection of light whose incident polarization is 45° will produce a 45° phase shift at internal incidences near 55°. Two such reflections produce a 90° phase shift. As we shall shortly see, Fresnel was well aware of the phase shifts, but he did not interpret them in quite the same way as he did later, and as we do today.

16. In, for example, Fresnel (1818e, 528) he used energy conservation to demonstrate the necessity of an extra half-wavelength path difference between the sets of rays that make up the two images in the analyzer (a necessity demanded empirically by the complementarity of their colors). Take an incident wave polarized at 45° to the lamina's principal section and then set the analyzer section parallel to the incident polarization. Taking the incident intensity as unity, then one of the two images in the analyzer will have amplitude:

$$\sqrt{\frac{1}{4} + \frac{1}{4} + \frac{1}{2}\cos\left(2\pi\frac{\delta}{\lambda}\right)} = \sqrt{\frac{1}{2} + \frac{1}{2}\cos\left(2\pi\frac{\delta}{\lambda}\right)}$$

where δ is the path difference between the rays that form the images in the lamina. Now, Fresnel continued, energy conservation requires that the square of this expression added to the square of the amplitude of the other image must be equal to one, the original intensity. Consequently the other image's amplitude must be:

$$\sqrt{\frac{1}{2} + \frac{1}{2}\cos\left[2\pi\left(\frac{\delta + \frac{1}{2}\lambda}{\lambda}\right)\right]}$$

Energy conservation alone demands the extra half-wavelength path difference. This is also the first trace in the documentation of Fresnel's formulas for combining waves with different phases.

17. The usual formula of the day for "refractive power" set $n^2 - 1$ proportional to the density, not simply n^2. However, Fresnel adopted the expression, which dates back to Newton, for the ratio of speeds of propagation of waves in media that are equally elastic but unequally dense. Consequently, even in this early deduction Fresnel was already assuming that isotropic media differ optically only in respect to density and not in respect to elasticity.

18. Nevertheless, it passed into one of the standard texts of the wave theory, Airy's *Tracts* (Airy 1858, 351–52), primarily because it is so simple.

19. Fresnel also obtained the ratios for the refracted wave, but he explicitly provided a formula only for the angle of polarization of the refraction:

$$\tan(\alpha_{refr}) = \frac{1}{2}\cot(\alpha)\left(\frac{\sin(2i) + \sin(2e)}{\sin(i + e)}\right)$$

This can be written more simply as:

$$\tan(\alpha_{refr}) = \cot(\alpha)\cos(i - e)$$

20. No doubt because:

$$\cos(e) = \sqrt{1 - \sin^2(e)} = \sqrt{1 - \frac{1}{n^2}\sin^2(i)} = \frac{1}{n}\sqrt{n^2 - \sin^2(i)}$$

If n is less than one (internal incidence), then for i greater than arcsine(n) cos(e) becomes imaginary.

21. In his *Mathematical Tracts* Airy (1858, 357–58n) supposed that Fresnel's interpretation derived from mathematics rather than from physics: "In several geometrical cases, the occurrence of an imaginary quantity indicates a change of 90° in the position of the line whose length is multiplied by $\sqrt{-1}$." Airy's assumption was repeated many times in later years, most notably by Preston (1901, 359).

Aimé Argand's geometric interpretation of imaginary quantities was printed in 1806, but according to Kline it was not influential (Kline 1972, 630–31), whereas Carl Gauss's 1831 publications on the subject did have a major impact. It is hardly likely that Fresnel would even have known Argand's work. But it is very likely that he would have looked for a physical clue to the meaning of an imaginary number, and he had one here in the fact that the light is totally reflected just when the ratios become complex. Since the amplitude could not change, it followed that the phase had to change, and a changed phase meant one had to deploy the quarter-wave decomposition for purposes of calculation. It was an easy and natural step to the conclusion that the real and imaginary parts represented the quarter-wave amplitudes.

22. Actually Fresnel first found not this expression but its negative for the cosine of the phase difference. The latter results from the tangent law in the form $-[\tan(i - e)/\tan(i + e)]$, which in fact results from Fresnel's boundary condition. The negative sign is, however, empirically incorrect—or at least, if the sine law has $-\sin(i - e)$, then the tangent law must have $+\tan(i - e)$ in order to produce the deviations in the plane of polarization Fresnel had long ago observed. Fresnel was well aware of this problem. He argued that, when the oscillations occur in the plane of incidence, then the boundary condition must be interpreted with regard to sign, and careful consideration shows that the result is to alter the sign of v (Fresnel 1823h, 787–89), and so in the end of the tangent law itself.

23. This is the generalization of Biot's original angle i to cases in which the ray does not pass perpendicular to the optic axis.

24. Fresnel himself remarked only that metals fail to polarize light well (Fresnel 1822c, 98). Although he never used the phrase "elliptical polarization," he did demonstrate that the light emerging from thin crystals consists in general of oscillations that follow an elliptical path (Fresnel 1827, 503). The first sixteen sections of this memoir, the second on double refraction, were separately printed in 1824 (Fresnel 1824c). They contain Fresnel's most extensive discussion of compounding oscillations in different directions and were very heavily relied upon by Herschel in his 1827 "Light."

25. Biot picked up the topic, which greatly angered Brewster, who complained:

"I soon learned . . . from M. Biot, that he meant to treat the subject in his Traité de Physique; and though I remonstrated against this as a breach of courtesy, I had the mortification to see the discovery, to which I perhaps attached too much importance, published for the first time in a foreign work." Biot, like Brewster, linked metallic reflection to chromatic polarization.

26. Brewster's reasoning on this point almost certainly shows the selectionist understanding of chromatic polarization. If metals, like thin crystals, separate the rays in a polarized beam into new subsets, then all conditions being the same, a second reflection should leave the subsets unaltered in polarization and affect only their ray ratios—their tints—just as thin crystals do.

27. Brewster did not first put the difference this way. He initially remarks that, for a fixed incident azimuth of +45° and a fixed first reflection at 75°, the angle of incidence necessary in the second metallic reflection to restore the plane polarization varies with the angle between the planes of first and second incidence. He next remarked that the restored azimuth itself also varies. Consequently, for a fixed angle of first incidence and a varying incident azimuth there are corresponding pairs of second incidences and azimuths to restore the plane polarization. Brewster built his ellipse on the functional relationships between the azimuths.

28. In "rectilinear" polarization the ellipse loses its meaning (there being no angle of restoration, since the light remains plane polarized after the first reflection), which Brewster interprets as the degeneration of the ellipse into a line.

29. Brewster seems to have calculated the "index" (which he gives as 3.732 for steel) by applying the formula to light once reflected at 75° with an incident azimuth of 45°.

30. The procedure is a bit more complicated than it might appear. In Brewster's table we have only integers for the total number of reflections necessary to move from plane polarization back to plane polarization. However, the number of "reflections" he uses to calculate the phase is not always integral; it occasionally has a fractional part of $\frac{1}{2}$. If, for example, the total number of reflections is 5, then Brewster calculates the number necessary to generate elliptical polarization as $2\frac{1}{2}$. He interprets this as meaning "that the ray must have acquired its elliptic polarization in the middle of the second and third reflexion; that is, when it had reached its greatest depth within the metallic surface. It then begins to resume its state of polarization in a single plane, and recovers it at the end of 3, 5, and 7 reflexions" (Brewster 1830c).

This explains why, in these cases, the phase as calculated from his formula differs slightly from what one obtains by dividing 90° by the half of the total number of reflexions.

31. Because the angle to which the analyzer must be turned in order to minimize within it the extraordinary beam produced by elliptically polarized light is not in or normal to the plane of reflection, which at once means that the phase difference cannot be 90°. Herschel's formulas make this very clear (1827, 460). Herschel was writing about the composition of orthogonal oscillations, but the principle holds just as well if we simply consider the compounding beams to be orthogonally polarized, project both onto a given direction of analyzer principal section by Malus's formulas, and then compute the resultant by interference of these two.

32. The significant section of Herschel's "Light" in this context is 621, entitled "Case of Interference of Rectilinear Vibrations." Here Herschel considers a pair of orthogonal oscillations in the same plane and deduces that the coordinates (x,y) in that plane satisfy, eliminating the time:

$$\left(\frac{x}{a}\right)^2 + \left(\frac{y}{b}\right)^2 - 2\cos(p - q)\frac{xy}{ab} = \sin(p - q)^2$$

where a,b are the amplitudes and p,q are the phase constants.

Brewster was quite capable of realizing that the axes of this ellipse will be parallel to the coordinate axes only if the phase difference $p - q$ is 90°. It is not at all necessary to construe the ellipse as the path traced out by an oscillation to conclude that it governs the intensity in an analyzer—all we need is that the polarizations of the two beams that are combined by the analyzer are orthogonal, that they are brought to the same plane by the analyzer, and that Fresnel's interference formulas then govern the resultant intensity.

33. To be precise, Fresnel demonstrates that Biot's plane of polarization for the ordinary ray is perpendicular to one of the two semiaxes cut in the ellipsoid by a plane through its center and normal to the ray. Consequently that direction in the ellipsoid must define the "ordinary" ray, and the orthogonal direction then defines the polarization for the extraordinary ray (taking, as usual, Fresnel's requirement that the plane of polarization must be normal to the direction of oscillation).

References

Dates in boldface denote articles not printed at the time. Dates in italics also denote articles not printed at the time, but those dates are also uncertain.

Airy, G. B.

1831 On the nature of the light in the two rays produced by the double refraction of quartz. *Cambridge Phil. Soc. Trans.* 4:79–123, 199–208.

1858 *Mathematical tracts on the lunar and planetary theories, the figure of the earth, precession and nutation, the calculus of variations, and the undulatory theory of optics.* 4th ed. Cambridge: Macmillan.

Ampère, A. M.

1828 Mémoire sur la détermination de la surface des ondes lumineux dans un milieu dont l'élasticité est différent suivant les trois directions principales. *Annal. Chimie* 39:113–45.

1835 Note de M. Ampère sur la chaleur et sur la lumière considérées comme résultant de mouvements vibratoires. *Annal. Chimie* 58:432–44.

Arago, F.

1811a Mémoire sur les couleurs des lames minces. In *Oeuvres,* 10:1–35. Read 18 February 1811. Originally published in *Mém. Soc. Arc.,* 1817.

1811b Mémoire sur la polarisation colorée. In *Oeuvres,* 10:36–74. Read 11 August 1811. Originally published as Mémoire sur une modification remarquable qu'éprouvent les rayons lumineux dans leur passage à travers certains corps et sur quelques autre phénomènes d'optique. *Mém. Inst.* 11:93–134.

1812a Notes sur les phénomènes de la polarisation de la lumière. In *Oeuvres,* 10:74–84. Note deposited spring 1812.

1812b Mémoire sur plusieurs nouveaux phénomènes d'optique. In *Oeuvres,* 10:85–97. Read 14 December 1812.

1812c Quatrième mémoire sur plusieurs phénomènes d'optique. In *Oeuvres,* 10:98–131. Read 28 December 1812.

1816 Remarques sur l'influence mutuelle de deux faisceaux lumineux qui se croisent sous un très-petit angle. In Fresnel, *Oeuvres,* 1:123–28. Originally published in *Annal. Chimie* 1:332.

1819 Rapport fait par M. Arago à l'Académie des Sciences au nom de la commission qui avait été chargée d'examiner les mémoires envoyés au concours pour le prix de la diffraction. In Fresnel, *Oeuvres,* 1:229–46. Originally printed in *Annal. Chimie* 11:5.

1821 Examen des remarques de M. Biot. Originally published in *Annal. Chimie*. Fresnel, *Oeuvres,* 1:591–600.

1822 Rapport fait à l'Académie des Sciences, dans la séance du 19 aout 1822 sur un mémoire de M. Fresnel relatif à la double réfraction. In Fresnel, *Oeuvres,* 2:459–64.

1824 Notice sur la polarisation de la lumiére. In *Oeuvres,* 7:291–446. Written for *Encyclopaedia Britannica.*
1830 Eloge historique d'Augustin Fresnel. In Fresnel, *Oeuvres,* 3:475–526.
1854 Histoire de ma jeunesse. In *Oeuvres,* 1:1–102.
1854–58 *Oeuvres de François Arago.* 14 vols. Paris: Gide.
1857 Malus. In *Biographies of distinguished scientific men.* 2 vols. Trans. W. H. Smyth, Baden Powell, and R. Grant, 2:117–70. London: Longman, Brown, Green.

Arago, F., and A. M. Ampère
1821 Rapport fait à l'Académie des Sciences, le lundi 4 juin 1821, sur un mémoire de M. Fresnel, relatif aux couleurs des lames cristallisées douées de la double réfraction. In Fresnel, *Oeuvres,* 1:553–68.

Arago, F., and A.-T. Petit
1816 Mémoire sur les puissances réfractives et dispersives de certains liquides et des vapeurs qu'ils forment. In Arago, *Oeuvres,* 10:123–31. Read 11 December 1813. Originally published in *Annal. Chimie.*

Arago, F., and L. Poinsot
1816 Rapport fait à la première classe de l'Institut, le 25 Mars 1816. . . . In Fresnel, *Oeuvres,* 1:79–87.

Arnold, D.
1978 The *Mécanique physique* of Siméon Denis Poisson: The evolution and isolation in France of his approach to physical theory (1800–1840). Ph.D. diss., University of Toronto.

Baker, B. B., and E. T. Copson
1939 *The mathematical theory of Huygens' principle.* Oxford: Clarendon Press.

Bartholin, E.
1669 *Experimenta crystalli islandici disdiaclastici quibus mira et insolita refractio detegitur.* Copenhagen: Hafniae.

Baumgartner, A.
1824 *Die Naturlehre nach ihrem gegenwartigen Zustande mit Rücksicht auf mathematische Begrundung.* Vienna: J. G. Heubner.
1831 *Die Naturlehre . . . Supplementband* (mathematical and experimental). First part printed in 1826. Vienna: J. G. Heubner.
1845 *Die Naturlehre. . . .* 8th ed. Vienna: Carl Gerold.

Bechler, Z.
1974 Newton's search for a mechanistic model of colour dispersion: A suggested interpretation. *Arch. Hist. Ex. Sci.* 11:1–37.

Bertrand, A.
1824 *Lettres sur la physique.* Paris and Leipzig: Bossange Frères.

Beudant, F. S.
1824 *Essai d'un cours élémentaire et général des sciences physiques.* 3d ed. Paris: Verdière.

Biot, J.-B.
1811 Sur de nouveaux rapports qui existent entre la réflexion et la polarisation de la lumière par les corps cristallisés. *Mém. Inst.* 12:134–280.
1812 Sur un nouveau genre d'oscillation que les molécules de la lumière éprouvent en traversant certains cristaux. *Mém. Inst.* 13:1–371.
1814a Observations sur la nature des forces qui partagent les rayons lumineux dans les cristaux doués de la double réfraction. Read 26 December 1814. *Mém. Inst.* 14(1813–15): 221–27.

1814b *Recherches expérimentales et mathématiques sur les mouvements des molécules de la lumière autour de leur centre de gravité*. Paris.

1815a Observations sur la nature des forces qui produisent la double réfraction. Read 2 January 1815. *Mém. Inst.* 14(1813–15): 228–34.

1815b Phénomènes de polarisation successive observées dans les fluides homogènes. *Bull. Soc. Philom.* [Paris], 190–92.

1816 *Traité de physique expérimentale et mathématique*. 4 vols. Paris: Deterville.

1817 *Précis élémentaire de physique expérimentale*. 2 vols. Paris: Deterville.

1818 Mémoire sur les lois générales de la double réfraction et de la polarisation dans les corps régulièrement cristallisées. *Mém. Acad. Sci.* 3:177–384.

1821 Remarques de M. Biot, sur un rapport lu, le 4 juin 1821, à l'Académie des Sciences, par MM. Arago et Ampère. In Fresnel, *Oeuvres,* 1:569–90. Originally published in *Annal. Chimie*.

Biot, J.-B., and F. Arago.

1806 Mémoire sur les affinités des corps pour la lumière et particulièrement sur les forces réfringentes des différents gaz. *Mém. Inst.* 7:301–85.

Born, M., and E. Wolf

1975 *Principles of optics,* 5th ed. Oxford: Oxford University Press.

Breithaupt, A.

1820 Ueber die lichtwandlung des schörls. *Ann. Phys. Chem.* 64:424–26.

Brewster, D.

1813 *Treatise on new philosophical instruments for various purposes in the arts and sciences with experiments on light and colours*. Edinburgh: John Murray; London: William Blackwood.

1818 On the laws of polarization and double refraction in regularly crystallized bodies. *Phil. Trans.* 108:199–272.

1830a On the law of partial polarization of light by reflexion. Read 4 February 1830. *Phil. Trans.* 120:69–84.

1830b On the laws of the polarization of light by refraction. *Phil. Trans.* 120:133–44.

1830c On the phenomena and laws of elliptic polarization, as exhibited in the action of metals upon light. Read 22 April 1830. *Phil. Trans.* 120:287–326.

1831 *Optics*. [A volume in Dionysius Lardner's] *The cabinet cyclopedia*. London: Longman, Rees, Orme, Brown and Green; and John Taylor. Revised American ed. A. D. Bache, 1833; reprinted Philadelphia: Lea and Blanchard, 1850.

1834 Report on the recent progress of optics. *BAAS Rep.*, 308–22.

Briot, C.

1864 *Essais sur la théorie mathématique de la lumière*. Paris: Mallet-Bachelier.

Brunot, A., and R. Coquand

1982 *Le Corps des Ponts et Chausées*. Paris: Editions du CNRS.

Buchwald, J. Z.

1980a Optics and the theory of the punctiform ether. *Arch. Hist. Ex. Sci.* 21:245–78.

1980b Experimental investigations of double refraction from Huygens to Malus. *Arch. Hist. Ex. Sci.* 21:311–73.

1983 Fresnel and diffraction theory. *Arch. Int. Hist. Sci.* 33:36–111.

1985 *From Maxwell to microphysics: Aspects of electromagnetic theory in the last quarter of the nineteenth century.* Chicago: University of Chicago Press.

Buffon, C.
1786 *Histoire naturelle des minéraux.* Paris.

Cantor, G.
1970 The changing role of Young's ether. *Brit. J. Hist. Sci.* 5:44–62.
1975 The reception of the wave theory of light in Britain: A case study illustrating the role of methodology in scientific debate. *Hist. Stud. Phys. Sci.* 6:109–32.
1978 The historiography of "Georgian" optics. *Hist. Sci.* 16:1–21.
1983 *Optics after Newton: Theories of light in Britain and Ireland, 1704–1840.* Manchester: Manchester University Press.
1984 Brewster on the nature of light. In Morrison-Low and Christie, "*Martyr of Science,*" 67–76.

Challis, J.
1875 *Remarks on the Cambridge Mathematical Studies and their relation to modern physical science.* Cambridge: Deighton, Bell.

Chappert, A.
1977 *Etienne Louis Malus (1775–1812) et la théorie corpusculaire de la lumière.* Paris: Vrin.

Cohen, M. R., and I. E. Drabkin
1975 *Source book in Greek science.* Harvard: Harvard University Press.

Crosland, M.
1967 *The Society of Arcueil.* London: Heinemann.

Cuvelier, P.
1977 Les experimenta crystalli islandici disdiaclastici d'Erasme Bartholin. *Rev. Hist. Sci.* 30:193–224.

Delambre, J.
1812 Notice sur la vie et les ouvrages de Malus et Lagrange. *Mém. Inst.* 13:xxvii–xxxiii.

Despretz, C.
1825 *Traité élémentaire de physique.* Paris: Méquignon-Marvis.

Ecole Polytechnique
1799 Rapport sur la situation de l'Ecole Polytechnique présenté au Ministre de l'Intérieur par le Conseil de Perfectionnement établi en exécution de la loi du 25 frimaire an 8.

Eichhorn, D.
1986 Die äussere und innere konische Refraktion, letzter überzeugender Baustein zum endültigen Durchbruch der Undulationstheorie des Lichts gegenüber der Emissionstheorie. Ph.D. diss., University of Munich.

Eisenlohr, F.
1853 *Einleitung in die höhere Optik.* Braunschweig: F. Vieweg.

Euclid
1895 *Opera omnia.* Ed. J. L. Heiberg and H. Menge. 8 vols. Leipzig: Teubner.

Fabry, C.
1938 La vie et l'oeuvre scientifique d'Augustin Fresnel. In *Oeuvres choisies,* 633–53. Paris: Gauthier-Villars.

Flaugergues, H.
1812–13 Sur la diffraction de la lumière. *J. Phys.* 74:125–29, 75:16–29, 76:142–54.

1819 Supplément à différents mémoires sur la diffraction de la lumière, publiés dans le *Journal de Physique. J. Phys.* 89:161–86.

Fox, R.
1974 The rise and fall of Laplacian physics. *Hist. Stud. Phys. Sci.* 4:81–136.

Frankel, E.
1972 Jean-Baptiste Biot: The career of a physicist in nineteenth century France. Ph.D. diss., Princeton University.
1974 The search for a corpuscular theory of double refraction: Malus, Laplace and the prize competition of 1808. *Centaurus* 18:223–45.
1976 Corpuscular optics and the wave theory of light: The science and politics of a revolution in physics. *Soc. Stud. Sci.* 6:141–84.
1977 J. B. Biot and the mathematization of experimental physics in Napoleonic France. *Hist. Stud. Phys. Sci.* 8:33–72.

Fresnel, A. J.
1815a Premier mémoire sur la diffraction de la lumière. In *Oeuvres,* 1:9–34.
1815b Complément au [premier] mémoire sur la diffraction de la lumière. In *Oeuvres,* 1:41–60.
1816a [Deuxième] mémoire sur la diffraction de la lumière. *Annal. Chimie* 1:239–81. In *Oeuvres,* 1:89–122.
1816b Supplément au deuxième mémoire sur la diffraction de la lumière. Presented 15 July 1816. In *Oeuvres,* 1:129–70.
1816c Mémoire sur l'influence de la polarisation dans l'action que les rayons lumineux exercent les uns sur les autres. In *Oeuvres,* 1:385–439 (two versions).
1817a Sur l'influence de la chaleur dans les couleurs développées par la polarisation. *Annal. Chimie* 4:298–300. In *Oeuvres,* 2:637–39.
1817b Mémoire sur les modifications que la réflexion imprime à la lumière polarisée. Presented 10 November 1817. In *Oeuvres,* 1:441–85.
1818a Sur l'influence du mouvement terrestre dans quelques phénomènes d'optique. *Annal. Chimie* 9:57–66, 286. In *Oeuvres,* 2:627–36.
1818b Note sur la théorie de la diffraction. Sealed note deposited 20 April 1818. In *Oeuvres,* 1:171–82.
1818c Supplément au mémoire [1817b] sur les modifications que la réflexion imprime à la lumière polarisée. Presented 19 January 1818. In *Oeuvres,* 1:487–508.
1818d Mémoire sur les couleurs développées dans les fluides homogènes par la lumière polarisée. *Mém. Acad. Sci.* 20(1849): 163–94; *Annal. Chimie* 17(1846): 172–99; *Ann. Phys. Chem.* 72(1848): 332–55.
1818e Note sur la théorie des couleurs que la polarisation développe dans les lames minces. In *Oeuvres,* 1:523–32.
1819a Mémoire sur l'action que les rayons de lumière polarisés exercent les uns sur les autres [with Arago]. *Annal. Chimie* 10:288–306; *Quart. J. Sci.* 11(1821): 381. In *Oeuvres,* 1:509–22.
1819b Mémoire sur la diffraction de la lumière [couronné par l'Académie des Sciences]. *Mém. Acad. Sci.* 5(1821–22; printed in 1826): 339–475; *Annal. Chimie* 11:246–96, 337–38 (does not include the introduction and section 1 of the mémoire couronné); *Ann. Phys. Chem.* 30(1836): 100–254, 255–61. In *Oeuvres,* 1:247–363.
1819c Résumé d'un mémoire sur la réflexion de la lumière. *Annal. Chimie* 15:379–86; *Edinburgh Phil. J.* 2:150–53; *Bull. Soc. Philom.* [Paris] (1820), 113–16; *Ann. Phys. Chem.* 30(1836): 255–61. In *Oeuvres,* 1:685–90.

1819d Mémoire sur la réflexion de la lumière. Presented 15 November 1819. *Mém. Acad. Sci.* 20(1849): 195–20; *Annal. Chimie* 17(1846): 316–38; *Ann. Phys. Chem.* 72(1848): 332–55. In *Oeuvres,* 1:691–712.

1821a Note sur l'expérience des franges produites par deux rhomboides de chaux carbonatée. In *Oeuvres,* 1:538–41.

1821b Note sur la polarisation mobile. In *Oeuvres,* 1:542–44.

1821c Note sur le calcul des teintes que la polarisation développe dans les lames cristallisées. *Annal. Chimie* 17:102–12, 167–96, 312–16. In *Oeuvres,* 1:609–53.

1821d Note sur les remarques de M. Biot relatif aux phénomènes de couleurs que produisent les lames minces. *Annal. Chimie* 17:393–412. In *Oeuvres,* 1:601–8.

1821e Premier mémoire sur la double réfraction. Presented 19 November 1821. In *Oeuvres,* 2:261–308.

1821f Mémoire sur la double réfraction: Extrait. Read 26 November 1821. In *Oeuvres,* 2:309–29.

1821g Note sur la double réfraction dans les cristaux à deux axes. Inserted in the *Monitor* for 12 December 1821. In *Oeuvres,* 2:331–34.

1821h Explication de la réfraction dans le système des ondes. *Bull. Soc. Philom.* [Paris], 152–60; *Annal. Chimie* 21(1822): 213–29.

1821i Note sur les interférences des rayons polarisés. In *Oeuvres,* 1:545–50.

1821j Note sur l'application du principe des interférences à l'explication des couleurs des lames cristallisées. In *Oeuvres,* 1:551–52.

1822a Notes sur la double réfraction du verre comprimé. *Bull. Soc. Philom.* [Paris], 139–42; *Annal. Chimie* 21:260–63. In *Oeuvres,* 1:713–18.

1822b [Extrait d'un mémoire] sur la double réfraction particulière que présente le cristal de roche dans la direction de son axe. *Bull Soc. Philom.* [Paris], 191–98 [extrait]; *Annal. Chimie* 20:376–82 [extrait] and 28(1825): 147–61, 263–79 [mémoire]; *Froriep, Notizien* 3(1823): 321–23 [extrait]; *Quart. J. Sci.* 15(1823): 165–66 [extrait]; *J. Phys. Chimie* 95:314–17 [extrait]; *Ann. Phys. Chem.* 21(1831): 276–89 [extrait]. In *Oeuvres,* 1:719–29 [extrait] and 1:731–51 [mémoire].

1822c De la lumière. Extract from the supplement to the French translation of Thomas Thomson's *Chemistry*. In *Oeuvres,* 2:3–146.

1822d Note sur les accès de facile réflexion et de facile transmission des molécules lumineuses dans le système de l'émission. In *Oeuvres,* 2:147–65.

1822e Extrait du supplément au mémoire sur la double réfraction. Read at the Academy 13 January 1822. In *Oeuvres,* 2:335–42.

1822f Supplément au mémoire sur la double réfraction. Dated 13 January 1822, presented 22 January 1822. In *Oeuvres,* 2:343–67.

1822g Second supplément au mémoire sur la double réfraction. Dated 31 March 1822, presented 1 April 1822. In *Oeuvres,* 2:369–442.

1822h Note sur l'accord des expériences de MM. Biot et Brewster avec la loi des vitesses donnée par l'ellipsoïde. Dated 27 May 1822. In *Oeuvres,* 2:443–58.

1822i Extrait du second mémoire sur la double réfraction. *Bull. Soc. Philom.* [Paris], 63–71; *Annal. Chimie* 28(1825): 263. In *Oeuvres,* 2:465–78.

1823a [Extrait d'un mémoire] sur la loi des modifications imprimées à la lumière polarisée par sa réflexion totale dans l'intérieur des corps transparents. Read 7 January 1823. *Bull. Soc. Philom.* [Paris], 29–36; *Annal. Chimie* 29(1825): 175–87. In *Oeuvres,* 1:753–62.

1823b Note sur la polarisation circulaire. In *Oeuvres,* 1:763–66.

1823c Quelques observations sur les principales objections de Newton contre le système des vibrations lumineuses et sur les difficultés que présente son hypothèse des accès. *Bibl. Univ.* 22:73–99. In *Oeuvres,* 2:167–82.

1823d Réponse de M. A. Fresnel à la lettre de M. Poisson: "Sur les vibrations d'une onde lumineuse. *Annal. Chimie* 23:32–49, 113–23. In *Oeuvres,* 2:215–38.

1823e Note sur le phénomène des anneaux colorés. *Annal. Chimie* 23:129–35; *Ann. Phys. Chem.* 12(1828): 599–603. In *Oeuvres,* 2:247–54.

1823f Sur les propriétés optiques de la tourmaline. *Bull. Soc. Philom.* [Paris], 91–92. In *Oeuvres,* 2:640–42.

1823g Sur les dilatations inégales qu'un même cristal peut éprouver dans différentes directions par l'effet de la chaleur. *Bull. Soc. Philom.* [Paris], 181–83. In *Oeuvres,* 2:644–45.

1823h Mémoire sur la loi des modifications que la réflexion imprime à la lumière polarisée. Read 7 January 1823. *Mém. Acad. Sci.* 11(1832): 393–433; *Annal. Chimie* 46(1831): 225–64; *Ann. Phys. Chem.* 22(1831): 90–126. In *Oeuvres,* 1:767–99.

1823i Réponse de M. A. Fresnel à la lettre de M. Poisson. *Annal. Chimie* 23:32, 113. Dated May and June. In *Oeuvres,* 2:215–38.

1824a Sur la direction des axes de double réfraction dans les cristaux. *Bull. Soc. Philom.* [Paris], 40. In *Oeuvres,* 2:643.

1824b Sur les contractions produites par la chaleur dans les cristaux. *Bull. Soc. Philom.* [Paris], 40. In *Oeuvres,* 2:646.

1824c Considérations théoriques sur la polarisation de la lumière. *Bull. Soc. Philom.* [Paris], 147–58; *Ann. Phys. Chem.* 22(1831): 68–90. Consists of the first sixteen sections of Fresnel (1827).

1826 Note en réponse aux questions de Sir John Herschel. In *Oeuvres,* 2: 647–60.

1827 Second mémoire sur la double réfraction. *Mém. Acad. Sci.* 7:45–176; *Ann. Phys. Chem.* 23(1831): 372–434, 494–556; Taylor, *Sci. Mem.* 5(1852): 238–333. In *Oeuvres,* 2:479–596.

1866–70 *Oeuvres complètes.* 3 vols. Ed. H. de Senarmont, E. Verdet, and L. Fresnel. Paris: Imprimérie Impériale.

Glazebrooke, R. T.

1879 An experimental determination of the values of the velocities of normal propagation of plane waves in different directions in a biaxal crystal, and a comparison of the results with theory. *Phil. Trans.* 170:287–377.

Good, G.

1982 J. F. W. Herschel's optical researches: A study in method. Ph.D. diss., University of Toronto.

Grattan-Guiness, I.

1981 Recent researches in French mathematical physics of the early nineteenth century: Essay review. *Ann. Sci.* 38:663–90.

Graves, R. P.

1882 *Life of Sir William Rowan Hamilton.* 3 vols. Dublin: Hodges, Figgis.

Hamilton, W R.

1833 On some results of a characteristic function in optics. *Papers,* 1:297–303.

1837 Third supplement to an essay on the theory of systems of rays. *Papers,* 1:164–294.
1931 *Mathematical papers,* ed. A. W. Conway and J. L. Synge. 3 vols. Cambridge: Cambridge University Press.

Hankins, T. L.
1980 *Sir William Rowan Hamilton.* Baltimore: Johns Hopkins University Press.

Haüy. R. J.
1788 Mémoire sur la double réfraction du spath d'Islande. *Mém. Acad. Sci.,* 34–61.
1803 *Traité élémentaire de physique.* 2 vols. Paris: Delance et Lesueur.
1807 *An elementary treatise on natural philosophy.* 2 vols. Trans. O. Gregory. London.

Hawkins, T.
1975 Cauchy and the spectral theory of matrices. *Historia Mathematica* 9: 1–129.

Herivel, J. W.
1973 The influence of Fourier on British mathematics. *Centaurus* 17:40–57.

Herschel, J.
1820 On the action of crystallized bodies on homogeneous light, and on the causes of deviations from Newton's scale in the tints which many of them develop on exposure to a polarized ray. *Phil. Trans.* 110:45–100.
1827 Light. In *Encyclopaedia Metropolitana* (1849), 4:341–586.

Home, R. W.
1983 Poisson's memoirs on electricity: Academic politics and a new style in physics. *Brit. J. Hist. Sci.* 16:239–58.

Huygens, C.
1690 *Traité de la lumière.* In *Oeuvres,* 19:457–537.
1937 *Oeuvres complètes.* The Hague: Société Hollandaise des Sciences.
1950 *Treatise on light.* Trans. S. P. Thompson. 2d ed. Chicago: University of Chicago Press. Originally published 1690.

James. F. A. J. L.
1984 The physical interpretation of the wave theory of light. *Brit. J. Hist. Sci.* 17:47–60.

Kipnis, N.
1984 History of the principle of interference. Ph.D. diss., University of Minnesota.

Kline. M.
1972 *Mathematical thought from ancient to modern times.* New York: Oxford University Press.

La Hire, P. de
1710 Sur une espèce de talc qu'on trouve communément proche de Paris audessous des bancs de pierre de plâtre. *Mém. Acad. Sci.,* 341–52.

Lamé, G.
1836 *Cours de physique de l'Ecole Polytechnique.* 2 vols. Paris: Bachelier.

Langins, J.
1979 The Ecole Polytechnique (1794–1804): From encyclopaedic school to military institution. Ph.D. diss., University of Toronto.
1980 Sur la première organisation de l'Ecole Polytechnique: Texte de l'arreté du 6 frimaire an III. *Rev. Hist. Sci.* 33:290–313.

Laplace, P. S. de
1809 Sur les mouvements de la lumière dans les milieux diaphanes. Read 30 January 1808. *Mém. Inst.* 10:300–342.
1878–1912 *Oeuvres complètes.* 14 vols. Paris: Gauthiers-Villars.
Lindberg, D. C.
1971 Alkindi's critique of Euclid's theory of vision. *Isis,* 62:469–89.
1976 *Theories of vision from Al-Kindi to Kepler.* Chicago: University of Chicago Press.
Lloyd, H.
1833a On conical refraction. *BAAS Rep.,* 370.
1833b On the phaenomena presented by light in its passage along the axes of biaxal crystals. *Trans. Roy. Irish Acad.* 17:145–58.
1834 Report on the progress and present state of physical optics. *BAAS Rep.,* 295–413.
1841 *Lectures on the wave theory of light.* Dublin.
1877 *Miscellaneous papers connected with physical science.* London: Longmans, Green.
Lohne, J. A.
1977 Nova experimenta crystalli islandici disdiaclastici. *Centaurus* 21: 106–48.
Malus. E.
1807 Sur la mésure du pouvoir réfringent des corps opaques. *Mém. Sav. Etr.* 2(1811): 508–17; *J. Ecole Poly.* 8(1809): 219–28.
1809a Sur une propriété de la lumière réfléchie. *Mém. Soc. Arc.* 2:143–58.
1809b Sur une propriété des forces répulsives qui agissent sur la lumière. *Mém. Soc. Arc.* 2:254–67.
1809c Sur une propriété de la lumière réfléchie par les corps diaphanes. *Bull. Soc. Philom.* [Paris], 1:266–69.
1810 Traité d'optique analytique. *Mém. Sav. Etr.* 2:214–302.
1811a Mémoire sur les phénomènes qui accompagnent la réflexion et la réfraction de la lumière. *Mém. Inst.* 11:112–20; *Bull. Soc. Philom.* [Paris], 2:320–25; *J. Phys. Chimie* 73:5–11.
1811b Théorie de la double réfraction. *Mém. Sav. Etr.* 2:303–508.
1811c Mémoire sur de nouveaux phénomènes d'optique. *Mém. Inst.* 11:105–11; *Bull. Soc. Philom.* [Paris], 2:5–11.
Mascart, M. E.
1889 *Traité d'optique.* 3 vols. Paris: Gauthier-Villars et G. Masson.
1891 *Traité d'optique.* 3 vols. Paris: Gauthier-Villars.
Mauskopf, S. H.
1976 Crystals and compounds: Molecular structure and composition in nineteenth-century French science. *Trans. Am. Phil. Soc.* 66:5–82.
Mayer, J. T.
1812 Über die polarität des lichtes. *Göttingische gelehrte Anzeigen* 2:1977–88.
Moigno, l'Abbé
1847 *Répertoire d'optique moderne.* Paris: A. Franck.
Morrell, J., and A. Thackray, eds.
1982 *Gentlemen of science: Early years of the British Association for the Advancement of Science.* Oxford: Clarendon Press.

Morrison-Low, A. D., and J. R. R. Christie
1984 *"Martyr of science": Sir David Brewster, 1781–1868*. Edinburgh: Royal Scottish Museum Studies.
Muncke, G.
1817 Ueber Licht-polarisirung. *Ann. Phys. Chem.* 57:203–16.
Newton, I.
1704 *Opticks*. London.
Péclet, E.
1823 *Cours de physique*. Marseille: A. Ricard.
Poisson, S. D.
1823a Extrait d'un mémoire sur la propagation du mouvement dans les fluides élastiques. Dated 24 March. In Fresnel, *Oeuvres,* 2:192–205. Originally published in *Annal. Chimie* 22:250.
1823b Extrait d'une lettre de M. Poisson à M. A. Fresnel. In Fresnel, *Oeuvres,* 2:206–13. Originally published in *Annal. Chimie* 22:270.
Potter, R.
1856 *Physical optics; or, The nature and properties of light: A descriptive and experimental treatise, part I*. London: Walton and Maberly.
1859 *Physical optics, part II. The corpuscular theory of light: Discussed mathematically*. Cambridge: Deighton, Bell; London: Bell and Daldy.
Powell, B.
1841 *A general and elementary view of the undulatory theory as applied to the dispersion of light and some other subjects*. London: John Parker.
Preston, T.
1901 *The theory of light*. 3d ed. Ed. C. J. Joly. London: Macmillan.
Procès-verbaux
1804–7 *Procès-verbaux des séances de l'Academie des Sciences*. Vol. 3. Paris.
Ronchi, V.
1956 *Histoire de la lumière*. Trans. Juliette Taton. Paris.
Sabra, A. I.
1981 *Theories of light from Descartes to Newton*. Cambridge: Cambridge University Press.
Salmon, G.
1928 *A treatise on the analytic geometry of three dimensions*. 7th ed. London: Longmans.
Sarton, G.
1932 Discovery of conical refraction by William Rowan Hamilton and Humphrey Lloyd (1833), *Isis* 17:154–70.
Schuster, A.
1909 *An introduction to the theory of optics*. 2d ed. London: Edward Arnold.
Shapiro, A. E.
1974 Kinematic optics: A study of the wave theory of light in the seventeenth century. *Arch. Hist. Ex. Sci.* 11:134–266.
1975 Newton's definition of a light ray and the diffusion theories of chromatic dispersion. *Isis* 66:194–210.
1979 Newton's "achromatic" dispersion law: Theoretical background and experimental evidence. *Arch. Hist. Ex. Sci.* 21:92–128.
1980 The evolving structure of Newton's theory of white light and color. *Isis* 71:211–34.

Silliman, R. H.

1967 Augustin Fresnel (1788–1827) and the establishment of the wave theory of light. Ph.D. diss., Princeton University.

1975 Fresnel and the emergence of physics as a discipline. *Hist. Stud. Phys. Sci.* 4:137–62.

Smeaton, W. A.

1962 *Fourcroy, chemist and revolutionary, 1755–1809.* Cambridge: W. Heffer.

Stokes, G. G.

1862 Report on double refraction. *BAAS Rep.*, 253–82.

Tralles, J. G.

1809 Bemerkungen. *Ann. Phys. Chem.* 31:294–96.

Weber, E. H., and W. Weber

1825 *Wellenlehre auf Experimente gegründet oder über die Wellen tropfbarer Flüssigkeiten mit Anwendung auf die Schall- und Lichtwellen.* Leipzig: Gerhard Fleischer.

Whewell, W.

1837 *History of the inductive sciences from the earliest to the present time.* 3 vols. London: John W. Parker.

Whittaker, E.

1973 *A history of the theories of aether and electricity.* 2 vols. New York: Humanities Press. Originally published 1910.

Wilson, D. B.

1968 The reception of the wave theory of light by Cambridge physicists (1820–1850): A case study in the nineteenth-century mechanical philosophy. Ph.D. diss., Johns Hopkins University.

Wollaston, W. H.

1802a A method of examining refractive and dispersive powers by prismatic refraction. *Phil. Trans.* 92:365–80; *Annal. Chimie* 46(1803): 36–60.

1802b On the oblique refraction of Iceland crystal. *Phil. Trans.* 92:381–86; *Annal. Chimie* 46(1803): 63–74.

Young. T.

1802 The Bakerian lecture: On the theory of light and colors. *Phil. Trans.* 92:12–48.

1804 The Bakerian lecture: Experiments and calculations relative to physical optics. *Phil. Trans.* 94:1–16.

1809 Review of Laplace's memoir "Sur la loi de la réfraction extraordinaire dans les cristaux diaphanes. . . . *Quart. Rev.* 2:337. In *Works,* 1: 220–33.

1810 Review of the Mémoires de physique et de chimie de la Société d'Arcueil," vols. I and II. *Quart. Rev.* 3:462. In *Works,* 1:234–59.

1814 Review of Malus, Biot, Seebeck and Brewster on Light. *Quart. Rev.* 11:42. In *Works,* 1:260–78.

1817 Chromatics. In *Supplement* to the *Encyclopaedia Britannica.* In *Works,* 1:279–342.

1855 *Miscellaneous works.* 3 vols. London: John Murray.

Name Index

Subject Index